Complexity
and Real
Computation

Springer
New York
Berlin
Heidelberg
Barcelona
Budapest
Hong Kong
London
Milan
Paris
Santa Clara
Singapore
Tokyo

Lenore Blum
Felipe Cucker
Michael Shub
Steve Smale

Complexity
and Real
Computation

Foreword by Richard M. Karp

With 47 Illustrations and 3 Color Plates

 Springer

Lenore Blum
International Computer Science Institute
Berkeley, CA 94704-1198
USA
and
Department of Mathematics
City University of Hong Kong
Kowloon, Hong Kong

Felipe Cucker
Universitat Pompeu Fabra
Barcelona 08008
Spain
and
Department of Mathematics
City University of Hong Kong
Kowloon, Hong Kong

Michael Schub
IBM T.J. Watson Research Center
Yorktown Heights, NY 10598-0218
USA

Steve Smale
Department of Mathematics
City University of Hong Kong
Kowloon, Hong Kong

Library of Congress Cataloging-in-Publication Data
Complexity and real computation / Lenore Blum . . . [et al.].
    p.   cm.
   Includes bibliographical references and index.
   ISBN 0-387-98281-7 (hc : alk. paper)
   1. Computer science. 2. Computational complexity. 3. Real-time
data processing. 4. Computer algorithms. I. Blum, Lenore.
  QA76..C5474   1997
  511.3—dc21   97-22859

Printed on acid-free paper.

Production managed by Lesley Poliner; manufacturing supervised by Jeffrey Taub.
Photocomposed copy prepared using the authors' L^AT_EX files.
Printed and bound by R. R. Donnelley and Sons, Harrisonburg, VA.
Printed in the United States of America.

9 8 7 6 5 4 3 2 1

ISBN 0-387-98281-7 Springer-Verlag New York Berlin Heidelberg   SPIN 10633130

Foreword

Computational complexity theory provides a framework for understanding the cost of solving computational problems, as measured by the requirement for resources such as time and space. The objects of study are algorithms defined within a formal model of computation. Upper bounds on the computational complexity of a problem are usually derived by constructing and analyzing specific algorithms. Meaningful lower bounds on computational complexity are harder to come by, and are not available for most problems of interest.

The dominant approach in complexity theory is to consider algorithms as operating on finite strings of symbols from a finite alphabet. Such strings may represent various discrete objects such as integers or algebraic expressions, but cannot represent real or complex numbers, unless the numbers are rounded to approximate values from a discrete set. A major concern of the theory is the number of computation steps required to solve a problem, as a function of the length of the input string.

Complexity theory groups problems into *complexity classes* such as P, the class of problems that can be solved in polynomial time (i.e., in a number of steps bounded by a polynomial function of the length of the input), and NP, the class of problems for which solutions can be verified in polynomial time. Solvability in polynomial time can be equated roughly with computational tractability, and accordingly there is great interest in determining whether particular problems lie in P. A major open question is whether the classes P and NP are equal; it would be surprising if they were, since our experience suggests that checking a solution is much easier than finding one. Further evidence against the equality of P and NP is the plethora of problems that lie in NP but are not known to lie in P. A problem in NP is called NP-*complete* if every problem in NP can be reduced to it in polynomial-time. The classes P and NP are equal if and only if the NP-complete problems lie in P. This fact is taken as evidence that the NP-complete problems are intractable.

The classical theory of computation does not deal adequately with computations that operate on real-valued data. Most computational problems in the physical sciences and engineering are of this type. In developing algorithms for such problems a good strategy is to assume initially that operations on real numbers are exact, and then make adjustments for roundoff in a second stage of analysis. Researchers in mathematical programming have considered the complexity of network flow problems and of linear programming when exact real arithmetic is permitted. Penrose raised the question of whether the Mandelbrot set is computable, and pointed out that such a question only makes sense in the context of exact computation on real numbers. Because standard discrete computational models are ill suited to questions such as these, alternative models have been developed. Algebraic complexity theory considers algorithms that include exact rational operations on real numbers and exact comparisons of real numbers, but requires that the dimensionality of the data presented to any algorithm is fixed. Thus, in studying a problem such as matrix multiplication, a separate algorithm is required for each combination of matrix dimensions, even though what we really want is a *uniform* algorithm that will accept matrices of any size. Moreover, the tools of algebraic complexity theory are best suited to problems whose complexity is clearly polynomial-bounded, and thus cannot address questions about polynomial-time solvability. The field of information-based complexity also allows exact rational operations on reals, but mainly considers problems such as integration, in which the input is a function rather than a finite array of real numbers.

Starting in the late '80s the authors of the present volume initiated the first model of computation which postulates exact arithmetic on real numbers, permits uniform algorithms which can accept input data of arbitrary finite dimension, and is oriented toward characterizing the limits of polynomial-time computation. Although the primary motivation was to model computation over the reals, they set up the theory so that it would apply to computation over any ring or field. Their model of computation allows addition, subtraction, multiplication, and (in the case of fields) division as primitive operations, and equality testing and (in the case of ordered structures) comparison, as primitive tests. When the input data to an algorithm are of fixed dimension the algorithms representable within the model resemble the flowcharts familiar to all programmers; when the dimension is variable the setup is more elaborate, as provision must be made for arrays of real numbers whose dimension is not fixed in advance. With respect to an arbitrary ring or field the authors give natural definitions of the decidable sets and computable functions, the complexity classes P and NP, polynomial-time reducibility, and NP-completeness. The classes P and NP over an algebraic structure R are denoted P_R and NP_R. The traditional classes P and NP are captured by taking the field $\mathbb{Z}_2$ as the underlying structure. The new framework immediately suggests deep mathematical questions, such as whether $P_\mathbb{R}$ is equal to $NP_\mathbb{R}$, and whether $P_\mathbb{C}$ is equal to $NP_\mathbb{C}$, where $\mathbb{R}$ denotes the reals and $\mathbb{C}$ denotes the complex numbers.

The present volume is the first book-length exposition of this theory. By allowing computation over noncountable structures such as the reals the theory provides a bridge between complexity theory and well-developed areas of continuous mathematics such as algebraic geometry and differential topology. In addition to the topics related to $P_\mathbb{R}$ and $NP_\mathbb{R}$, fundamental topics such as average-case analysis of algorithms, randomized algorithms, and parallel algorithms are explored. Chapters are also devoted to such classical problems as solving linear systems of equations, finding zeros of polynomials, and linear programming. The key importance of the *condition number*, which, roughly speaking, measures the closeness of a problem instance to the manifold of ill-posed instances, is clearly developed. The role of the condition number is just one of the many issues discussed in this volume that cannot be adequately addressed within the traditional discrete models of computation.

Let us give some of the highlights of the theory as it applies to the fields $\mathbb{R}$ and $\mathbb{C}$. It is immediately apparent that, for problems with a fixed number of input variables, each algorithm permits only a countable number of computational paths, and that the set of inputs that follow a given path is a semi-algebraic set; i.e., the set of solutions to a finite number of polynomial equations and inequalities. It follows that, in the case of a decision problem, the set of accepted inputs is a countable union of semi-algebraic sets. This has the immediate consequence that a number of interesting sets are not decidable: these include the Mandelbrot set, the Julia set of $T(z) = z^2 + 4$, and the set of starting points from which Newton's method converges to a zero of a certain univariate polynomial.

Define the Hilbert Nullstellensatz (HN) as the problem of deciding if a finite set of polynomials in n variables has a common solution. HN is NP-complete over both $\mathbb{R}$ and $\mathbb{C}$. The following problem is NP-complete over $\mathbb{R}$: given a degree-4 polynomial in n variables with real coefficients, decide whether it has a real zero. The authors conjecture that HN over $\mathbb{C}$ is not solvable in polynomial time, and hence that $P_\mathbb{C} \neq NP_\mathbb{C}$. They give some highly specific conjectures that would imply this result. For example, for a positive integer m let $\tau(m)$ denote the number of additions, subtractions, and multiplications needed to construct the integer m, starting from the integer 1. If, for every sequence m_k of positive integers, $\tau(m_k k!)$ grows faster than every fixed power of $\log(k)$, then HN/$\mathbb{C}$ does not lie in $P_\mathbb{C}$, and hence $P_\mathbb{C} \neq NP_\mathbb{C}$.

It is interesting to speculate as to whether the questions of whether $P_\mathbb{R} = NP_\mathbb{R}$ and whether $P_\mathbb{C} = NP_\mathbb{C}$ are related to each other and to the classical P versus NP question, which in the present setting asks whether $P_{\mathbb{Z}_2} = NP_{\mathbb{Z}_2}$. I am inclined to think that the three questions are very different and need to be attacked independently. Nevertheless, by providing a bridge between complexity theory and the fields of analysis, geometry, and topology, the authors may well have set the stage for new attacks on the classical problems of complexity theory, which heretofore have been attacked using only the tools of logic and combinatorics.

Seattle, Washington, May 1997   Richard M. Karp

Preface

Our book is divided into three parts.

The first part gives an extensive introduction and then proves the fundamental NP-completeness theorems, both those of Cook–Karp and their extensions to more general number fields as the real and complex numbers. Moreover some results comparing complexity classes for different fields are proved ("transfer" theorems).

The second part is closer to numerical analysis. It organizes material of a complexity theory for finding zeros of polynomial systems, one and many variables. A geometric point of view prevails, with the condition number playing a central role. Probability is featured, again with geometric perspectives.

The third part of our book returns to themes of classical computer science. We take a structural approach to complexity over the reals. Objects of study are now classes of problems classified as to their solvability given prescribed resource allocations.

This work has been supported by a number of institutions, starting with the IBM T. J. Watson Research Center in 1987. Substantial support has also been given by the International Computer Science Institute (ICSI) in Berkeley, the University of California at Berkeley, the Universidad Pompeu Fabra in Barcelona, and the City University of Hong Kong. Others include Columbia University, the Universidad de la República at Montevideo, the Universidad de Buenos Aires, and especially the Centre de Recerca Matemàtica (CRM) in Barcelona and the Instituto de Matematica Pura e Aplicada (IMPA) in Rio de Janeiro. One or more of us has given a lecture series or course at each of these institutions on parts of the material which was to become this book.

Other support for the development of the book has been given by the National Science Foundation (NSF), the Research Grants Council of Hong Kong, the Ministerio de Educación y Ciencia of Spain, and the Generalitat de Catalunya. To all of these institutions we give our thanks.

We note how the work here fits well into the spirit of the new organization "Foundations of Computational Mathematics" (FoCM). FoCM has held its first international meetings (Park City, Utah, July 1995, and IMPA, Rio de Janeiro, January 1997) and the proceedings of these meetings [Renegar, Shub, and Smale 1996; Cucker and Shub 1997] contain a number of research papers extending and developing the ideas of this book.

Throughout this book, the square □ denotes the end of a proof or its absence.

Hong Kong, March 1997

Lenore Blum
Felipe Cucker
Michael Shub
Steve Smale

Contents

II Some Geometry of Numerical Algorithms

III   Complexity Classes over the Reals

Part I

Basic Development

1
Introduction

1.1 Aim

The classical theory of computation had its origins in the work of logicians —of Gödel, Turing, Church, Kleene, Post, among others— in the 1930s. The model of computation that developed in the following decades, the Turing machine, has been extraordinarily successful in giving the foundations and framework for theoretical computer science.

The point of view of this book is that the Turing model (we call it "classical") with its dependence on 0s and 1s is fundamentally inadequate for giving such a foundation to the theory of modern scientific computation, where most of the algorithms —with origins in Newton, Euler, Gauss, et al.— are *real number algorithms*. Our viewpoint is not new. Already in 1948, John von Neumann, in his Hixon Symposium lecture, articulated the need for "a detailed, highly mathematical, and more specifically analytical theory of automata and of information." In this lecture, von Neumann was particularly critical of the limitations imposed on the "theory of automata" by its foundations in formal logic:

> There exists today a very elaborate system of formal logic, and specifically, of logic as applied to mathematics. This is a discipline with many good sides, but also serious weaknesses. . ... Everybody who has worked in formal logic will confirm that it is one of the technically most refractory parts of mathematics. The reason for this is that it deals with rigid, all-or-none concepts, and has very little contact with the continuous concept of the real or of the complex number, that is, with mathematical analysis. Yet analysis is the technically

most successful and best-elaborated part of mathematics. Thus formal logic, by the nature of its approach, is cut off from the best cultivated portions of mathematics, and forced onto the most difficult part of the mathematical terrain into combinatorics.

The theory of automata, of the digital, all-or-none type as discussed up to now, is certainly a chapter in formal logic. It would, therefore, seem that it will have to share this unattractive property of formal logic. It will have to be, from the mathematical point of view, combinatorial rather than analytical.

The goal of our work is to develop a formal theory of computation that integrates major themes of the classical theory and builds on the classical foundations, yet at the same time is more mathematical, perhaps less dependent on logic, and more directly applicable to problems in mathematics, numerical analysis, and scientific computing.

We attempt to develop systematically a theory of *real computation*. We do this in a way that preserves the Turing theory as a special case of the new theory, that is, by an appropriate choice of the fields admitted. In this way, results from computer science give insight to numerical analysis and the reverse holds as well.

1.2   Seven Examples

To give some motivation and flavor of what is to come, and to make the formal definitions of Chapters 2 and 3 easier to digest, we discuss in a preliminary way seven examples which we return to again and again throughout our work:

1.2.1   The Mandelbrot Set
1.2.2   A Julia Set
1.2.3   Newton's Method
1.2.4   The Knapsack Problem
1.2.5   The Hilbert Nullstellensatz as a Decision Problem
1.2.6   4-Feasibility
1.2.7   Linear Programming, Integer Programming.

1.2.1   Is the Mandelbrot Set Decidable?

The British mathematical physicist, Roger Penrose, in *The Emperor's New Mind*, p. 124, writes:

> Now we witnessed. . . a certain extraordinarily complicated looking set, namely the Mandelbrot set (see Figure A on page 5 and color insert). Although the rules which provide its definition are surprisingly simple, the set itself exhibits an endless variety of highly elaborate structures.

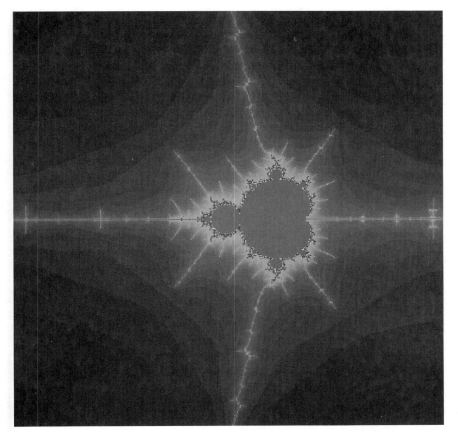

Figure A　The Mandelbrot set.

> Could this be an example of a non-recursive [i.e., undecidable] set,
> truly exhibited before our mortal eyes?

It is known that the boundary of the Mandelbrot set has a rich and complex structure. (See, for example, Figure B on page 6 and color insert, where a part of this boundary is shown.) Hence Penrose's query seems reasonable. Penrose is motivated to ask this question to make an argument against artificial intelligence. Although we do not find this use of mathematics compelling, the question of the decidability of the Mandelbrot set has another justification. It can partly answer and give insight to the question: can one decide if a differential equation is chaotic?

The Mandelbrot set $\mathcal{M}$ is defined as the set of complex numbers c such that the sequence $c, c^2 + c, (c^2 + c)^2 + c, \ldots$ remains bounded.

Figure B   On the boundary in Seahorse Valley. *Richard F. Voss/IBM Research.*

More formally, for $c \in \mathbb{C}$, the complex numbers, let $p_c(z) = z^2 + c$ and let $p_c^n(z)$ be the nth iterate of p_c applied to z. That is,

$$p_c^n(z) = p_c(\dots p_c(p_c(p_c(z)))), n \text{ times.}$$

So $p_c(0) = c$, $p_c^2(0) = c^2 + c$, ... Then $\mathcal{M}$ is the complement of the set

$$\mathcal{M}' = \{c \in \mathbb{C} \mid p_c^n(0) \to \infty \text{ as } n \to \infty\}.$$

The set $\mathcal{M}$ may also be described as the set of all inputs c that do not halt for the flowchart in Figure C.

This is because if ever the sequence c, $c^2 + c$, $(c^2 + c)^2 + c$ escapes the disk of radius 2, it will go off to infinity.[1]

---

[1]This fact is utilized in designing computer algorithms for drawing "pictures" of the Mandelbrot set. Let N be a large integer. For given point c, generate up to N elements of the sequence c, $c^2 + c$, $(c^2 + c)^2 + c$, ... along with their magnitudes. If and when some magnitude is greater than two, color c white, else color c black. Note that white points are *definitely* in $\mathcal{M}'$ whereas black points are *possibly* in $\mathcal{M}$ with our confidence level partly dependent on N. (For more sophisticated algorithms, see the book by Peitgen and Saupe [1988]).

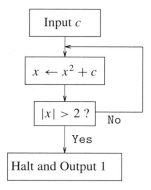

Figure C A flowchart associated with the Mandelbrot set.

To answer Penrose's query, one needs a "machine" or "algorithm" that, given input c, a complex number, will decide in a finite number of steps whether or not c is in $\mathcal{M}$. (See Figure D.)

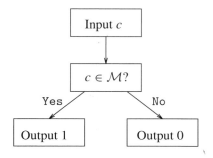

Figure D Desired decision machine for $\mathcal{M}$.

After asking his question, Penrose acknowledges being somewhat inexact (p. 125). The classical theory of computation presupposes that all the underlying sets are countable and hence ipso facto cannot handle these questions about subsets which are uncountable.

Next, Penrose seeks ways to bypass this problem. One way is to use computable real numbers (to describe the appropriate complex numbers). This would be the approach of *recursive analysis*, an area originating with early work of Turing. Here one might imagine a Turing machine being input a real number bit by bit.

Using its internal instructions, the machine operates on what it sees, possibly every so often outputting a bit. The resulting sequence, if any, would be considered in the limit the (binary expansion of the) real output.

Problems arise here when one wants to decide if two numbers are equal and so Penrose rejects this approach. As he points out on p. 126, "One implication of this is that even with such a simple set as the unit disc, ..., there would be no algorithm for deciding for sure ... whether the computable number $x^2 + y^2$ is actually equal to 1 or not, this being the criterion for deciding whether or not the computable complex number $x + iy$ lies on the unit circle. .... Clearly that is not what we want."

Another tack might be to consider the rational or algebraic skeleton of the problem. Thus, we could rephrase Penrose's question: given a complex number c whose real and imaginary parts are rational, or algebraic, decide whether or not c is in $\mathcal{M}$. Indeed, this has been a tack used by theoretical computer scientists to deal with problems whose natural underlying spaces are the real numbers (such as the linear programming problem) or the complex numbers. However, this approach is also problematical. For example, the curve $x^3 + y^3 = 1$ has no rational points with both x and y positive. So, the rational skeleton provides no useful information about the given curve.

After exploring several such approaches, Penrose (p. 129) concludes: "one is left with the strong feeling that the correct viewpoint has not yet been arrived at."

Thus Penrose's question, "Is the Mandelbrot set decidable?" makes no sense!

Now note that the flowchart of Figure C could be interpreted as a machine with a "halting set" which is precisely the complement $\mathcal{M}'$ of $\mathcal{M}$. (The set $\mathcal{M}'$ might be said to be "semidecidable.") This machine has the power to accept complex numbers, perform basic arithmetic operations on complex numbers, and to compare magnitudes. It is an example of a machine (to be defined formally later) over the real numbers $\mathbb{R}$ *not* $\mathbb{C}$ since it uses the real comparison $|z| > 2$.

What is not clear, is whether there is a similar kind of machine with $\mathcal{M}$ as its halting set.

In the next chapter we make precise and formal sense of the suggestions here. Penrose's question becomes well-defined and we will answer it.

1.2.2   Example of the Julia Set of $T(z) = z^2 + 4$

We are looking at a polynomial map $T : \mathbb{C} \to \mathbb{C}$ from the point of view of iteration or as a complex dynamical system. Thus we write $T^2(z) = T(T(z))$ and T^k for the composition of T with itself k times. Let us specify $T(z) = z^2 + 4$.

Observe that if $|z| > 2$, $|T^k(z)| \to \infty$ as $k \to \infty$.

Consider the flowchart in Figure E.

Call this machine M. In M there are 4 nodes (the boxes) which are called, as we descend in the diagram: input node, computation node, branch node, and output node. Again we have an example of a machine over the real numbers $\mathbb{R}$ since its branching depends on real inequality comparisons.

The halting set Ω_M of M is the set of inputs $z \in \mathbb{C}$ such that by following the flow of the flowchart, we eventually halt (or output). For example Ω_M contains the set of all z with $|z| > 2$. Moreover $0, \pm 1, \pm 2$ are all in Ω_M. However, fixed points of T (so $T(z) = z$, or $z^2 + 4 = z$) are not in the halting set. In fact any periodic point of T (so $T^k(z) = z$ some $k = 1, 2, 3, \ldots$) is not in Ω_M.

A little thought shows that Ω_M is an open set of complex numbers, so that $J = \mathbb{C} - \Omega_M$ must contain the closure of the set of periodic points of T. J is

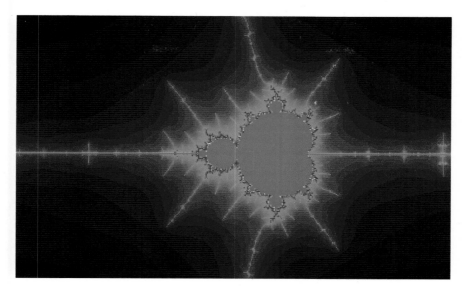

Figure A: The Mandelbrot Set.

Figure B: On the boundary in Seahorse Valley. *Richard F. Voss/IBM Research.*

Figure I: The dynamics of the Newton endomorphism for $f(z) = (z^2 - 1)(z^2 + 0.16)$. *Courtesy of S. Sutherland and Springer-Verlag.*

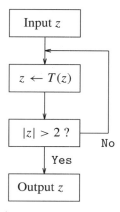

Figure E　A Julia set flowchart.

the Julia set of T in the terminology of complex dynamical systems and it can be proved that J is the closure of the set of periodic points of T and is homeomorphic to a Cantor set.

A question again suggested by the classical theory of computation is: is J decidable or, equivalently, is there a real machine with halting set J?

To see the equivalence of these questions, note that a decision machine for J can be converted into a machine with halting set J as indicated in Figure F.

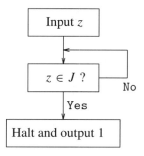

Figure F　J is a halting set (supposing J is decidable).

On the other hand, suppose J were the halting set of some machine M_J. Recall, the flowchart machine M given in Figure F has halting set $\mathbb{C} - J$. We construct a decision machine for J by hooking together M and M_J (see Figure G). To decide if $z \in J$, input z into both M and M_J and run the two machines in tandem. One and only one of these machines will halt; the one that does decides the membership of z. (Schematically, we have indicated a parallel process. This could be turned into a sequential process by alternating in turn between operations of M and M_J.)

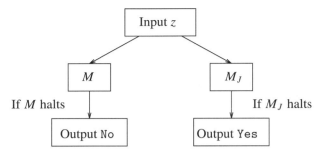

Figure G J is decidable (supposing J is a halting set).

When we have at hand our formal development of machines over $\mathbb{R}$, we will be able to answer the preceding questions (the answer is, J is not a halting set and hence not decidable).

1.2.3   Newton's Method

The previous two examples raised questions concerning the existence of machines that would *decide* Yes or No to queries of the form: given $z \in \mathbb{C}$, is $z \in \mathcal{M}$ (or J)?

On the other hand, often we want algorithms that *search* for solutions to problems of the form: given a polynomial f, find ζ such that $f(\zeta) = 0$.

Newton's method is the "search algorithm" sine qua non of numerical analysis and scientific computation. Here we briefly recall Newton's method for finding (approximate) zeros of polynomials in one variable.

Given a one-variable polynomial $f(z)$ over the complex numbers $\mathbb{C}$, define the *Newton endomorphism* $N_f : \mathbb{C} \to \mathbb{C}$ by

$$N_f(z) = z - \frac{f(z)}{f'(z)}. \tag{1.1}$$

This map is defined as long as $f'(z) \neq 0$.

Now for Newton's method. Pick an initial point $z_0 \in \mathbb{C}$ and generate the *orbit*

$$z_0, z_1 = N_f(z_0), z_2 = N_f(z_1), \ldots, z_{k+1} = N_f(z_k) = N_f^{k+1}(z_0), \ldots \, .$$

Some stopping rule such as "stop if $|f(z_k)| < \varepsilon$ and output z_k" is given. In practice, if the procedure has not stopped with an output after a certain number of iterates, or it becomes undefined at some stage, a new initial point is chosen.

We may represent Newton's method schematically as in Figure H. In this simple machine, we assume that f, N_f, and ε are built in. An initial point $z \in \mathbb{C}$ is "input" to the machine. Later we consider machines that allow f and ε to be input as well.

Proposition 1 *(a)* $f(\zeta) = 0$ if and only if $N_f(\zeta) = \zeta$ and *(b)* $N_f(\zeta) = \zeta$ implies $|N_f'(\zeta)| < 1$.

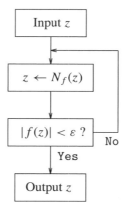

Figure H The Newton machine for f.

Part (a) of the proposition follows directly from the formula defining N_f (and noting that there are polynomials g and h such that $f/f' = g/h$ and if $f(\zeta) = 0$, then $g(\zeta) = 0$ but $h(\zeta) \neq 0$). To show (b) we observe that, for ζ a zero of f of multiplicity m,

$$|N'_f(\zeta)| = \frac{m-1}{m}.$$

To see this, note that $N'_f = ff''/(f')^2$ and evaluate it using the Taylor expansion of f about ζ,

$$f(z) = a_m(z - \zeta)^m + \text{ higher order terms} \quad a_m \neq 0.$$

Thus the *zeros* of f are the *fixed points* of N_f and the fixed points of N_f are *attracting*. This implies there are local neighborhoods about the zeros of f that contract under (iterates of) N_f. Hence, any point $z \in \mathbb{C}$ that, under the action of N_f, eventually enters one of these contracting neighborhoods will eventually approach a zero of f. This is the basis of Newton's method.

If ζ is a simple zero of f, then $N'_f(\zeta) = 0$; that is, ζ is a *superattracting* fixed point of N_f. It follows that there is an open set of points whose orbits under the Newton endomorphism eventually converge *quadratically* to ζ. That is, starting at any of the points in this open set, Newton's method will eventually converge to ζ, and moreover, at some stage, the precision of the approximation will double with each successive iteration. Analyzing "complexity" issues such as the speed of convergence of Newton's method is an important focus of this book.

Decidability questions also arise in this context. It is well known that Newton's method is not *generally convergent*. The main obstruction to general convergence is the existence of attracting periodic points of period at least 2. For example, consider the cubic polynomial

$$f(z) = z^3 - 2z + 2.$$

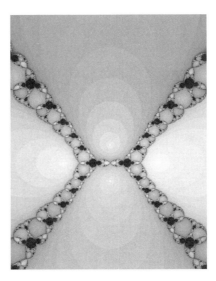

Figure I   The dynamics of the Newton endomorphism for $f(z) = (z^2 - 1)(z^2 + 0.16)$.
Courtesy of S. Sutherland and Springer-Verlag.

Here $N_f(z) = z - (z^3 - 2z + 2)/(3z^2 - 2)$. So $N_f(0) = 1$ and $N_f(1) = 0$; that is, 0 is a point of period 2 under the Newton endomorphism. Also, by the chain rule we see $(N_f^2)'(0) = 0$. So there is a neighborhood of points about 0 whose orbits under the Newton map fail to converge to a zero of f. These points are *super*attracted to the periodic orbit: $0, 1, 0, 1, \ldots$. So, given f, it is natural to ask: is the set of "good" starting points for Newton's method decidable?

Here we say that a point $z \in \mathbb{C}$ is good if its orbit under N_f converges to a zero of f. It can be shown that the set of good starting points coincides with the halting set of some Newton machine M (with built-in ε that depends on f). So as before we can rephrase our question: is the set of bad points a halting set? We return to this question more formally in the next chapter. Yet we include here a picture indicating the good and bad points for Newton's method applied to the polynomial $f(z) = (z^2 - 1)(z^2 + 0.16)$. In this picture (see Figure I and color insert), bad points are colored black and the set of them includes the Julia set for the Newton endomorphism.

1.2.4   The Knapsack Problem

Now suppose R is a commutative ring with unit. For specificity, one may suppose R is the ring of integers $\mathbb{Z}$, the rationals $\mathbb{Q}$, the reals $\mathbb{R}$, or the complex numbers $\mathbb{C}$. Consider the following problem.

Given $x_1, \ldots, x_n \in R$ decide if there is a nonempty subset $S \subset \{1, \ldots, n\}$ such that $\sum_{i \in S} x_i = 1$.

We call this problem the Knapsack Problem (KP) over R. In the classical theory, and more generally, if R is not a field, it is considered in the following form, also known as the Subset Sum Problem.

Given positive integers $x_1, \ldots, x_n, c$ decide if there is a subset $S \subset \{1, \ldots, n\}$ such that $\sum_{i \in S} x_i = c$.

Here we imagine c to be the capacity of a knapsack, x_i the weights of given items, and the question is: can one fill the knapsack to capacity with some subset of the items?

Note that in our formulation of the Knapsack Problem, the ring R need not be ordered. Thus, the machines we consider here branch only on equality comparisons over R.

In many ways, the Knapsack Problem is similar to our previous decidability problems. Let K_n be the Knapsack set which we write in the following form.

$$K_n = \{x \in R^n \mid \exists b \in \{0, 1\}^n \text{ such that } \sum b_i x_i = 1\}. \tag{1.2}$$

Now we are seeking a machine that will decide, given $x \in R^n$, if $x \in K_n$.

But unlike our previous problems, KP is easily seen to be decidable. Given $x \in R^n$, successively enumerate the nonzero elements of $\{0, 1\}^n$ and evaluate the corresponding $\sum b_i x_i$. If and when an evaluation is 1, halt and output 1 (yes). Otherwise, halt and output 0 (no). (See Figure J.) Given input $x \in R^n$ this algorithm (or machine) will stop with a correct decision in at most $2^n - 1$ enumerations.

Fixing n, we can view the input space for our algorithm as the finite-dimensional space R^n. However, since the algorithm is "uniform" in n, we can naturally view the input space as R^∞, the infinite disjoint union of R^n (essentially, the space of finite, but unbounded sequences over R). Thus, letting n vary, we have really sketched an algorithm to decide membership in $K = \bigcup K_n$. In the finite-dimensional case, the polynomial tests at the branch nodes can be built into the machine. In the "infinite-dimensional" case, the tests at the branch nodes are computed by subroutines.

The exhaustive search algorithm does not seem at all satisfactory. We may very well be unlucky and have to go through an exponential number of iterations before we halt with an answer. The big question is: can we do better? Is there an algorithm for deciding the Knapsack Problem in a polynomial number of "steps"? By this we mean, is there a uniform machine M and a positive integer c such that for each n and $x \in R^n$, M will decide if $x \in K$ in less than n^c steps? In our formal development we are more precise as to how to define a uniform machine and how to count steps. For the moment, it is the number of nodes we traverse (counting multiple visits) in a flowchart machine from input to output, given input x.

Notice that the exhaustive search algorithm does not use multiplication. A major unsolved problem is: can multiplication speed up the process?

Suppose R has no zero divisors. Let $k_n(x) \in R[x_1, \ldots, x_n]$ be the polynomial

$$k_n(x) = \prod_{b \in \{0,1\}^n} \left(\sum b_i x_i - 1 \right) \tag{1.3}$$

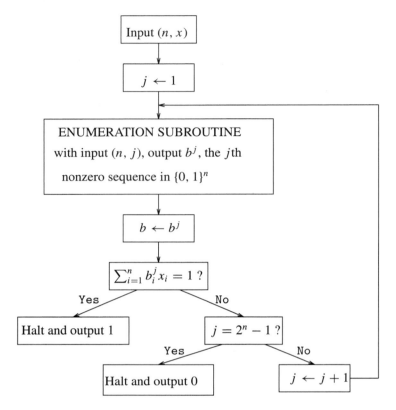

Figure J   Exhaustive search machine for solving the Knapsack Problem.

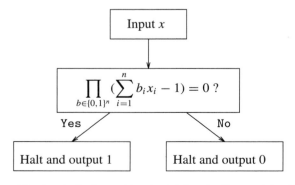

Figure K   An algebraic machine for solving the Knapsack Problem.

and $V_{k_n} = \{x \in R^n \mid k_n(x) = 0\}$. Then $V_{k_n} = K_n$. So, for each n we could construct a machine with k_n built in. (See Figure K.)

The input space of this machine is R^n. It decides if $x \in K$ immediately (in one step) by evaluating the polynomial $k_n(x)$ and checking if the result is 0. The degree of $k_n(x)$ is $2^n - 1$. So we have traded an exponential search for an evaluation of a polynomial whose degree is exponential in n. We show later that, for the case of nonuniform (and uniform) machines for deciding KP over $\mathbb{Z}$, $\mathbb{Q}$, $\mathbb{R}$, and $\mathbb{C}$ that branch only on equality comparisons, this tradeoff is intrinsic. If R is ordered, a big question remains: do order comparisons enable significant speedups?

Although it may be hard to decide in general if $x \in K$ we can quickly verify it is, if we are given a good "witness" $b \in \{0, 1\}^n$. Just sum up the corresponding x_is. We express this property of the Knapsack Problem by saying KP is in the class "NP." Over the integers, KP is a universal problem with this property —it is "NP-complete over $\mathbb{Z}$."

In the next chapters we develop these notions more formally over a ring R. We also return to the Knapsack Problem.

1.2.5   The Hilbert Nullstellensatz as a Decision Problem

Newton's method tackles the search problem.

> Given a polynomial f over $\mathbb{C}$, *find* a zero of f.

Here we consider the decision problem.

> Given a finite set of polynomials f_i in n variables over $\mathbb{C}$, *decide* if the f_i have a common zero.

By the Fundamental Theorem of Algebra, for one polynomial in one variable, for nonconstant polynomials the answer is always yes. This is not true for several polynomials in several variables.

We call this decision problem the Hilbert Nullstellensatz over $\mathbb{C}$ and we denote it by HN/$\mathbb{C}$. Thus, one seeks an algorithm, in fact an algebraic algorithm over $\mathbb{C}$, which on input $f = \{f_1, \dots, f_k\}$ produces yes if and only if there is a $\zeta \in \mathbb{C}^n$ such that $f_i(\zeta) = 0$ for all i.

By an algebraic algorithm we have in mind a machine whose computations, like Newton's method, involve the basic arithmetic operations, but whose branching now depends only on equality comparisons and not on order, that is, a machine now *over* $\mathbb{C}$.

The input f to such a machine can be thought of as the vector of coefficients of the f_i in $\mathbb{C}^N$ where N is given by the formula

$$N = \sum_{i=1}^{k} \binom{n + d_i}{n}, \quad d_i = \deg f_i, \quad i = 1, \dots, k.$$

Thus, this N represents the size $S(f)$ of the input f.

There are algorithms that accomplish this task. In the linear case, that is, when $\deg f_i = 1$ for $i = 1, \ldots, k$, this is a simple linear algebra problem. In the general case, Hilbert has shown that the answer is no if and only if there exist polynomials $g_1, \ldots, g_k$ in n variables with the property

$$\sum_{i=1}^{k} g_i f_i = 1, \tag{1.4}$$

where the equality is equality as polynomials. Bounds on the degrees of the g_i have been proved and in fact we may take $\deg g_i \leq D^n$, where $D = \max\{3, d_1, \ldots, d_n\}$. Thus, equation (1.4) becomes a finite-dimensional linear algebra problem: to find the coefficients of the g_i. As one can readily see, the number of Gaussian elimination steps required is exponential in the size $S(f)$ of the input vector. Equation (1.4) together with the bounds are known as the effective Nullstellensatz.

This suggests a problem which we formulate as a main conjecture.

Conjecture. The Hilbert Nullstellensatz over $\mathbb{C}$ is intractable.

By this we mean there is no algorithm that solves HN/$\mathbb{C}$ with the number of arithmetic operations $\mathcal{A}(f)$ satisfying the bound

$$\mathcal{A}(f) \leq S(f)^c,$$

where c is a universal constant. But a precise formulation of the conjecture awaits our formal definition of a machine over $\mathbb{C}$.

Continuing to anticipate our mathematical development, we show that HN/$\mathbb{C}$ is in NP over $\mathbb{C}$ by noting that given f and a test point $\zeta \in \mathbb{C}^n$, we may test ζ for a zero, that is, whether $f_i(\zeta) = 0$ for $i = 1, \ldots, k$, in a number of arithmetic operations that are polynomial in $S(f)$. Moreover HN/$\mathbb{C}$ is shown to be universal with this property so that any problem in NP over $\mathbb{C}$ can be quickly reduced to HN/$\mathbb{C}$. That is to say HN/$\mathbb{C}$ is NP-complete over $\mathbb{C}$.

From these considerations we deduce the following.

> The Hilbert Nullstellensatz over $\mathbb{C}$ is intractable if and only if P $\neq$ NP over $\mathbb{C}$.

Thus, we are able to take the NP out of the purely complexity-theoretic question, "Is P $\neq$ NP? over $\mathbb{C}$" and replace it by a simpler and more algebraic question, "Is HN/$\mathbb{C}$ intractable over $\mathbb{C}$?". These ideas are developed in Chapter 5 and pursued in Chapter 7.

The preceding sketches our ideas on formulating complexity issues in an algebraic framework. How does this relate to the P $\neq$ NP problem of classical complexity theory?

By abuse of notation write HN/$\mathbb{Z}_2$ for the following problem.

> Given a finite set of polynomials f_i in n variables over $\mathbb{Z}_2$, decide if the f_i have a common zero in $(\mathbb{Z}_2)^n$.

By considerations similar to the preceding, HN/$\mathbb{Z}_2$ is seen to be intractable if and only if P $\neq$ NP (in the classical sense). Since our algorithms in the previous formulation are defined in terms of algebra (characteristic 2 algebra), it could be said that we are placing classical complexity into an algebraic setting in addition to extending it to new problems such as P $\neq$ NP over $\mathbb{C}$. These new problems are interesting in their own right. But also, by posing problems such as P $\neq$ NP within a broader framework, we may be able to employ new mathematical tools to study the classical case as well as gain new insights by analogy or by direct connections.

1.2.6 Feasibility of Real Polynomials

Let us replace the field of the complex numbers by the real numbers in the preceding problem and consider polynomials $f_1, \ldots, f_k \in \mathbb{R}[X_1, \ldots, X_n]$. The problem at hand now is to decide if there exists a common root $\xi \in \mathbb{R}^n$. Since the reals have a natural order, it is natural here to consider algorithms that branch on order comparisons.

A particular feature of the real case, not shared by the complex one, enables us to consider the same problem with only one polynomial at the cost of slightly increasing the degree. We associate with the polynomials $f_1, \ldots, f_k$ the single polynomial $g = \sum_{i=1}^{k} f_i^2$. Now g has the property that for every $\xi \in \mathbb{R}^n$, ξ is a common root of all the f_i if and only if $g(\xi) = 0$. Therefore, solving our problem for the f_i turns out to be equivalent to solving it for g. Moreover, if not all the f_i are linear, then the degree of g is at least 4. Let us restrict our attention to degree-4 polynomials and consider the following problem which we denote by 4-FEAS.

> Given a degree-4 polynomial in n variables with real coefficients, decide whether it has a real zero.

Again, the input g for this problem can be seen as a vector in $\mathbb{R}^N$, where

$$N = \binom{n+4}{4}$$

is the size $S(g)$ of this input.

An algorithm for solving this problem was first given by Tarski in the context of exhibiting a decision procedure for the theory of real numbers. In the context of complexity theory, Tarski's algorithm is highly intractable. The number of arithmetic operations performed by this algorithm grows in the worst case by an exponential tower of n 2s. Later on, Collins devised another algorithm that solved 4-FEAS within a number of arithmetic operations bounded by

$$2^{2^{S(g)}}.$$

More recent algorithms achieve single exponential bounds. These algorithms are quite elaborate and we do not sketch them at this point.

Again, this suggests a problem which we formulate as another conjecture.

Conjecture. The 4-FEAS problem is intractable.

That is, we conjecture that there is no algorithm that solves 4-FEAS with the number of arithmetic operations $\mathcal{A}(f)$ satisfying the bound

$$\mathcal{A}(f) \leq S(f)^c,$$

where c is a universal constant. But again, a precise formulation of the conjecture awaits our formal definition of a machine over $\mathbb{R}$.

As with the Hilbert Nullstellensatz, 4-FEAS is seen to be in NP over $\mathbb{R}$. We also show it to be universal with this property; that is, 4-FEAS is NP-complete over $\mathbb{R}$.

1.2.7 Linear Programming and Integer Programming

Over the reals consider the following problem.

Given a set of m linear inequalities in n variables

$$A_i x \geq b_i \quad i = 1, \ldots, m, \tag{1.5}$$

where $A_i x = \displaystyle\sum_{j=1}^{n} a_{ij} x_j$, $a_{ij} \in \mathbb{R}$ and $b_i \in \mathbb{R}$, decide if there is a point $x \in \mathbb{R}^n$ satisfying (1.5).

We call this problem the (real) *linear programming feasibility* (LPF) problem. We now write the system of inequalities (1.5) as

$$Ax \geq b,$$

where A is the $m \times n$ matrix whose ith row is A_i and b the m-vector whose ith entry is b_i. Then we may rewrite LPF/$\mathbb{R}$ as follows:

Given the $m \times n$ real matrix A and $b \in \mathbb{R}^m$, decide if there is an $x \in \mathbb{R}^n$ such that $Ax \geq b$.

This problem is simpler than HN/$\mathbb{R}$ in that the functions are linear, but more complicated by the fact that we consider inequalities. The set of solutions of $Ax \geq b$ is called a *polyhedron*.

The (real) *linear programming optimization* (LPO) problem is:

with input (A, b, c) *minimize* $c \cdot x$ subject to $Ax \geq b$, where A is an $m \times n$ real matrix, $b \in \mathbb{R}^m$, and $c \in \mathbb{R}^n$, or decide no minimum exists.

Replacing the reals by the integers or the rationals in LPF we have the *integer programming* (IPF) and the rational linear programming *feasibility* problems.

Given an $m \times n$ matrix A with integer (respectively, rational) entries a_{ij} and $b \in \mathbb{Z}^m$ (respectively, $\mathbb{Q}^m$), determine if there is an $x \in \mathbb{Z}^n$ (respectively, $\mathbb{Q}^n$) such that $Ax \geq b$.

The corresponding integer programming (IPO) and rational linear programming *optimization* problems are:

$$\text{minimize } c \cdot x$$
$$\text{subject to } Ax \geq b$$
or determine that no minimum exists.

Here A is an $m \times n$ integer (or rational) matrix, $b \in \mathbb{Z}^m$ (or $\mathbb{Q}^m$), $c \in \mathbb{Z}^n$ (or $\mathbb{Q}^n$), and $x \in \mathbb{Z}^m$ (or $\mathbb{Q}^m$).

Algorithms are known that solve all six of these problems, but there is a great deal of difference in what is known about the efficiency of algorithms which solve them.

Integer programming is set apart from real or rational linear programming by the fact that the solutions of linear equations with integer coefficients are not necessarily integers; for example, $2x = 1$. But it is also set apart from the reals by the notion of the size of the input. For the reals, the input size $S(A, b)$ or $S(A, b, c)$ of the problem is the number of real variables involved, $mn + m$ for the feasibility problem and $mn + m + n$ for the optimization problem. No algorithm is known for either the feasibility or the optimization problem with total number of arithmetic operations $\mathcal{A}(A, b)$ or $\mathcal{A}(A, b, c)$ satisfying the following complexity bounds

$$\mathcal{A}(A, b) \leq S(A, b)^d \text{ or}$$
$$\mathcal{A}(A, b, c) \leq S(A, b, c)^d$$

for a universal constant d. Thus an outstanding problem is: is linear programming tractable over $\mathbb{R}$? Again, to make this precise one needs a formal definition of an algorithm, or machine, over $\mathbb{R}$.

Complexity estimates for algorithms for integer programming traditionally take the binary lengths of the integers used into account both in the input size of a problem "instance" and the "cost" of the algorithm in that instance.

The *height* of an integer x, $\mathrm{ht}(x)$, is the first integer greater than or equal to $\log(|x| + 1)$.

For the feasibility problem we take input size $S_{\mathrm{ht}}(A, b)$ to be equal to be $S(A, b)$ times the maximum height of all the integers a_{ij}, $i = 1, \ldots, m$, $j = 1, \ldots, n$ and b_i, $i = 1, \ldots, m$, where a_{ij} are the entries of the matrix A and b_i the components of the vector b.

For the optimization problem we take input size $S_{\mathrm{ht}}(A, b, c)$ to be equal to $S(A, b, c)$ times the maximum height of all the integers a_{ij}, $i = 1, \ldots, m$, $j = 1, \ldots, n$, b_i, $i = 1, \ldots, m$, and c_j, $j = 1, \ldots, n$.

The cost of an algorithm for the problem instance (A, b) (respectively, (A, b, c)), $C_{\mathrm{ht}}(A, b)$ (respectively, $C_{\mathrm{ht}}(A, b, c)$), is similarly adjusted by multiplying the number of algebraic operations $\mathcal{A}(A, b)$ (respectively, $\mathcal{A}(A, b, c)$) by the maximum height of any integer appearing in the computation. The integer programming problems are NP-complete in the classical model (we formulate and establish this in Chapter 6 for the feasibility problem). The intractability of either one is equivalent to $P \neq NP$ in the classical case, where now by intractability we mean there is no algorithm and constant $d > 0$ for either problem such that

$$C_{ht}(A, b) \leq S_{ht}(A, b)^d \text{ or } C_{ht}(A, b, c) \leq S_{ht}(A, b, c)^d.$$

The situation for the rationals, rational linear programming feasibility, and optimization, is dramatically different. The height $ht(x)$ of a rational number $x = p/q$ where p and q are relatively prime is defined as $\max\{ht(p), ht(q)\}$. Otherwise the definitions of the input sizes $S_{ht}(A, b)$, $S_{ht}(A, b, c)$, and costs $C_{ht}(A, b)$, $C_{ht}(A, b, c)$ are the same as in integer programming. In Chapter 15 we prove that there is an algorithm for the rational linear programming feasibility problem and a constant $d > 0$ such that

$$C_{ht}(A, b) \leq S_{ht}(A, b)^d.$$

1.3   The Classical Theory of Computation

As we have noted, the classical theory of computation had its origins in work of logicians in the 1930s. Of course at that time, there were no computers as we know them. Although this work, in particular Turing's (1936), clearly anticipated the development of the modern digital general-purpose computer, a primary motivation for the logicians was to formulate and understand the concept of *decidability*, or of a *decidable set*. In particular, the aim was to make sense of such questions as, "Is the set of true sentences of arithmetic decidable?" or "Is the set of diophantine equations with integer solutions decidable?" [2]

Intuitively, a set S is decidable if there is an "effective procedure" that given any element u of U (some natural universe containing S) will decide in a finite number of steps whether or not u is in S, that is, if the characteristic function of S (with respect to U) is "computable." To put the first query in this format, U would be the set of arithmetic sentences, S the true ones. For the second, U would be the set of polynomials with integer coefficients and S the subset of those with integer solutions.

The models of computation designed by these logicians were intended to capture the essence of this concept of effective procedure or computation. The idea was to design theoretical machines with operations, and finitely described rules for proceeding step by step from one operation to the next, so simple and constructive that it would be self-evident that the resulting computations were effective.

---

[2]The latter question is known as Hilbert's Tenth Problem, posed by David Hilbert (along with 22 other seminal problems) at the Second International Congress of Mathematicians in Paris on August 8, 1900. It was originally taken for granted, by mathematicians in general, and Hilbert in particular, that the answers to the preceding questions were both affirmative. The queries were actually posed as tasks: produce decision procedures for the given sets. The incompleteness/undecidability results of Gödel in 1931 in the first place, and of Matiyasevich in 1971 on the unsolvability of Hilbert's Tenth Problem in the second, show such tasks cannot be carried out. In Part IV where we deal with questions of logic we give a proof of Gödel's Theorem within our more algebraic setting.

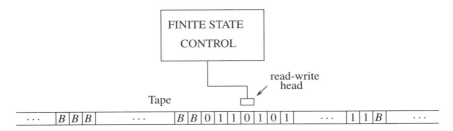

Figure L   A Turing machine.

A number of distinct formal models of computation were proposed. A primary example is the *Turing machine*. (See Figure L.)

Here we have a *finite state control* device with a *read-write head* and a two-way infinite *tape* consisting of an infinite number of *cells*. The control device is regulated by a *program* which is a finite set of instructions of the form (q, s, o, q'). Here q and q' belong to a finite set $\{q_0, \ldots, q_N\}$ called the set of *states* of the machine, s is a symbol 0,1, or B (for blank), and o is one of the following *operations*: R (move right one cell), L (move left one cell), 0 (print 0), 1 (print 1), or B (print B). The instruction is interpreted as follows: if the device is in state q with read-write head scanning a cell containing symbol s, then the device performs operation o and goes into state q'. (If o is a print operation, then it is implicit that the head erases the current symbol before printing.) We assume the program is *consistent*; that is, for each q and s there is at most one instruction starting with the pair (q, s).

The machine operates as follows: given an *input* string x, a finite sequence of 0s and 1s written on consecutive tape cells (with Bs everywhere else), the head is placed over the leftmost symbol of x. (If x is the empty string, the head is placed on any cell.) The control device is started in *initial state q_0* and proceeds according to the program instructions until it can no longer proceed (i.e., it reaches a state q while scanning a symbol s for which there is no instruction starting with the pair q, s). If and when this occurs the *output* is the string of 0s and 1s starting at the current scanned cell (and going right) until the first occurrence of a B (which may be the current cell, in which case the output is the empty string). One might consider input and output strings as natural numbers written in binary (a convention here could be that the empty string is interpreted as 0 and a nonempty input string always has 1 in its leftmost place).

Here, and in each formalism for computation, a function f from the natural numbers $\mathbb{N}$ to $\mathbb{N}$ is defined to be *computable* if it is the *input–output* map of some such machine. Thus we can now say formally: a set of natural numbers is *decidable* if its characteristic function is computable, in this case by a Turing machine. Recall that $\mathbb{N} = \{0, 1, 2, 3, \ldots\}$.

A fundamental object of study is the *halting set* of a machine. This is the set of all inputs for which the machine *halts*, that is, produces an output. It is clear that the halting sets are exactly the *semidecidable sets*: a set S of natural numbers is semidecidable if there is a machine that outputs 1 when input an element of S, and otherwise outputs 0 or does not halt. A little "programming" shows that S is decidable if and only if both it and its complement are semidecidable. (Schematically, see Figures F and G.)

This notion of computability can be naturally extended to the integers $\mathbb{Z}$, the rational numbers $\mathbb{Q}$, or any domain that can be "effectively encoded" in $\mathbb{N}$. Thus, for example, by "Gödel" coding sentences (of a first-order language) as natural numbers, one can begin to formally ask (and answer) within the formalism questions about the decidability of the set of true sentences of various mathematical theories.

It is quite remarkable that even though the formalism we just described and the others proposed were often markedly different, in each case, the resulting class of computable functions —and hence decidable (as well as semidecidable) sets— was exactly the same. Thus, the class of computable functions appears to be a natural class, independent of any specific model of computation.[3] And consequently, the answers to the basic questions of decidability are independent of formalism.

This gives one a great deal of confidence in the theoretical foundations of the theory of computation. Indeed, what is known as *Church's thesis* is an assertion of belief that the classical formalisms completely capture our intuitive notion of computable function.[4] Compelling motivation clearly would be required to justify yet a new model of computation.

1.4   Toward a Mathematical Foundation of Numerical Analysis

There is a sense in which this work is a study of the laws of computation. Thus we write not from the point of view of the engineer who looks for a good algorithm that solves the problem at hand, or who wishes to design a faster computer. The perspective is more like that of a physicist, trying to understand the laws of scientific computation. Idealizations are appropriate, but such idealizations should carry basic truths.

Scientific computation is the domain of computation that is based mainly on the equations of physics. For example, from the equations of fluid mechanics, scientific computation helps in the better design of airplanes, or assists in weather prediction. The theory underlying this side of computation is called numerical analysis.

---

[3] In classical terminology, these functions are often called the *recursive functions*, decidable sets are the *recursive sets*, and semidecidable sets are the *recursively enumerable sets*.

[4] Thus, for example, in the light of Church's thesis, the negative solution to Hilbert's Tenth Problem can be gotten by showing there is no Turing machine to decide the solvability in integers of diophantine polynomials.

There is a substantial conflict between theoretical computer science and numerical analysis. These two subjects with common goals have grown apart. For example, computer scientists are uneasy with calculus, whereas numerical analysis thrives on it. On the other hand numerical analysts see no use for the Turing machine.

The conflict has at its roots another age-old conflict, that between the continuous and the discrete. Computer science is oriented by the digital nature of machines and by its discrete foundations given by Turing machines. For numerical analysis, systems of equations and differential equations are central and this discipline depends heavily on the continuous nature of the real numbers.

The developments described in the previous section (and the next) have given a firm foundation to computer science as a subject in its own right. Use of Turing machines yields a unifying concept of the algorithm well formalized. Thus this subject has been able to develop a complexity theory that permits discussion of lower bounds of all algorithms without ambiguity.

The situation in numerical analysis is quite the opposite. Algorithms are primarily a means to solve practical problems. There is not even a formal definition of algorithm in the subject. One is reminded of how the development of the definition of differentiable manifold was so important in the history of differentiable topology. The history of algebraic geometry gives us a similar lesson.

Thus we view numerical analysis as an eclectic subject with weak foundations; this certainly in no way denies its great achievements through the centuries.

A major obstacle to reconciling scientific computation and computer science is the present view of the machine, that is, the digital computer. As long as the computer is seen simply as a finite or discrete object, it will be difficult to systematize numerical analysis. We believe that the Turing machine as a foundation for real number algorithms can only obscure concepts.

Toward resolving the problem we have posed, we are led to expanding the theoretical model of the machine to allow real numbers as inputs. There has been great hesitation to do this because of the digital nature of the computer. Here we might learn a lesson from the history of science. In particular, Isaac Newton was faced with an analogous problem in writing his *Principia*. At the time of Newton, scientists assumed that the world was atomistic, as viewed by the ancient Greek, Democritus. Newton accepted that picture according to which all matter is composed of indivisible particles, a finite number in each bounded region. On the other hand, Newton's mathematics was continuous as was Euclid's. Moreover, the differential equations Newton needed for his theory involved calculus and the continuum, contrasting with the corpuscular view of the universe. It was a substantial problem for Newton to reconcile the discrete world with the continuous mathematics. The resolution was produced by analyzing the effect of replacing an object (e.g., the earth) by a finite number of particles, then making a better approximation with a larger number of particles.

In the limit, the mathematics becomes continuous. Thomas Kuhn in *The Copernican Revolution* (1957) writes:

> In 1685 [Newton] proved that, whatever the distance to the external corpuscle, all the earth corpuscles could be treated as though they were located at the earth's centre. That surprising discovery, which at last rooted gravity in the individual corpuscles, was the prelude and perhaps the prerequisite to the publication of *Principia*.

And Kuhn adds:

> At last it could be shown that both Kepler's Law and the motion of a projectile could be explained as the result of an innate attraction between the fundamental corpuscles of which the world machine was constructed.

Now our suggestion is that the modern digital computer could be idealized in the same way that Newton idealized his discrete universe. The machine numbers are rational numbers, finite in number, but they fill up a bounded set of real numbers (e.g., between -1000 and 1000) sufficiently densely that viewing the computer as manipulating real numbers is a reasonable idealization, at least in a number of contexts.

Moreover, if one regards computer-graphical output such as our picture of the Mandelbrot or Julia sets with their apparently fractal boundaries and asks to describe the machine that made these pictures, one is driven to the idealization of machines that work on real or complex numbers in order to give a coherent explanation of these pictures. For a wide variety of scientific computations the continuous mathematics that the machine is simulating is the correct vehicle for analyzing the operation of the machine itself.

These reasonings give some justification for taking as a model for scientific computation a machine model that accepts real numbers as inputs. Of course a great many issues such as roundoff error must be dealt with. Moreover the ultimate justification is: does the model developed this way give new insights and understanding to the use of the big machines? To attempt to show this is a major goal of our work.

1.5   Classical Complexity Theory and Its Extension

A cornerstone of classical complexity theory is the theory of NP-completeness and the *fundamental* P $\neq$ NP? problem.

A main goal of this book is to extend this theory to the real and complex numbers and, in particular, to pose and investigate the fundamental problem within a broader mathematical framework.

The foundations for such a theory are developed in the next chapters. But here we give some background and briefly and informally introduce some of the classical notions.

Since the 1930s, much of the work of logicians focused on identifying and classifying decidable and undecidable problems. A prevailing view was that once a problem was known to be decidable (or solvable), then by and large, it was not terribly deep or interesting. In contrast, there was a great deal of interest and activity designed to untangle and understand the rich hierarchy amongst the undecidable problems (the "degrees of unsolvability").

To relate the notion of solvable problem to our earlier discussion of decidability (in Section 1.3), we can view a *decision problem* as a pair (X, X_{yes}). Here X is the set of *problem instances*, and X_{yes} the subset of *yes-instances*. Thus X plays the role of the universe U and X_{yes} of the subset S. The problem is *decidable* (or *solvable*) if X_{yes} is decidable (by a machine that on input $x \in X$ will output 1 if x is in X_{yes} and 0 if not). So, for example, for Hilbert's Tenth Problem, X would be the set of diophantine equations (polynomial equations with integer coefficients) and X_{yes} the subset of those with integer solutions.

With the advent of the digital computer, and its promise of solving hitherto intractable problems, interest perked in the realm of the solvable with the quest for efficient algorithms. Although there were many successes, it soon became apparent that a number of problems (such as the famous Traveling Salesman Problem) although solvable in principle, defied efficient solution. These problems seemed in essence intractable. Thus, amongst the solvable, there appeared to be yet another rich and natural hierarchy, with the dichotomy of tractability/intractability mirroring the earlier dichotomy of decidability/undecidability. And so, the theory and field of *computational complexity* was born.[5]

The foundation of this theory was developed in the 1960s, primarily by researchers originally trained in mathematics and logic but who found more hospitable environments for these interests in the newly emerging computer science departments. The theory began in an abstract setting with the formulation of axiomatic complexity measures yielding surprising speedup theorems, and then became more concrete with the NP-completeness results in the early 1970s. It

---

[5]Again in his Hixon Symposium lecture, von Neumann voiced the need for such a theory:

> Throughout all modern logic, the only thing that is important is whether a result can be achieved in a finite number of elementary steps or not. The size of the number of steps which are required, on the other hand, is hardly ever a concern of formal logic. Any finite sequence of correct steps is, as a matter of principle, as good as any other. It is a matter of no consequence whether the number is small or large, or even so large that it couldn't possibly be carried out in a lifetime, or in the presumptive lifetime of the stellar universe as we know it. . . . [On the other hand] in the case of an automaton the thing which matters is not only whether it can reach a certain result in a finite number of steps at all but also how many such steps are needed.

A primary concern here for von Neumann was his conviction that the cumulative effect of the small but nonzero probability of component failure "may (if unchecked) reach the order of magnitude of unity —at which point it produces, in effect, complete unreliability." In fact, to the contrary, the phenomenon of error build-up due to computer failure has not posed difficulties anywhere near the magnitude posed by the (apparent) intractability phenomenon.

is primarily this latter work, showing the equivalence of literally thousands of often seemingly unrelated difficult problems, that has captured the attention of researchers from many fields. These problems have the property that an efficient solution to any one can be easily converted to an efficient solution to any other.

The formalisms of classical complexity theory are founded on the models and formalisms of classical computation theory. Formal measures of complexity are intended to indicate various degrees of difficulty inherent in problems. These difficulties could be measured by the amount of information necessary to describe a problem (*descriptional* or *informational complexity*), the power of the language needed (*descriptive complexity*), or the amount of resources, such as time or space, required to *solve* the problem. In this book we primarily follow the tradition of *computational complexity* which studies the cost of computation with regard to time, or number of steps, for solution. The complexity of a problem is then measured in terms of the complexity of machines for solving it. Paramount here is that *complexity* is given as a function of *input word size L*, classically measured in bits.

A machine M is said to be a *polynomial time machine* if there are positive integers c and q such that for all inputs x,

$$\text{cost}_M(x) < c(\text{size}(x))^q.$$

Here $\text{cost}_M(x)$ denotes the number of *basic* operations performed by machine M from input x to output. A decision problem (X, X_{yes}) is in *class* P, or solvable in *polynomial time*, if it is decidable by a polynomial time machine.

Polynomial-time is an attempt to capture a notion of tractability and is what is meant in this discussion when we use qualifiers such as "quick," "efficient," "short," and "fast."

Note that to give upper bounds on complexity or to show a problem is tractable it is sufficient to demonstrate one appropriate machine. On the other hand, to claim a lower bound g for complexity or that a problem is not in *class* P (and hence intractable) is more problematic. For now we must demonstrate that *every* machine for solving it has a complexity function that grows faster than g or, in the latter case, faster than any polynomial.

The more subtle concept of *class* NP is meant to capture the notion that some problems have the property that for each yes-instance there exists a quick verification, or short proof, of this fact. Moreover, the proof system is sound in the sense that no proof exists for a no-instance.

Since a quick decision also serves as a quick verification, we see that class P is contained in class NP. It is natural to ask the converse: if a yes-instance has a short proof, can we find some such proof quickly? This is the essence of the fundamental P = NP ? problem.

Again, as in the case of decidability and computability, for all this to be reasonable and natural, we must have some degree of assurance that these notions and classes are independent of most "reasonable" formalisms.

To illustrate some of these ideas, we consider probably the most well-known problem of classical complexity theory, the Traveling Salesman Problem (TSP):

given n cities, the distances (a_{ij}) between them, and a positive number
k, does there exist a tour through all the cities with total distance less
than or equal to k?

and the related Shortest Path Problem (SPP):

given n cities, the distances (a_{ij}) between them, two specified cities l
and m, and a positive number k, does there exist a path from l to m
with total distance less than or equal to k?

The SPP is solvable in order n^2 operations.[6] On the other hand, the TSP appears
not at all to be tractable. All known solutions essentially require us to enumerate
the $(n - 1)!$ possible tours.[7]

Let us look at these problems a bit more formally. First, we can easily pose them
as decision problems in the form (X, X_{yes}). For example, for the TSP let

$$X = \{(A, k) \mid A = (a_{ij}) \text{ is an } n \times n \text{ matrix of } \textit{distances}, k > 0\}$$

and

$$X_{\text{yes}} = \{(A, k) \in X \mid \text{there is a } \textit{tour } \tau \text{ with Dist}(A, \tau) \leq k\}.$$

Here τ is a cyclic permutation of $\{1, 2, \ldots, n\}$ and

$$\text{Dist}(A, \tau) = \sum_{i=1}^{n-1} a_{\tau_i \tau_{i+1}} + a_{\tau_n \tau_1}.$$

Notice that X is the set of $\textit{all}$ problem instances, for all n. This reflects the fact
that we are interested in solving problems uniformly.

Notice also that, in stating these particular problems, we have made no assumption that the distances are integers; it makes perfectly good sense to talk about
these particular problems over the reals or any ring with order.

Over $\mathbb{R}$, a natural measure of the size of a TSP or SPP instance would be n^2 (the
number of entries in the matrix A of distances). Over $\mathbb{Z}$, a more natural measure
would be $n^2 b$ where b is the maximum of the heights (or binary lengths) of the
distances (a_{ij}) and k. This size roughly reflects the number of symbols needed to
describe the instance (or its bit length) and is essentially the classical measure.

The classical measure of cost, the bit cost, is the number of Turing machine
operations for solution. Thus the bit costs of the preceding solutions for SPP and

---

[6]We indicate a solution to the special case when the distances between distinct cities are
either 1 or $k + 1$. At stage 0, label city l with the number 0. At stage $s + 1$, label all unlabeled
cities that are distance 1 from the cities labeled s by the number $s + 1$. If no such cities
exist, terminate process and answer "yes" if city m is labeled by a number $\leq k$, otherwise
answer "no."

[7]By Sterling's formula, $n!$ is asymptotically equal to $(n/e)^n \sqrt{2\pi n}$ which is exponential
in n.

TSP are of order $n^2 b$ and $(n-1)!b$, respectively. Over the reals, a natural measure of cost could be the number of arithmetic computations and comparisons, and so for the preceding solutions to SPP and TSP, of order n^2 and $(n-1)!$, respectively. Thus, over $\mathbb{R}$ or over $\mathbb{Z}$, the cost of these solutions (as a function of size of instance) is linear in the case of SPP and exponential in the case of the TSP.

Now the TSP, although not known to be in class P over any ordered ring, is seen to be in class NP in any reasonable sense. Although we may not be able to easily tell if a TSP instance has a "good" tour (i.e., one of total distance bounded by k), if we are handed a good one we can quickly check it out: first check that it is indeed a tour and then sum up the n distances along the tour and compare with k.

The TSP is NP-*complete* over $\mathbb{Z}$; that is, it is universal for NP problems over $\mathbb{Z}$. If (X, X_{yes}) is a problem in NP over $\mathbb{Z}$, then *problem instances* $x \in X$ can be efficiently encoded as Traveling Salesman instances T_x such that $x \in X_{\text{yes}}$ if and only if T_x has a good tour. Thus an efficient solution to the TSP will yield an efficient solution to any other NP problem. Hence the importance of the TSP in classical complexity theory —not only because it is one of the ubiquitous problems of discrete optimization, but also because of its NP-completeness!

The Knapsack Problem (KP) introduced in Section 1.2.4 can also be posed as a decision problem in the preceding form. Let

$$X = R^\infty = \bigcup_{n \geq 0} R^n,$$

$$X_{\text{yes}} = \{x \in X \mid \exists b \in \{0, 1\}^n \text{ such that } \sum b_i x_i = 1\}. \tag{1.6}$$

Over any ring R, KP is in class NP in any reasonable sense. Moreover, the Knapsack Problem is also NP-complete over $\mathbb{Z}$ and hence equivalent to the TSP with respect to complexity.

We develop the theory of NP and NP-completeness over an arbitrary ring more formally and fully in the next chapters. In this general framework, we get a number of NP-completeness results, both classical and new. We define the classical Satisfiability Problem (SAT) and show it is NP-complete over $\mathbb{Z}_2$. We show that the Integer Programming Problem (IP) is NP-complete over $\mathbb{Z}$. Extending the theory to nonclassical domains, we show that the Hilbert Nullstellensatz is NP-complete over $\mathbb{C}$ and that 4-FEAS is NP-complete over $\mathbb{C}$ or $\mathbb{R}$. We do not know if TSP or KP are NP-complete over $\mathbb{R}$ or $\mathbb{C}$ (we suspect not).

In Part III, we describe a variety of complexity classes over the reals. In addition, we pursue a taxonomy of computational problems in this setting according to which of these classes they belong.

1.6　Complexity Theory in Numerical Analysis

It is natural to be skeptical about machines using exact arithmetic in numerical analysis. Most numerical problems can only be solved to within an accuracy of ε. Roundoff error is an important fact in the use of actual machines for solving scientific problems. Does it make sense to try to extend the complexity theory of computer science to numerical analysis?

We recall that computer scientists say that an algorithm defined by a machine M is *tractable* (or polynomial time, or in P) if the computation time $T(x)$ associated with input x satisfies the bound

$$T(x) \leq c(\text{size } (x))^q \quad \text{for all inputs } x, \tag{1.7}$$

where the constants c and q depend only on M. Here time is the number of Turing machine operations and size is the number of bits. A *problem is tractable* if there is a tractable algorithm solving it.

Some of the algorithms of numerical analysis are quite immediately tractable in a natural extension of this definition. Consider the problem of solving a linear system of equations $Ax = b$. The input of the problem is a nonsingular $n \times n$ matrix A and a vector $b \in \mathbb{R}^n$. Gaussian elimination produces an output x, solving this problem in less than cn^3 arithmetic operations. Therefore one can speak of the "tractability" of Gaussian elimination where Turing operators are replaced by arithmetic operations (and comparisons). Also the size of the input now becomes naturally the number of input variables. Thus

$$T(A, b) \leq c(\text{size}(A, b))^{3/2}$$

for Gaussian elimination.

More generally in numerical analysis it is important to take into account the desired accuracy ε of an approximate solution. This is because most problems cannot be solved exactly, even using exact arithmetic. Thus one must modify the concept of tractable and one way to do this is to consider $\varepsilon < 1$ as an additional (special) input to the problem. Then one demands that the time T of computation satisfy

$$T(\varepsilon, x) \leq (|\log \varepsilon| + \text{size}(x))^q, \ \varepsilon < 1. \tag{1.8}$$

Much recent work on solving nonlinear equations fits into this framework, where even sometimes $|\log \varepsilon|$ is replaced by $\log |\log \varepsilon|$ in (1.8).

Frequently among the set of inputs to a problem, there is a subset of "ill-posed" problems where the main algorithms fail and may even fail in principle. In general as an input gets closer to this ill-posed set the time of computation becomes larger.

A "condition number," a function on the input, has been defined traditionally to deal with this phenomenon. If the condition number of a certain input is large, then the time of computation can be expected to become large and the effects of round-off error to become substantial. It has often been shown that there is a relation between the condition number of an input and the reciprocal of the distance to the ill-posed set. Then the desired complexity results have the form

$$T(\varepsilon, x) \leq (|\log \varepsilon| + \log \mu(x) + \text{size}(x))^q,$$

where μ is the condition number of x.

The preceding remarks are all given a substantial development in Part II, with the main examples from the complexity analysis of algorithms in nonlinear equations and systems.

1.7   Summary

As we have said, our goal is to develop a theory of machines that will take real numbers as inputs.

Generally speaking, mathematical theories are built on plausible abstractions and simplifications intended to capture the essence of, rather than precisely describe, phenomena they are attempting to model. We hope that the basic assumptions reflect fundamental underlying principles, and that the results inferred from these assumptions reveal new truths. Justification for our model ultimately depends on how well this last task is accomplished.

The basic arithmetic operations $(+,-,\times,/)$ are taken as primary in the structures of computation that are described. This point of view bestows an algebraic emphasis so that it becomes natural to suppose that the inputs and states of the machines are numbers (or finite sequences of numbers) in a field (mathematical sense of the word). In the main case this is the field of real numbers. But certainly the field of complex numbers is also important.

There are natural situations where division cannot be done as within the integers $\mathbb{Z}$. So to cover those cases we extend our model of computation to machines over a ring.

Now formulating a theory of computation in this manner, that is, over a field K, we are able to include and extend the classical theory by taking $K = \mathbb{Z}_2$ (the field of two elements). In this way, the classical theory takes on an algebraic setting. By choosing K to be the real numbers $\mathbb{R}$, we are able to obtain a setting that provides a foundation of numerical analysis. The notion of an algorithm over $\mathbb{R}$ becomes well-defined as a mathematical object in its own right. So we have developed an *extension* of the classical theory to a new theory which can be specialized to the study of real number algorithms. This theory by the nature of our development is primarily algebraic. More precisely, when the field K is an ordered field as is the case of $\mathbb{R}$, the comparisons include $\leq$, and the geometry becomes what is called semi-algebraic. As we show later, the classical algorithms of mathematics and of computer science naturally fit into this framework.

It is important to remark that a fundamental property of classical computation is that the machines are finite objects, even though they operate on inputs that have no a priori bound on size. This property is satisfied by the machines developed here.

There is much sophisticated mathematics involved in many of the subjects of computation theory that this book addresses. In each case we attempt to illustrate the principles involved at the most elementary level we can. For example, we analyze Newton's method first for one variable and prove Bézout's Theorem first for one homogeneous equation. But we also hope to introduce, as gently as we can, the pleasures of a myriad of mathematical methods from elementary algebraic geometry and mathematical logic, to elementary applications of integral geometry, Morse theory, and Lie groups.

1.8   Brief History and Comparison with Other Models of Computation

The ideas presented in this chapter are at the confluence of different traditions in mathematics and computer science.

On the one hand, there is the work of classical computability and complexity. The initial motivating force here was —as we have already remarked— the question of the decidability of the arithmetic, and also the tenth problem posed by Hilbert at the Second International Congress of Mathematicians in 1900. A common characteristic of these problems is the possibility of expressing their underlying objects (arithmetic sentences and diophantine equations) in a language over a finite alphabet. Not surprisingly, the host of theoretical computational models that were subsequently proposed to formalize the notion of decidability were designed to act on finite strings over a finite alphabet.[8]

This was the case with the general recursive functions of Kleene [1936], the λ-computable functions of Church [1936], the computable functions of Turing [1936], and the canonical systems of Post [1943], to mention just the most influential models. Perhaps less expectedly, all models were equivalent in the sense that they defined the same class of computable functions. This gave rise to *Church's thesis*, discussed in Section 1.3.

On the other hand, there is a long-standing tradition of decidability results in algebra and analysis that we refer to as the *numerical tradition*. This theory led to algorithms —such as Newton's method discussed in Section 1.2 and Gaussian elimination for solving linear systems of equations— as well as to several undecidability results. A paradigm here is Galois' result on the nonsolvability by radicals of polynomial equations of degree-5 or more. It is important to notice that these algorithms manipulate real numbers in much the way proposed in this chapter.

With the arrival of the digital computer, attention shifted from decidability to complexity issues, and the first of the traditions previously described produced a sophisticated theory of complexity of which the P versus NP question described in Section 1.5 became central. We owe to this research concepts and tools that enable us to classify computational problems into complexity classes reflecting different resource requirements, and then to discover structural relations among these classes.

During the 1960s, Rabin [1960b], Hartmanis and Stearn [1965], and Blum [1967] developed the notion of measuring the complexity of a problem in terms of the number of steps required to solve it with an algorithm. This led in a natural way to the association by Cobham [1964], Edmonds [1965], and Rabin [1966] of the concept of "feasible" or "tractable" problems to the class P. Simultaneously, it

---

[8]Furthermore, at the turn of the century, with recent discoveries of paradoxes in the foundation of mathematics well in mind, there was an understandable preoccupation with questions of consistency. This surely was a factor in stipulating, in the early computational models, that the simplest operations were to be performed on the simplest objects.

was observed that a large class of search problems seemed to defy the existence of algorithms significantly better than brute force. At the beginning of the 1970s Cook [1971] defined a reduction for a class of decision problems corresponding to these search problems and proved the existence of complete problems. Concretely, he proved the completeness of the Satisfiability Problem of propositional logic. Independently Levin [1973] obtained similar results for a class of search problems and proved the existence of six complete problems, including the one of finding a satisfying truth assignment for a given propositional formula. This is the search version of Cook's completeness result for decision problems. Shortly afterwards, Karp [1972] considered the class of decision problems dealt with by Cook and coined the name NP for it. Moreover he showed that a series of familiar decision problems from different areas of discrete mathematics were also complete, coining the name NP-complete. This gave strong impetus to the subject that was reflected in work exhibiting hundreds of NP-complete problems and, on the other hand, in attempts to prove the inequality P $\neq$ NP leading to results on the structure of the class NP.

A lively exposition on the P versus NP question (containing a large list of NP-complete problems) can be found in the already classic book by Garey and Johnson [1979]. A survey of the state of the art of this question is given in [Sipser 1992]. In this latter article, a recently discovered letter of Gödel to von Neumann dated 1956 is reproduced in which Gödel stated the P versus NP question in the form of the time required by a Turing machine to test whether a formula of the predicate calculus has a proof of a given length.

The rise of complexity issues in the numerical tradition is less attached to the advent of the digital computer. Early in 1937, in a short note of Scholz [1937], complexity questions arose under the form of the number of additions needed to produce a given integer starting from 1. Seventeen years later Ostrowski [1954] conjectured the optimality of Horner's rule for evaluating univariate polynomials. In order to do so, he defined a formal model of computation and associated with it an idea of cost. This was followed by a flow of results concerning lower bounds (including the proof of Ostrowski's conjecture by Pan [1966]) for computational models with the following two characteristics:

(i) they take their inputs from R^n where R is a ring, and

(ii) their basic operations are arithmetic and complexity is measured by how many such operations are performed.

In most cases, the ring R was chosen to be the field of real numbers $\mathbb{R}$ and this choice, together with the second characteristic, reflected the kind of computations done in numerical analysis. However, these models were essentially nonuniform. This fact, useful for the search of lower bounds, becomes an obstruction to developing a theory of complexity for general-purpose algorithms. Two very influential papers at the end of the 1960s were those of Winograd [1967] and Strassen [1969]. They helped to make this search for lower bounds in algebraic problems an independent subject of study, now known as *algebraic complexity*. Some central examples of algebraic computational models along with lower bounds for them are given by Steele and Yao [1982], Ben-Or [1983], and Smale [1987]. Two early books on algebraic complexity are the ones by Borodin and Munro [1975] and by Winograd [1980]. A recent survey of the subject can be found in [Strassen 1990]. For a comprehensive book see [Bürgisser, Clausen, and Shokrollahi 1996].

Complexity issues are at the forefront of current research related to designing algorithms for finding zeros of polynomials and determining the solvability of polynomial systems. Amongst the major references here are: Collins [1975]; Schönhage [1982]; Shub and Smale [1993a, 1993b, 1993c, 1996, 1994]; Ben-Or, Kozen, and Reif [1986]; Grigoriev and Vorobjov [1988]; Renegar [1987a, 1992a]; Pan [1987, 1995]; Canny [1988]; and Heintz, Roy, and Solerno [1990]. This work may be considered the modern counterpart to algorithmic investigations begun earlier in the century by Hermann [1926], Van der Waerden [1949], and Tarski [1951], in particular related to elimination theory for real closed fields.

Although the history of numerical analysis provides us with a great motivating force towards our efforts here, we only give the reference [Goldstine 1977].

The computational model sketched in this chapter, and whose formal definition and properties are at the core of this book, was first introduced in [Blum, Shub, and Smale 1989]. It meets both traditions since it incorporates the universality of the classical ones (existence of universal machines and of NP-complete problems) while keeping the assumptions of the numerical one (real numbers given as an entity and unit cost of arithmetical operations) that make it suitable for modeling numerical analysis algorithms. Here again, there is a growing body of work —by Cucker [1993], Koiran [1993], Meer [1990, 1992, 1993], Michaux [1989, 1991], and Poizat [1995] among others— some of which is incorporated into later chapters of this book. A recent survey is [Meer and Michaux 1997].

In addition to the work already described, there are many more contributions by mathematicians and computer scientists that predate and relate to our model. We proceed now to review some of them.

Close to the classical approach, Rabin [1960a] developed a theory of computable algebra and fields in which the underlying domains can be effectively coded by natural numbers and are thus, necessarily countable.

On the other hand, the theories of computation over abstract structures are perhaps more general than ours. See, for example, [Engeler 1967] (contained also in [Engeler 1993]), [Friedman 1971] (or as discussed by Shepherdson in [Harrington, Morley, Seedrov, and Simpson 1985]), [Tiuryn 1979], and [Moschovakis 1986]. These general approaches both exploit and explore the logical properties of procedures. But, when applied to specific structures such as the reals, they do not yield the concrete mathematical results (such as the undecidability of the Mandelbrot set, NP-completeness of the Hilbert Nullstellensatz) that will quite naturally follow from the model we develop here.

There is yet another possible approach to the complexity of real valued problems known as recursive analysis, originating with Turing's seminal paper [Turing 1936]. Indeed, in this paper Turing introduced the notion of computable real numbers *before*, and as a means to, defining computation over the integers.[9] Here the machine model is the classical Turing machine and one deals with real numbers that, roughly speaking, are fed to the machine bit by bit. This contrasts with the numerical tradition where real numbers are viewed not as their decimal (or binary) expansion, but rather as mathematical entities. Text references for recursive analysis are [Ko 1991] for complexity matters and [Weihrauch 1987] for computability issues. Other references are [Friedman and Ko 1982], [Pour-El and Richards 1983], [Hoover 1987], and [Kreitz and Weihrauch 1982].

More closely related are the register machines of Shepherdson and Sturgis [1963] and the RAMs or random access machines. These were originally defined over the integers (see [Aho, Hopcroft, and Ullman 1974] for a definition) but with an algebraic character in their ground operations. An extension of the RAM to the real numbers is suggested in the book of Preparata and Shamos [1985]. The goal of the model is primarily to describe algorithms in computational geometry and the formal development of a theory of computability or complexity is not pursued. A different algebraic/logical approach to computability, called the combinatory programme, has been developed by Engeler and his students in [Engeler 1995]. Also, in the book previously mentioned by Borodin and Munro [1975], the authors state that their underlying model of computation will be the RAM. However, immediately afterwards they say that this "code can be 'unwound' and separate programs can be written for each 'degree' of the desired class of functions," justifying therefore the subsequent use of models of fixed dimension. Again, the formal development of a theory of computability and complexity over fields is not pursued.

Perhaps closest to our approach is the work of Herman and Isard [1970] on computability over arbitrary fields. Here, some finite-dimensional problems over the reals are shown to be undecidable in a manner similar to our proof of the undecidability of the Mandelbrot set. Also close is the work of Tucker [1980] and Tucker and Zucker [1992] who employ the theory of computing over abstract

---

[9]This fact does not seem well known, so it is of considerable historical interest to examine the very first paragraph of Turing's paper.

> The "computable" numbers may be described briefly as the real numbers whose expressions as a decimal are calculable by finite means. Although the subject of this paper is ostensibly the computable numbers, it is almost equally easy to define and investigate computable functions of an integral variable or a real or computable variable, computable predicates, and so forth. The fundamental problems involved are, however, the same in each case, and I have chosen the computable numbers for explicit treatment as involving the least cumbrous technique. I hope shortly to give an account of the relations of the computable numbers, functions, and so forth to one another. This will include a development of the theory of functions of a real variable expressed in terms of computable numbers. According to my definition, a number is computable if its decimal can be written down by a machine.

structures to obtain computability and noncomputability results in line with ours. Friedman and Mansfield [1992] have also specialized the abstract theory to specific structures to good avail.

Another model, again close in spirit, is a theory of real Turing machines outlined by Abramson [1971]. The machine model developed here can operate on arbitrarily long vectors of real numbers. The main thrust of the article is to develop a hierarchy of noncomputable functions according to their use of a greater-lower-bound operation.

Yet another approach, meeting ours in some points, is information-based complexity, developed by Traub, Wasilkowski, and Woźniakowski [1988]. A paradigm problem whose complexity is analyzed here is: given a function f of class C^p in $[0, 1]^n$, compute $\int_{[0,1]^n} f$. As one notes, inputs for this problem cannot be given in general by a finite vector of real numbers. So one must assume the existence of a routine that given $x \in [0, 1]^n$ returns $f(x)$. The complexity is evaluated in terms of the operations done as well as in terms of the number of times this routine is used. Again we are in the realm of the numerical tradition since the arithmetic is performed on real numbers at a constant cost and the main issue is the search for lower and upper bounds.

We close this section with some general references to the topics introduced in this chapter.

A good reference for the Mandelbrot and Julia sets is [Devaney 1989]. For the undecidability of the Mandelbrot set see [Blum and Smale 1993]. An undecidability result, related to chaotic dynamics as well, but very different in methodology and spirit is in [da Costa and Doria 1990]. A major reference for Hilbert's Tenth Problem is [Matiyasevich 1993].

For the Nullstellensatz, see [Lang 1993a], [Kendig 1977], and [Kunz 1985]. More advanced books in algebraic geometry are [Eisenbud 1995], [Hartshorne 1977], and [Shafarevich 1977], the first of them also containing a proof of the Nullstellensatz. These references, however, only deal with the qualitative aspect of the Nullstellensatz. Exponential bounds for the degrees in the Nullstellensatz were proved by Brownawell [1987] and refined by Kollar [1988] and Caniglia, Galligo, and Heintz [1988]. Related results are in [Krick and Pardo 1996]. Real polynomials, real algebraic sets, and semi-algebraic sets are exposed in the monographs of Benedetti and Risler [1990] and by Bochnak, Coste, and Roy [1987].

For Newton's method see Smale's survey article [Smale 1985]. A classical reference for linear and integer programming is the book of Schrijver [1986].

For the classical theory of computability and Turing machines see the books by Davis [1965], Rogers [1967], and Cutland [1980]. Classical complexity theory is a younger subject. The books by Balcázar, Díaz, and Gabarró [1988, 1990] or by Papadimitriou [1994] offer very good introductions to its achievements. Other references for this chapter are [Blum 1991; Blum 1990], [Smale 1988; Smale 1990], and [Shub 1993a]. In particular [Blum 1990] and [Traub and Woźniakowski 1982] discuss the question of polynomial time for the linear programming for real machines and related questions of conditioning. For a related discussion see [Barvinok and Vershik 1993].

The quotations from von Neumann, Penrose, and Kuhn are taken from [von Neumann 1963], [Penrose 1991], and [Kuhn 1957].

A version of this chapter appears in [Blum, Cucker, Shub, and Smale 1996b].

2
Definitions and First Properties of Computation

We begin the development of our theory of computation with the definition of a *finite-dimensional machine* over a ring. Although it is necessary to define a *machine* to be a more general object in order to fully develop a uniform theory —in particular with regard to complexity issues— a great deal can already be gleaned from the finite-dimensional case.

Indeed, in addition to being able to easily introduce in this setting the basic notions of computability and decidability and the fundamental tools (the computing endomorphism, register equations, and canonical path construction) that are used time and again, a number of major themes of this book emerge here. We show the undecidability of the Mandelbrot set, give lower bounds (on certain resources needed to solve the Knapsack Problem in the algebraic case), demonstrate the role of semi-algebraic sets, and initiate the theory of NP-completeness. In fact, the proof of our Main Theorem yielding the basic NP-completeness results in Chapter 5 are shown to be contained essentially in the succinct descriptions we obtain here for time-T halting sets.

Hence, this chapter is somewhat longer, with ideas perhaps more interleaved, than in subsequent chapters.

Although our primary interest is in developing foundations for a theory of computing —and complexity— over the reals, we first consider the somewhat more general framework of computing over a ring R, focusing on the basic arithmetic operations of addition and multiplication. We generally assume R is a commutative ring with unit, usually without zero divisors. If we wish to allow division, we suppose R is a field. Some important examples are: the ring of integers $\mathbb{Z}$; the integers mod p, $\mathbb{Z}_p$ ($p = 2, 3, 5, \ldots$); the rationals $\mathbb{Q}$; the reals $\mathbb{R}$; and the complex numbers $\mathbb{C}$. With $R = \mathbb{Z}$ (or $\mathbb{Z}_2$), we are also able to get much of the classical theory of computation (and complexity) as a special case. At the same time it is not hard to imagine how to proceed in an even more abstract fashion. But we do not take this route. We wish to emphasize that quite often the special character of the theory depends on the algebraic or geometric properties of the specific underlying structure and much is lost when the viewpoint is too abstract.

2.1   The Model of Computation: The Finite-Dimensional Case

Recall the simple Newton machine of Section 1.2 (see Figure A). We would like a model of computation in which Newton's method could be represented as naturally as in this machine, and in which its salient features would be as apparent. A Turing machine for implementing Newton's method, by reducing all operations to bit operations, would wipe out its basic underlying mathematical structure.

We note some essential features of the Newton machine. The machine is represented by a directed graph with four types of nodes —*input, computation, branch*, and *output*— each with associated functions and conditions on incoming and outgoing edges. For example, associated with the computation node is the Newton endomorphism N_f where

$$N_f(z) = z - \frac{f(z)}{f'(z)}, \tag{2.1}$$

and associated with the branch node is the stopping rule $|f(z)| < \epsilon$. The computation node has one outgoing edge and the branch node has two.

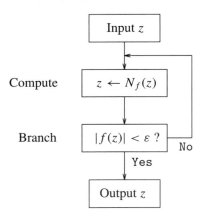

Figure A   A simple Newton machine.

Associated with the Newton machine is a natural input–output map Φ computed by "following the flow of the flowchart" as shown next.

Input $z \in \mathbb{C}$ into the machine. Compute $N_f(z)$ and replace "state" z by the new state $N_f(z)$, again called z. Test if $|f(z)| < \epsilon$. If Yes, then halt and output z. If No, return to the computation node and repeat the process. Continue in this fashion. If and when the machine halts, the output is the value of the input–output map. So with input $z_0 \in \mathbb{C}$, the output is the first point z_T (if such exists) in the orbit of z_0 under iterates of N_f for which $|f(z_T)| < \epsilon$. If $N_f(z_k)$ is undefined at some stage, the machine "crashes" and there is no output.

The machine will not in general halt on all inputs (e.g., for some inputs it may go into an infinite loop) and thus defines only a partial map. Let $\Omega \subset \mathbb{C}$ be the set of all inputs for which the machine halts with an output. The set Ω is called the *halting set* of the machine and the input–output map Φ is naturally defined on Ω with values in $\mathbb{C}$; that is, $\Phi : \Omega \to \mathbb{C}$.

We observe that

$$\Omega = \bigcup_{0 < T < \infty} \Omega_T,$$

where

$$\Omega_T = \{z \in \mathbb{C} \mid |f(N_f^{T'}(z))| < \epsilon \text{ for some } T', 0 \leq T' \leq T\}. \qquad (2.2)$$

And for $z \in \Omega_T$, $\Phi(z) = N_f^{T'}(z)$, where T' is the least $T > 0$ such that $z \in \Omega_T$. The set Ω_T is called the *time-T halting set* of the machine.

The Newton machine is really a machine over $\mathbb{R}$, not $\mathbb{C}$, since it tests the real order relation $|f(z)| < \epsilon$. So it is natural to view $\mathbb{C}$ as $\mathbb{R}^2$ and to consider $\mathbb{R}^2$ as being the "input space," "state space," and "output space" of this machine.

The Newton endomorphism associated with the computation node is given by a rational function, that is, a quotient of two polynomials, over $\mathbb{C}$. Over $\mathbb{R}$, the Newton endomorphism can be viewed as a map

$$g = (g_1, g_2) : \mathbb{R}^2 \to \mathbb{R}^2,$$

where

$$g_1(x, y) = \mathrm{Re} N_f(x + iy) \text{ and } g_2(x, y) = \mathrm{Im} N_f(x + iy).$$

Here Re and Im denote the real and imaginary parts, respectively, of a complex number. It can easily be checked that the real and imaginary parts of a polynomial f over $\mathbb{C}$ are given by polynomials over $\mathbb{R}$, denoted also by Re f and Im f, respectively. Similarly, if f is a rational function over $\mathbb{C}$, then Re f and Im f are rational functions over $\mathbb{R}$. Thus, each $g_i : \mathbb{R}^2 \to \mathbb{R}$, $i = 1, 2$, is a rational function over $\mathbb{R}$, and so g is a rational map over $\mathbb{R}$.

By expressing the stopping rule associated with the branch node as $|f(z)|^2 < \epsilon^2$, we can replace the stopping rule by an equivalent relation of the form $h(x, y) < 0$, where $h = f_1^2 + f_2^2 - \epsilon^2 : \mathbb{R}^2 \to \mathbb{R}$ is a polynomial over $\mathbb{R}$. Here $f_1 = \mathrm{Re} f$ and $f_2 = \mathrm{Im} f$.

Thus, the sets Ω_T in (2.2) are given by "semi-algebraic conditions" and the halting set Ω is seen to be a countable union of semi-algebraic sets over $\mathbb{R}$. A *semi-algebraic set* is a Boolean combination of sets defined by polynomial equalities and inequalities. More is said about this later.

We modify our Newton machine to reflect these changes (see Figure B). This derived Newton machine is an example of a *finite-dimensional machine* over a ring, or field in this case, which we formally define in this chapter. Some, but not all, of our theory of computation over a ring is captured in the finite-dimensional case. To obtain universal machines, and to develop a general theory of complexity for uniform algorithms, a more general model of machine is needed. We define such a model in the next chapter.

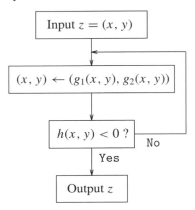

Figure B   Derived Newton machine.

We focus on rings because their basic operations —addition, subtraction, and multiplication— are the basic arithmetic operations. Polynomials are built from these basic operations. Fields add division and rational functions to our repertoire, but also some complications. So for simplicity of exposition, we sometimes define concepts first for rings, modifying them later for fields. Later one could add "subroutines" or "oracle" nodes for other functions or operations. Mostly, we assume our rings have an order reflecting the basic comparisons allowed over the integers, rationals, and reals. However, if we wish to focus on the *algebraic* character of a structure, or structures without order such as the finite fields or the complex numbers, we do not assume order.

So now let R be an ordered commutative ring (or field) with unit. Unless otherwise stated, we assume our rings are without zero divisors; that is, that all rings are *integral domains*. To obtain much of our goal, a reader may suppose R is the real field.

Let R^n be the "vector space" of n-tuples of elements of R.

Definition 1   A *finite-dimensional machine M over R* consists of a finite directed connected graph with four types of nodes: *input, computation, branch*, and *output*. The unique input node has no incoming edges and only one outgoing edge. All other nodes have possibly several incoming edges. Computation nodes have only one outgoing edge, branch nodes exactly two, Yes and No, and output nodes none.

In addition the machine has three spaces: an *input space* $\mathcal{I}_M$, *state space* $\mathcal{S}_M$, and *output space* $\mathcal{O}_M$ of the form R^n, R^m, R^ℓ, respectively, where n, m, and ℓ are positive integers. Associated with each node of the graph are maps of these spaces and *next node* assignments.

(1) Associated with the *input node* is a linear map $I : \mathcal{I}_M \to \mathcal{S}_M$ and unique next node β_1.

(2) Each *computation node* η has an associated *computation map*, a polynomial (or rational) map $g_\eta : \mathcal{S}_M \to \mathcal{S}_M$, and unique next node β_η. If R is a field, g_η can be a rational map.

(3) Each *branch node* η has an associated *branching function*, a nonzero polynomial function $h_\eta : \mathcal{S}_M \to R$. The next node along the Yes outgoing edge, β_η^+, is associated with the condition $h_\eta(z) \geq 0$ and the next node along the No outgoing edge, β_η^-, with $h_\eta(z) < 0$.

(4) Finally, each *output node* η has an associated linear map $O_\eta : \mathcal{S}_M \to \mathcal{O}_M$ and no next node.

A polynomial (or rational) map $g : R^m \to R^m$ is given by m polynomials (or rational functions) $g_j : R^m \to R$, $j = 1, \ldots, m$. If g is a rational map associated with a computation node (in the case R is a field), we assume each g_j is given by a *fixed* pair of polynomials (p_j, q_j), where $g_j(x) = (p_j(x))/(q_j(x))$.

Remark 1 It is worth mentioning here that there are a number of *standard* ways to *represent* polynomials in m variables of degree d over R. One way is as a sum of monomials $a_\alpha x^\alpha$ where $a_\alpha \in R$, $\alpha = (\alpha_1, \ldots, \alpha_m) \in \mathbb{N}^m$, $\sum_{i=1}^{m} \alpha_i \leq d$ and $x^\alpha = x_1^{\alpha_1} \ldots x_m^{\alpha_m}$. The terms in the sum can be ordered with respect to the lexicographic order on α. Usually, we have this *dense* representation (where every monomial is specified), or a variant, in mind. (*Sparse* representations disregard terms where $a_\alpha = 0$. Sometimes polynomials are given by *straight-line programs*.)

To a finite-dimensional machine M one can attach a function from a subset of the input space to the output space called the *input–output* map Φ_M. Heuristically Φ_M is defined by "following the flow" of the flowchart. Roughly one uses the orientation to pass from one node to the next node and then perform the "instruction" associated with that node. An important point is that instructions are carried out step by step. The formal definition is given in Section 2.2.

In this chapter, when we say "machine over R" we mean finite-dimensional machine over R. We say finite-dimensional since, in contrast to the general case, the input space, state space, and output space are each finite-dimensional.

If R is a ring or field without order, for example, the complex numbers $\mathbb{C}$ or the integers mod p $\mathbb{Z}_p$, we modify the definition of machine by stipulating in (3) that β_η^+ is associated with the condition $h_\eta(z) = 0$ and β_η^- with $h_\eta(z) \neq 0$.

Of course, in the ordered case, we can test if $h(z) = 0$ by a subroutine that uses the comparisons $h(z) \geq 0$ and $-h(z) \geq 0$. (See Figure C.)

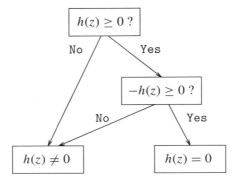

Figure C   Testing for equality.

The Newton machine, as well as the flowcharts for the Mandelbrot and Julia sets (Figures C and E of Chapter 1) are examples of finite-dimensional machines over $\mathbb{R}$. Also, for fixed n, the exhaustive search and algebraic machines for solving the Knapsack Problem (Figures J and K of Chapter 1) are also finite-dimensional. Notice that the algebraic machine for the n-dimensional Knapsack Problem, unlike the other examples, has no loops; that is, its graph is a tree. The paths in the tree give a particularly simple "path decomposition" of R^n into the union of the sets $V_1 = \{x \in R^n \mid k_n(x) \neq 0\}$ and $V_2 = \{x \in R^n \mid k_n(x) = 0\}$, where $k_n(x) = \prod_{b \in \{0,1\}^n}(\sum b_i x_i - 1)$. For fixed n, the exhaustive search machine can be "unwound" to produce a machine with no loops, again giving a path decomposition of R^n into a finite (but now exponential in n) number of semi-algebraic sets. The other examples of machines are not in general "time-bounded." In the Newton example, an analogous unwinding produces a countable path decomposition of the halting set into semi-algebraic sets:

$$\Omega = \bigcup_{0 < T < \infty} V_T,$$

where

$$V_T = \{z \in \mathbb{C} \mid |f(N_f^T(z))| < \epsilon \text{ and } |f(N_f^k(z))| \geq \epsilon \text{ for } 0 < k < T\}.$$

We end this section with several comments and observations about our definition of machines.

Remark 2

(1) If R is a field, a computation map associated with a computation node may not always be defined due to the vanishing of some denominator. However, without loss of generality to our theory, we are able to suppose that a computation map is always defined for every input. This can be done as follows. Insert subroutines into the given machine that check for the vanishing of relevant denominators before entering a computation node. If some denominator is zero, then the machine goes into a loop. Otherwise, the machine proceeds into the computation node as originally specified.

For example, the computation map N_f (2.1) associated with the computation node of the Newton machine may not be defined for some inputs. The machine in Figure D is equivalent to the Newton machine but does not have this problem. From now on, we suppose our machines are of the latter type; that is, computation maps are defined for all inputs.

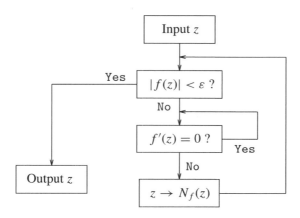

Figure D A Newton equivalent.

(2) It is often convenient to stipulate (again without loss of generality to our theory) that a machine M over R has a *unique* output node. This can be accomplished using a little programming device that collapses all output nodes to one. We can then identify the nodes of the machine with the *labels* $1, \ldots, N$ where 1 denotes the input node and N the output node. We assume such an identification unless otherwise indicated.

(3) Note that associated with a machine M over R is a finite set of *machine constants*

$$\mathcal{C}_M = \{1, c_1, \ldots, \overset{\bullet}{c}_{k_M}\} \subset R.$$

These are the coefficients of the maps associated with the nodes of the machine. Thus M is also a machine over a subring of R generated by $\mathcal{C}_M$.

(4) Also associated with a machine M is its *degree* D_M, the maximum of the degrees of all polynomials occurring in the maps associated with the nodes of M.

(5) One can think of the coordinate indices of the state space $\mathcal{S}_M = R^n$ as addresses of *registers* of the machine M. Thus, a finite-dimensional machine has a finite number of registers and, saying that M is in state $x = (x_1, \ldots, x_m)$ is to say that for $i = 1, \ldots, m$, x_i is in register with address i.

(6) We generally omit the subscript M for the machine when there is no confusion. On the other hand, we include it sometimes for emphasis. Similar remarks apply to the words "over R." In addition, we sometimes say that M is a machine over $(R, <)$ to emphasize that R is ordered; if we wish to emphasize that M branches only on equality comparisons, we say that M is a machine over $(R, =)$.

(7) We sometimes blur the distinction between variables x, y, z, ... and elements x, y, z, ... The intended meaning should be clear from the context.

2.2   The Input–Output Map, Halting Set, and the Computing Endomorphism

Let R be a ring and M a machine over R. Let $\mathcal{N} = \{1, \ldots, N\}$ be the set of nodes of M (with 1 the input node and N the output node) and $\mathcal{S}$ its state space. We call the space of node/state pairs $\mathcal{N} \times \mathcal{S}$ the *full state space* of the machine.[1]

Associated with M is the natural *computing endomorphism*

$$H : \mathcal{N} \times \mathcal{S} \to \mathcal{N} \times \mathcal{S} \tag{2.3}$$

of the full state space to itself. That is, H maps each node/state pair (η, x) to the unique *next node/next state* pair (η', x') determined by the graph of M and its associated maps.

To describe H explicitly, it is convenient to have *next node* assignments and *computation maps* defined for each node $\eta \in \mathcal{N}$. Thus we let

$$\beta_\eta = N \quad \text{for } \eta = N \text{ and}$$

$$g_\eta(x) = x \quad \text{for } \eta = 1, N, \text{ or a branch node.}$$

Now let $\mathcal{B} \subset \mathcal{N}$ be the subset of branch nodes of M and $\mathcal{C}$ be $\mathcal{N} - \mathcal{B}$. Then,

$$H(\eta, x) = (\beta_\eta, g_\eta(x)) \text{ for } \eta \in \mathcal{C}, \text{ and}$$

$$H(\eta, x) = \left\{ \begin{array}{l} (\beta_\eta^-, g_\eta(x)) \text{ if } h_\eta(x) < 0 \\ (\beta_\eta^+, g_\eta(x)) \text{ if } h_\eta(x) \geq 0 \end{array} \right\} \text{ for } \eta \in \mathcal{B}. \tag{2.4}$$

In case R is a ring without order, then

$$H(\eta, x) = \left\{ \begin{array}{l} (\beta_\eta^-, g_\eta(x)) \text{ if } h_\eta(x) = 0 \\ (\beta_\eta^+, g_\eta(x)) \text{ if } h_\eta(x) \neq 0 \end{array} \right\} \text{ for } \eta \in \mathcal{B}.$$

---

[1] Our terminology is influenced somewhat by that of dynamical systems. In line with classical nomenclature, we might also call $\mathcal{N} \times \mathcal{S}$ the *configuration space* of M.

The computing endomorphism is our main technical as well as conceptual tool. We use it to define basic notions such as the *halting-time function* T_M, the *input–output map* Φ_M, and the *halting set* Ω_M of a machine M over R as follows.

With input $x \in \mathcal{I}_M$, let $x^0 = I(x) \in \mathcal{S}_M$. Then with *initial point* $z^0 = (1, x^0) \in \mathcal{N} \times \mathcal{S}$, generate the *computation*, $z^0, z^1, z^2, \ldots, z^k, \ldots$, that is, the orbit of z^0 under iterates of H; that is,

$$z^0 = (1, x^0), z^1 = H(z^0), \ldots, z^k = (\eta^k, x^k) = H(z^{k-1}), \ldots. \qquad (2.5)$$

Let $\pi_{\mathcal{N}} : \mathcal{N} \times \mathcal{S} \to \mathcal{N}$ be the projection of the full state space onto $\mathcal{N}$. Then the sequence of nodes

$$\eta^0 = 1, \eta^1, \ldots, \eta^k, \ldots, \quad \text{where } \eta^k = \pi_{\mathcal{N}}(z^k) \text{ for } k = 0, 1, \ldots \qquad (2.6)$$

is the *computation path* γ_x *traversed* by input x.

Similarly, the *state trajectory* of input x is the sequence of states

$$x^0 = I(x), x^1, \ldots, x^k, \ldots, \quad \text{where } x^k = \pi_{\mathcal{S}}(z^k), \; k = 0, 1, \ldots. \qquad (2.7)$$

Here $\pi_{\mathcal{S}} : \mathcal{N} \times \mathcal{S} \to \mathcal{S}$ is the projection onto $\mathcal{S}$.

The computation *halts* if (ever) there is a *time* T such that $z^T = (N, u)$ for some $u \in \mathcal{S}$. If this is the case, the finite sequence

$$(z^0, z^1, \ldots, z^T) \in (\mathcal{N} \times \mathcal{S})^{T+1} \qquad (2.8)$$

is called a *halting computation*, and the finite sequence of nodes

$$(\eta^0, \eta^1, \ldots, \eta^T = N) \in \mathcal{N}^{T+1}, \qquad (2.9)$$

a *halting path traversed by* x. The *halting time* $T_M(x)$ is the least such T (we say M *halts* on input x in *time* T), and we define

$$\Phi_M(x) = O(x^T) \in \mathcal{O}_M. \qquad (2.10)$$

If there is no such T (i.e., M *does not halt* on input x), then $\Phi_M(x)$ is not defined and we let $T_M(x) = \infty$. Thus, Φ_M the *input–output map* is a partial map of the input space to the output space.

The *halting set* of M, Ω_M, is the set of all inputs on which M halts. That is,

$$\Omega_M = \{x \in \mathcal{I}_M \mid T_M(x) < \infty\}. \qquad (2.11)$$

Thus, $T_M : \Omega_M \to \mathbb{Z}^+$ and $\Phi_M : \Omega_M \to \mathcal{O}_M$.

We can extend these notions to the case R is a field. Although H may now only be a partial map, orbits of initial points $z^0 = (1, x^0)$ under iterates of H are always defined (see Remark 2), and so the previous definitions all make sense.

The complements of the Mandelbrot and Julia sets, as well as the good starting points for Newton's method, are examples of halting sets over $\mathbb{R}$.

With the machinery developed, we can begin to formalize other concepts and questions introduced in Chapter 1. For example, we define a map

$$\varphi : X \to R^\ell, \quad X \subset R^n \qquad (2.12)$$

to be *f.d. computable over R* if it is the input–output map of some finite-dimensional machine M over R; that is, if $R^n = \mathcal{I}_M$, $R^\ell = \mathcal{O}_M$, $X = \Omega_M$, and $\varphi = \Phi_M$ on X. We say *M computes φ*.

Remark 3

(1) In the next chapter we define a machine over R to be a more general object than a finite-dimensional machine. It can be shown that if a map as in (2.12) is the input–output map of a machine over R (in the general sense), then it is the input–output map of a finite-dimensional machine over R. See Section 2.8 for a reference. It follows that, if $X \subset R^n$ is the halting set of a machine over R (in the general sense), then it is a halting set of a finite-dimensional machine over R. Hence we may omit the qualifier f.d. in the preceding definition of computability. As long as we restrict ourselves to finite-dimensional spaces, the implicit qualifier "finite dimensional" in our subsequent definitions and discussion in this chapter are nonrestrictive.

(2) The preceding theory reduces to the classical (i.e., Turing) theory when $R = \mathbb{Z}$. That is, the f.d. computable functions over $\mathbb{Z}$ are exactly the classical recursive functions (see Section 1.3). Therefore this model of computation is sufficiently powerful to develop the classical theory of computation. (By Church's thesis, we would have cause for concern had we produced more functions computable over $\mathbb{Z}$.) References for this equivalence are given in Section 2.8.

For complexity theory, the comparison to the classical theory makes sense only after we have defined our general model of computation. Then we show that, both in terms of computation and complexity, the new theory naturally reduces to the classical when $R = \mathbb{Z}_2$. (See Chapters 3 and 4.)

(3) In the classical theory it is easily seen that the halting sets are exactly the output sets, in fact the output sets of machines with input space $\mathbb{N}$. This is why classically halting sets are called the *recursively enumerable* sets.

For machines that branch on inequality ($<$) comparisons over $\mathbb{R}$, or an arbitrary real closed field, it is also true that the halting sets are exactly the output sets. But now the proof uses Tarski's elimination theory. Indeed, for a subring $R \subseteq \mathbb{R}$ of infinite transcendence over $\mathbb{Z}$, a necessary and sufficient condition that "halting sets = output sets" is that R is a real closed field. Recall, a *real closed field* is an ordered field k in which positive elements have square roots and every odd degree polynomial with coefficients in k has a root in k. For subrings of $\mathbb{R}$ of finite transcendence, the satisfiability of "halting sets = output sets" is completely determined by the Dedekind cuts of members of a transcendence base. For references, see Section 2.8.

Definition 2 A set $S \subset R^n$ is *decidable over R* if its characteristic function $\chi_S : R^n \to R$ is computable over R where

$$\chi_S(x) = \begin{cases} 1 \text{ if } x \in S \\ 0 \text{ otherwise.} \end{cases}$$

Otherwise S is *undecidable over R*.

In this setting, Penrose's question of Chapter 1 may thus be posed formally as follows.

Is the Mandelbrot set $\mathcal{M}$ decidable over $\mathbb{R}$?

That is, does there exist a machine M over $\mathbb{R}$ with input space $\mathbb{C} (= \mathbb{R}^2)$ such that

$$\Phi_M(x) = \begin{cases} 1 \text{ if } x \in \mathcal{M} \\ 0 \text{ otherwise.} \end{cases}$$

But before addressing this, it is worth noting that $\mathcal{M}'$, the complement of $\mathcal{M}$, is *semidecidable* over $\mathbb{R}$. That is, there is a machine M' over $\mathbb{R}$ with input space $\mathcal{I}$ ($\mathbb{R}^2$ in this case) such that

$$\Phi_{M'}(x) = \begin{cases} 1 & \text{if } x \in \mathcal{M}' \\ 0 \text{ or undefined} & \text{otherwise.} \end{cases} \tag{2.13}$$

A semidecision machine for $\mathcal{M}'$ is given essentially by the flowchart in Figure C of Chapter 1. It is easy to see that the semidecidable sets over R are exactly the halting sets over R.

Proposition 1 *A set $S \subset R^\ell$ is decidable over R just in case both S and its complement $R^\ell - S$ are semidecidable (i.e., halting sets) over R.*

So Penrose's question reduces to the following.

Is the Mandelbrot set $\mathcal{M}$ a halting set over $\mathbb{R}$?

Proof of Proposition 1.   It is clear that the decidable sets are semidecidable. Now suppose, without loss of generality, that M is a semidecision machine for S with $\Omega_M = S$ and M' is a semidecision machine for S' with $\Omega_{M'} = S'$. We construct a decision machine M^* for S. A schema for such a machine is given in Figure G of Chapter 1 by replacing M_J by M'. More formally, let

$$\mathcal{I}_{M^*} = \mathcal{I}_M = \mathcal{I}_{M'}$$
$$\mathcal{S}_{M^*} = \mathcal{S}_M \times \mathcal{S}_{M'}$$
$$\mathcal{O}_{M^*} = R.$$

For $x \in \mathcal{I}_{M^*}$, let $I_{M^*}(x) = (I_M(x), I_{M'}(x))$. For $y^* \in \mathcal{S}_{M^*}$, let $O_{M^*}(y^*)$ be the first coordinate of y^*.

The nodes of M^* and their associated maps are defined recursively with

$$\mathcal{N}_{M^*} \subseteq \{0, 1\} \times \mathcal{N}_M \times \mathcal{N}_{M'} \cup \{1^*, N^*\}.$$

Here 1^* and N^* are distinguished symbols.

Let 1^* be the input node of M^*, N^* be the output node of M^*, and $\beta_{1^*} = (1, 1, 1) \in \mathcal{N}_{M^*}$.

Suppose $\eta^* = (s, \eta, \eta') \in \mathcal{N}_{M^*}$, where $s \in \{0, 1\}$, $\eta \in \mathcal{N}_M$, and $\eta' \in \mathcal{N}_{M'}$. We consider cases.

(a) Suppose $s = 1$ and η is an input or computation node of M. Then η^* is a computation node of M^* with $g_{\eta^*}(y^*) = (g_\eta(y), y')$, and $\beta_{\eta^*} = (0, \beta_\eta, \eta') \in \mathcal{N}_{M^*}$. Here $y^* = (y, y')$, where $y \in \mathcal{S}_M$ and $y' \in \mathcal{S}_{M'}$.

Similarly, if $s = 0$ and η' is an input or computation node of $\mathcal{N}_{M'}$, then η^* is a computation node with $g_{\eta^*}(y^*) = (y, g_{\eta'}(y'))$ and $\beta_{\eta^*} = (1, \eta, \beta_{\eta'}) \in \mathcal{N}_{M^*}$.

(b) If $s = 1$ and η is a branch node, then η^* is a branch node with g_{η^*} the identity map, $\beta_{\eta^*}^- = (0, \beta_\eta^-, \eta')$, and $\beta_{\eta^*}^+ = (0, \beta_\eta^+, \eta')$.

Similarly, if $s = 0$ and η' is a branch node, then η^* is a branch node with g_{η^*} the identity map, $\beta_{\eta^*}^- = (1, \eta, \beta_{\eta'}^-)$, and $\beta_{\eta^*}^+ = (1, \eta, \beta_{\eta'}^+)$.

(c) If $s = 1$ and η is an output node of M, then η^* is a computation node of M^* with $g_{\eta^*}(y^*) = (1, \ldots, 1) \in \mathcal{S}_{M^*}$ and $\beta_{\eta^*} = N^*$.

On the other hand, if $s = 0$ and η' is an output node of M', then η^* is a computation node of M^* with $g_{\eta^*}(y^*) = (0, \ldots, 0) \in \mathcal{S}_{M^*}$ and $\beta_{\eta^*} = N^*$.

Thus, M^* on input x alternates between simulating machine M and machine M'. It outputs 1 if M halts, 0 if M' halts. $\qquad\square$

2.3   Halting Sets and Computable Maps

We are now ready to take a closer look at halting sets and computable maps. The main result of this section, Theorem 1, states that halting sets are countable unions of semi-algebraic sets and computable maps are piecewise polynomial or piecewise rational maps.

We retain the assumptions and the notation of the previous sections. In particular, we assume machine M over R is in *normal form*. That is, we assume:

1. the set of nodes $\mathcal{N}$ is $\{1, \ldots, N\}$, where 1 denotes the input node and N the (unique) output node;

2. iterates of the computing endomorphism H are always defined on initial point $z^0 = (1, I(x))$, $x \in \mathcal{I}$.

In addition, we assume:

3. all branch nodes are *standard*, that is, that $h_\eta(z) = z_1$ for each branch node η and $z \in \mathcal{S}$.

We have already remarked that the first two assumptions do not restrict our theory. Neither does the third.

Proposition 2 *For each machine M over R there is a machine M' over R with standard branches only such that $\Phi_{M'} = \Phi_M$; that is, both M and M' have the same input–output maps. Moreover, we may suppose $S_{M'} = R \times S_M$, $T_{M'}(x) \leq 2T_M(x)$, $C_{M'} = C_M$, and $D_{M'} = D_M + 1$. Recall, C and D denote the set of machine constants and the machine degree, respectively.*

Proof.   Modify M by adding an initial coordinate space to its state space. Then before each branch node η insert a computation node η' with $g_{\eta'}(y_0, y) = (h_\eta(y), y)$ for $y_o \in R$, $y \in S_M$. Now replace η by a standard branch.   □

Now suppose M is a machine over R. Let $x \in \mathcal{I}_M$ and let $z^0, z^1, \ldots, z^k, \ldots$ be the computation with initial point $z^0 = (1, I(x))$. So,

$$z^k = (\eta^k, x^k) = H^k(1, I(x)),$$

where H^k is the kth iterate of H, $k = 0, 1, \ldots$.

Let $\gamma \,(= \gamma_x)$ be the computation path $\eta^0 = 1, \eta^1, \ldots, \eta^k, \ldots$, and let $\gamma(k)$ be the *initial computation path* $(\eta^0, \eta^1, \ldots, \eta^k) \in \mathcal{N}^{k+1}$ of γ of *length* k. Define the *initial path set*

$$\mathcal{V}_{\gamma(k)} = \{x' \in \mathcal{I}_M \mid \gamma_{x'}(k) = \gamma(k)\}. \tag{2.14}$$

Thus $\mathcal{V}_{\gamma(k)}$ is the set of points in the input space whose computation paths coincide with γ for the first k *steps*.

At each step k in the path γ, M *evaluates* a polynomial (or rational) map

$$G_{\gamma(k)} : \mathcal{I}_M \to \mathcal{S} \tag{2.15}$$

defined by $G_{\gamma(k)} = g_{\eta^k} \circ \ldots \circ g_{\eta^0} \circ I$, the composition of the computation maps along the initial path $\gamma(k)$ with I. Thus, for $x \in \mathcal{V}_{\gamma(k)}$, we have

$$G_{\gamma(k)}(x) = x^{k+1} = \pi_\mathcal{S} H^{k+1}(1, I(x)),$$

where $\pi_\mathcal{S} : \mathcal{N} \times \mathcal{S} \to \mathcal{S}$ is the projection onto $\mathcal{S}$.

Likewise, at each branch step k in path γ (i.e., η_k is a branch node) M *evaluates* the *step-k branching function*

$$f_{\gamma(k)} : \mathcal{I}_M \to R \tag{2.16}$$

defined by $f_{\gamma(k)} = \pi_1 \circ G_{\gamma(k)}$, where $\pi_1 : \mathcal{S}(= R^m) \to R$ is the projection onto the first coordinate. If R is a field, the branching function $f_{\gamma(k)}$ may be a rational function, otherwise it is a polynomial.

We note that $\mathcal{V}_{\gamma(k)}$ is determined by the *branching conditions* along the path $\gamma(k)$. In particular, let

$$L_{\gamma(k)} = \{f_{\gamma(k')} \mid k' < k, \ k' \text{ a branch step in } \gamma, \text{ and } \eta^{k'+1} = \beta^-(\eta^{k'})\}$$

and

$$R_{\gamma(k)} = \{f_{\gamma(k')} \mid k' < k, \ k' \text{ a branch step in } \gamma, \text{ and } \eta^{k'+1} = \beta^+(\eta^{k'})\}.$$

Then, in case R is ordered, we have

$$\mathcal{V}_{\gamma(k)} = \{x \in \mathcal{I}_M \mid f(x) < 0, g(x) \geq 0, f \in L_{\gamma(k)}, g \in R_{\gamma(k)}\}. \qquad (2.17)$$

Or, in case R is unordered,

$$\mathcal{V}_{\gamma(k)} = \{x \in \mathcal{I}_M \mid f(x) \neq 0, g(x) = 0, f \in L_{\gamma(k)}, g \in R_{\gamma(k)}\}. \qquad (2.18)$$

This description of $\mathcal{V}_{\gamma(k)}$ motivates our interest in *semi-algebraic* and *quasi-algebraic sets*. Suppose R is a ring or field and f and $g \in R[x_1, \ldots, x_n]$. We call $f(x) = g(x)$ a polynomial *equation* (or *equality*) *over* R, and $f(x) \neq g(x)$, a polynomial *inequality over* R. If R has an order, then the following are polynomial *inequalities over* R.

$$f(x) \neq g(x), \ f(x) < g(x), \ f(x) \geq g(x), \ f(x) > g(x), \ f(x) \leq g(x).$$

Definition 3 A set $S \subset R^n$ is *basic semi-algebraic* over R in the ordered case (or *basic quasi-algebraic*, in the unordered case) if S is the set of elements in R^n that satisfy a finite *system* of polynomial equalities and inequalities over R. A *semi-algebraic* (*or quasi-algebraic*) set is a finite union of basic semi-algebraic (or basic quasi-algebraic) sets.

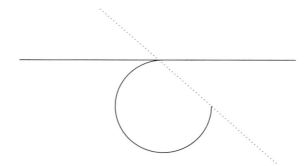

Figure E   A semi-algebraic set in $\mathbb{R}^2$: $S = (S_1 \cap S_2) \cup S_3$, where $S_1 = \{(x, y) \in \mathbb{R}^2 \mid x^2 + y^2 = 1\}$, $S_2 = \{(x, y) \in \mathbb{R}^2 \mid x + y \leq 1\}$, and $S_3 = \{(x, y) \in \mathbb{R}^2 \mid y = 1\}$.

Semi-algebraic sets are the fundamental objects of real algebraic geometry. The semi-algebraic subsets of $\mathbb{R}$ are finite unions of intervals —bounded or unbounded, open, closed (including single point sets), or half open. In higher dimensions, they provide more interesting geometric configurations such as those that arise in robotics and motion planning. (See Figure E for an example in $\mathbb{R}^2$.) In logic, (basic) quasialgebraic sets are called (basic) *constructible* sets. In algebraic geometry, basic quasi-algebraic sets are called *quasi-projective*. The quasi-algebraic subsets of $\mathbb{C}$ are the finite sets and their complements.

We remark that there are various equivalent *normal forms* for describing basic semi-algebraic (or basic quasi-algebraic) sets. For example, a basic semi-algebraic set S can be described as

$$S = \{x \in \mathbb{R}^n \mid f(x) < 0, \ g(x) \geq 0, \ \text{for } f \in F, \ g \in G\},$$

for some finite sets F and G of polynomials over R. Clearly, finite intersections of basic sets are basic. We also note that the semi-algebraic (or quasi-algebraic) sets are closed under Boolean operations.

Lemma 1

(1) *If R is an ordered ring or field, then $\mathcal{V}_{\gamma(k)}$ is basic semi-algebraic (or basic quasi-algebraic in the unordered case).*

(2) *If $\gamma_1(k) \neq \gamma_2(k)$, then $\mathcal{V}_{\gamma_1(k)} \cap \mathcal{V}_{\gamma_2(k)} = \emptyset$.*

Proof. If the branching functions evaluated by the machine are polynomials, then the first part of the lemma follows directly from the descriptions (2.17) and (2.18) of $\mathcal{V}_{\gamma(k)}$.

Otherwise, we note that by our stipulation that machines check for vanishing denominators before entering computation nodes (Remark 2), if a rational function p/q belongs to $L_{\gamma(k)} \cup R_{\gamma(k)}$, then $q(x) \neq 0$ for all $x \in \mathcal{V}_{\gamma(k)}$. Thus we may assume without loss that $q(x) \neq 0$ is one of the conditions defining $\mathcal{V}_{\gamma(k)}$.

Now for $x \in R^n$, if $q(x) \neq 0$, then

$$p(x)/q(x) = 0 \text{ if and only if } p(x) = 0.$$

If in addition R is ordered, then

$$p(x)/q(x) < 0 \text{ if and only if } p(x)q(x) < 0$$

and

$$p(x)/q(x) \geq 0 \text{ if and only if } p(x)q(x) \geq 0.$$

Using these correspondences, we can easily rewrite (2.17) and (2.18) to see that $\mathcal{V}_{\gamma(k)}$ is basic semi-algebraic or basic quasi-algebraic.

For the second part of the lemma, note that due to the bifurcation at branching steps, we have $\mathcal{V}_{\gamma_1(k')} \cap \mathcal{V}_{\gamma_2(k')} = \emptyset$ at the first step k' where γ_1 and γ_2 differ. □

Now let

$$\Gamma_T = \{\gamma_x(T) \mid T_M(x) \leq T \text{ for } x \in \mathcal{I}_M\} \tag{2.19}$$

be the set of *time-T halting paths*. Note that Γ_T is finite of size at most 2^T.

For each halting path $\gamma = (\eta^0, \eta^1, \ldots, \eta^T) \in \Gamma_T$ and $k \leq T$ it makes sense to define $\gamma(k)$ to be the *initial computation path* $(\eta^0, \eta^1, \ldots, \eta^k)$ of γ, and so we can adapt to time-T halting paths definitions given previously for computation paths

—such as $\mathcal{V}_{\gamma(k)}$, $G_{\gamma(k)}$, and $f_{\gamma(k)}$. In particular, $\mathcal{V}_\gamma = \mathcal{V}_{\gamma(T)}$, and for each $\gamma \in \Gamma_T$, the input–output map Φ_M restricted to $\mathcal{V}_\gamma$ is given explicitly by

$$\Phi_M|_{\mathcal{V}_\gamma} = O \circ G_{\gamma(T-1)} \tag{2.20}$$

and so is a polynomial, or rational, map.

Let

$$\Omega_T = \{x \in \mathcal{I} \mid T_M(x) \le T\} \tag{2.21}$$

be the *time-T halting set* of M. Then $\Omega_T = \bigcup_{\gamma \in \Gamma_T} \mathcal{V}_\gamma$ is a finite disjoint union.

Let

$$\Gamma_M = \bigcup_{T < \infty} \Gamma_T \tag{2.22}$$

be the *set of halting paths* of M and let

$$\Gamma'_M = \{\gamma \in \Gamma_M \mid N \text{ occurs only once in } \gamma\} \tag{2.23}$$

be the *set of minimal halting paths*. Note that each $x \in \Omega_M$ belongs to one and only one $\mathcal{V}_{\gamma'}$, $\gamma' \in \Gamma'_M$. In particular, $x \in \mathcal{V}_{\gamma'}$ where γ' is the minimal length halting path traversed by x. So, $\Omega_M = \bigcup_{\gamma \in \Gamma'_M} \mathcal{V}_\gamma$ is a countable disjoint union.

Combining Lemma 1 and the previous discussion, we have the following result.

Theorem 1 (Path Decomposition Theorem.) *For any machine M over R the following properties hold.*

(1) *For any $T > 0$, $\Omega_T = \bigcup_{\gamma \in \Gamma_T} \mathcal{V}_\gamma$ is a finite disjoint union of basic semi-algebraic sets (respectively, basic quasi-algebraic sets, in the unordered case).*

(2) *$\Omega_M = \bigcup_{\gamma \in \Gamma'_M} \mathcal{V}_\gamma$ is a countable disjoint union of basic semi-algebraic (respectively, basic quasi-algebraic) sets.*

(3) *For $\gamma \in \Gamma_M$, $\Phi_M|_{\mathcal{V}_\gamma}$ is a polynomial map, or a rational map if R is a field.*

□

Thus, the halting set of a machine M over R is a countable union of semi-algebraic (or quasi-algebraic) sets; the input–output map Φ_M is a piecewise polynomial or rational map. These properties are useful in demonstrating that specific sets are not halting sets and that specific functions are not computable. We do so in the next section. But first, we discuss some other consequences of Theorem 1 and related matters.

Definition 4 A *semi-algebraic formula* $\varphi(x)$ (or *quasi-algebraic formula* in the unordered case) is a finite combination of polynomial equalities and inequalities over R linked by the *logical connectives* $\wedge$ ("and"),[2] and $\vee$ ("or"). Here x denotes the sequence of variables $x_1, \ldots, x_n$.

---

[2]We sometimes denote this connective by &.

Remark 4

(1) Implicit in much of our discussion in this section is a natural correspondence between semi-algebraic (or quasi-algebraic) sets and semi-algebraic (or quasi-algebraic) formulas.

For example, if γ is a computation path and R is an ordered ring, the formula $\varphi_{\gamma(k)}(x)$ given by

$$\left(\bigwedge_{f \in L_{\gamma(k)}} f(x) < 0 \right) \wedge \left(\bigwedge_{f \in R_{\gamma(k)}} f(x) \geq 0 \right)$$

is a basic semi-algebraic formula and we have

$$\mathcal{V}_{\gamma(k)} = \{x \in R^n \mid \varphi_{\gamma(k)}(x) \text{ holds over } R\}.$$

Indeed, semi-algebraic (or quasi-algebraic) sets are exactly those sets *defined* by semi-algebraic (or quasi-algebraic) formulas φ, and are thus of the form

$$S_\varphi = \{x \in R^n \mid \varphi(x) \text{ holds over } R\}.$$

This is due to the natural correspondence between Boolean operations on sets and logical connections of formulas, namely,

$$S_{\varphi \wedge \psi} = S_\varphi \cap S_\psi \quad \text{and} \quad S_{\varphi \vee \psi} = S_\varphi \cup S_\psi.$$

We can also consider *semi-algebraic (or quasi-algebraic) systems* $\Phi(x) = (\varphi_1(x), \ldots, \varphi_k(x))$ where each φ_i, $i = 1, \ldots, k$ is a semi-algebraic (or quasi-algebraic) formula. Then Φ *defines* the set $S_\Phi \subseteq R^n$ where $S_\Phi = S_\varphi$ for $\varphi = \varphi_1 \wedge \ldots \wedge \varphi_k$.

(2) Thus, the time-T halting set Ω_T is defined by the semi-algebraic (or quasi-algebraic) formula $\psi_T(x)$ given by

$$\bigvee_{\gamma \in \Gamma_T} \varphi_\gamma(x).$$

We remark that, in general, this formula contains a number of polynomial equalities and inequalities over R that are exponential in T.

(3) We are sometimes interested in the inputs that output a fixed value v, for example, the inputs that output the value 1. Thus, we are interested in the sets

$$\Omega_M(v) = \{x \in \mathcal{I} \mid \Phi_M(x) = v\}$$

and

$$\Omega_T(v) = \{x \in \Omega_T \mid \Phi_M(x) = v\}, \tag{2.24}$$

where Φ_M is the input–output map of machine M. Note that

$$\Omega_T(v) = \bigcup_{\gamma \in \Gamma_T} \mathcal{V}_\gamma(v),$$

where

$$\mathcal{V}_\gamma(v) = \{x \in \mathcal{V}_\gamma \mid \Phi_M(x) = v\}, \text{ for } \gamma \in \Gamma_T.$$

Also,

$$\mathcal{V}_\gamma(v) = \mathcal{V}_\gamma \bigcap \{x \in R^n \mid O \circ G_{\gamma(T-1)}(x) = v\}.$$

So, $\Omega_T(v)$ is a semi-algebraic (or quasi-algebraic) set over R.

(4) Theorem 1 can be strengthened in several ways. First, since there are only a finite number of machine constants, all the relevant sets and maps can be defined by polynomials with coefficients from the same finitely generated extension of the prime ring or field. Secondly, in Part (2), "countable union" can be replaced by "effective union" in the classical Turing sense. Also, this condition is sufficient. That is, for any finite subset C of R, an effective union of basic semi-algebraic (respectively, quasi-algebraic) sets in R^n, defined by polynomials with coefficients from C, is the halting set of some machine M over R with constants from C. See Section 2.8.

(5) Finally, we remark that Theorem 1 has little to say about computing over $\mathbb{Z}$ since all subsets of $\mathbb{Z}^n$ are countable and singleton sets are semi-algebraic over the prime ring $\mathbb{Z}$. Indeed, halting sets over $\mathbb{Z}$ are exactly the effective (i.e., recursively enumerable) sets.

We finish this section with an application of some notions introduced thus far. The result, Proposition 3, though simple, is illustrative of an underlying goal of this book.

Let M be a finite-dimensional machine over $(\mathbb{R}, <)$; that is, M branches on inequality $(<)$ comparisons. The number of halting paths of M of length k is bounded by 2^k. This follows since the next node function is determined for every node other than a branching node and for the latter only two next nodes are possible. For a halting path γ denote by $B(\gamma)$ the number of branch nodes of γ. The same argument shows that the number of minimal halting paths γ' with $B(\gamma') \leq k$ is also bounded by 2^k.

For a given $x \in \mathcal{I}_M = \mathbb{R}^n$, let γ'_x denote the minimal halting path traversed by x. The number of branch nodes $B(\gamma'_x)$ performed during the computation of M with input x can be seen both as a trivial lower bound for the total number of steps performed during this computation and as a measure of an obstruction for Φ_M to be rational.

Our goal is to prove a lower bound for the amount of branching needed to compute the *floor function*. This is the function

$$\lfloor \ \rfloor : \mathbb{R} \to \mathbb{Z}$$

$$x \to \lfloor x \rfloor,$$

where $\lfloor x \rfloor$ is the largest integer less than or equal to x. We begin with a lemma whose simple proof is omitted.

Lemma 2 *The semi-algebraic subsets of $\mathbb{R}$ are exactly the finite union of intervals.*
□

Proposition 3 *Let M be a f.d. machine over $\mathbb{R}$ that computes the function $x \to \lfloor x \rfloor$. Then $\{x \in \mathbb{R} \mid B(\gamma_x') \geq \log x\}$ is an unbounded subset of $\mathbb{R}$. Here γ_x' denotes the minimal halting path traversed by x.*

Proof.   Consider the interval $[0, \ell]$ for $\ell \in \mathbb{N}$ and suppose that $T_M(x) \leq T_\ell$ for all $x \in [0, \ell]$. Thus, there are at most 2^{T_ℓ} computation paths associated with the elements in $[0, \ell]$. Associated with each such path is a rational function on the input variable $\varphi_\gamma : \mathbb{R} \to \mathbb{R}$ which coincides with the $\lfloor \ \rfloor$ function over the set $\mathcal{V}_\gamma$.

Suppose $r \in \mathbb{N}$, $r \leq \ell$ and consider the interval $(r - 1, r]$. Since the sets $\mathcal{V}_\gamma$ are semi-algebraic subsets of $\mathbb{R}$ it follows from Lemma 2 that there is a path γ such that $\mathcal{V}_\gamma$ contains an open subset of $(r - 1, r]$. Thus, φ_γ is the constant function r on some open subset of $\mathbb{R}$ from which we deduce that φ_γ is a constant function. It follows that $2^{T_\ell} \geq \ell$ which in turn implies the proposition.   □

2.4   The Mandelbrot Set Is Undecidable

We can now give precise answers to the questions about the decidability of the Mandelbrot set and other sets considered in Chapter 1. These answers, however, rely on results from the theory of complex dynamics lying outside the realm of this book and thus we omit their proofs. Relevant references can be found in Section 2.8.

Theorem 2 *The Mandelbrot set is not decidable over $\mathbb{R}$.*

Theorem 2 follows from Lemma 3 which in turn follows from the fact that the boundary of the Mandelbrot set has Hausdorff dimension 2.

Lemma 3 *The Mandelbrot set is not the countable union of semi-algebraic sets over $\mathbb{R}$.*
□

We can now prove Theorem 2.

Proof of Theorem 2.   From Proposition 1 it is sufficient to show that the Mandelbrot set is not a halting set of any machine. Now use Theorem 1 and Lemma 3 to see that, indeed, this is the case.   □

Similarly, "most" Julia sets are not decidable over $\mathbb{R}$ since, from the theory of complex analytic dynamical systems, we know that most Julia sets have fractional Hausdorff dimension. But one of the consequences of Theorem 1 is that halting sets must have integral Hausdorff dimension.

Indeed, for hyperbolic rational maps of the Riemann sphere, the following can be proved.

Theorem 3 *A Julia set is decidable over* $\mathbb{R}$ *if and only if it is*
  1. *a round circle,*
  2. *an arc of a round circle, or*
  3. *the whole sphere.* □

Consider the third example of Section 1.2 of deciding whether a real number x is a good starting point for Newton's method. It is sufficient to use the cubic $f(x) = x^3 - 2x + 2$ in that example to obtain the undecidability result.

It is known that the set of points that do not converge to a root of this f under iteration by Newton's method is exactly a Cantor set. But a Cantor set cannot be the countable union of semi-algebraic sets. In fact a Cantor set is uncountable, yet contains no intervals except for points. Thus, we conclude the following.

Theorem 4 *The set of points that converge under Newton's method is generally undecidable over* $\mathbb{R}$. □

2.5 The Canonical Path Construction

Upper bound results tell us that certain tasks can be accomplished if we have sufficient means. Lower bounds give us information about necessary resources. Undecidability results, like those in the previous section, can be considered ultimate lower bounds —no resources are enough! Proposition 3 is an example of a lower bound result. In this section we give a lower bound on the computational resources needed to solve the Knapsack Problem in the *algebraic case*, that is, when only equality comparisons are allowed. We use a construction, called the *canonical path construction*, that proves useful again later for other lower bound results.

In this section we assume that R is an infinite field.

Consider a machine M over $(R, =)$; that is, M branches only on equality comparisons. Let $\gamma = \eta^0, \eta^1, \ldots, \eta^k, \eta^{k+1}, \ldots$ be a computation path of M.

Recall from (2.16) that associated with each branch step k in γ is a rational branching function $f_{\gamma(k)}$ evaluated by M. We say that the branch step k is *exceptional* if $f_{\gamma(k)}$ is 0 on $\mathcal{V}_{\gamma(k)}$, and *nonexceptional*, otherwise. If k is an exceptional branch step, then $\eta^{k+1} = \beta^+(\eta^k)$ and $\mathcal{V}_{\gamma(k+1)} = \mathcal{V}_{\gamma(k)}$. Let

$$\mathcal{B}_\gamma = \{k \mid k \text{ is a nonexceptional branch step in } \gamma\}. \tag{2.25}$$

Note that if $\gamma(T)$ is a halting path for some $T > 0$, then $\mathcal{B}_\gamma$ is finite, and of size less than T.

Definition 5 The path γ is a *canonical path* of M if $\mathcal{B}_\gamma \neq \emptyset$ and for each $k \in \mathcal{B}_\gamma$, $\eta^{k+1} = \beta^-(\eta^k)$.

Thus, canonical paths are traversed by inputs that always take the No (i.e., $\neq 0$) outgoing edge at nonexceptional branch steps. Note that M has at most one canonical path.

For the rest of this section we suppose that M is *time bounded*; that is, there is a $T > 0$ such that the halting time $T_M(x) \leq T$ for all $x \in \mathcal{I}_M$. In this case, if M has no canonical path, then the input–output map Φ_M is a rational map. So, for example, by Lemma 4, if Φ_M takes on exactly $n > 1$ values, then M must have a canonical path γ.

Now suppose γ is a canonical path of M. Let

$$F_\gamma = \prod_{k \in \mathcal{B}_\gamma} f_{\gamma(k)}. \qquad (2.26)$$

The following is clear.

Proposition 4 *The path set*

$$\mathcal{V}_\gamma = \{x \in \mathcal{I}_M \mid f_{\gamma(k)}(x) \neq 0 \text{ for each } k \in \mathcal{B}_\gamma\} = \{x \in \mathcal{I}_M \mid F_\gamma(x) \neq 0\}.$$

Hence, since R is infinite, $\mathcal{V}_\gamma$ is Zariski dense in R^n. □

Recall a set $X \subseteq R^n$ is *Zariski dense* in R^n if only the zero polynomial vanishes on X. (See Appendix A at the end of Part I for a definition of the Zariski topology as well as for some other basic notions of algebraic geometry.) The next lemma states an important property of Zariski dense subsets of R^n.

Lemma 4 *Let $X \subseteq R^n$ be dense with respect to the Zariski topology in R^n and suppose $\varphi : X \to R$ is a rational function. If the image of φ is a finite set, then it consists of a single point.*

Proof. Let $\{a_1, \ldots, a_m\}$ be the image of φ and consider $X_i = \{x \in X \mid \varphi(x) = a_i\}$ for $1 \leq i \leq m$. Since $\cup X_i = X$ we have that $\cup \overline{X_i} = R^n$, where the bar denotes Zariski closure. But since R^n is not the union of a finite number of proper algebraic subsets (which can be seen by induction on n) there is an $i \leq m$ such that $\overline{X_i} = R^n$.

Consider now the rational function $\widetilde{\varphi} = \varphi - a_i$. Restricted to X_i the function $\widetilde{\varphi}$ is identically zero. But if a rational function vanishes on a set, it vanishes in its Zariski closure. Thus, $\widetilde{\varphi}$ is identically zero on R^n. From here we deduce that φ is the constant function a_i over R^n and a fortiori over X. □

Theorem 5 (Canonical Path Theorem.) *Let $S \subset R^n$. Suppose M is a decision machine for S over R and that γ is its canonical path. If S is not Zariski dense in R^n, then $S \cap \mathcal{V}_\gamma = \emptyset$. Hence*

(a) $\Phi_M|_{\mathcal{V}_\gamma} \equiv 0$, *and*

(b) $F_\gamma|_S \equiv 0$.

Proof. If $S \cap \mathcal{V}_\gamma \neq \emptyset$ then, by Lemma 4, $\Phi_M|_{\mathcal{V}_\gamma} \equiv 1$, and so $\mathcal{V}_\gamma \subseteq S$. Hence, by Proposition 4, S would be Zariski dense in R^n. □

We now proceed to the lower bound for the Knapsack Problem. The proof relies on a Nullstellensatz which we prove in the next proposition. The dependency of some lower bounds on various forms of Nullstellensätze is a theme which appears more in this book.

Proposition 5 *Suppose $f, h \in R[x_1, \ldots, x_n]$, and $\deg h = 1$. If f vanishes on the set of zeros of h in R^n, then f is a multiple of h.*

Proof.   After a change of coordinates we can suppose that h is the monomial x_n. Write f as

$$a_d x_n^d + \ldots + a_1 x_n + a_0$$

with $a_i \in R[x_1, \ldots, x_{n-1}]$ for $i = 0, \ldots, d$. Since f vanishes on the zeros of h we have that for every $y \in R^{n-1}$, $f(y, 0) = 0$; that is, $a_0(y) = 0$. But since R is infinite this implies $a_0 \equiv 0$ and thus, f is a multiple of h.   □

Theorem 6 *Let R be an infinite field and M a time bounded f.d. machine over $(R, =)$ deciding the n-Knapsack Problem. Let γ be the canonical path of M and B_γ the set of nonexceptional branch steps of γ. Then, for any representation of $f_{\gamma(k)}$ as a quotient p_k/q_k of polynomials, $\sum_{k \in B_\gamma} d_k \geq 2^n - 1$, where d_k denotes the degree of p_k.*

Proof.   Recall from Section 1.2 that the n-Knapsack Problem is the problem of deciding the set

$$K_n = \{x \in R^n \mid \exists b \in \{0, 1\}^n \text{ such that } \sum_{i=1}^n b_i x_i = 1\}.$$

Consider for any $b \in \{0, 1\}^n$, $b \neq 0$, the linear polynomial

$$h_b = \sum_{i=1}^n b_i x_i - 1$$

and its zero set S_b in R^n. The set K_n is the union of the $2^n - 1$ sets S_b.

Each S_b, being the zero set of a nontrivial polynomial is not Zariski dense in R^n. Therefore, neither is K_n.

Now consider a machine M deciding K_n as in the statement of the theorem. It follows from Theorem 5(b) that K_n is included in the set of zeros of $P_\gamma = \prod_{k \in B_\gamma} p_k$. Thus, according to Proposition 5 for any $b \in \{0, 1\}^n$, $b \neq 0$, the linear polynomial h_b divides P_γ. Since the h_b are relatively prime it follows that $\deg P_\gamma \geq 2^n - 1$.   □

2.6   The Register Equations and Succinct Descriptions

In this section we develop the *register equations* derived from the computing endomorphism. These equations give us explicit and succinct mathematical formulas for describing fundamental objects such as halting computations and halting sets and so play a central role in our development of complexity theory and in explorations into decidability.

We continue with the finite-dimensional case, retaining all the definitions, notations, and assumptions of Sections 2.1 through 2.3. In particular, we assume machine M over R is in normal form. Thus, the set of nodes is $\mathcal{N} = \{1, \ldots, N\}$, where 1 denotes the input node and N the output node, iterates of the computing endomorphism H are always defined on initial point $z^0 = (1, I(x))$, $x \in \mathcal{I}$, and branching depends only on comparing the first coordinate z_1 of $z \in \mathcal{S}$ to zero.

As remarked earlier, it follows from Theorem 1 that the time-T halting set Ω_T is described by a semi-algebraic (or quasi-algebraic) formula whose length is exponential in T. We are seeking a more succinct description.

To say that $x \in \Omega_T$ is to say that there is a sequence $(z^0, z^1, \ldots, z^T) \in (\mathcal{N} \times \mathcal{S})^{T+1}$ satisfying the *next state conditions*

$$z^k = H(z^{k-1}), \quad k = 1, \ldots, T \tag{2.27}$$

and the *initial* and *terminal conditions*

$$z^0 = (1, I(x)) \quad \text{and} \quad z^T = (N, u) \tag{2.28}$$

for some $u \in \mathcal{S}$.

Equations (2.27) and (2.28) represent the First Form of our (time-T) *register equations*. We examine them more closely in this section.

It is useful to make explicit the coordinate maps of H, namely, the *next node map* $\beta : \mathcal{N} \times \mathcal{S} \to \mathcal{N}$ and the *next state map* $g : \mathcal{N} \times \mathcal{S} \to \mathcal{S}$. Again, let $\mathcal{B}$ be the set of branch nodes, and $\mathcal{C} = \mathcal{N} - \mathcal{B}$. Then

$$\beta(\eta, x) = \beta_\eta, \text{ for } \eta \in \mathcal{C},$$

$$\beta(\eta, x) = \left\{ \begin{array}{l} \beta_\eta^- \text{ if } x_1 < 0 \ (x_1 = 0) \\ \beta_\eta^+ \text{ if } x_1 \geq 0 \ (x_1 \neq 0) \end{array} \right\} \text{ for } \eta \in \mathcal{B}, \text{ and} \tag{2.29}$$

$$g(\eta, x) = g_\eta(x) \text{ for all } \eta \in \mathcal{N}.$$

Thus, $H(\eta, x) = (\beta(\eta, x), g(\eta, x))$.

So, to say that $x \in \Omega_T$ is to say there is a sequence $(\eta^0, x^0), \ldots, (\eta^T, x^T) \in (\mathcal{N} \times \mathcal{S})^{T+1}$ such that

$$\eta^k = \beta(\eta^{k-1}, x^{k-1}) \text{ and } x^k = g(\eta^{k-1}, x^{k-1}), \quad k = 1, \ldots, T \tag{2.30}$$

$$(\eta^0, x^0) = (1, I(x)) \quad \text{and} \quad \eta^T = N. \tag{2.31}$$

Equations (2.30) and (2.31) are the register equations in their Second Form.

In order to view these equations as equations over R, we extend the maps β, g, and H to maps of vector spaces over R. We do this by considering the injection $\mathcal{N} \hookrightarrow R^N$ given by $j \to e_j$, where e_j is the jth coordinate vector, for $j = 1, \ldots, N$.

Now we can define $\widehat{\beta} : R^N \times S \rightarrow R^N$, $\widehat{g} : R^N \times S \rightarrow S$ (and hence $\widehat{H} = (\widehat{\beta}, \widehat{g}) : R^N \times S \rightarrow R^N \times S$) as follows,

$$\widehat{\beta}(\alpha, x) = \sum_{j=1}^{N} \alpha_j e_{\beta(j,x)} \text{ and } \widehat{g}(\alpha, x) = \sum_{j=1}^{N} \alpha_j g(j, x). \qquad (2.32)$$

Once again, we rewrite the register equations collecting terms to one side. We thus obtain their Third Form

$$\alpha^k - \widehat{\beta}(\alpha^{k-1}, x^{k-1}) = 0 \text{ and } x^k - \widehat{g}(\alpha^{k-1}, x^{k-1}) = 0, \ k = 1, \dots, T \qquad (2.33)$$

$$(\alpha^0, x^0) - (e_1, I(x)) = 0 \text{ and } \alpha^T - e_N = 0. \qquad (2.34)$$

As before, x varies over R^n, and the x^k, $k = 0, \dots, T$ vary over R^m, where n and m are, respectively, the dimensions of the input and state spaces of machine M. The α^k, $k = 0, \dots, T$ are also variables, now ranging over R^N.

Notation. Let $\mathcal{R}_T(x, z)$ denote the (time-T) register equations (2.33) and (2.34), where $x = (x_1, \dots, x_n)$ and $z = (\alpha^0, x^0, \dots, \alpha^T, x^T)$. Here z ranges over R^t, where $t = (N + m)(T + 1)$. Thus, we may write z as $(z_1, \dots, z_t)$, where each variable z_j ranges over R.

Remark 5 The time-T halting set Ω_T can now be described succinctly in terms of $\mathcal{R}_T(x, z)$. Namely,

$$x \in \Omega_T \Leftrightarrow \mathcal{R}_T(x, z) \text{ is solvable over } R.$$

That is,

$$\Omega_T = \{x \in R^n \mid \exists z \in R^t \text{ such that } \mathcal{R}_T(x, z) \text{ holds over } R\}.$$

Here $t = (N + m)(T + 1)$.

Much of our understanding about computation and halting sets comes from analyzing the time-T register equations $\mathcal{R}_T(x, z)$ and deriving *equivalent* systems with nice structural properties.

Remark 6

(1) Note that in equations (2.33), $\widehat{g}$ is a polynomial or rational map, but that $\widehat{\beta}$ generally is not even continuous, so we must handle these equations with care!

(2) Note that in $\mathcal{R}_T(x, z)$ the n input variables $x = (x_1, \dots, x_n)$ *only* appear in the m linear equations given by $x^0 - I(x) = 0$ of (2.34).

Theorem 7 *Suppose M is a machine over R with $\mathcal{I} = R^n$, $S = R^m$, $\mathcal{N} = \{1, \dots, N\}$, and suppose the degree of M is D_M.*

Then for any time bound $T > 0$, the system of register equations $\mathcal{R}_T(x, z)$ is equivalent to a semi-algebraic (or quasi-algebraic) system $\Phi_T(x, z)$, where

(1) $x = (x_1, \ldots, x_n)$, $z = (z_1, \ldots, z_t)$, $t = (N + m)(T + 1)$, and the variables x_i and z_j range over R;

(2) $\Phi_T(x, z)$ contains at most $4(N + m)T$ polynomial equations and $2T$ linear inequalities; and

(3) the equations have degree at most $D_M + 1$, or $ND_M + 1$ if R is a field.

We prove Theorem 7 by closely examining the register equations. But first we state two important corollaries. The first gives us a succinct semi-algebraic description of Ω_T.

Corollary 1 For any machine M over R and any time $T > 0$, the time-T halting set $\Omega_T \subseteq R^n$ is the projection of a semi-algebraic (or quasi-algebraic) set $S \subseteq R^{n+t}$ defined by a "small" semi-algebraic (or quasi-algebraic) system Φ_T.

In particular, $\Omega_T = \{x \in R^n \mid \exists z \in R^t$ such that $\Phi_T(x, z)$ holds over $R\}$, where Φ_T is a semi-algebraic (or quasi-algebraic) system and the bounds —on t, on the number of equations and inequalities in Φ_T, and on their degrees— are the same as in Theorem 7.   □

Remark 7 We are also interested in describing the inputs that output a fixed value v, say 1. Recall that $\Omega_T(v) = \{x \in \Omega_T \mid \Phi_M(x) = v\}$. To describe $\Omega_T(v)$, we add to the register equations the ℓ linear equations

$$O(x^T) - v = 0. \tag{2.35}$$

Here ℓ is the dimension of the output space of M.

Corollary 2 The set $\Omega_T(v) \subset R^n$ is the projection of a semi-algebraic (or quasi-algebraic) set $S_v \subset R^{n+t+\ell}$ defined by the semi-algebraic (or quasi-algebraic) system $\Phi_{T,v}(x, z)$ which is the system $\Phi_T(x, z)$ of Theorem 7 plus the ℓ linear equations (2.35).   □

Proof of Theorem 7.   Our task is to analyze the number and kinds of equations in the system given by (2.33) and (2.34). Let n and m be, respectively, the dimensions of the input and state spaces of M, N the number of nodes, and let D_M be the degree of M, that is, the maximum of the degrees of all polynomials occurring in the maps associated with the nodes of M. Then:

1. There are $(T + 1)(N + m)$ variables in the system of equations (2.33). Considering $x = (x_1, \ldots, x_n)$ as variables, system (2.34) adds n more.

2. System (2.34) is a system of $2N + m$ linear equations.

3. If R is a ring, then the subsystem

$$x^k - \widehat{g}(\alpha^{k-1}, x^{k-1}) = 0, \ k = 1, \ldots, T$$

of (2.33) is a system of Tm polynomial equations of degree at most D_M+1. If R is a field, then these equations may contain rational functions. By clearing denominators, the system can be converted into an equivalent polynomial system of degree at most $ND_M + 1$.

4. The subsystem

$$\alpha^k - \widehat{\beta}(\alpha^{k-1}, x^{k-1}) = 0, \quad k = 1, \ldots, T \tag{2.36}$$

of (2.33) is more subtle and we analyze it in stages.

Our first step in analyzing (2.36) is to rewrite it explicitly as a semi-algebraic (or quasi-algebraic) formula.

For each $j \in C$, let $\beta_j^- = \beta_j^+ = \beta_j$ and define the linear maps $\widehat{\beta}^-, \widehat{\beta}^+ : R^N \to R^N$ by

$$\widehat{\beta}^-(\alpha) = \sum_{j=1}^{N} \alpha_j e_{\beta_j^-} \quad \text{and} \quad \widehat{\beta}^+(\alpha) = \sum_{j=1}^{N} \alpha_j e_{\beta_j^+}. \tag{2.37}$$

Then, if R is ordered, $\alpha' = \widehat{\beta}(\alpha, x)$ if and only if

$$x_1 < 0 \ \text{ and } \ \alpha' = \widehat{\beta}^-(\alpha) \ \text{ or } \ x_1 \geq 0 \ \text{ and } \ \alpha' = \widehat{\beta}^+(\alpha).$$

So let $B(\alpha', \alpha, x)$ be the semi-algebraic formula

$$(x_1 < 0 \ \wedge \ \alpha' - \widehat{\beta}^-(\alpha) = 0) \ \vee \ (x_1 \geq 0 \ \wedge \ \alpha' - \widehat{\beta}^+(\alpha) = 0), \tag{2.38}$$

and replace the system (2.36) by the semi-algebraic system

$$B(\alpha^k, \alpha^{k-1}, x^{k-1}), \ k = 1, \ldots, T. \tag{2.39}$$

If R is unordered, use the quasi-algebraic formulas $B_=(\alpha^k, \alpha^{k-1}, x^{k-1})$ gotten from the formulas $B(\alpha^k, \alpha^{k-1}, x^{k-1})$ by replacing $<$ by $\neq$ and $\geq$ by $=$.[3]

The semi-algebraic system (2.39) contains $2TN$ linear equations and $2T$ linear inequalities. In the unordered case, (2.39) contains $2TN + T$ linear equations and T linear inequalities.

Putting together the preceding pieces completes the proof.   □

---

[3]It is not hard to see that the semi-algebraic formula (2.38) is equivalent to the semi-algebraic formula $\widehat{B}(\alpha', \alpha, x)$,

$$(x_1 \geq 0 \ \vee \ \alpha' - \widehat{\beta}^-(\alpha) = 0) \ \wedge \ (x_1 < 0 \ \vee \ \alpha' - \widehat{\beta}^+(\alpha) = 0).$$

So we can replace the system (2.36) by the semi-algebraic system

$$\widehat{B}(\alpha^k, \alpha^{k-1}, x^{k-1}), \ k = 1, \ldots, T.$$

Similarly, if R is unordered, use $\widehat{B}_=(\alpha^k, \alpha^{k-1}, x^{k-1})$.

Remark 8 We note that if R is a ring, then the coefficients of the polynomials in the semi-algebraic (quasi-algebraic) system $\Phi_T(x, z)$ belong to the set of machine constants (up to a factor of -1). If R is a field, then the coefficients may be N-fold products of machine constants (again up to a factor of -1).

2.7   More Succinct Descriptions

We now further refine the register equations for a number of important cases.

For this section suppose M is a machine over $(R, =)$, where R is a field (e.g., $\mathbb{C}$ or $\mathbb{Z}_2$) or M is a machine over $(R, <)$, where R is $\mathbb{Z}$, $\mathbb{Q}$, or $\mathbb{R}$. Again, n and m are the dimensions of the input and state spaces of M, and N and D_M are the number of nodes and degree of M, respectively.

Definition 6 We say that a system $\mathcal{R}(x, z)$ is *n-equivalent* to a system $\Phi(x, w)$ if $\pi_n(S_\mathcal{R}) = \pi_n(S_\Phi)$, where

$$S_\mathcal{R} = \{(x, z) \in R^{n+t} \mid \mathcal{R}(x, z) \text{ holds over } R\}$$
$$S_\Phi = \{(x, w) \in R^{n+s} \mid \Phi(x, w) \text{ holds over } R\}$$

and $\pi_n : R^{n+k} \to R^n$ is the projection onto the first n coordinate spaces.

Theorem 8 *The system of register equations $\mathcal{R}_T(x, z)$ (2.33) and (2.34) is n-equivalent to a system of quadratic equations*

$$q_1(x, w) = 0, \ldots, q_k(x, w) = 0,$$

where $w = (w_1, \ldots, w_s)$ and $s, k \leq (n+mT)^{c_M}$. Here c_M is a constant depending on N and D_M —and not on n, m, or T.

Remark 9 Suppose $\Omega = \pi_n(S_\mathcal{R})$; that is,

$$\Omega = \{x \in R^n \mid \exists z \in R^t \text{ such that } \mathcal{R}(x, z) \text{ holds over } R\}$$

and suppose $\mathcal{R}$ is n-equivalent to a system Φ. Then $\Omega = \pi_n(S_\Phi)$; that is,

$$\Omega = \{x \in R^n \mid \exists w \in R^s \text{ such that } \Phi(x, w) \text{ holds over } R\}.$$

Thus, the time-T halting set Ω_T is the projection of an *algebraic set* in R^{n+s} defined by the k quadratic equations in Theorem 8; that is,

$$\Omega_T = \{x \in R^n \mid \exists w \in R^s \text{ such that } q_1(x, w) = 0, \ldots, q_k(x, w) = 0\}.$$

Remark 10 If R is any ordered ring or field (e.g., $\mathbb{Z}$, $\mathbb{Q}$, or $\mathbb{R}$), $q_i : R^m \to R$ for $i = 1, \ldots, k$, and $z \in R^m$, then $q_1(z) = 0, \ldots, q_k(z) = 0$ if and only if

$$\sum_{i=1}^{k} q_i(z)^2 = 0.$$

So we have the following.

Corollary 3 *If M is a machine over $(R, <)$, where R is $\mathbb{Z}$, $\mathbb{Q}$, or $\mathbb{R}$, then the register equations $\mathcal{R}_T(x, z)$ (2.33) and (2.34) can be replaced by a single degree-4 polynomial equation $p(x, w) = 0$, where $w = (w_1, \ldots, w_s)$ and $s < (n + mT)^{c_M}$ with c_M a constant depending only on N and D_M.* $\qquad\square$

Proof of Theorem 8.   Our primary goal is to show that the register equations are equivalent to a "small" (with respect to T, m, and n) system of polynomial equations.

To do so, we must show that the subsystem

$$\alpha^k - \widehat{\beta}(\alpha^{k-1}, x^{k-1}) = 0, k = 1, \ldots, T \tag{2.40}$$

of the register equations (2.33) is equivalent to a "small" polynomial system. Here we must handle the branching conditions.

Define a map $\chi : R \to R$ by

$$\chi(x) = \begin{cases} -1 & \text{if } x < 0 \ (x \neq 0) \\ 0 & \text{if } x = 0 \ (x = 0) \\ 1 & \text{if } x > 0 \end{cases} \tag{2.41}$$

and let Γ_χ be the graph of χ; that is, $\Gamma_\chi = \{(x, y) \in R^2 \mid \chi(x) = y\}$.

Lemma 5 *For $(R, =)$, where R is a field or $(R, <)$, where R is $\mathbb{Z}$, $\mathbb{Q}$, or $\mathbb{R}$, there is an algebraic set $\Lambda \subset R^2 \times R^r$ defined by e polynomial equations of degree d with $r, e, d \leq 5$ such that $\Gamma_\chi = \pi_2(\Lambda)$.*

As in Definition 6, $\pi_2 : R^2 \times R^r \to R^2$ is the projection onto the first two coordinate spaces.

In other words, under the hypotheses of Lemma 5, there are e polynomials $p_1, \ldots, p_e$ of degree d in $2 + r$ variables such that

$$\Lambda = \{(x, y, u) \in R^2 \times R^r \mid p_1(x, y, u) = 0, \ldots, p_e(x, y, u) = 0\}$$

and

$$\chi(x) = y \ (\text{i.e., } (x, y) \in \Gamma_\chi) \text{ if and only if } \exists u \in R^r \text{ such that } (x, y, u) \in \Lambda.$$

Before we prove Lemma 5 we use it to achieve our primary goal, (2.43).
Let

$$S_\beta = \{(\alpha', \alpha, x) \in R^N \times R^N \times R^m \mid \alpha' - \widehat{\beta}(\alpha, x) = 0\}.$$

Corollary 4 *Under the hypothesis on R of Lemma 5, there is an algebraic set $S_P \subseteq R^{2N+m} \times R^{1+r}$ defined by $e + 1$ polynomials of degree d with $r, e, d \leq 5$ such that $S_\beta = \pi_{2N+m}(S_P)$.*

Proof of Corollary 4. Suppose R is field. For $j \in \mathcal{N}$, let $\psi_j : R \to R^N$ be the unique quadratic map (or linear in the case of equality branching) taking

$$\left.\begin{array}{rcl} -1 & \to & e_{\beta_j^-} \\ 0 & \to & e_{\beta_j^+} \\ 1 & \to & e_{\beta_j^+} \end{array}\right\} \quad \text{(in case of equality branching)}.$$

Recall, for $j \in \mathcal{C} = \mathcal{N} - \mathcal{B}$, $\beta_j^- = \beta_j^+ = \beta_j$.

So, for $x \in R^m$, $\psi_j(\chi(x_1)) = e_{\beta(j,x)}$. So (see (2.32))

$$\widehat{\beta}(\alpha, x) = \sum_{j=1}^N \alpha_j \psi_j(\chi(x_1)). \tag{2.42}$$

If R is not a field (i.e., if $R = \mathbb{Z}$), ψ_j is a map over its quotient field (i.e., over $\mathbb{Q}$).

In this case, modify the last equation to be $2\widehat{\beta}(\alpha, x) = 2 \sum_{j=1}^N \alpha_j \psi_j(\chi(x_1))$.

Now we construct the polynomials defining S_P. For $i = 1, \ldots, e$ let $\widehat{p}_i(\alpha', \alpha, x, y, u)$ be the polynomial $p_i(x_1, y, u)$ of Lemma 5 defining Λ. Here x_1 is the first coordinate of x.

Let $\widehat{p}_{e+1}(\alpha', \alpha, x, y, u) = \mu\alpha' - \mu \sum_{j=1}^N \alpha_j \psi_j(y)$, where $\mu = 2$ if $R = \mathbb{Z}$ and $\mu = 1$ otherwise.

Let

$$S_P = \{(\alpha', \alpha, x, y, u) \in R^{2N+m+1+r} \mid \widehat{p}_i(\alpha', \alpha, x, y, u) = 0, i = 1, \ldots, e+1\}.$$

Now by (2.42), $\alpha' = \widehat{\beta}(\alpha, x)$ if and only if there is a $y \in R$ such that $\chi(x_1) = y$ and $\alpha' = \sum_{j=1}^N \alpha_j \psi_j(y)$. By Lemma 5 and the definition of the $\widehat{p}_i, i = 1, \ldots, e$, and $\widehat{p}_{e+1}$, this holds if and only if there is a $y \in R$ and $u \in R^r$ such that $\widehat{p}_i(\alpha', \alpha, x, y, u) = 0$ for $i = 1, \ldots, e+1$. $\qquad \square$

Thus, the subsystem (2.40) is equivalent to a system

$$p_i(\alpha^k, \alpha^{k-1}, x^{k-1}, y^k, u^k) = 0, \ i = 1, \ldots, e+1, \ k = 1, \ldots, T$$

of at most $6T$ polynomial equations, each of degree at most 5.

Combining this result with Theorem 7 we meet our primary goal. Let $x = (x_1, \ldots, x_n)$, $z = (z_1, \ldots, z_t) = (\alpha^0, x^0, \ldots, \alpha^T, x^T)$, and $w = (w_1, \ldots, w_s) = (y^1, u^1, \ldots, y^T, u^T)$ where the x_i, z_j, and w_k range over R. Then:

> The system of register equations $\mathcal{R}_T(x, z)$ is $(n + t)-$
> equivalent to an *algebraic* system $\Phi'_T(x, z, w)$ in $n + t + s$
> variables ranging over R, where $t = (N + m)(T + 1)$ (2.43)
> and $s \le 6T$, consisting of at most $4(N + m)T + 6T$
> polynomial equations of degree at most $ND_M + 3$.

Proof of Lemma 5.   In each case we define Λ by polynomial equations $p_1 = 0, \ldots, p_e = 0$.

For the case $(R, =)$, R a field, we use the equivalence

$$x \neq 0 \iff \exists u \, (ux = 1). \tag{2.44}$$

Here we let

$$p_1(x, y, u) = (y + 1)x, \quad p_2(x, y, u) = xu - y.$$

In this case $r = 1$, $e = 2$, and $d = 2$.

For the case $(\mathbb{R}, <)$, or any ordered field where positive elements have square roots, we use the equivalence

$$x > 0 \iff \exists u \, (xu^2 = 1). \tag{2.45}$$

Now let

$$p_1(x, y, u) = (y + 1)x(y - 1), \quad p_2(x, y, u) = xu^2 - y.$$

Here $r = 1$, $e = 2$, and $d = 3$.

For the case $(\mathbb{Q}, <)$ we modify the last set of polynomials by replacing u^2 by $u_1^2 + u_2^2 + u_3^2 + u_4^2$ producing a system of two cubic polynomials now in the variables x, y, u_1, u_2, u_3, u_4. So in this case $r = 4$, $e = 2$, and $d = 3$. Here we are using Lagrange's theorem for $\mathbb{Q}$ and $\mathbb{Z}$:

$$u \geq 0 \iff \exists u_1, u_2, u_3, u_4 \, (u = u_1^2 + u_2^2 + u_3^2 + u_4^2). \tag{2.46}$$

Finally, for $(\mathbb{Z}, <)$ we use the additional equivalence

$$x > 0 \iff \exists u \geq 0 \, (x = 1 + u) \tag{2.47}$$

and let

$$p_1 = (x + (1 + u_1^2 + u_2^2 + u_3^2 + u_4^2))x(x - (1 + u_1^2 + u_2^2 + u_3^2 + u_4^2))$$
$$p_2 = (y + 1)x(x - (1 + u_1^2 + u_2^2 + u_3^2 + u_4^2))$$
$$p_3 = y(x + (1 + u_1^2 + u_2^2 + u_3^2 + u_4^2))(x - (1 + u_1^2 + u_2^2 + u_3^2 + u_4^2))$$
$$p_4 = (y - 1)(x + (1 + u_1^2 + u_2^2 + u_3^2 + u_4^2))x.$$

Here $r = 4$, $e = 4$, and $d = 5$. $\square$

The proof of Theorem 8 is completed combining (2.43) with the next lemma. Here we note that the number of monomials in a polynomial in n variables of degree D is bounded above by $(n + 1)^D$.

Lemma 6   *Suppose R is any ring or field. Then any system of polynomial equations $p_1(x) = 0, \ldots, p_t(x) = 0$ in n variables of degree D and with at most K monomials per equation (so $K \leq (n + 1)^D$) is n-equivalent to a quadratic polynomial system $q_1(x, y) = 0, \ldots, q_{t+t'}(x, y) = 0$ in $n + n'$ variables with $n', t' \leq KD$. Here $x = (x_1, \ldots, x_n)$ and $y = (y_1, \ldots, y_{n'})$.*

Proof. Suppose $2 < d \leq D$. For each $I = (i_1, \ldots, i_d)$ with $i_j \in \{1, \ldots, n\}$ and $i_j \leq i_{j+1}$, let $x_I = x_{i_1} \cdot \ldots \cdot x_{i_d}$ and let $y_{I_1}, \ldots, y_{I_{d-2}}$ be $d - 2$ new variables.

Replace each monomial of the form $a_I x_I$, $a_I \in R$, from the original system by the quadratic monomial $a_I x_{i_1} y_{I_1}$ and add to the system the $d - 2$ quadratic polynomial equations

$$y_{I_1} - x_{i_2} y_{I_2} = 0$$
$$y_{I_2} - x_{i_3} y_{I_3} = 0$$
$$\vdots$$
$$y_{I_{d-2}} - x_{i_{d-1}} x_{i_d} = 0.$$

$\square$

Remark 11 In the next chapters, in particular with regard to complexity considerations, it is important at times to keep track of coefficients as systems get transformed one to another. To follow up on Remark 8, we now note that, in transforming the semi-algebraic (quasi-algebraic) system $\Phi_T(x, z)$ to the algebraic system $\Phi'_T(x, z, w)$ (2.43) and then to the quadratic system (Lemma 6), the set of coefficients essentially remained the same (perhaps some got multiplied by 2). In the final transformation to the degree-4 polynomial, in case R is ordered (Remark 10), the new coefficients are now products $c_1 c_2$ of the old.

2.8 Additional Comments and Bibliographical Remarks

The main reference for this chapter is [BSS], that is, [Blum, Shub, and Smale 1989], where for example, the equivalence of the theory of computation over $\mathbb{Z}$ with the classical one is given.

Proposition 3 appears in [BSS]. It is a simple example of a lower bound for *topological complexity* which, roughly speaking, measures the amount of branching needed to solve a problem. Another example, also taken from [BSS], bounds the topological complexity of the TSP over $\mathbb{R}$.

Proposition 6 *Let* TSP_n *be the* n *dimensional Traveling Salesman Problem over* $\mathbb{R}$. *If M decides* TSP_n, *then the number of minimal halting paths of M is at least* $(n - 1)!/2$. $\square$

An exponential bound on the number of minimal halting paths of any machine deciding the Knapsack Problem is also mentioned in [BSS]. A more complicated example of lower bounds for topological complexity is given in [Smale 1987] where the following problem is considered. Given a complex polynomial $f = x^d + a_{d-1} x^{d-1} + \ldots + a_0$ and an $\varepsilon > 0$, compute ε-approximations of all the roots of f. The main result of [Smale 1987] states that any machine solving this problem has at least $(\log d)^{2/3}$ minimal halting paths. Subsequently Vassiliev [1992] refined this result proving that the number of paths lies in the interval

$$[d - \min D_p(d), d - 1],$$

where $D_p(d)$ is the sum of digits in the p-adic decomposition of the number d, and the minimum is taken over all primes p. Here machines are considered over $\mathbb{R}$ and the input coefficients of f are given as pairs of real numbers.

Theorem 1 appears in [BSS]. Theorem 2 appears in [Blum and Smale 1993]. For the mathematics of the Mandelbrot set $\mathcal{M}$ see [Douady and Hubbard 1984, 1985]. The proof that the boundary of $\mathcal{M}$ has Hausdorff dimension 2 is given by Shishikura [1994]. A direct proof of Lemma 3 is folklore following discussions with Michel Herman, Adrian Douady, John Hubbard, and Dennis Sullivan.

For Theorem 3 see [BSS] and for the fact that the "bad" initial points for the cubic $f(x) = x^3 - 2x + 2$ is a Cantor set see [Barna 1965 and Smale 1985].

For semialgebraic sets and their geometry see the books Benedetti and Risler 1990 and Bochnak, Coste, and Roy 1987]. The latter may also be used as a reference for the theory of real closed fields. Another reference is [Lang 1993a].

For the characterization of rings satisfying the property "halting sets = output sets" see [Byerly 1993 and Michaux 1991]. An interesting line of research is to determine which results of the classical theory generalize and which do not. See [Friedman and Mansfield 1992] for some examples as well as an effective version of Theorem 1.

3
Computation over a Ring

A pillar of the classical theory of computation is the existence of a *universal* Turing machine, a machine that can compute any computable function. This theoretical construct foretold and provides a foundation for the modern general-purpose computer. Classical constructions of universal machines generally utilize computable encodings of finite sequences of integers by a single integer in finite time. These codings also ensure that our theory of finite-dimensional machines over $\mathbb{Z}$ is sufficient to capture the full classical theory. However, such computable codings are not possible over the real numbers. Nor, from a strictly algebraic point of view, are they necessarily desirable. If we wish to construct universal machines over the reals, and to develop a general theory of computation, we are led naturally to consider machines that can handle finite but *unbounded* sequences. This in fact is closer to Turing's original approach.

Likewise, from a complexity perspective, we are led to similar considerations. A goal of computational complexity is to quantify the intrinsic difficulty of solving problems. The problems we have dealt with thus far have been finite-dimensional in that the *problem instances* have been given by a fixed finite number of parameters. Thus, for example, the decision problem for the Mandelbrot set is 2-dimensional over $\mathbb{R}$; each problem instance (to decide) is a point in $\mathbb{R}^2$. Now if we wish to have a natural framework for handling *uniform* procedures for solving or deciding problem instances of arbitrary dimension, say for solving polynomials of arbitrary degree or for deciding the Knapsack problem for all n, we are again led to consider machines that handle unbounded sequences. This framework enables us to formalize in Chapter 4 the idea of a *tractable* problem by means of the class P over a ring and then, in Chapter 5, the more subtle notion of the class NP over a ring. We are then able to formulate the fundamental "P = NP?" question over the reals, as well as in more general settings.

With these considerations in mind we are motivated to define our *machines* over a ring R.

3.1 R^∞ and R_∞

Suppose R is a commutative ring (or field) with unit.

To accommodate finite but unbounded sequences we consider the spaces R^∞ and R_∞.

Here R^∞ is the disjoint union

$$R^\infty = \bigsqcup_{n \geq 0} R^n,$$

where for $n > 0$, R^n is the standard n-dimensional space over R and R^0 is the 0-dimensional space with just one point $\mathbf{0}$. The space R^∞ is a natural one to represent problem instances of arbitrarily high dimension. For $x \in R^n \subset R^\infty$, we call n the *length* of x.

We denote by R_∞ the bi-*infinite direct sum* space over R. Elements of R_∞ have the form

$$x = (\ldots, x_{-2}, x_{-1}, x_0 . x_1, x_2, \ldots),$$

where $x_i \in R$ for all integers i, $x_k = 0$ for $|k|$ sufficiently large, and $\,.\,$ is a distinguished marker between x_0 and x_1. For each i, we say that x_i is the ith *coordinate* of x. Let 0 denote the sequence with all elements 0.

We can define polynomial and rational functions and maps on R_∞ as follows.

Suppose $h : R^m \to R$ is a polynomial (or rational) function of degree d over R. Then h *defines* a *polynomial* (or *rational*) function

$$\widehat{h} : R_\infty \to R$$

on R_∞ of *dimension m* and *degree d* by letting $\widehat{h}(x) = h(x_1, \ldots, x_m)$ for each $x \in R_\infty$.

Suppose $g_i : R^m \to R$, $i = 1, \ldots, m$ are polynomial (or rational) functions of maximum degree d over R. Then the g_i, $i = 1, \ldots, m$, *define* a *polynomial* (or *rational*) map on R_∞

$$\widehat{g} : R_\infty \to R_\infty$$

of *dimension m* and *degree d* by letting $(\widehat{g}(x))_i = \widehat{g}_i(x)$, $i = 1, \ldots, m$ and $(\widehat{g}(x))_i = x_i$ for $i < 1$ or $i > m$. These defined functions and maps naturally inherit the algebraic properties of their defining functions and maps by, for example, letting $\widehat{f} + \widehat{g} = \widehat{f + g}$, for polynomials $f, g : R^n \to R$.

The space R_∞ has natural *shift* operations, *shift left* σ_l and *shift right* σ_r, where

$$\sigma_l(x)_i = x_{i+1} \text{ and } \sigma_r(x)_i = x_{i-1}.$$

Shift left has the effect of shifting each element of the sequence x one coordinate to the left (thus shifting the distinguished marker one coordinate to the right). Shift right is the inverse operation. These operations combined with the polynomial and rational functions on R_∞ make R_∞ a natural state space for our machines.

We relate the spaces R^∞ and R_∞, by defining maps $I_\infty : R^\infty \to R_\infty$ and $O_\infty : R_\infty \to R^\infty$ as follows.

$$I_\infty(x) = (\ldots, 0, 0, 0, \widehat{n}. x_1, x_2, \ldots, x_n, 0, 0, 0, \ldots) \text{ for } x \in R^n, \tag{3.1}$$

where for $n > 0$, $\widehat{n}$ denotes the sequence of n 1s and $\widehat{0} = 0$, and

$$O_\infty(\ldots, x_0. x_1, \ldots, x_\ell, \ldots) = \begin{cases} 0 \in R^0 & \text{if } \ell = 0 \\ (x_1, \ldots, x_\ell) \in R^\ell & \text{otherwise,} \end{cases} \tag{3.2}$$

where $\ell = \min_{i \geq 0}\{x_{-i} = 0\}$.

3.2 Machines over R

Definition 1 A *machine M* over R is a finite connected directed graph, containing now five types of nodes, four as in the finite-dimensional case —*input, computation, branch*, and *output*— plus a set Σ_M of *shift nodes*, with associated maps. The space R^∞ is both the underlying *input space* $\mathcal{I}_M$ and *output space* $\mathcal{O}_M$ of M, and R_∞ is the *state space* $\mathcal{S}_M$.

The input, computation, branch, and output nodes and associated maps are as in the definition of a finite-dimensional machine. But now we use the notion of defined polynomial (or rational) map on R_∞ for computation nodes, the notion of defined polynomial on R_∞ for branch nodes, and stipulate that $I_M = I_\infty$ and that $O_\eta = O_\infty$, for each output node η.

Shift nodes are similar to computation nodes. Each such node has (possibly several) incoming edges and one outgoing edge. Associated with each shift node η is a map $g_\eta \in \{\sigma_l, \sigma_r\}$ of the state space to itself and a unique next node β_η.

The *dimension* K_M and *degree* D_M of M are, respectively, the maximum of the dimensions and the degrees of all maps associated with its computation and branch nodes.

As in the finite-dimensional case, we can assume, without loss of generality to our theory, that computations associated with computation nodes are always defined for every input. Indeed for simplicity, we sometimes describe our theory over a ring without division. The corresponding concepts and results for fields, where divisions may take place, are obtained by suitably replacing polynomials by rational functions.

Remark 1

(1) If M had R_∞ as its state space but only the four previous type nodes, it would still essentially be a finite-dimensional machine. The increased power comes from the shift nodes and maps that enable the accessing of coordinates of arbitrarily high dimension, or *registers* of arbitrary high address, and allow the use of as much work and storage space as needed. Indeed, the expansion of the state space to R_∞ together with the addition of the shift nodes provide the ability to implement uniform procedures. Thus, we sometimes refer to our machines as *uniform machines*.

(2) A finite-dimensional machine can be considered as a special case of machine, one without shift nodes. A machine M without *loops* (or cycles), that is, a machine whose graph is a tree, is essentially a finite-dimensional machine without loops (in fact, a f.d. machine of dimension at most $K_M + s$, where s is the number of shift nodes in M). Finite-dimensional machines without loops are known as *tame* machines, special cases of which are *algebraic decision trees* and *algebraic computation trees*. Such machines are considered in Chapter 16.

(3) Note that a classical Turing machine is a machine over $\mathbb{Z}_2$. Thus a machine over $\mathbb{R}$ might be considered to be a "real" Turing machine.

(4) Indeed, a crucial aspect of our theory of computation is that a machine is in essence a *finite* object with a finite description (that includes a finite number of constants from R specifying the defined polynomials and maps), even though inputs can have arbitrary large length and be sequences of elements from R, or in case $R = \mathbb{R}$, sequences of real numbers.

As before, we often assume, again without loss of generality to our theory, that our machines are in *normal form*. That is, as before: there is a unique output node; the set of nodes $\mathcal{N}$ is $\{1, \ldots, N\}$, where 1 denotes the input node and N the output node; computation maps are always defined for legitimate inputs to their associated nodes; and all branch nodes are standard; that is, $h_\eta(x) = x_1$ for each branch node η. We also assume that the dimension of all maps associated with the computation nodes is K_M.

It is convenient again to let $\beta_N = N$ and $g_\eta(x) = x$ for $\eta = 1, N$, or a branch node. And as before, we let $\mathcal{B}$ be the set of branch nodes of M and $\mathcal{C} = \mathcal{N} - \mathcal{B}$; thus $\mathcal{C}$ contains the shift nodes. We omit subscripts M when there is no confusion.

Most definitions of the last chapter can be adapted here by replacing R^n and R^ℓ by R^∞ and by replacing R^m by R_∞. In particular, the *full state space* $\mathcal{N} \times \mathcal{S}$, the *computing endomorphism* $H : \mathcal{N} \times \mathcal{S} \to \mathcal{N} \times \mathcal{S}$, *computations*

$$z^0 = (1, I(x)), \ldots, z^k = (\eta^k, x^k) = H^k(z^0), \ldots,$$

and thus, computation paths $\eta^0 = 1, \eta^1, \ldots, \eta^k, \ldots$, *state trajectories* $x^0 = I(x), x^1, \ldots, x^k, \ldots$, *halting computations, halting paths* $\gamma \in \Gamma_M$, *minimal halting paths* $\gamma' \in \Gamma'_M$, *halting time* $T_M(x)$, *halting sets* Ω_M, *input–output maps* Φ_M, *computable maps, decidable, undecidable,* and *semidecidable sets* are all defined exactly as before with the appropriate adaptations. The same holds for *path sets*

$\mathcal{V}_y$, *time-T halting paths* Γ_T, and *time-T halting sets* Ω_T, and the First Form of the register equations (2.27) and (2.28).

Most results about finite-dimensional machines carry over to the general case. Some have essentially identical proofs, possibly with appropriate modifications. For example, we can use a "pairing function" $\varphi_2 : R^\infty \times R^\infty \to R^\infty$ (see Section 3.5) to adapt the proof that a set is decidable just in case both it and its complement are halting sets from the finite-dimensional setting to the general case. Other results require some care. For example, since the shift maps are not polynomial or rational, care need be taken in describing the path set decomposition of the halting sets (see Section 3.3) as well as in analyzing the register equations (see Section 3.4).

We end this section with some conventions and assumptions that are helpful in designing and analyzing machines. It can be seen with a little "programming" that these assumptions do not alter our theory of computation.

Remark 2

(1) For convenience, when the characteristic of R is zero, we sometimes suppose $I_\infty(x) = (\dots, 0, n \cdot x_1, x_2, \dots, x_n, 0, \dots)$ for $x \in R^n \subseteq R^\infty$. This is justified since, in this case, it is easy to convert an element $(\dots, 0, \widehat{n} \cdot x_1, x_2, \dots) \in R_\infty$ to $(\dots, 0, n \cdot x_1, x_2, \dots) \in R_\infty$ using a subroutine such as Init (see Figure A). Recall $\widehat{n}$ is the sequence of n ones.

(2) It is sometimes convenient to define computations that involve nonpositive coordinates. For example, we may wish to associate a function such as

$$g(x) = (\dots, x_{-4}, x_{-3}, \frac{x_0}{2}, \frac{x_0}{2}, \frac{x_0}{2} \cdot x_1, x_2, \dots)$$

with some computation node of a machine M. In fact, we do so in Section 4.4. Again, with a little programming, such a machine M can be converted to a formal machine M'. Let $s_M \geq 0$ be the smallest integer such that only coordinates $k > -s_M$ are called on in any computation node of M. We convert machine M to M' by first adding s_M to all coordinate indices in all the computation maps of M, then inserting immediately after the input node a subroutine to shift right s_M times, and finally inserting directly before the output node a subroutine to shift left s_M times.

(3) In many important situations, the input–output map Φ_M has values in R^ℓ for some fixed $\ell > 0$. This is the case, for example, when Φ_M is a 0–1 valued function. In such cases, it is convenient to consider the output space of M to be the finite-dimensional space R^ℓ and the map associated with the output node to be $O_\ell : R_\infty \to R^\ell$, where $O_\ell(x) = (x_1, \dots, x_\ell)$ for each $x \in R_\infty$.

(4) Finally, we say a machine M over R is *proper* if for all $T, n > 0$, if $x \in R^n \subset R^\infty$ is in the time-T halting set of M, then $(x^T)_{-r} \in \{0, 1\} \subseteq R$ for $r \geq 0$. From now on, unless otherwise indicated, we assume all machines are proper. Again, this is without loss of generality to our theory.

$$(\ldots, 0, \widehat{n} \cdot x_1, x_2, \ldots)$$

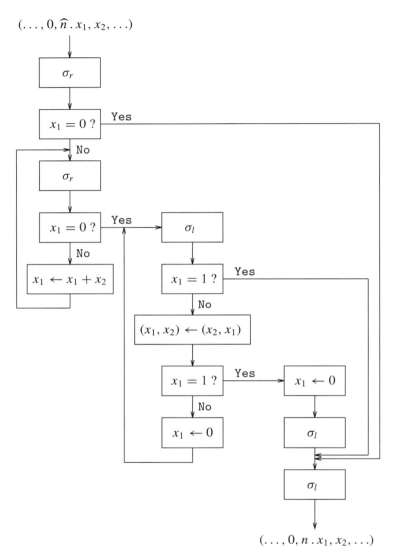

Figure A   The subroutine `Init`.

3.3   Path Decomposition and Reduction to the Finite-Dimensional Case

Let R be a commutative ring with unit. For $A \subset R^\infty$, let $A^n = A \cap R^n$ be the *n-dimensional component* of A.

Given a machine M over R of dimension K_M we wish to describe, by ordinary polynomial equalities and inequalities, the n-dimensional components of path sets, time-T halting sets, and halting sets, namely, the sets

$$\mathcal{V}_\gamma^n, \quad \Omega_T^n, \quad \text{and} \quad \Omega_M^n.$$

We observe that each initial path $\gamma(k)$ of a computation path can contain at most k shift nodes; that is, M can perform at most k shift operations in *time k*. Thus, in time k, M can perform computations utilizing at most $K_M + k$ coordinates of points in the state space $\mathcal{S} = R_\infty$. To obtain complete information on a computation with input $x \in R^n$ in time k, it is sufficient to observe at most $r + 2k$ coordinates of points in $\mathcal{S}$, k to the left of the distinguished symbol and $r + k$ to the right. Here $r = \max(K_M, n)$.

Note that restricted to a finite set of coordinates or "registers," the shift operations, as well as the computation maps of M, are given by ordinary polynomials. Thus, in time k, and on inputs $x \in R^n \subset R^\infty = \mathcal{I}$, the machine acts essentially as a finite-dimensional machine, that is, a machine with a finite number of registers. This fact enables us to develop algebraic machinery to describe the behavior of machines completely analogous to that developed for the finite-dimensional case.

For $m > 0$ let $\mathcal{S}_m = \{(x_{-(m-1)}, \ldots, x_0, x_1, \ldots, x_m) \mid x_i \in R\} = R^{2m}$.

It is convenient to have at hand the injection $\tilde{\iota} : \mathcal{S}_m \to \mathcal{S}$ defined by

$$\tilde{\iota}(x_{-(m-1)}, \ldots, x_0, x_1, \ldots, x_m) = (\ldots, 0, x_{-(m-1)}, \ldots, x_0 . x_1, \ldots, x_m, 0, \ldots),$$

the projection $\tilde{\pi} : \mathcal{S} \to \mathcal{S}_m$ defined by

$$\tilde{\pi}(x) = (x_{-(m-1)}, \ldots, x_0, x_1, \ldots, x_m),$$

and the projection $\tilde{\pi}_1 : \mathcal{S}_m \to R$ onto the "first" coordinate; that is, $\tilde{\pi}_1(x_{-(m-1)}, \ldots, x_0, x_1, \ldots, x_m) = x_1$.

Here and in what follows, all maps marked with a "tilde" have an implicit dependency on m which we make explicit if necessary.

Now, we associate with the input map I the the *modified input map*

$$\tilde{I} = \tilde{\pi} \circ I : \mathcal{I} \to \mathcal{S}_m,$$

and with each g_η the *modified computation map*

$$\tilde{g}_\eta = \tilde{\pi} \circ g_\eta \circ \tilde{\iota} : \mathcal{S}_m \to \mathcal{S}_m.$$

We observe that, restricted to R^n, $\tilde{I}$ is linear and that the maps $\tilde{g}_\eta$ are polynomial, all in the ordinary sense.

In particular:

1. If $\eta = 1$, N or a branch node, $\tilde{g}_\eta$ is the identity.

2. If η is a computation node and $m \geq K_M$, the dimension of M, then

$$\tilde{g}_\eta(x_{-(m-1)}, \ldots, x_0, x_1, \ldots, x_m) =$$
$$(x_{-(m-1)}, \ldots, x_0, \underbrace{\hat{g}_1(x_1, \ldots, x_{K_M}), \ldots, \hat{g}_{K_M}(x_1, \ldots, x_{K_M})}_{K_M}, \underbrace{\ldots, x_m}_{m - K_M}),$$

where $\widehat{g}_1, \ldots, \widehat{g}_{K_M}$ are the ordinary polynomials defining g_η.

3. If η is a shift node with $g_\eta = \sigma_l$, then

$$\widetilde{g}_\eta(x_{-(m-1)}, \ldots, x_0, \ldots, x_m) = (x_{-(m-2)}, \ldots, x_0, \ldots, x_m, 0),$$

and if η is a shift node with $g_\eta = \sigma_r$, then

$$\widetilde{g}_\eta(x_{-(m-1)}, \ldots, x_0, \ldots, x_m) = (0, x_{-(m-1)}, \ldots, x_0, \ldots, x_{m-1}).$$

We also associate with the output map O_∞ the *modified output map*

$$\widetilde{O} = O_\infty \circ \widetilde{\iota} : \mathcal{S}_m \to R^\infty.$$

This is not a polynomial map.

Now suppose γ is a computation path $\eta^0, \eta^1, \ldots, \eta^k, \ldots$. As in (2.15), at each step k in path γ, M evaluates a map $G_{\gamma(k)} : \mathcal{I} \to \mathcal{S}$, now defined by the composition of the computation and shift maps along the initial path $\gamma(k)$ with I.

For each step $k' \leq k$, we define the *modified evaluation map*

$$\widetilde{G}_{\gamma(k')} : \mathcal{I} \to \mathcal{S}_m$$

by $\widetilde{G}_{\gamma(k')} = \widetilde{g}_{\eta_{k'}} \circ \cdots \circ \widetilde{g}_{\eta_0} \circ \widetilde{I}$, where $\widetilde{g}_{\eta_i} : \mathcal{S}_m \to \mathcal{S}_m$ for $i = 0, \ldots, k'$. Restricted to R^n, $\widetilde{G}_{\gamma(k')}$ is an ordinary polynomial map.

If $\eta_{k'}$ is a branch node, we define the *modified step-k' branching function*

$$\widetilde{f}_{\gamma(k')} : \mathcal{I} \to R$$

by $\widetilde{f}_{\gamma(k')} = \widetilde{\pi}_1 \circ \widetilde{G}_{\gamma(k')}$. Again, restricted to R^n, $\widetilde{f}_{\gamma(k')}$ is an ordinary polynomial function.

We now have the tools to apply the arguments of Chapter 2 to see that $\mathcal{V}^n_{\gamma(k)}$, the n-dimensional component of the path set $\mathcal{V}_{\gamma(k)}$, is basic semi-algebraic (or basic quasi-algebraic in the unordered case), described by polynomial equalities and inequalities using the polynomials $\widetilde{f}_{\gamma(k')}$, $k' < k$.

Letting $k = T$ we obtain a path decomposition theorem as in the previous chapter.

Theorem 1 (Path Decomposition Theorem) *For any machine M and positive integers n and T, the following properties hold.*

(1) $\Omega^n_T = \bigcup_{\gamma \in \Gamma_T} \mathcal{V}^n_\gamma$. *That is, the n-dimensional component of the time-T halting set is a finite disjoint union of basic semi-algebraic sets (respectively, basic quasi-algebraic sets, in the unordered case).*

(2) $\Omega^n_M = \bigcup_{\gamma \in \Gamma'} \mathcal{V}^n_\gamma$, *where Γ' is the set of minimal halting paths of M. That is, the n-dimensional component of the halting set is a countable disjoint union of basic semi-algebraic (or basic quasi-algebraic) sets.*

(3) *For $\gamma \in \Gamma_T$, $\Phi_M|_{\mathcal{V}^n_\gamma} = \widetilde{O} \circ \widetilde{G}_{\gamma(T-1)}$. In other words, the input–output map Φ_M, restricted to the n-dimensional component of the path set $\mathcal{V}_\gamma$, is a polynomial map composed with the modified output map $\widetilde{O}$.* $\square$

As remarked earlier, $\widetilde{O}$ is not a polynomial. However, if the machine outputs are contained in R^{ℓ}, for some $\ell < \infty$, then in Part (3) of the theorem, $\widetilde{O}$ can be replaced by a projection. This is the case, for example, when M is a decision machine, that is, when Φ_M is a characteristic function.

Remark 3 We observe that to describe the halting path sets V_{γ}^n for $\gamma \in \Gamma_T$, we need only consider the *basic active state space* $\mathcal{S}_m$, where $m = K_M + T$, independent of n, and polynomial equalities and inequalities utilizing polynomials f of the form

$$\widetilde{\pi}_1 \circ \widetilde{g}_{\eta_{k'}} \circ \cdots \circ \widetilde{g}_{\eta_0} \circ \widetilde{I}, \quad \text{for some } k' < T.$$

Sometimes, however, to describe the computation on inputs from R^n, we also consider the *active state space* $\mathcal{S}_m$, where $m = K_M + T + n$.

We finish this section by stating a proposition that follows from the previous discussion and is useful in the next chapters. It stresses the close relationship between finite-dimensional and uniform machines in the presence of a time bound.

Proposition 1 *Let M be a machine over R of degree d and $t : \mathbb{N} \to \mathbb{N}$ be a function such that for every $x \in R^n$ the halting time $T_M(x)$ corresponding to the computation of M with input x is bounded above by $t(n)$. Then, for every $n \in \mathbb{N}$ there exists a f.d. machine M_n of degree d computing $\Phi_M|_{R^n}$ with the same property; that is, for every $x \in R^n$, $T_{M_n}(x) \leq t(n)$.* $\square$

3.4 The Register Equations

In this section we develop and investigate the *register equations* for machines over a ring R. We closely follow the development given in Sections 2.6 and 2.7 for finite-dimensional machines.

Let $H : \mathcal{N} \times R_{\infty} \to \mathcal{N} \times R_{\infty}$ be the computing endomorphism of a machine M over R defined exactly as in Section 2.2. Recall that the state space $\mathcal{S}$ now is R_{∞}.

The First Form of the register equations is exactly (2.27) and (2.28).

Again, exactly as in the finite-dimensional case, we make explicit the coordinate maps

$$\beta : \mathcal{N} \times R_{\infty} \to \mathcal{N} \quad \text{and} \quad g : \mathcal{N} \times R_{\infty} \to R_{\infty}$$

and then extend to maps

$$\widehat{\beta} : R^N \times R_{\infty} \to R^N \quad \text{and} \quad \widehat{g} : R^N \times R_{\infty} \to R_{\infty}$$

in an analogous way (see (2.29) and (2.32)).

The register equations describe the halting computation of points in the time-T halting set Ω_T. As we have observed in Remark 3, to do so, we need only describe how M operates on the basic active state space $\mathcal{S}_m = R^{2m}$, where $m = K_M + T$.

Thus, we can directly apply the development and results of Section 2.6 to the case at hand.

To say $x \in \Omega_T^n$ is to say there is a sequence $(\alpha^0, x^0), \ldots, (\alpha^T, x^T)$ in $R^N \times R^{2m}$ such that

$$\alpha^k - \widetilde{\beta}(\alpha^{k-1}, x^{k-1}) = 0 \text{ and } x^k - \widetilde{g}(\alpha^{k-1}, x^{k-1}) = 0 \quad k = 1, \ldots T \qquad (3.3)$$

and

$$(\alpha^0, x^0) - (e_1, \widetilde{I}(x)) = 0 \text{ and } \alpha^T - e_N = 0. \qquad (3.4)$$

These equations (3.3) and (3.4), where

$$x = (x_1, \ldots, x_n) \text{ and } z = (\alpha^0, x^0, \ldots, \alpha^T, x^T),$$

are the (time-T) *register equations* $\mathcal{R}_T(x, z)$ in their Third Form (see (2.33) and (2.34)), where

$$\widetilde{\beta} : R^N \times R^{2m} \to R^N \text{ and } \widetilde{g} : R^N \times R^{2m} \to R^{2m}$$

are the appropriate modifications of the maps $\widehat{\beta}$ and $\widehat{g}$.

Remark 4 We note that there is considerable uniformity in the register equations. In particular, they depend essentially on $m = K_M + T$ and hardly on n. Indeed, the variables $x = (x_1, \ldots, x_n)$ only appear in the linear equation $x^0 - \widetilde{I}(x) = 0$ of (3.4).

At times we add conditions on x^T such as

$$x_0^T = x_1^T = 1 \text{ and } x_{-1}^T = 0. \qquad (3.5)$$

This says the output value is 1. Thus, $x \in \Omega_T^n(1)$ if and only if there are solutions over R to the equations (3.3) through (3.5).

Except for the first set of equations in (3.3), the rest are all ordinary polynomial equations, clearing denominators if necessary. We replace the first set of equations in (3.3) by the pair of semi-algebraic (or quasi-algebraic) formulas

$$\widetilde{\pi}_1(x^{k-1}) \overset{(=0)}{\geq} 0 \ \vee (\alpha^k - \widehat{\beta}^-(\alpha^{k-1}) = 0), \ \widetilde{\pi}_1(x^{k-1}) \overset{(\neq 0)}{<} 0 \ \vee (\alpha^k - \widehat{\beta}^+(\alpha^{k-1}) = 0),$$

where $\widehat{\beta}^-, \widehat{\beta}^+ : R^N \to R^N$ are as in (2.37).

We can now directly apply the results of Sections 2.6 and 2.7 to get equivalent semi-algebraic (or quasi-algebraic) and algebraic forms for the register equations, and hence we get the corresponding succinct descriptions of the time-T halting sets Ω_T^n. In particular, let $x = (x_1, \ldots, x_n)$ and $z = (z_1, \ldots, z_t) = (\alpha^0, x^0, \ldots, \alpha^T, x^T)$, where the x_i and z_j range over R. Let $\mathcal{R}_T(x, z)$ be the system (3.3) and (3.4). Then:

Theorem 2

(1) *The register equations $\mathcal{R}_T(x, z)$ of machine M over R are equivalent to a semi-algebraic (or quasi-algebraic) system $\Phi_T(x, z)$, where*

- *the number of R-variables is at most $n + cT^2$, (i.e., $t \leq cT^2$),*
- *the number of polynomial equations is $\leq cT^2$, each of degree $\leq c$, and*
- *the number of inequalities (all linear) is $\leq 2T$.*

Here c is a constant depending on N, D_M, and K_M, the number of nodes of M, the degree of M, and the dimension of M, respectively, and is independent of n and T. (See Theorem 7 of Chapter 2.)

(2) *Over $(\mathbb{C}, =)$, $(\mathbb{Z}_2, =)$, or any field $(F, =)$, as well as over $(\mathbb{Z}, <)$, $(\mathbb{Q}, <)$, and $(\mathbb{R}, <)$, we have:*

(a) *$\mathcal{R}_T(x, z)$ is n-equivalent to an algebraic system $\Phi'_T(x, w)$ (i.e., a system of polynomial equations and no inequalities) with the same qualitative bounds as in (1). The constant c again depends only on N, D_M, and K_M. (See (2.43).)*

(b) *And in turn, $\Phi'_T(x, w)$ is n-equivalent to a quadratic system $Q_T(x, w')$, where both the length of w' and the number of equations are bounded above by $(n + T^2)^{c_M}$, where c_M is a constant depending only on N, D_M, and K_M. (See Theorem 8 of Chapter 2.)*

(3) *Over $(\mathbb{Z}, <)$, $(\mathbb{Q}, <)$, and $(\mathbb{R}, <)$, the register equations $\mathcal{R}_T(x, z)$ are n-equivalent to a single degree-4 polynomial equation $p_T(x, w)$, where $w = (w_1, \ldots, w_s)$, the variables w_i range over R, $s \leq (n + T^2)^{c_M}$, and c_M is a constant depending only on N, D_M, and K_M. (See Corollary 3 of Chapter 2.)*

Remark 5 Our earlier observations about the coefficients of the transformed systems in the finite-dimensional case (see Remarks 8 and 11 of Chapter 2) continue to apply here.

3.5 Product Spaces, Codes and Universal Machines

Consider the following problems over R.

(1) Given input, an $n \times m$ matrix A, and a vector v in R^m, compute Av.

(2) Given input, a polynomial p in n variables of degree 4, and a point $x \in R^n$, evaluate $p(x)$.

(3) Given input, a machine M over R, and a point $x \in R^\infty$, compute the value $\Phi_M(x)$.

A common characteristic of these problems is that their inputs are taken from product spaces. In the first one the input is given by $m(n + 1)$ elements of R. However, the list of these elements is not enough to describe the pair (A, v). One needs to specify where A ends and v begins. This can be done by giving n and m (or $\widehat{n}$ and $\widehat{m}$). Similarly, in the second problem one needs to specify n, a list of the coefficients of p, and x. Thus, in these two problems, the input can naturally be considered as an element of R^∞.

The situation regarding the third problem is more complicated. Here the input is a pair $(\pi(M), x)$, the first element of which describes the machine M. The description $\pi(M)$ is a finite sequence of descriptions of the nodes of M, each of which in turn may contain a description of a polynomial or rational map. Thus, we have two related issues to address here. One is how to construct nodal descriptions; the other is how to represent finite sequences of such descriptions as elements of R^∞. We consider the latter issue first.

In the classical theory of computability, that is, with R equal to $\mathbb{Z}$ or $\mathbb{Z}_2$, there are various ways of dealing with finite sequences. They generally make use of some injective computable function

$$\varphi : \mathbb{N}^\infty \to \mathbb{N}$$

that codes a sequence of natural numbers into a single one, whose inverse also is computable, and with $\varphi(\mathbb{N})$ decidable. Such "Gödel" encodings however do not necessarily exist over an arbitrary ring or field R and, in particular, they do not exist over $\mathbb{R}$. Instead, we utilize encodings that are more natural over R.

In the following we suppose the characteristic of R is zero. The case of the positive characteristic can be handled with slight modifications.

Consider the *pairing function*

$$\varphi_2 : R^\infty \times R^\infty \to R^\infty$$
$$(a, b) \to (n, m, a_1, \ldots, a_n, b_1, \ldots, b_m),$$

where n and m are the lengths of a and b, respectively.

It is easy to extend φ_2 to $R^\infty \times \ldots \times R^\infty$, a product of k copies of R^∞. Actually, one can use the same idea and define

$$\widetilde{\varphi} : (R^\infty)^\infty \to R^\infty$$
$$(a^1, \ldots, a^k) \to (k, n_1, \ldots, n_k, a^1, \ldots, a^k),$$

where n_i denotes the length of $a^i \in R^\infty$ for $i = 1, \ldots, k$.

We freely suppose such encodings of sequences in the rest of this book.

An important example is a *standard encoding* of rational maps. A polynomial of degree d in n variables can be described by the sequence $(d, n, \bar{a})$, where $\bar{a}$ denotes the sequence of its coefficients given, say, in lexicographic order using a standard representation of polynomials (see Remark 1 of Section 2.1). A rational function is given by an ordered pair of polynomials and a rational map is given by an appropriate finite sequence of rational functions. Utilizing φ_2 and $\widetilde{\varphi}$, rational functions and rational maps can thus be considered as elements of R^∞.

Now let us return to our third problem and the issue of descriptions.

A machine node can be described by its label η, its type, its next node β_η (or a pair $(\beta_\eta^+, \beta_\eta^-)$ in the case of a branching node), and its associated map g_η. To be specific, here $\eta, \beta_\eta \in \{1, \ldots, N\}$, and for type we can suppose the numbers 1 to 5 denote an input, computation, branch, output, and shift node, respectively. An associated rational map can be described as previously, and for shift nodes we can use 0 to denote a right shift and 1 a left shift. The description, or *program*, of M is then given by the sequence $\pi(M)$ of such nodal descriptions. Using $\widetilde{\varphi}$ we can view $\pi(M) \in R^\infty$. And, for $x \in R^\infty$, we can view $(\pi(M), x) \in R^\infty$ as well. Thus, in the language of programming, a program is its own code.

A *universal machine* over R is a machine U which, when input a pair $(\pi(M), x)$ with $x \in R^\infty$, will output $\Phi_M(x)$. Thus, a universal machine solves Problem (3). Such a machine U can be constructed so that given input $(\pi(M), x)$, U systematically reads the program $\pi(M)$ and carries out the instructions at hand applied to x, starting with instruction labeled 1, proceeding to instruction labeled β_1, and so forth. If U ever arrives at instruction labeled N, it halts and outputs exactly what M would output. An essential component of U is a universal polynomial evaluator (UPE), a subroutine which takes as input (f, x) in R^∞, with $f \in R[x_1, \ldots, x_n]$ and $x \in R^n$, and outputs $f(x) \in R$. There are various ways of constructing a UPE depending on the different methods of representing and evaluating polynomials. Shift nodes play an essential role here as they did in enabling U to read the program $\pi(M)$.

3.6   Additional Comments and Bibliographical Remarks

Again the basic reference for this chapter is [BSS]. However, the approach here to machines over the real numbers and other rings is closer to that of Turing machines. Thus the objects are the same as in [BSS], the definitions equivalent, but the development here we feel is more natural and elegant. Moreover this approach includes the case of computation over a finite field.

A construction of a universal machine with universal polynomial evaluator is given in [BSS]. In this reference there is also a sketch of a proof that a computable function $f : \mathbb{R}^n \to \mathbb{R}^m$ with $n, m \in \mathbb{N}$ is computable by a finite-dimensional machine. A different proof is given by Michaux [1989]. As a consequence of his proof it follows that every decidable subset of $\mathbb{R}^\infty$ can be decided by a machine that uses a number of registers which is linear in the input length.

Finally we remark that the formal definitions of machine are intended to capture informal notions of algorithm. On the other hand, we sometimes use the word "algorithm" when we informally talk about formal machines in subsequent chapters. See also Section 4.4.

4

Decision Problems and Complexity over a Ring

Classical complexity theory deals primarily with combinatorial (discrete, integer) problems. We extend the theory here to consider a wider class of problems.

At this point, as has been traditional, we focus on *decision problems*. These are problems with "yes/no" answers, for example, to questions of the form, "Is x a member of S?" or, "Given a polynomial system $f = 0$ over R, does it have a solution over R?" These problems are classified into *complexity classes* such as P or NP indicating measures of their difficulty.

In this chapter we formally define the notion of *cost*, of *class* P, and of *polynomial time reductions* over a ring R. Polynomial time reductions enable us to transfer complexity results proved about one decision problem over R to a seemingly different one over R. We focus here on a number of basic feasibility problems, including the feasibility of semi-algebraic and algebraic systems over R, and establish fundamental reductions amongst them. Problems of particular interest are over $\mathbb{Z}_2$, $\mathbb{Z}$, $\mathbb{Q}$, $\mathbb{R}$, and $\mathbb{C}$.

4.1 Decision Problems

Definition 1 A *decision problem* over R is a set $S \subseteq R^\infty$.

The decidability/complexity of a decision problem S is measured by the computability/complexity of its *characteristic function* χ_S. Recall that for $x \in R^\infty$,

$$\chi_S(x) = \begin{cases} 1 & \text{if } x \in S \\ 0 & \text{otherwise.} \end{cases}$$

We are most often interested in *structured* problems, that is, when the *instances* x have a special form.

Definition 2 A *structured decision problem* over R is a pair (X, X_{yes}) with

$$X_{yes} \subset X \subset R^\infty.$$

Here X is the set of *problem instances* and X_{yes}, the set of *yes-instances*. Let X_{no} denote $X - X_{yes}$, the set of *no-instances*.

The relevant function now is the *restricted* characteristic function χ_{yes} where $\chi_{yes} = \chi_{X_{yes}}|x$.

Of course, a decision problem S is structured with $X = R^\infty$ and $X_{yes} = S$. On the other hand, if we suppose, as we do unless otherwise stated, that the set of problem instances X is decidable over R, then χ_{yes} is computable if and only if $\chi_{X_{yes}}$ is. Thus, we freely use *decision problem* to denote either a structured or an unstructured problem.

We focus our attention primarily on the following decision problems. The underlying ring or field R is specified or clear from context.

- SA-FEAS: The Feasibility of Semi-Algebraic Systems

- QA-FEAS: The Feasibility of Quasi-Algebraic Systems

- HN: The Hilbert Nullstellensatz

- QUAD: The Feasibility of Quadratic Systems

- 4-FEAS: The Feasibility of Degree-4 Polynomials

- LPF: The Linear Programming Feasibility Problem

- IPF: The Integer Programming Feasibility Problem

- TSP: The Traveling Salesman Problem

- KP: The Knapsack Problem

- SAT: The Satisfiability Problem

Most of these problems have already been introduced in Chapter 1. Thus, HN/R is the problem of deciding if a finite system of polynomial equations over R has a solution over R and SAT is the problem of deciding if a propositional formula is satisfiable, the seminal problem of the classical theory of NP-completeness. The problem SA-FEAS (similarly, QA-FEAS) is the problem of deciding for each semi-algebraic (quasi-algebraic) system Φ whether or not $S_\Phi \neq \emptyset$ (see Remark 3 of Chapter 2). Also, QUAD/R is the problem of deciding if a finite system of quadratic equations over R has a solution over R.

For the problems SA-FEAS, LPF, and TSP, R is ordered, for IPF, $R = \mathbb{Z}$, and for SAT, $R = \mathbb{Z}_2$.

In Section 1.5 we represented KP and TSP as decision problems in the form (X, X_{yes}). The other problems can also be represented in this form. For example, to represent HN, let

$$X = \{f \in R[x_1, \ldots, x_n]^m \mid m, n \in \mathbb{Z}^+\}, \tag{4.1}$$

where $R[x_1, \ldots, x_n]^m$ is the m-fold product of $R[x_1, \ldots, x_n]$, that is, the set of tuples $(f_1, \ldots, f_m)$ with $f_i \in R[x_1, \ldots, x_n]$ and

$$X_{yes} = \{f \in X \mid \exists z \in R^n \text{ such that } f(z) = 0\}.$$

Similarly, to represent 4-FEAS, let

$$X = \{f \in R[x_1, \ldots, x_n] \mid \text{degree } f \leq 4\}$$

and

$$X_{yes} = \{f \in X \mid \exists z \in R^n \text{ s.t. } f(z) = 0\}.$$

We suppose that here, and in each of our examples, $X \subseteq R^\infty$ via a standard encoding (see Section 3.5).

For LPF over $(R, <)$, let

$$X = \{(f, b) \mid f = (f_1, \ldots, f_m) : R^n \to R^m \text{ is a linear map and } b \in R^m\}$$

and

$$X_{yes} = \{(f, b) \in X \mid \exists z \in R^n \text{ such that } f(z) \geq b\}.$$

For SA-FEAS we utilize the following normal form representation of semi-algebraic systems. Let

$$X = \{\Phi = (\varphi_1, \ldots, \varphi_m) \mid \varphi_j = (f_{j1}, \ldots, f_{j\ell}, g_{j1}, \ldots, g_{jr}),$$
$$f_{ji}, g_{jk} \in R[x_1, \ldots, x_n]\}.$$

Here φ_j represents the following semi-algebraic formula which we also denote by φ_j.

$$f_{j1}(x) > 0 \vee \ldots \vee f_{j\ell}(x) > 0 \vee g_{j1}(x) = 0 \vee \ldots \vee g_{jr}(x) = 0.$$

Let

$$X_{yes} = \{\Phi \in X \mid \cap_{j=1}^m S_{\varphi_j} \neq \emptyset\}.$$

Thus, letting $\varphi = \wedge_{j=1}^m \varphi_j$, we have $\Phi \in X_{yes}$ if and only if $S_\varphi \neq \emptyset$.

Note that the system of register equations of machine M over R in their semi-algebraic form, Φ_T (see Theorem 2 of Chapter 3), can easily be rewritten in the normal form given previously. In particular, each formula in Φ_T can be represented as (f, g), where f is either x_i or 0 and g is a polynomial of degree bounded by a constant (depending only on M).

To represent SAT see Section 5.5.

Remark 1 By abuse of notation, we sometimes identify a decision problem with its set of yes-instances. Thus, for example, if we write $f \in$ 4-FEAS/R we mean f is a degree-4 polynomial in n variables over R that has a zero in R^n.

Remark 2 For complexity theory we are primarily interested in problems that are decidable, that is, for which the characteristic function χ_{yes} is computable over R. Over $\mathbb{R}$, $\mathbb{C}$ and $\mathbb{Z}_2$ all the preceding relevant problems are decidable. Over $\mathbb{Z}$, the problems IPF, TSP, and KP are decidable but 4-FEAS, and hence QUAD, HN, QA-FEAS, and SA-FEAS are not. Over $\mathbb{Q}$, LPF, TSP, and KP are decidable but it is not known if 4-FEAS or, equivalently, if HN is decidable or not. For references, see Section 4.5.

Remark 3 Closer to the classical Hilbert Nullstellensatz, although not as immediately relevant in what follows as HN, is $HN_{R,k}$, where R and k are rings, $R \subseteq k$, X is given by (4.1), but now

$$X_{yes} = \{f \in X \mid \exists z \in k^n \text{ such that } f(z) = 0\}.$$

Important examples are $HN_{\mathbb{Z},\mathbb{C}}$ and $HN_{\mathbb{Z}_2,\overline{\mathbb{Z}_2}}$, where $\overline{\mathbb{Z}_2}$ denotes the algebraic closure of $\mathbb{Z}_2$. These problems are decidable over $\mathbb{Z}$ and $\mathbb{Z}_2$, respectively. By the Hilbert Nullstellensatz, $HN_{\mathbb{Z},\mathbb{C}} = HN_{\mathbb{Z},\overline{\mathbb{Q}}}$.

4.2　Complexity and the Class P

The *complexity* of a function is measured by the optimal *cost* of computing it. The cost of computing a function by a machine reflects the number of fundamental operations performed from input to output. As in the classical tradition, cost is a function of input *size*.

We say a machine over $\mathbb{R}$ works in *polynomial time* if for each input $x \in \mathbb{R}^n \subset \mathbb{R}^\infty$

$$\mathrm{cost}(x) \leq cn^q$$

for some fixed $c, q \geq 1$, where $\mathrm{cost}(x)$ is the number of nodes traversed during the computation. Here n is the size of x. Let us proceed more systematically.

Classically, size is given by the bit length. But now, over an arbitrary ring R, size depends on a *height* function ht_R defined on R and on the vector length.

To reflect the algebraic character of our machines, we generally take ht_R to be the *unit height*; that is, $ht_R(x) = 1$ for all $x \in R$. Unit height is natural for developing a complexity theory over $\mathbb{R}$ and $\mathbb{C}$ that embodies *algebraic* complexity. It also corresponds to the notion of bit over $\mathbb{Z}_2$.

Over $\mathbb{Z}$, if we wish to reflect classical bit complexity, it is natural to consider *logarithmic* or *bit height*. Here $ht_{\mathbb{Z}}(x) = \lceil \log(|x| + 1) \rceil$ for all $x \in \mathbb{Z}$. Over $\mathbb{Q}$, logarithmic height is then defined by $ht_{\mathbb{Q}}(x) = \max(ht_{\mathbb{Z}}(p), ht_{\mathbb{Z}}(q))$ for $x \in \mathbb{Q}$, where $x = p/q$ and p and q are relatively prime integers.

Sometimes, we may wish to consider unit height on $\mathbb{Z}$ and $\mathbb{Q}$ (e.g., if we wish to focus on the algebraic character of computation) or logarithmic height on $\mathbb{R}$ (if we wish cost to take magnitudes of numbers into account). Over algebraic number fields, it may be natural to relate ht_R to the number-theoretic notion of height over R.

So, now suppose we have a function ht_R defined on R with values in the non-negative integers. For $x = (x_1, \ldots, x_n) \in R^n$, let $\mathrm{ht}_R(x) = \max \mathrm{ht}_R(x_i)$.

Definition 3 For $x \in R^n \subset R^\infty$, define $\mathrm{length}(x) = n$ and $\mathrm{size}(x) = n \cdot \mathrm{ht}_R(x)$.

Thus over $\mathbb{R}$, $\mathbb{C}$ or $\mathbb{Z}_2$ with unit height, the size of the input is the dimension, that is, vector length, of the input. Over $\mathbb{Z}_2$, this is also the bit length of the *string* x. And over $\mathbb{Z}$ with logarithmic height, size roughly reflects the bit length of the sequence x.

Throughout this book, unless otherwise stated, we suppose ht_R is the unit height for all fields and rings except $\mathbb{Z}$ and $\mathbb{Q}$, where we assume ht_R is logarithmic height. We refer to this as our *basic assumption* about height. Notice that in the former case, $\mathrm{length}(x)$, and hence $\mathrm{size}(x)$, is input into the machine along with x. In the latter case, it is convenient to also input the same information. To do so, we alter the input map slightly

$$I_\infty(x) = (\ldots, 0, \widehat{\mathrm{ht}_R}(x), 0, \widehat{n} \cdot x_1, x_2, \ldots, x_n, 0, 0, \ldots) \quad \text{for } x \in \mathbb{Z}^n.$$

Likewise, if our input space $\mathcal{I}$ is a product space (see Section 3.5), for example, if $\mathcal{I} = R^\infty \times R^\infty$, we suppose that information about the size of each "factor" of $y \in \mathcal{I}$ is input into the machine. So, for example, if $y = (x, w) \in \mathcal{I} = R^\infty \times R^\infty$, then information about both $\mathrm{size}(x)$ and $\mathrm{size}(w)$ are input into the machine.

Now suppose M is a machine over a ring R with a height function ht_R defined.

Definition 4 For $x \in R^n \subset R^\infty$,

$$\mathrm{cost}_M(x) = T(x) \times \mathrm{ht}_{\max}(x),$$

where $T(x)$ is the halting time of M on input x and $\mathrm{ht}_{\max}(x)$ is the maximum height of any element occurring in the computation of M on input x. More precisely,

$$\mathrm{ht}_{\max}(x) = \max_{k \leq T(x)} \mathrm{ht}_R(\widetilde{x}^k),$$

where $\widetilde{x}^0, \cdots, \widetilde{x}^k, \ldots$ is the *modified* state trajectory in S_m of input x. Here $m = \max(K_M, n) + T(x)$ and $\widetilde{x}^k = \widetilde{\pi}(x^k)$, where $\widetilde{\pi} : S \to S_m$ and $x^0, \ldots, x^k, \ldots$ is the state trajectory of x in S. (See Sections 3.2 and 3.3.)

Over the reals or any ring with unit height, cost is just the halting time and reflects the number of basic algebraic operations (as well as the number of shifts) from input to output, whereas over the integers or rationals with logarithmic height, cost reflects the number of bit operations. We call the former measure *unit* or *algebraic cost* and the latter (which is essentially the classical complexity measure), *logarithmic* or *bit cost*. In Chapter 6 we show that complexity theory over $\mathbb{Z}_2$ with algebraic cost is equivalent to complexity theory over $\mathbb{Z}$ with bit cost. Thus, "classical complexity theory" is captured by complexity theory over $\mathbb{Z}_2$ with algebraic cost.

Our basic assumption about height means that, unless otherwise stated, we suppose unit cost for all fields and rings except $\mathbb{Z}$ and $\mathbb{Q}$ where we assume bit cost. At times though, we may want cost to take into account other features affecting computational complexity such as the degrees of polynomials evaluated along computation paths (see Chapter 20).

Definition 5 A machine M over R is a *polynomial time* machine on $X \subset R^\infty$ if there are positive integers c and q such that

$$\text{cost}_M(x) \leq c(\text{size}(x))^q, \quad \text{for all } x \text{ in } X.$$

Polynomial time is meant to formalize the notion of tractability.

Remark 4 Closer to classical notation, for $s \in \mathbb{N}$ we can let

$$\text{cost}_{M,X}(s) = \sup\{\text{cost}_M(x) \mid x \in X \text{ and size}(x) = s\}. \tag{4.2}$$

Then M is polynomial time on X if there are positive integers c and q such that for all sizes s,

$$\text{cost}_{M,X}(s) \leq cs^q.$$

Remark 5 Much of complexity theory is concerned with cost "up to a polynomial." Indeed, when we have asserted that certain assumptions are without loss to our theory, in terms of complexity, this has meant that the relevant cost functions are polynomially related. Thus, for example, in Section 3.2, the cost of converting from $\widehat{n}$ to n is linear in n and the process of simulating computations involving negative coordinates increases the complexity only by an additive constant.

Remark 6 Polynomial time machines on X are examples of *time bounded* machines, that is, machines where for all x in X, $\text{cost}_M(x) \leq f(\text{size}(x))$ for some bounding function $f : \mathbb{N} \to \mathbb{N}$.

Definition 6 A map $\varphi : X \to Y \subset R^\infty$ is said to be a *p-morphism* over R, or *polynomial time computable*, if φ is computable by a polynomial time machine on X.

Now we are in a position to formally define the class P over R.

Definition 7 A decision problem $S \subseteq R^\infty$ is in *class* P *over* R ($S \in P_R$) if its characteristic function χ_S is a *p*-morphism over R.

Definition 8 A structured decision problem (X, X_{yes}) is in *class* P *over* R if its restricted characteristic function χ_{yes} is a *p*-morphism over R. In this case we write $(X, X_{\text{yes}}) \in P_R$.

Remark 7 If (X, X_{yes}) is a structured decision problem, we assume unless otherwise stated, that the decision problem $X \in P_R$. This means that we can decide in polynomial time given an element of R^∞ whether it is an *admissible* problem instance. All the problems in Section 4.1 have this property. Under this assumption, $(X, X_{\text{yes}}) \in P_R$ if and only if $X_{\text{yes}} \in P_R$.

Remark 8 Polynomial time is meant to capture an intuitive notion of tractability. So, problems in class P over R are sometimes called *tractable* over R and those not in class P over R are called *intractable*.

Remark 9 Let us return to the problems listed in Section 4.1. Clearly, if a problem is in class P over R, then it is decidable over R. So, by virtue of undecidability (see Section 4.5), the problems 4-FEAS, QUAD, HN, QA-FEAS, and SA-FEAS are not in class P over $\mathbb{Z}$ no matter which cost measure we may consider. They are also unlikely to be in class P over $\mathbb{Q}$.

The problem LPF is in class P over $\mathbb{Q}$ with bit cost (we prove this in Chapter 15) but not known to be in class P over $\mathbb{Q}$ with unit cost.

For the remaining problems in the list, none is known to be in class P_R and none is known *not* to be in class P_R. However, all are decidable in *exponential time*. The same is true for $\text{HN}_{\mathbb{Z},\mathbb{C}}$ and $\text{HN}_{\mathbb{Z}_2,\overline{\mathbb{Z}_2}}$.

Definition 9 A decision problem $S \subseteq R^\infty$ is in class EXP over R ($S \in \text{EXP}_R$) if its characteristic function χ_S is computable by an *exponential time* machine over R, that is, if there is a machine M that computes χ_S over R and positive integers c and q, such that

$$\text{cost}_M(x) \leq 2^{c(\text{size}(x))^q}$$

for all $x \in R^\infty$.

Remark 10 The classes P_R and EXP_R are two important complexity classes. We consider other complexity classes $\mathcal{C}_R$ as well throughout this book. Consistent with classical notation, we generally write class $\mathcal{C}$ to denote a class $\mathcal{C}_{\mathbb{Z}_2}$.

Remark 11 With the intent of preserving classical terminology insofar as possible, we are using expressions such as "polynomial time" or "exponential time" to denote what might more properly be called "polynomial cost" or "exponential cost" in our development. In the case of unit height, which is mainly used in this book, cost is just the halting time and so there is no confusion here. In addition, with respect to both realizations of the classical theory within our development — namely, complexity over $\mathbb{Z}_2$ with unit height or complexity over $\mathbb{Z}$ with logarithmic height— our notion of cost is essentially the same as the classical notion of time.

On the other hand, we note that, in the case of $\mathbb{Z}$ with logarithmic height, open questions arise regarding the relationship between halting time (as a function of input size) and complexity classes with respect to bit cost. In particular, let $P_{T,<}$ be the class of problems over $(\mathbb{Z}, <)$ decidable by machines with halting time bounded by polynomials in the input size measured in bits. Then clearly, $P_{(\mathbb{Z},<)} \subseteq P_{T,<}$, where $P_{(\mathbb{Z},<)}$ is the classical class P (see Chapter 6). An open question is: is $P_{T,<} \subseteq P_{(\mathbb{Z},<)}$?

4.3   Polynomial Time Reductions

Now, assume the ring R and complexity measure are fixed.

Definition 10 We say a decision problem S is *p-reducible* to a decision problem S' and write

$$S \hookrightarrow_p S'$$

if there is a p-morphism $\varphi : R^\infty \to R^\infty$ such that $\varphi(S) \subseteq S'$ and $\varphi(R^\infty - S) \subseteq R^\infty - S'$. The morphism φ is called a *polynomial time reduction*, or simply, a *p-reduction*.

Thus, if φ is a p-reduction then, for all $x \in R^\infty$,

$$x \in S \text{ if and only if } \varphi(x) \in S'.$$

Remark 12 Suppose $S \hookrightarrow_p S'$. Then S' is at least as "hard" as S. Any decision procedure for S' can be easily converted into a decision procedure for S by means of a p-reduction φ. To decide if problem instance $x \in R^\infty$ is a yes-instance, simply convert x into problem instance $\varphi(x) \in R^\infty$ using a polynomial time machine for φ and then decide if $\varphi(x) \in S'$. The complexity of the converted procedure is no worse (up to a polynomial) than the original. So, for example, if $S' \in P_R$, then $S \in P_R$.

Remark 13 Clearly, p-reducibility is transitive.

Again we have a corresponding definition for structured problems.

Definition 11 Let (X, X_{yes}) and (Y, Y_{yes}) be structured decision problems. We say that (X, X_{yes}) is *p-reducible* to (Y, Y_{yes}) if there is a p-morphism $\varphi : X \to Y$ such that $\varphi(X_{\text{yes}}) \subseteq Y_{\text{yes}}$ and $\varphi(X_{\text{no}}) \subseteq Y_{\text{no}}$. In this case we write

$$(X, X_{\text{yes}}) \hookrightarrow_p (Y, Y_{\text{yes}}).$$

Remark 14 Suppose $(X, X_{\text{yes}}) \subseteq (Y, Y_{\text{yes}})$; that is, $X_{\text{yes}} \subseteq Y_{\text{yes}}$ and $X_{\text{no}} \subseteq Y_{\text{no}}$. Thus, we get $(X, X_{\text{yes}}) \hookrightarrow_p (Y, Y_{\text{yes}})$ via the identity map. So, for example, QUAD $\hookrightarrow_p$ HN.

Theorem 1 *Suppose either unit or bit cost.*

(1) *Over any ring or field R,* HN $\hookrightarrow_p$ QUAD.

(2) *Over any ordered ring or field R,* QUAD $\hookrightarrow_p$ 4-FEAS.

(3) *Over* $(\mathbb{Z}, <)$,

    (a) BHN $\hookrightarrow_p$ BQUAD $\hookrightarrow_p$ B4FEAS.

    (b) BQUAD $\hookrightarrow_p$ BIP $\hookrightarrow_p$ IPF.

Here BHN, BQUAD, B4FEAS, and BIP are the bounded counterparts of HN, QUAD, 4-FEAS, and IPF, respectively. For example, BHN is given by

$$X = \{(f, k) \mid f \in \mathbb{Z}[x_1, \dots, x_n]^m \text{ and } n, m, k \in \mathbb{Z}^+\}$$

and

$$X_{\text{yes}} = \{(f, k) \in X \mid \exists z \in \mathbb{Z}^n, |z_i| \leq k, i = 1, \dots, n \text{ and } f(z) = 0\}.$$

To get BIP, replace (f, k) in the preceding X by (f, b, k) where f is a linear system and $b \in \mathbb{Z}^m$ and replace, in the preceding X_{yes}, the condition $f(z) = 0$ by the condition $f(z) \geq b$.

Proof of Theorem 1.   For each case our task is to construct a machine M that computes an appropriate p-morphism. A formal construction and analysis would be quite cumbersome. So we proceed informally. Justification for this approach is discussed in the next section.

For (1) and (2) the constructions of the requisite p-reductions are essentially contained in Lemma 6 and Remark 10, respectively, of Chapter 2. The observations concerning the coefficients of the transformed systems in Remark 11 of Chapter 2 assures that these constructions are p-morphisms with respect to bit cost as well as unit cost.

For the first reduction in (3a) we convert a problem instance (f, k) of BHN to a problem instance (f', k') of BQUAD as follows. Let $f' = (\varphi_1(f), B_k)$, where φ_1 is the p-reduction given in (1) and B_k is the following system of quadratic polynomials,

$$(k^2 - x_i^2) - (1 + v_{i1}^2 + \dots + v_{i4}^2), \quad i = 1, \dots, n.$$

Let $k' = k^{D-1}$ where D is the degree of f. Note that $\text{ht}_{\mathbb{Z}}(k') \leq (\text{size}(f, k))^2$ and that B_k introduces $4n$ new variables. So size(f', k') is polynomial in size(f, k).

The second reduction is exactly as in (2). Here the bound k remains unchanged.

For (3b), we first show that BQUAD $\hookrightarrow_p$ BIP. To do this, we convert a problem instance $x = (f, k)$ of BQUAD to an equivalent problem instance $x' = (f', b', k')$ of BIP in polynomial time.

We start by replacing each quadratic polynomial $\sum_{i \leq j \leq n} a_{ij} x_i x_j$ in x by a pair of linear polynomials,

$$\sum_{i \leq j \leq n} a_{ij} z_{ij} \quad \text{and} \quad - \sum_{i \leq j \leq n} a_{ij} z_{ij} \tag{4.3}$$

and a system of quadratic polynomial equations,

$$z_{ij} = x_i x_j \quad \text{for } i \leq j \leq n. \tag{4.4}$$

We now show that, in the case of bounded variables, multiplication can be defined in terms of linear inequalities. We first consider the case of positive integers.

Lemma 1 *Let $q \in \mathbb{Z}^+$. For $0 \leq x, y < 2^{q+1}$, $z = xy$ is defined over $(\mathbb{Z}, <)$ by the following system of linear equations and linear inequalities.*

$$
\left.
\begin{aligned}
x = \sum_{i=0}^{q} a_i 2^i, \quad y = \sum_{j=0}^{q} b_j 2^j, \quad z = \sum_{\ell=0}^{2q} c_\ell 2^\ell \\
c_\ell = \sum_{i+j=\ell} w_{ij} \qquad\qquad \text{for } 0 \leq \ell \leq 2q \\
0 \leq a_i, b_j, w_{ij} \leq 1 \qquad \text{for } -q \leq i, j \leq q \\
a_i + b_j - 1 \leq w_{ij} \qquad \text{for } -q \leq i, j \leq q \\
3w_{ij} \leq 2a_i + 2b_j \qquad \text{for } -q \leq i, j \leq q.
\end{aligned}
\right\}
\qquad (4.5)
$$

Here a_i, b_j, w_{ij}, and c_ℓ are new variables for $0 \leq i, j \leq q$, and $0 \leq \ell \leq 2q$.

In other words, over $(\mathbb{Z}, <)$, the system $\{0 \leq x, y < 2^{q+1}, z = xy\}$ is 3-equivalent to the system (4.5) together with the bounds $0 \leq x, y < 2^{q+1}$.

Proof. Suppose $0 \leq x, y < 2^{q+1}$. Then, given the binary representation of x, y, and z, the product $z = xy$ is defined by the first two lines of (4.5) provided $w_{ij} = a_i b_j$ for $0 \leq i, j \leq q$. □

Lemma 2 *If $a_i, b_j, w_{ij} \in \mathbb{Z}$ satisfy the inequalities in the last three lines of (4.5), then $w_{ij} = a_i b_j$.*

Proof. For $a_i, b_j, w_{ij} \in \{0, 1\}$, just check the matrix of possibilities. □

We now consider the general case. Let $x, y \in \mathbb{Z}$ such that $-2^{q+1} < x, y < 2^{q+1}$. Then there exist x^+, x^-, y^+, y^- with $0 \leq x^+, x^-, y^+, y^- < 2^{q+1}$ such that $x = x^+ - x^-$ and $y = y^+ - y^-$. If $z = xy$ we have that

$$
z = z_1 - z_2 - z_3 + z_4, \qquad (4.6)
$$

where

$$
z_1 = x^+ y^+, \; z_2 = x^- y^+, \; z_3 = x^+ y^-, \; z_4 = x^- y^-. \qquad (4.7)
$$

Now let q be the least integer such that $2^{q+1} \geq k$. Replace each quadratic equation in (4.4) by an equation as in (4.6) and for each equation as in (4.7) a corresponding system of linear equations and linear inequalities as given in (4.5) plus the conditions $0 \leq x_i^+, x_i^-, x_j^+, x_j^- < 2^{q+1}$. Then, in the resulting system, replace each equation of the form $g = h$ by the pair of polynomials $g - h$ and $h - g$ and replace all inequalities $g \geq h$ (or $h \leq g$) by $g - h$. Let f' be the resulting system of linear polynomials together with the linear polynomials in (4.3). It is not hard to check that f' is "small" with respect to size(x).

Let $b' = (0, \ldots, 0)$ and let the new bound $k' = 2^{2(q+1)}$. This completes the first reduction in (3b).

The second reduction BIP $\hookrightarrow_p$ IPF is easy. Given problem instance $x = (f, b, k)$ in BIP we produce an equivalent instance $x' = (f', b')$ in IPF by adding to f the polynomials $x_i + k$ and $-x_i + k$ for each variable x_i in f, and appending $2n$ zeros to b, thus ensuring the conditions $x_i + k \geq 0$ and $-x_i + k \geq 0$ that express the original bounds $|x_i| \leq k$ for $i = 1, \ldots, n$. □

The p-reductions in the preceding proposition are used in the next chapters.

Remark 15 The arrows in (1), (2), and (3a) are all reversible. So the corresponding problems are *polynomial time equivalent*. The first arrow in (3b) can be reversed noting that a system of linear inequalities $f(z) \geq b$ over $\mathbb{Z}$ can be transformed to a system of quadratic equations

$$(f_j(z) - b_j) - (u_{j1}^2 + \ldots + u_{j4}^2) = 0, \ j = 1, \ldots, n.$$

The reversibility of the second arrow in (3b) is not at all clear but is shown in Chapter 6.

For another example, clearly HN $\hookrightarrow_p$ SA-FEAS, but again the reverse direction is not as apparent. Let SA-FEAS$|_c$ be the restricted problem where each instance is a system of simple semi-algebraic formulas of the form $f > 0 \vee g = 0$, where f is a variable x, or 0, and g is a polynomial of degree bounded by c. Over $\mathbb{Z}$, $\mathbb{Q}$, and $\mathbb{R}$, we can construct p-reductions SA-FEAS$|_c \hookrightarrow_p$ HN using the tools in the proof of Lemma 5 of Chapter 2. For example, over $\mathbb{R}$, the formula $f > 0 \vee g = 0$ transforms to the polynomial equation

$$(fu^2 - 1)g = 0,$$

which has one new variable and degree bounded by $c + 3$.

A direct p-reduction for the full problem is more subtle —since care must be taken to control the size of the transformed problem— but follows as a corollary to Theorem 1 of Chapter 5. Similar remarks can be made about QA-FEAS over any field (e.g., $\mathbb{Z}_2$, $\mathbb{C}$).

4.4 Management, Machines and Algorithms

In this section we discuss a rationale for constructing machines and computable functions in an informal fashion. We begin with an example.

Consider the problem of computing the dot product of two vectors; that is, given the vectors $v = (v_1, \ldots, v_n) \in \mathbb{R}^n$ and $w = (w_1, \ldots, w_n) \in \mathbb{R}^n$ compute

$$v \cdot w = \sum_{i=1}^{n} v_i w_i.$$

If we were to write an informal computer program $\mathcal{P}$ to solve this problem we might write something with the following simple form.

```
input v, w ∈ ℝⁿ
z := 0
for i = 1 to n do
    z := z + vᵢwᵢ
end do
output z.
```

Here the expression $z := x$ reads as "z is set equal to x" and the expression **return** z means "output z."

We now give a formal, detailed construction of a machine $M_{\mathcal{P}}$ over $\mathbb{R}$ which solves the same problem.

Given input $(v, w) \in \mathbb{R}^\infty$, the machine $M_{\mathcal{P}}$ produces the initial state in its state space

$$(\ldots, 0, \widehat{2n} \,.\, v_1, \ldots, v_n, w_1, \ldots, w_n, 0, \ldots)$$

which is then converted, using the initial subroutine $\texttt{Init}$ (see Section 3.2), to the state

$$(\ldots, z_{-1}, z_0 \,.\, z_1, z_2, \ldots) = (\ldots, 0, 2n \,.\, v_1, \ldots, v_n, w_1, \ldots, w_n, 0, \ldots).$$

The components of this "vector" are then arranged so that the products $v_i w_i$ can be performed by $M_{\mathcal{P}}$. This is done first with a few machine operations to produce the state

$$(\ldots, 0 \,.\, n, n, n, v_1, \ldots, v_n, w_1, \ldots, w_n, 0, \ldots)$$

and then a sequence of operations to produce the sequence of states:

$$(\ldots, 0.n, n - 1, n, v_1, v_2, \ldots, v_n, w_1, \ldots, w_n, 0, \ldots)$$
$$(\ldots, 0, v_1.n, n - 1, n - 1, v_2, \ldots, v_n, w_1, \ldots, w_n, 0, \ldots)$$
$$(\ldots, 0, v_1, v_2.n, n - 1, n - 2, \ldots, v_n, w_1, \ldots, w_n, 0, \ldots)$$
$$\vdots$$
$$(\ldots, 0, v_1, v_2, \ldots, v_{n-1}, v_n.n, n - 1, 0, w_1, \ldots, w_n, 0, \ldots)$$
$$(\ldots, 0, v_1, v_2, \ldots, v_{n-2}, v_{n-1}.n, n - 2, 0, w_1, v_n, w_2, \ldots, w_n, 0, \ldots)$$
$$(\ldots, 0, v_1, v_2, \ldots, v_{n-3}, v_{n-2}.n, n - 3, 0, w_1, v_{n-1}, v_n, w_2, \ldots, w_n, 0, \ldots)$$
$$\vdots$$
$$(\ldots, 0, v_1.n, 0, 0, w_1, v_2 \ldots, v_n, w_2, \ldots, w_n, 0, \ldots).$$

At each step here $M_{\mathcal{P}}$ exchanges the contents of its first, second, third, and fourth registers, performs a substraction, and then shifts one position.

A few more operations yield

$$(\ldots, 0, v_1 w_1 \,.\, n - 1, n - 1, n - 1, v_2, \ldots, v_n, w_2, \ldots, w_n, 0, \ldots).$$

Now proceeding as before, the machine produces in turn the states:

$$v_1 w_1 + v_2 w_2.n - 2, n - 2, n - 2, v_3, \ldots, v_n, w_3, \ldots, w_n$$
$$v_1 w_1 + v_2 w_2 + v_3 w_3.n - 3, n - 3, n - 3, v_4, \ldots, v_n, w_4, \ldots, w_n$$
$$\vdots$$
$$v_1 w_1 + v_2 w_2 + \ldots + v_n w_n.0, 0, \ldots,$$

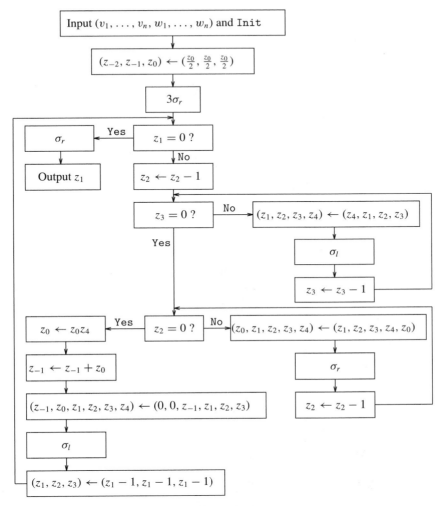

Figure A   The machine $M_{\mathcal{P}}$ implementing $\mathcal{P}$.

where we have omitted the 0s at the beginning and end of these sequences. Finally, the machine shifts right and outputs the contents of its first register.

A machine $M_{\mathcal{P}}$ that performs these computations is given in Figure A. It is quite cumbersome and not immediately illuminating.

Between the simple program $\mathcal{P}$ we wrote at the beginning of this section and the machine $M_{\mathcal{P}}$ there are a number of differences. We focus on one important one.

If we understand cost to be the number of operations performed to solve the problem, there is a gap between $\mathcal{P}$ and $M_{\mathcal{P}}$. Program $\mathcal{P}$ performs a number of arithmetic operations that are linear in $2n$, the input size. On the other hand, the cost of solving the same problem by machine $M_\mathcal{P}$, that is, the halting time, is quadratic in n. The reason for this difference is clear. The computation of $M_\mathcal{P}$ uses a quadratic number of steps to move information from distant registers to an initial few where the arithmetic operations are performed.

If we view the arithmetic operations performed by $\mathcal{P}$ as *essential*, then the additional steps performed by $M_\mathcal{P}$ can be viewed as the *management* done by $M_\mathcal{P}$ in order to execute the essential arithmetic operations. This management is responsible for the difference between the intuitive complexity understood as the number of essential operations and the running time of the machine $M_\mathcal{P}$. The increase in cost is only polynomial. Quite generally, if the number of essential operations performed in solving a problem is polynomial in the input size, then there is a machine that solves this problem with polynomial cost (see also Remark 5).

Since for much of our work we are concerned with classes of problems closed under polynomial increases in cost, the preceding remark liberates us from the burden of exhibiting machines as defined in Chapter 3. Instead, to prove that a function f is computable and to estimate complexity, we describe in some informal yet unambiguous way a sequence of essential operations for computing f. We use the word *algorithm* for such a description. The program $\mathcal{P}$ for computing the dot product is an example. Cost is then estimated in terms of the number of operations executed by such an algorithm.

We close this section by recalling standard notation of classical complexity theory which is also useful in our development.

Definition 12 Let $f, g : \mathbb{N} \to \mathbb{R}$. We say that f *is of order* g and we write $f = O(g)$ if there exist constants $n_0 \in \mathbb{N}$ and $c \in \mathbb{R}$, $c > 0$, such that $|f(n)| \leq c|g(n)|$ for all $n \geq n_0$.

The purpose of denoting $f = O(g)$ is to express the fact that f grows at most as fast as g. This comparative growth rate disregards multiplicative constants and thus enables us to omit them. A similar situation is common in mathematics when we write $f \sim g$ to denote the fact that $\lim_{n \to \infty} (f(n))/(g(n)) = 1$. Here, the relation $\sim$ allows us to eliminate smaller additive factors.

4.5   Additional Comments and Bibliographical Remarks

The decidability of SA-FEAS over $\mathbb{R}$ follows from the effective elimination theory of Tarski [1951]. The geometric content of this theory is that projections of semi-algebraic sets over $\mathbb{R}$ are semi-algebraic. The algorithmic content is an effective procedure for going from a semi-algebraic set to its projection. The resulting decision algorithm is highly intractable. Singly exponential time decision algorithms for these problems have more recently been given by a number of researchers. See Section 18.6 for a fuller discussion and references.

The decidability of HN over $\mathbb{C}$, $HN_{\mathbb{Z},\mathbb{C}}$ over $\mathbb{Z}$, and $HN_{\mathbb{Z}_2,\overline{\mathbb{Z}_2}}$ over $\mathbb{Z}_2$ (in exponential time) follow from the effective Nullstellensatz mentioned in Section 1.2 and proved in [Brownawell 1987; Kollár 1988; Caniglia, Galligo, and Heintz 1988].

The undecidability of HN, and the like, over $\mathbb{Z}$ follows from Matiyasevich's theorem [Matiyasevich 1993] on the unsolvability of diophantine equations, that is, of Hilbert's Tenth Problem.

It is perhaps worth saying a few words at this point about how the classical decidability/undecidability results of logic relate to our development. For example, what do Tarski's results implying that "the reals are decidable" and Gödel's that "the integers are not" mean in our context?

We say that a set $S \subset R^n$ is *definable* over $(R, <)$ if S is derived from a semialgebraic set over R by a finite sequence of projections and complements. Using this terminology, Tarski [1951] implies:

$$\text{every definable set over } \mathbb{R} \text{ is decidable over } \mathbb{R} \text{ (in bounded time),} \qquad (4.8)$$

whereas Gödel [1931] implies:

$$\text{there exist definable undecidable sets over } \mathbb{Z}. \qquad (4.9)$$

We give a brief outline of the proof of undecidabilty of the integers, that is, of (4.9).

Over $\mathbb{Z}$ (or more generally, over any ring of algebraic integers).

(a) For each $n \leq \infty$, there exist undecidable halting sets in $\mathbb{Z}^n$. (This is proved using a Cantor style diagonal argument on the halting set of a universal machine.)

(b) For each $n < \infty$, time-T halting sets in $\mathbb{Z}^n$ are definable over $\mathbb{Z}$ (via the register equations, see Corollary 1 of Chapter 2).

And so, by Gödel's sequencing lemma, we have the following.

(c) For each $n < \infty$, halting sets in $\mathbb{Z}^n$ are definable over $\mathbb{Z}$.

Thus (4.9) follows from (a) and (c).

Matiyasevich's theorem, stated in this format is as follows.

(c′) For each $n < \infty$, halting sets in $\mathbb{Z}^n$ are diophantine, that is, projections of algebraic sets over $\mathbb{Z}$. (The converse is easily seen to be the case.)

The undecidability of HN over $\mathbb{Z}$ follows from (a) and (c′).

Now, if $\mathbb{Z}$ were the projection of an algebraic set over $\mathbb{Q}$, then we would also have the undecidability of HN over $\mathbb{Q}$. This is not known. However, $\mathbb{Z}$ is definable in $\mathbb{Q}$ [Robinson 1949] and so "the rationals are undecidable." That is,

$$\text{there exist definable undecidable sets over } \mathbb{Q}.$$

We now return to the reals. We note that statement (a) (and also statement (b)) is true when $\mathbb{Z}$ is replaced by $\mathbb{R}$. So, by (4.8), (c) cannot be true over $\mathbb{R}$. Neither can

$\mathbb{Z}$ be definable over $\mathbb{R}$. For again by (4.8), we have that definable subsets of $\mathbb{R}$ are semi-algebraic, and thus are finite unions of intervals with algebraic endpoints. On the other hand, $\mathbb{Z}$ is decidable over $\mathbb{R}$, as the following simple algorithm shows.

> input $x \in \mathbb{R}$
> $z := |x|$
> **while** $z \geq 0$ **do** $z := z - 1$ **end do**
> $z := z + 1$
> **if** $z = 0$ output 1, **else** output 0.

And indeed, it is very simple to show (see Proposition 2 of Chapter 22) that each subset $S \subseteq \mathbb{Z}$ can be decided by a machine over $(\mathbb{R}, <)$ with a built-in constant c_S coding the characteristic function χ_S. Such machines over $\mathbb{R}$ can be considered as a type of "oracle" Turing machine. A formal model of an oracle machine is defined in Section 21.3. We develop in more detail the question of the power of real machines over binary inputs in Chapter 22.

The situation over $\mathbb{C}$ is in some ways similar to $\mathbb{R}$; that is,

every definable set over $\mathbb{C}$ is decidable over $\mathbb{C}$,

where definable now means derived from quasi-algebraic sets. But over $\mathbb{C}$, decidable sets in $\mathbb{C}^n$ are decidable by time-bounded machines and hence are quasi-algebraic sets.[1] This (and an extension to subsets of $\mathbb{C}^\infty$) is proved in [Cucker and Rosselló 1993]. Thus, unlike over $\mathbb{R}$,

every decidable set in $\mathbb{C}^n$ is definable over $\mathbb{C}$,

and $\mathbb{Z}$ cannot be decided over $\mathbb{C}$.

For further elaboration of these ideas, particularly with respect to Gödel's theorem and decidability over a ring, see [Blum and Smale 1993]. See Section 18.6 for additional comments on elimination theory.

In subsequent chapters of this book we encounter algorithms and bibliographical references for the remaining complexity results mentioned in Section 4.2. The "big Oh" notation, which is ubiquitous in complexity theory, was introduced by Bachman [1894]. See [Knuth 1976] for other notions of comparative rates of growth such as "little oh," "big Theta," and "big Omega." We meet the latter in Chapter 16.

---

[1] This was pointed out to Blum by D. Marker and B. Poizat in personal communication.

5
The Class NP and NP-Complete Problems

Although it may not be at all obvious given a polynomial over R how to decide whether it has a zero over R, it is a straightforward procedure to verify a solution that may be presented to us. Just plug the purported solution into the polynomial and evaluate it. Is this verification tractable in our model of computation? An affirmative answer will depend on the underlying mathematical properties of the ring or field, as well as our measure of complexity, and is at the core of the notion of NP.

A key impetus for the development of classical complexity theory was the discovery of a large number of NP-*complete* problems, that is, NP problems that are universal for the class NP. These problems have no known polynomial time decision algorithms, and their universality amounts to the fact that the existence of such an algorithm for one of them would provide polynomial time decision algorithms for all problems in NP. Thus, the fundamental "P = NP ?" question reduces to ascertaining the complexity of any one of the NP-complete problems.

The seminal result in the classical setting was Cook's theorem stating that the Satisfiability Problem is NP-complete.

In our expanded framework, we obtain analogous results for the real and complex numbers. It is the goal of this chapter to prove such results. We prove that the Hilbert Nullstellensatz and the 4-Feasibility problems introduced in Section 1.2 are NP-complete over $\mathbb{C}$ and $\mathbb{R}$, respectively. The completeness of these problems follows quite naturally from the register equations developed in Chapter 3. If the base ring is $\mathbb{Z}_2$, we recover Cook's theorem from these equations.

5.1   The Class NP

We begin this section with a formal definition of NP_R.

Definition 1 A decision problem $S \subseteq R^\infty$ is in *class* NP over R ($S \in NP_R$) if there is a machine M over R with $\mathcal{I}_M = R^\infty \times R^\infty$ and positive integers c and q such that

(1) if $x \in S$, then there exists some $w \in R^\infty$ (called a *guess* or *witness* for a solution) such that $\Phi_M(x, w) = 1$ and $\mathrm{cost}_M(x, w) \leq c(\mathrm{size}(x))^q$;

(2) if $x \notin S$, then there is no $w \in R^\infty$ such that $\Phi_M(x, w) = 1$.

We are assuming in this definition that if R is considered with an order $<$, then M is a machine over $(R, <)$. Otherwise M is a machine over $(R, =)$.

The machine M is called a *nondeterministic polynomial time* (NP_R) decision machine over R for the problem S. Property (1) reflects the nondeterministic aspect of this notion; that is, for each yes-instance, we just require that some polynomial time verifiable witness *exist*, not necessarily that one can be found. What is important here is that the cost of verification is polynomial in the *size* of x. Property (2) requires that the verification process have integrity; that is, there is no witness to a *no-instance*.

When $R = \mathbb{Z}_2$ we have the definition of the classical NP. Thus, we simply write NP when we mean $NP_{\mathbb{Z}_2}$.

Remark 1 We note that for each x we need only consider potential witnesses w satisfying the bounds $\mathrm{size}(x, w) \leq c(\mathrm{size}(x))^q + K_M$ and $\mathrm{ht}_R(w) \leq c(\mathrm{size}(x))^q$.

We continue with our basic assumptions about height and cost (see Section 4.2). Since information about size (length and height) is input to the machine, we can assume, without changing the class NP_R, that our nondeterministic machine M satisfies the additional property,

(3) for all $x \in X$, $w \in R^\infty$, $\mathrm{cost}_M(x, w) \leq c(\mathrm{size}(x))^q$ and $\Phi_M(x, w) \in \{0, 1\}$.

Remark 2 We continue our practice of describing machines by means of informal algorithms (see Section 4.4). In the case of nondeterministic machines, as is common in classical complexity theory, we also use the expression

$$\text{guess } w \in R^\infty$$

to denote the input into the algorithm of the witness w. This expression can be placed at any stage of the algorithm and has a cost of $\mathrm{size}(w)$. In this way we differentiate the input of the witness w from the input of the instance x.

Proposition 1 *If $S \hookrightarrow_p S'$ and $S' \in NP_R$, then $S \in NP_R$.*

Proof.   Let φ be a polynomial time reduction and let M' be an NP_R decision machine for S'. The following algorithm shows $S \in NP_R$.

> input x
> compute $y := \varphi(x)$
> guess $w \in R^\infty$ and compute $z := \Phi_{M'}(y, w)$
> output z.

$\square$

Again, for structured problems (X, X_{yes}), we have a related definition for the class NP over R. We leave the precise formulation to the reader. As before, since we are assuming $(R^\infty, X) \in P_R$, there is a natural correspondence between problems and structured problems in class NP_R.

It is possible to give a more algebraic formulation of class NP_R in terms of the class P_R.

Definition 2 A computable map $\varphi : R^\infty \to R^\infty$ is said to be *honest* on $V \subset R^\infty$ if there are positive integers c and q such that for all $v \in V$, one has $\text{size}(v) \leq c(\text{size}(\varphi(v)))^q$.

Thus, up to a polynomial, maps honest on V do not collapse sizes of elements in V.

Proposition 2 *A decision problem $S \in NP_R$ if and only if*

$$S = \varphi(V)$$

for some decision problem $V \in P_R$ and p-morphism $\varphi : R^\infty \to R^\infty$ which is honest on V.

Proof. First suppose $S \in NP_R$ with M its NP_R decision machine and p_M the corresponding polynomial cost bound. Let $X = R^\infty \times R^\infty$ and π be the "projection" onto the first coordinate, that is, for $(x, w) \in X$, $\pi(x, w) = x$. Let

$$V = \{(x, w) \in X \mid x \in S, \ \text{size}(x, w) \leq p_M(\text{size}(x)) + K_M$$
$$\text{and } \Phi_M(x, w) = 1\}.$$

Clearly, π is a p-morphism, it is honest on V and, by nondeterminism and Remark 1, $\pi(V) = S$.

We now show $V \in P_R$. Let M^* be defined by the following algorithm.

> input $(x, w) \in X$
> **if** $\text{size}(x, w) > p_M(\text{size}(x)) + K_M$ **then** output 0
> **else** output $\Phi_M(x, w)$.

The second line of this algorithm requires some attention. Note that in the **if** condition there is an integer polynomial to be computed and an inequality to decide. If R is ordered, this can be carried out in a straightforward fashion using a computation node followed by a branch node. If R is not ordered, the right side of the inequality can be computed by a subroutine that outputs its integer value in binary (i.e., as a vector of 0s and 1s) or unary (i.e., as a vector of 1s). This vector can then be compared to the left-hand side when represented in the same fashion. All this can be done in polynomial time on X.

So, M^* is polynomial time on X and $\Phi_{M^*}(V) = \{1\}$. But also, $\Phi_{M^*}(X - V) = \{0\}$. For if $(x, w) \notin V$, then either $x \notin S$, in which case $\Phi_M(x, w) = 0$, or $\text{size}(x, w) > p_M(\text{size}(x)) + K_M$, or $\Phi_M(x, w) = 0$. So in each case, $\Phi_{M^*}(x, w) = 0$.

Now, for the other direction, suppose $V \in P_R$ and $S = \varphi(V)$ for some p-morphism $\varphi : R^\infty \to R^\infty$, honest on V. Let p_φ and p_V be the polynomial bounds on the cost of computing the p-morphism and on the collapsing size, respectively. Let M be a decision machine for V with polynomial cost bound p_M. We construct an NP_R decision machine M^* for S via the following algorithm.

> input $x \in R^\infty$
> guess $v \in R^\infty$
> **if** $\text{size}(v) > p_V(\text{size}(x))$ **then** output 0
> **else** compute $\Phi_M(v)$
>   **if** $\Phi_M(v) = 0$ **then** output 0
>   **else** compute $\varphi(v)$
>   **if** $\varphi(v) = x$ **then** output 1
>   **else** output 0.

The **if** condition in the third line here can be handled as in the previous algorithm.

We first observe that M^* is polynomial time in the size of x: the cost of the first check is polynomial in $\text{size}(x)$. Then, if $\text{size}(v) \leq p_V(\text{size}(x))$, the cost of computing $\Phi_M(v)$ is bounded by $p_M(p_V(\text{size}(x)))$ and the cost of computing $\varphi(v)$ is bounded by $p_\varphi(p_V(\text{size}(x)))$.

Next we see that M^* is a nondeterministic decision machine for S: if $x \in S$, then there is a $v \in V$ such that $\varphi(v) = x$. Thus, $\Phi_M(v) = 1$ and, by honesty, $\text{size}(v) \leq p_V(\text{size}(x))$. So, $\Phi_{M^*}(x, v) = 1$. Conversely, if $\Phi_{M^*}(x, v) = 1$, then by construction of M^*, both $\Phi_M(v) = 1$ and $\varphi(v) = x$. So, $v \in V$ and hence $x \in S$. □

Corollary 1 *For any ring R one has $P_R \subseteq NP_R$.*

Proof. Let φ be the identity map. □

Remark 3 Over $\mathbb{Z}_2$, it is easy to see that $NP \subseteq EXP$; that is, problems in NP are decidable in exponential time. This is because the number of potential witnesses w with $\text{size}(w) \leq c(\text{size}(x))^q + K_M$ is at most $2^{c'(\text{size}(x))^q}$ for some integer $c' > 0$. Similarly, over $\mathbb{Z}$ with bit cost, $NP_\mathbb{Z} \subseteq EXP_\mathbb{Z}$. Over $\mathbb{R}$ and $\mathbb{C}$, or $\mathbb{Z}$ with unit cost, the number of such potential witnesses is infinite and so the analogous inclusion of complexity classes is not clear. Nor is it clear in this case if problems in NP_R are in general decidable over R. We return to this question later in Section 5.4.

In our general setting, it is natural to ask, as we do in the classical setting, does $P = NP$ over R? Over $(\mathbb{R}, <)$ or $(\mathbb{C}, =)$, this question presents us with compelling

new open problems. We show that here, as in the classical case, there exist natural
NP problems not known to be in class P.

5.2   Important NP Problems

We now refer to the list of decision problems in Section 4.1. In the following, if the
problem (X, X_{yes}) is defined using an order relation we assume we are working
over $(R, <)$. If R is not naturally ordered, we assume we are working over $(R, =)$.
For the remaining cases, unless otherwise specified, we assume our statements
hold over both $(R, <)$ and $(R, =)$.

An important component of many constructions here is a *universal polynomial
system evaluator* (UPSE) over R. The space of admissible inputs for such a machine
or "subroutine" is

$$W = \{(f, z) \mid f \in R[x_1, \ldots, x_n]^m, z \in R^n, \text{ and } m, n \in \mathbb{Z}^+\}.$$

Given input $(f, z) \in W$, UPSE outputs $f(z)$. Note that $W = X \times R^\infty$, where X is
the space of problem instances of HN.

Any standard method for evaluating polynomials can be used to construct a
UPSE that is polynomial time on W with respect to either bit or unit cost, and in
the latter case, with the cost depending *only* on the size of f. We now assume such
a UPSE is at hand.

Proposition 3   QA-FEAS, SA-FEAS, HN, QUAD, *and* 4-FEAS *are all in class*
NP_R *with respect to unit cost.*

Proof.   For each of these problems, let X denote the appropriate space of problem
instances. For the last three problems, an NP_R decision machine is constructed as
follows.

> input f
> guess $z \in R^n$
> use UPSE to compute $y := f(z)$
> **if** $y = 0$ output 1
> **else** output 0.

For the other two problems, replace the branching instruction by a subroutine
to check if the appropriate equalities or inequalities hold.   □

Important special cases include $HN \in NP_\mathbb{C}$, 4-FEAS $\in NP_\mathbb{R}$, and QUAD $\in NP_{\mathbb{Z}_2}$.

Remark 4   Another example of a problem in NP over R is the linear programming
feasibility problem LPF over $(R, <)$. Here the complexity measure plays a crucial
role in the proof. In the unit cost case, the proof is straightforward, similar to
the proof of Proposition 3. However, for the case $R = \mathbb{Z}$ with bit cost (i.e., for
the integer programming feasibility problem IPF), the proof is considerably more
subtle. Here one shows that if an integer programming problem instance is feasible
over $\mathbb{Z}$, then it also has a "small" integer solution. This is done in the next chapter.

For the next two propositions we suppose either bit or unit cost.

Proposition 4 BSA, BHN, BQUAD, B4FEAS, *and* BIP *are all in class* NP$_{\mathbb{Z}}$.

Proof. By p-reducibility and Proposition 1, we need just show BIP $\in$ NP$_{\mathbb{Z}}$. The nondeterministic decision machine operates on inputs $x = (f, b, k) \in X$, and guesses $z \in \mathbb{Z}^{\infty}$, by first checking if $|z_i| \leq k$. If not, it outputs 0; if yes, it utilizes UPSE to evaluate $f(z)$. If $f(z) \geq b$, it outputs 1; if not, it outputs 0.

If $x \in X_{\text{yes}}$, there is a bounded solution and hence a verifiable witness z such that $\text{ht}_{\mathbb{Z}}(z) \leq \text{ht}_{\mathbb{Z}}(k)$. With this witness, the cost of the first check, the evaluation, and the final check are each polynomial in size(x). $\quad\square$

Proposition 5 KP *and* TSP *are in class* NP$_R$.

Proof. For both KP and TSP a witness is just a subset of indices indicating which coordinates of the problem instance x to add up. The cost of this summing is linear in size(x). The cost of a final test to check the appropriate equality or inequality is also linear (constant in the case of unit cost). For TSP an additional linear test is necessary to determine if the indices define a tour. $\quad\square$

5.3   NP-Completeness

Definition 3 A decision problem $\widehat{S}$ is said to be NP-*hard* over R (or NP$_R$-hard) if every $S \in$ NP$_R$ is p-reducible to it.

Definition 4 A decision problem $\widehat{S}$ is said to be NP-*complete* over R (or NP$_R$-complete) if it is NP$_R$-hard and in class NP$_R$.

Thus, an NP$_R$-complete problem $\widehat{S}$ is the "hardest" NP$_R$ problem. The following two propositions are clear but fundamental.

Proposition 6 *Let* $\widehat{S}$ *be an* NP-*complete problem over* R. *Then* $\widehat{S} \in$ P$_R$ *if and only if* P$_R =$ NP$_R$. $\quad\square$

Proposition 7 *Suppose* $\widehat{S}$ *is* NP-*hard over* R *and* $\widehat{S} \hookrightarrow_p S$. *Then* S *is* NP-*hard over* R. $\quad\square$

5.4   NP-Complete Problems over $\mathbb{C}$, $\mathbb{R}$, and $\mathbb{Z}$

Our major NP-completeness results derive from the following Main Theorem.

Theorem 1

(1) *With respect to either unit or bit cost,*

(a) QA-FEAS *is NP-hard over* $(R, =)$ *for any ring or field R.*

(b) HN *and* QUAD *are NP-hard over* $(F, =)$, *for any field F (e.g.,* $\mathbb{Z}_2$, $\mathbb{C}$).

(2) *With respect to either unit or bit cost,*

(a) SA-FEAS *is NP-hard over* $(R, <)$ *for any ordered ring or field R.*

(b) HN, QUAD, *and* 4-FEAS *are NP-hard over* $(\mathbb{Z}, <)$, $(\mathbb{Q}, <)$, *and* $(\mathbb{R}, <)$ *(or any real closed field).*

(3) *With respect to bit cost,* BHN, BQUAD, B4FEAS, *and* BIP *are NP-hard over* $(\mathbb{Z}, <)$.

Proof.   We need only show the result for the first problem in each of (1a), (1b), (2a), (2b), and (3). The rest follow by the p-reductions given by Theorem 1 of Chapter 4.

So let (Y, Y_{yes}) denote one of the first problems and suppose $(X, X_{\text{yes}}) \in \text{NP}_R$ for the corresponding ring or field R. Our task is to construct a p-morphism $\varphi : X \to Y$ such that, for each problem instance $\mathbf{x} \in X$,

$$\mathbf{x} \in X_{\text{yes}} \text{ if and only if } \varphi(\mathbf{x}) \in Y_{\text{yes}}. \tag{5.1}$$

We do so utilizing the register equations of an NP_R machine M for (X, X_{yes}) with polynomial cost function defined by constants $c, q > 0$.

Let $x = (x_1, \ldots, x_\ell)$, $w = (w_1, \ldots, w_m)$, and let $n = \ell + m$. For $T > 0$, let $\mathcal{R}_T^{(1)}(x, w, z)$ be the time-T register equations (3.3) and (3.4) for M with input (x, w) together with the equations (3.5) stipulating that the output value is 1. Here, $z = (\alpha^0, x^0, \ldots, \alpha^T, x^T)$.

Let $\Phi_T^{(1)}(x, w, v)$ denote the system n-equivalent to $\mathcal{R}_T^{(1)}(x, w, z)$ —quasi-algebraic for (1a), semi-algebraic for (2a), and algebraic for (1b), (2b), and (3)— derived from Theorem 2 of Chapter 3 (parts (1) and (2a)). Thus, for each of our cases,

$$(\mathbf{x}, \mathbf{w}) \in \Omega_T(1) \text{ if and only if } \Phi_T^{(1)}(\mathbf{x}, \mathbf{w}, v) \text{ is feasible.} \tag{5.2}$$

Recall that $\Omega_T(1)$ is the set of elements in the time-T halting set of M that have output value 1.

We now construct the p-reduction for parts (1) and (2) of our theorem as follows.

For $\mathbf{x} \in X$, let $\varphi(\mathbf{x}) = \Phi_T^{(1)}(\mathbf{x}, w, v)$, where $T = c(\text{size}(\mathbf{x}))^q$

and $w = (w_1, \ldots, w_m)$ with $m = c(\text{size}(\mathbf{x}))^q + K_M$. $\tag{5.3}$

Let $n = \text{length}(\mathbf{x}) + m$. By the estimates given in Theorem 2 of Chapter 3, length(v) and the number of polynomial equations and inequalities in $\Phi_T^{(1)}(\mathbf{x}, w, v)$ are bounded by a polynomial in n and T which in turn are polynomials in size($\mathbf{x}$). Furthermore, the degrees of the polynomials are bounded by a constant depending only on M. In case of logarithmic height, the size of the coefficients grows negligibly (see Remarks 8 and 11 of Chapter 2). Hence, size($\Phi_T^{(1)}(\mathbf{x}, w, v)$) is bounded by a polynomial in size($\mathbf{x}$). The construction of the $\Phi_T^{(1)}$ is completely uniform. And so, φ is a p-morphism.

It is also a reduction. By the definition of class NP_R we have $\mathbf{x} \in X_{\text{yes}}$ if and only if there is a $\mathbf{w} \in R^\infty$ with length($\mathbf{w}$) $= m$ such that $\Phi_M(\mathbf{x}, \mathbf{w}) = 1$ and $\text{cost}_M(\mathbf{x}, \mathbf{w}) \leq T$, that is, such that $(\mathbf{x}, \mathbf{w}) \in \Omega_T(1)$. By (5.2), this is the case if and only if $\Phi_T^{(1)}(\mathbf{x}, \mathbf{w}, v)$ is feasible. Thus we have demonstrated (5.1).

For Part (3), the reduction must produce an algebraic system plus a stipulated bound on the magnitude of a solution. So for $\mathbf{x} \in X$, let $\varphi(\mathbf{x}) = (\Phi_T^{(1)}(\mathbf{x}, w, v), k)$, where $\Phi_T^{(1)}(\mathbf{x}, w, v)$ is as in (5.3) for Part (2b), and let $k = 2^{c(\text{size}(x))^q} + N$.

By the preceding, size($\Phi_T^{(1)}(\mathbf{x}, w, v)$) is bounded by a polynomial in size($\mathbf{x}$), and so is size(k). Hence φ is a p-morphism.

Again, as before, $\mathbf{x} \in X_{\text{yes}}$ if and only if $\Phi_T^{(1)}(\mathbf{x}, w, v)$ is feasible. We must show that feasibility implies feasibility with $|w_i|, |v_j| \leq k$.

Now, if $\mathbf{x} \in X_{\text{yes}}$, we may assume there is a witness $\mathbf{w}$ such that $\text{ht}_{\mathbb{Z}}(\mathbf{w}) \leq c(\text{size}(\mathbf{x}))^q$ —and hence $|w_i| \leq k$— and such that the heights of all elements in the computation of M with input $(\mathbf{x}, \mathbf{w})$ are also bounded by $c(\text{size}(\mathbf{x}))^q$. Now the variables v_j either represent states in the computation —and hence can be solved by elements with magnitude bounded by k— or are auxiliary variables of the type y, u_1, u_2, u_3, u_4 added to eliminate inequalities of the form $\pm \pi_1(z) > 0$, where z represents an element in the computation (see the proof of Lemma 5 of Chapter 2). The solutions for y have magnitude bounded by 1 and for the u_i bounded by $|\pi_1(z)|$, which in turn is bounded by k. □

Remark 5 We now return to the questions raised in Remark 3 concerning the decidability and complexity of NP problems with respect to unit cost. Over $(\mathbb{Z}, <)$, HN is not decidable. But with respect to unit cost, HN $\in$ NP (Proposition 3). Hence, in this case, NP problems are not in general decidable, nor is P equal to NP.

Over $(\mathbb{R}, <)$ and $(\mathbb{C}, =)$, HN $\in$ EXP (see Section 4.5 for references) and, by the Main Theorem, HN is at least as hard as any NP problem. So here, as in the classical case (but for quite different reasons), we have NP $\subseteq$ EXP. Thus, both over $(\mathbb{R}, <)$ and over $(\mathbb{C}, =)$, in analogy with the classical fundamental question, it is eminently reasonable for us to ask, "Is P = NP?"

Corollary 2 *All problems in* (1) *and* (2) *are* NP-*complete with respect to unit cost. All problems in* (3) *are* NP-*complete with respect to bit cost.*

Proof.   By the results of Section 5.2 each of these problems is in the appropriate class NP.   □

So, by Proposition 6, the fundamental "P = NP?" question is equivalent over $\mathbb{R}$ to "Is 4-FEAS in P?" and over $\mathbb{C}$ to "Is HN in P?" thus focusing our attention on these fundamental problems.

Remark 6 By Theorem 1 of Chapter 4, BQUAD $\hookrightarrow_p$ IPF, so IPF is also NP-hard over $\mathbb{Z}$ with respect to the bit cost. Recall that over $\mathbb{Z}$ with bounds, multiplication can be defined in terms of linear inequalities (see Lemma 1 of Chapter 4). This is fundamental to our p-reduction and hence for the NP-hardness of integer programming. In the next chapter we show that IPF $\in$ NP over $\mathbb{Z}$ with respect to bit cost. Hence IPF is NP-complete over $\mathbb{Z}$ with respect to bit cost.

Corollary 3

(1) *Over any field (e.g., $\mathbb{Z}_2$, $\mathbb{C}$), QA-FEAS $\hookrightarrow_p$ HN.*

(2) *Over $\mathbb{Z}$, $\mathbb{Q}$, and $\mathbb{R}$, SA-FEAS $\hookrightarrow_p$ HN.*   □

5.5   NP-Complete Problems over $\mathbb{Z}_2$

In this section we prove the seminal NP-completeness result of classical complexity theory.

Theorem 2 (Cook)   *The problem* 3-SAT *(of deciding if a 3-CNF propositional formula is satisfiable) is* NP-*complete.*

The task at hand is to understand the statement of this theorem and then derive it from the results of the last section.

Definition 5   A *literal* v_i is a variable x_i or the *negation* of a variable $\neg x_i$.

A *propositional formula* $\Phi(x)$ is a finite string of literals linked by the *logical connectives* $\wedge$ ("and") and $\vee$ ("or"). Here x denotes the variables $x_1, \ldots, x_n$ occurring in $\Phi(x)$.

A propositional formula $\Phi(x)$ is in *conjunctive normal form* ($\Phi(x) \in$ CNF) if for some positive integers ℓ and s,

$$\Phi(x) = \bigwedge_{k=1}^{\ell} \phi_k(x),$$

where $\phi_k(x) = v_{k_1} \vee \ldots \vee v_{k_{s'}}$, the v_{k_i}s are literals, $k_i \in \{1, \ldots, n\}$, and $s' \leq s$. The formula $\phi_k(x)$ is called a *clause*. We say ℓ is the *length* and s is the *width* of $\Phi(x)$. We say $\Phi(x) \in$ 3-CNF if $s = 3$, that is, if there are at most three literals per clause.

A CNF formula $\Phi(x)$ in n variables defines a map $\Phi : \{0, 1\}^n \to \{0, 1\}$ given by substituting $b \in \{0, 1\}^n$ for x in $\Phi(x)$ and evaluating using the basic *Boolean* rules (where the operations $\vee$ and $\wedge$ are associative and commutative):

$$\neg 1 = 0, \ \neg 0 = 1$$
$$1 \vee 0 = 1 \vee 1 = 1 \wedge 1 = 1$$
$$0 \wedge 1 = 0 \wedge 0 = 0 \vee 0 = 0.$$

Clauses are evaluated first and then, $\Phi(b) = \bigwedge_{k=1}^{\ell} \phi_k(b)$. Thus, $\Phi(b) = 1$ if and only if for each $k \in \{1, \ldots, \ell\}$, $\phi_k(b) = 1$.

Definition 6 A formula $\Phi(x)$ is *satisfiable* ($\Phi(x) \in$ SAT) if there is a $b \in \{0, 1\}^n$ such that $\Phi(b) = 1$.

We can now define the classical problem 3-SAT: given $\Phi(x) \in$ 3-CNF, is $\Phi(x) \in$ SAT?

Remark 7 The exhaustive search method for deciding whether $\Phi(x) \in$ SAT requires, in the worst case, checking if $\Phi(b) = 1$ for each of the 2^n elements $b \in \{0, 1\}^n$. All known methods are exponential in n.

We now reformulate the problem 3-SAT in our setting, in particular, as a decision problem over $R = \mathbb{Z}_2$.

With each 3-CNF formula $\Phi(x) = \bigwedge_{k=1}^{\ell} \phi_k(x)$ we naturally associate a system $P_\Phi = \{p_{\phi_k}(x)\}_{k=1}^{\ell}$ of ℓ simple polynomials over $\mathbb{Z}_2$ of degree at most 3 defined as follows. For each clause ϕ_k, $k = 1, \ldots, \ell$,

$$p_{\phi_k}(x) = w_{k_1} \cdot w_{k_2} \cdot w_{k_3},$$

where $k_i \in \{1, \ldots, n\}$ and

$$w_{k_i} = \begin{cases} x_{k_i} + 1 & \text{if } v_{k_i} \text{ is } x_{k_i} \\ x_{k_i} & \text{if } v_{k_i} \text{ is } \neg x_{k_i} \\ 1 & \text{otherwise.} \end{cases}$$

We call polynomials of the form $p_{\phi_k}(x)$, 3-*simple*, and finite systems of 3-simple polynomials over $\mathbb{Z}_2$, 3-CNF *systems*. The *length* of a 3-CNF system is the number ℓ of polynomials.

Note, for $b \in \{0, 1\}^n$, $\phi_k(b) = 1$ if and only if $p_{\phi_k}(b) = 0$. And so, $\Phi(b) = 1$ if and only if $P_\Phi(b) = 0$.

Thus, the satisfiability of the 3-CNF formula $\Phi(x)$ is *equivalent* to the feasibility of the 3-CNF system of equations $P_\Phi = 0$. So we can think of the system P_Φ as the *algebraic representation* of $\Phi(x)$. On the other hand, it is an easy exercise to see that every 3-CNF system P over $\mathbb{Z}_2$ naturally represents a 3-CNF formula $\Phi_P(x)$ of the same length. Hence, we have a natural algebraic formulation of 3-SAT.

Definition 7 The (decision) problem 3-SAT over $\mathbb{Z}_2$ is given by (X, X_{yes}) where

$$X = \{P \mid P \text{ is a 3-CNF system}\}$$

and

$$X_{\text{yes}} = \{P \in X \mid P = 0 \text{ is feasible over } \mathbb{Z}_2\}.$$

We now formally understand the statement of Theorem 2.

By the results of Section 5.4 it is proved once we establish the following two propositions, the first being immediate by Propositions 1 and 3.

Proposition 8 *The problem* 3-SAT *belongs to* NP. $\qquad\qquad\square$

Proposition 9 *The problem* QUAD *over* $\mathbb{Z}_2$ *is p-reducible to* 3-SAT.

Proof. We associate with each quadratic polynomial $q \in \mathbb{Z}_2[x_1, \ldots, x_n]$ a 3-CNF system Φ_q in the n variables $x_1, \ldots, x_n$ plus m additional variables, where $m \leq 2n^2$, and of length at most $4(m+1)$, such that for $b \in \{0,1\}^n$

$$q(b) = 0 \text{ if and only if } \exists b' \in \{0,1\}^m, \ \Phi_q(b, b') = 0.$$

It then follows that for each set of quadratic polynomials

$$Q = \{q_1, \ldots, q_r\} \subset \mathbb{Z}_2[x_1, \ldots, x_n]$$

we can associate a 3-CNF system Φ_Q in $n + rm$ variables and of length at most $4(m+1)r$ such that

$$Q = 0 \text{ is feasible over } \mathbb{Z}_2 \text{ if and only if } \Phi_Q = 0 \text{ is feasible over } \mathbb{Z}_2.$$

Here $\Phi_Q = \bigcup_{i=1}^{r} \Phi_{q_i}$. Thus, we have our p-reduction.

So now suppose $q \in \mathbb{Z}_2[x_1, \ldots, x_n]$ is a quadratic polynomial presented in *normal form*

$$\sum_{i=1}^{s} \sigma_i + \sum_{j=1}^{t} \tau_j + c,$$

where $\sigma_i = x_k x_l$ some $1 \leq k \leq \ell \leq n$, $\tau_j = x_k$ some $1 \leq k \leq n$, $c \in \mathbb{Z}_2$, and no terms are repeated.

Replace the quadratic equation $q = 0$ by the following n-equivalent quadratic system $P_q = 0$, where each equation has *at most* three terms and three variables.

$$S + T + c = 0 \tag{5.4}$$

$$\sigma_i + x_k x_l = 0 \ (i = 1, \ldots, s), \quad \tau_j + x_k = 0 \ (j = 1, \ldots, t) \tag{5.5}$$

$$S + (\sigma_1 + y_1) = 0, \ y_1 + (\sigma_2 + y_2) = 0, \ldots, y_{s-2} + (\sigma_{s-1} + \sigma_s) = 0 \tag{5.6}$$

$$T + (\tau_1 + w_1) = 0, \ w_1 + (\tau_2 + w_2) = 0, \ldots, w_{t-2} + (\tau_{t-1} + \tau_t) = 0. \tag{5.7}$$

The variables of this system are the xs plus S, T, the σs, τs, ys, and ws.

Note that equation (5.4) adds 2 new variables, the $s + t$ equations in (5.5) add $s + t$ new variables, the $s - 1$ equations in (5.6) add $s - 2$ new variables, and finally, the $t - 1$ equations in (5.7) add $t - 2$ more. Thus, P_q is a system of $m + 1$ equations in the variables $x_1, \ldots, x_n$ plus m new variables, where $m = 2(s + t) - 2 \leq 2n^2$. The n-equivalence is straightforward noting that $x = -x$ in $\mathbb{Z}_2$.

So, for $b \in \{0,1\}^n$,

$$q(b) = 0 \text{ if and only if } \exists b' \in \{0,1\}^m, \ P_q(b, b') = 0. \tag{5.8}$$

We now transform P_q to an equivalent 3-CNF system Φ_q with the same variables.

Now suppose the system P_q is given by the polynomials $p_1, \ldots, p_{m+1}$. For each polynomial p_i we construct a 3-CNF system Φ_i with the same variables as p_i and length at most 4 such that for $b^* \in \{0,1\}^{n+m}$,

$$p_i(b^*) = 0 \text{ if and only if } \Phi_i(b^*) = 0.$$

Note that each polynomial p_i is of the form $x + y + z$ or $x + y \cdot z$, where the xs, ys, and zs are either variables or constants 1 or 0. The 3-CNF-systems are constructed using the following transformations.

$$x + y + z \to \{xyz, \ x(y + 1)(z + 1), \ (x + 1)y(z + 1), \ (x + 1)(y + 1)z\} \quad (5.9)$$
$$x + y \cdot z \to \{x(y + 1), \ x(z + 1), \ (x + 1)yz\}.$$

Now, letting $\Phi_q = \bigcup_{i=1}^{m+1} \Phi_i$ we have, for $b^* \in \{0, 1\}^{n+m}$,

$$P_q(b^*) = 0 \ \text{ if and only if } \ \Phi_q(b^*) = 0. \quad (5.10)$$

Combining (5.8) and (5.10) we are done. $\qquad\square$

Remark 8 If we define 3-CNF formulas to have exactly three distinct literals per clause, as is sometimes done, the preceding proof on NP-completeness can be modified by changing the second transformation in (5.9) to

$$x + y \cdot z \to \{x(y+1)u, \ x(y+1)(u+1), \ x(z+1)w, \ x(z+1)(w+1), \ (x+1)yz\}.$$

Corollary 4 (Cook) *The problem* SAT *is* NP-*complete.* $\qquad\square$

We leave it to the reader to define the problem SAT.

5.6   Additional Comments and Bibliographical Remarks

The fundamental $P = NP$? question merges the two themes of concrete algorithms and the theory of algorithms. Classically, this is evident from the sheer number of NP-complete problems that have emanated from diverse areas of computer science, pure and applied. The tractability — or not— of any one of these problems determines the answer to the fundamental question. Over $\mathbb{R}$ and $\mathbb{C}$, the merging comes, not from the number of diverse problems (which is not as striking) but rather because the basic NP-complete problem here *is* HN, the fundamental problem of deciding the solvability of systems of polynomial equations.

We end this chapter with some historical comments and several remarks concerning the status of P and NP under various scenarios.

The original proof of the NP-completeness of 3-SAT is in [Cook 1971]. The NP-completeness of other problems over $\mathbb{Z}_2$, or over $\mathbb{Z}$ with bit cost, ultimately derives from the NP-completeness of 3-SAT via chains of p-reductions. The sequence of reductions to TSP, KP (in the form of Subset Sum), and IPF is in [Karp 1972]. The canonical reference, [Garey and Johnson 1979], is a veritable encyclopedia of classical NP-complete problems and reductions.

Over $(\mathbb{R}, <)$, neither TSP nor KP is known to be NP-complete, nor are they likely to be, unless $P = NP$. Note that binary witnesses are sufficient for these problems and that multiplications and divisions are not necessary. We return to these issues in Chapters 19 and 21.

Analogous to 3-SAT in the classical setting, 4-FEAS is the optimal NP-complete problem over $\mathbb{R}$ in the sense that 1, 2, and 3-FEAS are all in class P over $(\mathbb{R}, <)$ [Triesch 1990].

In the classical setting, the problems $HN_{\mathbb{Z},\mathbb{C}}$ and $HN_{\mathbb{Z}_2,\overline{\mathbb{Z}_2}}$ are NP-hard but not known to be in NP. $HN_{\mathbb{Z},\mathbb{C}}$ is in the *polynomial hierarchy* [Koiran 1996]. For a discussion of the polynomial hierarchy see Chapter 21.

Now as to the status of the P = NP? question, we have seen in Remark 5 that the undecidability of HN over $\mathbb{Z}$ implies P $\neq$ NP over $(\mathbb{Z}, <)$ *with respect to unit cost*. Indeed, Matiyasevich's Theorem implies that, in this setting, NP = HALT, the class of halting sets over $(\mathbb{Z}, <)$.

On the other hand, *with respect to bit cost* over $\mathbb{Z}$, NP $\subseteq$ EXP. So in this case, the undecidability of HN over $\mathbb{Z}$ implies HN $\notin$ NP. Thus for each $c > 0$ and $q > 0$, there is a finite system f of polynomials in $\mathbb{Z}[x_1, \ldots, x_n]$ such that for all $z \in \mathbb{Z}^n$, if $f(z) = 0$, then $\mathrm{size}(z) > c(\mathrm{size}(f))^q$. More generally, the undecidability of HN over $\mathbb{Z}$ implies there is no effective bound on the size of a witness for any effective method of verification. Similar remarks can be made about the other problems listed in Proposition 3.

Over $(\mathbb{Z}, =)$, P $\neq$ NP with either unit or bit cost. In the unit case, it follows by the undecidability of HN over $\mathbb{Z}$. But there is a more elementary argument. For by Lagrange's theorem (2.46), we see that $\mathbb{N}$ is in class NP with respect to either measure. In case of unit cost, $\mathbb{N} \notin$ P, for otherwise $\mathbb{N}$ would be decidable in constant time. For a proof that $\mathbb{N} \notin$ P in case of bit cost see [Shub 1993b].

We can also conclude that P $\neq$ NP over $(\mathbb{Q}, <)$ with respect to unit cost. Since we do not know that HN is undecidable over $\mathbb{Q}$, the argument now must take a tack different than the one previously used for $(\mathbb{Z}, <)$: if P = NP over $(R, <)$, then projections of semi-algebraic sets (which are in NP) must be in P and hence must be semi-algebraic. But then, by the reverse of Tarski's Theorem, R must be real closed [Macintyre, McKenna, and van den Dries 1983], which $\mathbb{Q}$ is not. The same argument demonstrates that NP $\not\subset$ BD, the class of problems decidable by time bounded machines. (The use of reverse Tarski in this context was pointed out to us by B. Poizat and C. Michaux. See also Section 7.9.) We can summarize the preceding discussion in the following table.

Status of P = NP?

RING	BRANCH	COST	HN $\in$ DEC	HN $\in$ NP	NP $\subseteq$ EXP	P = NP
$\mathbb{Z}$	$<$	unit	No	Yes	No	No
$\mathbb{Z}$	$<$	bit	No	No	Yes	?
$\mathbb{Z}$	$=$	unit	No	Yes	No	No
$\mathbb{Z}$	$=$	bit	No	No	Yes	No
$\mathbb{Q}$	$<$	unit	?	Yes	No	No
$\mathbb{R}$	$<$	unit	Yes	Yes	Yes	?
$\mathbb{C}$	$=$	unit	Yes	Yes	Yes	?
$\mathbb{Z}_2$	$=$	unit	Yes	Yes	Yes	?

For results concerning the relativization of the $P = NP$? question using oracles over the reals (or other ordered rings) see [Emerson 1994]. A more detailed discussion of this point is in Section 21.5.

6

Integer Machines

Computations with integers are at the core of a long-standing tradition in both computer science and number theory. A common feature is a measure of magnitude of an integer x given by the number of zeros and ones necessary to write the binary expansion of x (i.e., $\lceil \log(|x| + 1) \rceil$). The relevant complexity measure is the bit cost. In the first section of this chapter, we show that machines over $\mathbb{Z}$ with bit cost and over $\mathbb{Z}_2$ with unit cost are polynomially equivalent. Thus, we refer to either setting as classical.

We have seen in the preceding chapter how the register equations yield natural NP-complete problems in several settings: over the reals, over the complexes, and over finite rings. Likewise, we have seen that the integer programming feasibility problem (IPF) is NP-hard over the integers with bit cost.

The main result of this chapter (Theorem 4) shows that IPF is in NP and hence is NP-complete. To do this, we prove in Section 6.2 some basic results of linear inequalities and polyhedra in real Euclidean space that are of general interest.

6.1 Polynomial Equivalence of Machines over $\mathbb{Z}$ and $\mathbb{Z}_2$

For any real number z denote by $\lceil z \rceil$ the smallest integer greater than or equal to z. Then, we recall from Chapter 4 that the logarithmic or bit height of an integer x is defined to be $\text{ht}_{\mathbb{Z}}(x) = \lceil \log(|x| + 1) \rceil$. We also recall that if $(x_1, \ldots, x_k) \in \mathbb{Z}^k$, the size of $(x_1, \ldots, x_k)$ is defined to be the $\max_{i \leq k} \text{ht}_{\mathbb{Z}}(x_i)$ times k.

Note that the size of a vector $(x_1, \ldots, x_k)$ of integers, measured with respect to logarithmic height, differs from the size of the same vector when considered as a vector of reals. In the first case, the notion of size reflects the size of a representation of the vector $(x_1, \ldots, x_k)$ by the binary expansions of its components. In the second, it is the dimension k.

This chapter deals with bit cost for machines over $\mathbb{Z}$. Recall that if M is a machine over $\mathbb{Z}$, and $x \in \mathbb{Z}^\infty$, the bit cost of the computation of M over x is the product of $T_M(x)$, the halting time of the computation of M with input x, times the maximum height of any integer produced during the computation.

The main result of this section is that one can simulate machines over $\mathbb{Z}_2$ with machines over $\mathbb{Z}$ with only a polynomial slowdown and, conversely, we can simulate machines over $\mathbb{Z}$ with machines over $\mathbb{Z}_2$ with only a polynomial slowdown. This polynomial slowdown is with reference to the bit cost for integer machines and the unit cost for machines over $\mathbb{Z}_2$. A first step is to make their inputs and outputs correspond.

Consider the set inclusion $\mathbb{Z}_2^\infty \to \mathbb{Z}^\infty$ induced by the set inclusion $\mathbb{Z}_2 \to \mathbb{Z}$ sending 0 to 0 and 1 to 1. A machine M over $\mathbb{Z}$ is said to *compute a function* $f : \mathbb{Z}_2^\infty \to \mathbb{Z}_2^\infty$ when the restriction of the input–output function computed by M to $\{0, 1\}^\infty$ is f. Thus, it makes sense to say that a machine M over $\mathbb{Z}$ and a machine M_2 over $\mathbb{Z}_2$ compute the same function from $\mathbb{Z}_2^\infty$ to $\mathbb{Z}_2^\infty$.

Theorem 1   *For a machine M_2 over $\mathbb{Z}_2$ with (unit) cost bounded by t, there is a machine M over $\mathbb{Z}$ computing the same function on $\mathbb{Z}_2^\infty$ with (bit) cost also bounded by t.*

Proof.   We begin by noting that each arithmetic operation in $\mathbb{Z}_2$ can be simulated over $\mathbb{Z}$ by a single integer polynomial computation. For example, the addition of $x, y \in \mathbb{Z}^2$ is performed over $\mathbb{Z}$ by the polynomial $x + y - 2xy$ and the multiplication by xy. Therefore, for any polynomial $f \in \mathbb{Z}_2[X_1, \ldots, X_k]$, there exists a polynomial $\tilde{f} \in \mathbb{Z}[X_1, \ldots, X_k]$ such that for each vector x in $\{0, 1\}^k$ one has that f and $\tilde{f}$ return the same value when evaluated on x (considering x to be in $\mathbb{Z}_2^k$ in the first case and in $\mathbb{Z}^k$ in the second).

Given any machine M_2 over $\mathbb{Z}_2$ we consider a machine M over $\mathbb{Z}$ obtained by replacing the polynomials f in the computation nodes of M_2 by their corresponding $\tilde{f}$. The branch nodes of M_2 test an element in $\{0, 1\}$ for equality with zero. This test can be done by M as well. We conclude that the machine M computes the same function as M_2 when restricted to binary inputs. $\square$

For the converse we proceed differently. Now it is necessary to consider machines over $\mathbb{Z}_2$ computing functions $f : \mathbb{Z}^\infty \to \mathbb{Z}^\infty$. This is done by using base-2 expansions of integers. For a positive integer m denote its binary expansion by $m^1, \ldots, m^{\ell_m}$ where each m^i is 0 or 1. Also, for an integer m, let $s(m)$ be the sequence $(1, 0)$ if $m < 0$ and the empty sequence otherwise. Then define $\phi : \mathbb{Z}^\infty \to \mathbb{Z}_2^\infty$ by

$$\phi(\overline{m}) = (s(m_1), 0, m_1^1, 0, m_1^2, 0, \ldots, m_1^{\ell_{m_1}}, 1, 1, s(m_2), 0, m_2^1, 0, \ldots, 0, m_n^{\ell_{m_n}}),$$

where $\overline{m} = (m_1, \ldots, m_n) \in \mathbb{Z}^n$. Note that the function ϕ codes the bit 0 in the expansion of m_i by a pair $(0, 0)$, the bit 1 by a pair $(0, 1)$, a negative sign with a pair

(1, 0), and the pair (1, 1) separates coordinates in $\mathbb{Z}^\infty$. The function ϕ thus defined is injective. We say that a machine M_2 over $\mathbb{Z}_2$ computes a function $f : \mathbb{Z}^\infty \to \mathbb{Z}^\infty$ if the following diagram commutes

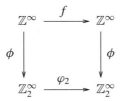

where φ_2 is the input–output function of M_2. In this way, it also makes sense to say that a machine M over $\mathbb{Z}$ and a machine M_2 over $\mathbb{Z}_2$ compute the same function from $\mathbb{Z}^\infty$ to $\mathbb{Z}^\infty$.

Theorem 2 *There exists a polynomial p such that for any machine M over $\mathbb{Z}$ with (bit) cost bounded by t, there is a machine M_2 over $\mathbb{Z}_2$ computing the same function with (unit) cost bounded by $p(t)$.*

We need a simple lemma.

Lemma 1 *There are machines M_+ and M_* over $\mathbb{Z}_2$ which on input a pair of integers (x, y) compute $x + y$ and xy, respectively, in polynomial time in the heights of x and y, that is, whose costs are bounded by a polynomial in the heights of x and y.*

Proof.   The primary school algorithms for addition and multiplication (that we can modify to work in base-2 instead of base-10) perform a number of elementary steps that are linear for addition and quadratic for multiplication as functions of the heights of x and y.   □

Remark 1 Note that although integer addition can be performed in a linear number of elementary steps, this does not imply that the machine M_+ over $\mathbb{Z}_2$ works in linear time. This is due to the fact that each of these steps might be preceded by some management steps devoted to moving the value of a coordinate in the state space to another coordinate. Although this does not affect the polynomial character of the running time, it can certainly destroy its linearity. A similar statement can be made about the quadratic bound for multiplication.

Proof of Theorem 2.   Consider the finitely many polynomials associated with the computation nodes of M. We replace each by a fixed sequence of arithmetic operations computing it. By the preceding lemma each of these operations can be performed in polynomial time in the length of its inputs. The bound t for the halting time of M imposes the same bound on the heights of the intermediate values of the computations. Therefore, each arithmetic operation of M can be done in time bounded by a polynomial in t.

Note that in the binary representation of integers, the sign needs to be coded with an additional binary value. It is this additional value that allows the branch nodes of M to be simulated in M_2.   □

Corollary 1 *The classes* P *over* $\mathbb{Z}$ *and* $\mathbb{Z}_2$ *correspond by the inclusion* $\mathbb{Z}_2^\infty \to \mathbb{Z}^\infty$ *and the binary representation given by* ϕ. *The same is true for the class* NP *and for its subclass of* NP-*complete problems.*   □

6.2   Some Results on Linear Inequalities

In this section we prove some results on linear inequalities over $\mathbb{R}$ that are used in several parts of this book. In the next section they are used to derive the existence of small solutions for feasible instances of the integer programming feasibility problem. We begin with a general theorem that gives a "nonnegative version" of one of the central theorems in linear algebra, namely, the fact that if a vector v is a linear combination of a set of vectors S, then there is a linearly independent subset S' of S such that v is a linear combination of the elements of S'. The nonnegativity condition added to this statement replaces "linear combination" by "linear combination with nonnegative coefficients." Consider vectors a_1, a_2, a_3, and b in $\mathbb{R}^2$ in the two configurations shown in Figure A. In situation (i), one sees that b is a linear combination of a_1, a_2, and a_3 with nonnegative coefficients (in fact, 0 for a_3 and positive for the other two). In situation (ii), one checks that this is not possible. On the other hand, in (i) one cannot find a line leaving all the as in one closed halfplane and b in the complementary open halfplane, whereas in (ii) this is clearly possible. The next theorem formalizes this phenomenon.

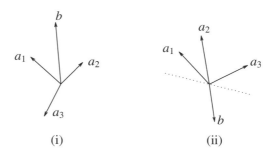

(i)   (ii)

Figure A   Two configurations of vectors in $\mathbb{R}^2$.

Theorem 3 *Let* $a_1, \ldots, a_m, b \in \mathbb{R}^n$. *Then, one and only one of the following statements holds.*

(a) *The vector* b *is a nonnegative linear combination of linearly independent vectors from the set* $\{a_1, \ldots, a_m\}$.

(b) *Let t be the rank of $\{a_1, \ldots, a_m, b\}$. Then there exists a hyperplane $\{x \in \mathbb{R}^n \mid c \cdot x = 0\}$ containing $t - 1$ linearly independent vectors from the set $\{a_1, \ldots, a_m\}$ such that $c \cdot b < 0$ and $c \cdot a_1, \ldots, c \cdot a_m \geq 0$.*

Proof.   We may assume that $a_1, \ldots, a_m$ span $\mathbb{R}^n$.

Clearly (a) and (b) exclude each other since otherwise, if $b = \lambda_1 a_1 + \ldots + \lambda_m a_m$ with $\lambda_1, \ldots, \lambda_m \geq 0$, we would have that

$$0 > c \cdot b = \lambda_1 c \cdot a_1 + \ldots + \lambda_m c \cdot a_m \geq 0. \tag{6.1}$$

Let us see that one of the two statements holds. Consider a set $D = \{a_{i_1}, \ldots, a_{i_n}\}$ of linearly independent elements from $a_1, \ldots, a_m$, and apply the following iterative process.

(i)   Let $b = \lambda_{i_1} a_{i_1} + \ldots + \lambda_{i_n} a_{i_n}$. If $\lambda_{i_1}, \ldots, \lambda_{i_n} \geq 0$, then (a) holds and we are done.

(ii)   Otherwise, let h be the smallest number among $i_1, \ldots, i_n$ such that $\lambda_h < 0$ and let $\{x \in \mathbb{R}^n \mid c \cdot x = 0\}$ be the hyperplane spanned by $D - \{a_h\}$ normalized such that $c \cdot a_h = 1$. Therefore we have that $c \cdot b = \lambda_h < 0$.

(iii)   If $c \cdot a_1, \ldots, c \cdot a_m \geq 0$, Statement (b) holds and we are done.

(iv)   Otherwise, choose the smallest s such that $c \cdot a_s < 0$. Then replace D by $(D - \{a_h\}) \cup \{a_s\}$ and start the iteration anew.

We now show that this process terminates, leading thus to either (a) or (b). For that, let D_k denote the set D at the begining of the kth iteration. If the process does not terminate, then for some $k, l \in \mathbb{N}$, $k < l$ we must have $D_k = D_l$. Let r be the highest index h for which a_h has been removed at the end of one of the iterations $k, k+1, \ldots, l-1$. Let p be this iteration. As $D_k = D_l$, we have that a_r must have been added in some iteration q with $k \leq q < l$. So,

$$D_p \cap \{a_{r+1}, \ldots, a_m\} = D_q \cap \{a_{r+1}, \ldots, a_m\}. \tag{6.2}$$

Let $D_p = \{a_{i_1}, \ldots, a_{i_n}\}$, $b = \lambda_{i_1} a_{i_1} + \ldots + \lambda_{i_n} a_{i_n}$, and let c' be the vector c found in (ii) of iteration q.

Then for these $i_1, \ldots, i_n$ the following inequalities hold.

(1) If $i_j < r$, then $\lambda_{i_j} \geq 0$ and $c' \cdot a_{i_j} \geq 0$. This follows from the facts that r is the smallest index i_j satisfying $\lambda_{i_j} < 0$ and that since a_r is added in iteration q one must have that $c' \cdot a_i \geq 0$ for all $i < r$.

(2) If $i_j = r$, then $\lambda_{i_j} < 0$ and $c' \cdot a_r < 0$. This is for the same reason.

(3) If $i_j > r$, then $c' \cdot a_{i_j} = 0$. By 6.2 one has that a_{i_j} also belongs to D_q but then, since $i_j \neq r$, and the choice of c' in Step (ii) of iteration q, one has that $c' \cdot a_{i_j} = 0$.

Finally, note that

$$0 > c' \cdot b = \lambda_{i_1} c' \cdot a_{i_1} + \ldots + \lambda_{i_n} c' \cdot a_{i_n} \geq 0,$$

where the first inequality follows from the definition of c' and the second inequality follows from (1), (2), and (3). But this is not possible and thus, the assumption that the described process does not terminate leads to a contradiction. $\qquad\square$

Several consequences can be derived from Theorem 3. The following result is useful later.

Corollary 2 (Carathéodory's Theorem) *Let $a_1, \ldots, a_m, b \in \mathbb{R}^n$. If b is a nonnegative linear combination of $a_1, \ldots, a_m$, then there exists a linearly independent subset $\{a_{i_1}, \ldots, a_{i_n}\} \subseteq \{a_1, \ldots, a_m\}$ such that b is a nonnegative linear combination of $a_{i_1}, \ldots, a_{i_n}$.*

Proof. If (a) in the Theorem 3 holds, we are done. Otherwise, equation (6.1) leads to a contradiction. $\qquad\square$

Several forms of Farkas's Lemma are easily deduced from Theorem 3. This lemma is a cornerstone of the theory of linear inequalities. Moreover, it has consequences for the complexity of real linear programming as we demonstrate shortly. In what follows we denote by yA the matrix multiplication $y^T A$ of a vector $y \in \mathbb{R}^m$ and an $n \times m$ matrix A.

Corollary 3 (Farkas's Lemma) *Let A be an $m \times n$ real matrix and let $b \in \mathbb{R}^m$.*

(i) *There exists $x \in \mathbb{R}^n$, $x \geq 0$ such that $Ax = b$ if and only if for each $y \in \mathbb{R}^m$ satisfying $yA \geq 0$ one has $y \cdot b \geq 0$.*

(ii) *There exists $x \in \mathbb{R}^n$ such that $Ax \geq b$ if and only if for each $y \in \mathbb{R}^m$, $y \geq 0$ satisfying $yA = 0$ one has $y \cdot b \leq 0$.*

(iii) *There exists $x \in \mathbb{R}^n$, $x \geq 0$ such that $Ax \geq b$ if and only if for each $y \in \mathbb{R}^m$, $y \geq 0$ satisfying $yA \geq 0$ one has $y \cdot b \leq 0$.*

Proof.

(i) Let $x \geq 0$ such that $Ax = b$ and let y satisfy $yA \geq 0$. Then $y \cdot b = y \cdot (Ax) = (yA) \cdot x \geq 0$. This proves one direction. To prove the other, suppose that there is no $x \geq 0$ making $Ax = b$. If $a_1, \ldots, a_m$ are the columns of A, this means that b is not a nonnegative linear combination of $a_1, \ldots, a_m$. By Theorem 3 this implies the existence of a $y \in \mathbb{R}^m$ such that $y \cdot b < 0$ and $yA \geq 0$.

(ii) Let A' be the matrix $[-I, A, -A]$ and consider the system $A'x' = b$ with $x' = (x^1, x^2, x^3)$, where $x^1 \in \mathbb{R}^m$ and $x^2, x^3 \in \mathbb{R}^n$. We claim that $Ax \geq b$ has a solution x if and only if $A'x' = b$ has a nonnegative solution x'.

Indeed, if x is a solution of $Ax \geq b$, then we can define for $j = 1, \ldots, m$ and $i = 1, \ldots, n$,

$$x_j^1 = A_j x - b_j, \qquad x_i^2 = \begin{cases} x_i & \text{if } x_i \geq 0 \\ 0 & \text{otherwise} \end{cases}$$

and

$$x_i^3 = \begin{cases} -x_i & \text{if } x_i < 0 \\ 0 & \text{otherwise} \end{cases}$$

and one has that the x' thus defined satisfies $A'x' = b$ and $x' \geq 0$. Conversely, given any $x' \geq 0$ satisfying $A'x' = -b$ we define $x_i = x_i^2 - x_i^3$ for $i = 1, \ldots, n$ and we have that $Ax \geq b$.

Now apply Part (i) to A' and b. We deduce that there is an $x' \in \mathbb{R}^{m+2n}$ such that $x' \geq 0$ and $A'x' = b$ if and only if for all $y \in \mathbb{R}^m$ with $yA' \geq 0$ one has $y \cdot b \geq 0$. But $yA' \geq 0$ is equivalent to $y \leq 0$ and $yA = 0$. The conclusion now follows.

(iii) Finally, let A' be the matrix $[-I, A]$. Then as in Part (ii) one proves that $Ax \geq b$ has a solution $x \geq 0$ if and only if $A'x' = b$ has a solution $x' \geq 0$ and applies Part (i) again. $\qquad\square$

There is a complexity theoretical consequence we can derive from Farkas's Lemma related to the linear programming feasibility problem over the reals.

Given a decision problem $S \subseteq \mathbb{R}^\infty$ we call the problem $\overline{S} = \mathbb{R}^\infty - S$ the *complement* of S. For example $\overline{4\text{-FEAS}}$ is the problem of deciding whether a real polynomial of degree 4 has no real roots.

It is easy to see that $P_\mathbb{R}$ is closed under complements; that is, if a decision problem S belongs to $P_\mathbb{R}$, then $\overline{S}$ also belongs to $P_\mathbb{R}$. One simply exchanges the outputs 0 and 1 for any machine deciding S in polynomial time. The closure under complements does not seem to be shared by $NP_\mathbb{R}$. For instance, it is not clear what witness one can use to prove that a real polynomial has no real roots.

Definition 1 We let $coNP_\mathbb{R}$ be the class of decision problems S over $\mathbb{R}$ such that $\overline{S} \in NP_\mathbb{R}$. A structured decision problem (X, X_{yes}) over $\mathbb{R}$ is in $coNP_\mathbb{R}$ if $(X, X - X_{\text{yes}}) \in NP_\mathbb{R}$.

A decision problem S is $coNP_\mathbb{R}$-*complete* if $S \in coNP_\mathbb{R}$ and every decision problem in $coNP_\mathbb{R}$ p-reduces to S.

Proposition 1 *Let S be a decision problem that belongs to both $NP_\mathbb{R}$ and $coNP_\mathbb{R}$. If S is $NP_\mathbb{R}$-complete or $coNP_\mathbb{R}$-complete, then $NP_\mathbb{R} = coNP_\mathbb{R}$.*

Proof. Let us suppose that S is $NP_\mathbb{R}$-complete and let us consider any problem T in $NP_\mathbb{R}$. Because of the completeness of S, we have that T reduces to S. The same reduction shows that the complement $\overline{T}$ of T reduces to the complement $\overline{S}$ of S. But since $\overline{S}$ belongs to $NP_\mathbb{R}$ the reduction composed with the $NP_\mathbb{R}$ algorithm for $\overline{S}$ provides an $NP_\mathbb{R}$ algorithm for $\overline{T}$ showing that T is in $coNP_\mathbb{R}$.

Conversely, if T belongs to $coNP_\mathbb{R}$, its complement belongs to $NP_\mathbb{R}$ and therefore reduces to S. This provides a reduction from T to $\overline{S}$ and as before, the composition of this reduction with an $NP_\mathbb{R}$ algorithm for $\overline{S}$ shows that T belongs to $NP_\mathbb{R}$.

The same proof applies if we suppose that S is $coNP_\mathbb{R}$-complete. $\qquad\square$

Proposition 2   *The linear programming feasibility problem over the reals belongs to both* $NP_\mathbb{R}$ *and* $coNP_\mathbb{R}$. *Consequently, if* LPF *over* $\mathbb{R}$ *is* $NP_\mathbb{R}$ *or* $coNP_\mathbb{R}$ *complete, then* $NP_\mathbb{R} = coNP_\mathbb{R}$.

Proof.   We have seen in Chapter 5 that LPF over $\mathbb{R}$ belongs to $NP_\mathbb{R}$. In order to see that it belongs to $coNP_\mathbb{R}$ consider an input (A, b) where A is an $m \times n$ real matrix and $b \in \mathbb{R}^m$. The nonexistence of an $x \in \mathbb{R}^n$ such that $Ax \geq b$ is equivalent by Corollary 3(ii) to the existence of a $y \in \mathbb{R}^m$ satisfying $y \geq 0$, $yA = 0$, and $y \cdot b > 0$. And this last property can be decided in nondeterministic polynomial time by just guessing the $y \in \mathbb{R}^m$ and checking the three inequalities. $\qquad\square$

The preceding result suggests that LPF is not $NP_\mathbb{R}$-complete. More facts supporting this belief are shown in Section 20.3. Proposition 2 also holds for LPF over $\mathbb{Q}$ since Farkas's Lemma is valid over $\mathbb{Q}$ (and even over any ordered field). As we have already anticipated, a more dramatic property holds in this case, namely, the existence of a polynomial time algorithm for LPF (see Chapter 15).

A last simple result follows from Theorem 3 that is helpful in providing good upper bounds for heights of solutions to the integer programming problem. It is a weak form of the well-known duality principle of linear programming.

Corollary 4   *Let A be an $m \times n$ real matrix and let $b \in \mathbb{R}^m$ and $c \in \mathbb{R}^n$. Now consider the sets*

$$M = \{c \cdot x \mid x \geq 0, \ Ax = b\}$$

and

$$N = \{y \cdot b \mid yA \geq c\}.$$

Then if one of these sets is empty, the other is either empty or unbounded.

Proof.   Suppose one of these sets of real numbers, say M, is bounded and nonempty. We show that in this case, N is nonempty. Let μ be the maximum of M (which exists since everything is linear) and let x_0 be such that $c \cdot x_0 = \mu$. Assume N is empty. One can then apply Corollary 3(ii) (with the role of rows and columns exchanged) to deduce the existence of a $z \geq 0$ such that $Az = 0$ and $c \cdot z > 0$. Therefore one has that $x_0 + z \geq 0$ and $A(x_0 + z) = Ax_0 + Az = b + 0 = b$. But $c \cdot (x_0 + z) = \mu + c \cdot z > \mu$ contradicting the maximality of μ.

A similar argument applies if one supposes N bounded and nonempty and M empty. $\qquad\square$

6.3 The Complexity of Integer Programming

The aim of this section is to prove that the integer programming feasibility problem introduced in Section 1.2 is NP-complete over the integers.

Theorem 4 *The integer programming feasibility problem is* NP-*complete.*

The NP-hardness of IPF follows from the register equations of integer machines (see Theorem 1 and Remark 6 of Chapter 5). Our effort is thus devoted to showing that IPF belongs to NP.

We first prove an easy lemma providing bounds for the values of integer determinants and then a theorem providing "small" rational points in closed polyhedra.

Lemma 2 *Let A be an n × n integer matrix whose entries have absolute value bounded by a. Then the absolute value of its determinant* $\det A$ *is less than* $(na)^n$.

Proof. The determinant is a sum of $n!$ factors, each smaller in absolute value than a^n. Since $n! \leq n^n$ the result follows. $\square$

Theorem 5 *Let P be a nonempty polyhedron in* $\mathbb{R}^n$ *defined by a system* $Ax \leq b$ *of N inequalities, where the entries of A and the components of b are integers with absolute value smaller than a. Then there is a rational point* $x \in P$ *whose components have the same denominator and whose numerators and denominator are bounded in absolute value by* $(na)^n$. *If the components of b are not assumed to be integers but arbitrary real numbers, then there exist* $I \subset \{1, \ldots, N\}$ *and* $y \in P$ *of the form*

$$y_i = \sum_{j \in I} u_{ij} b_j, \ i = 1, \ldots, n, \tag{6.3}$$

where $|I| \leq n$ *and the* u_{ij}*s are rationals such as those described for the components of* x.

Proof. Let z be a point in P. We define the matrix A_+ by changing the sign of the elements in column j of A when $z_j < 0$. Clearly, the point z_+ whose components are the absolute value of the components of z is a nonnegative solution of the system $A_+ x \leq b$. Since the coefficients of A_+ are those of A up to sign, we can assume without loss of generality that the system $Ax \leq b$ has a nonnegative solution. If now we consider the $N \times (n + N)$ integer matrix $A' = [A, I]$ we deduce that the system

$$A'x' = b, \ x' \geq 0 \tag{6.4}$$

is satisfiable in $\mathbb{R}^{n+N}$.

The matrix A' has rank N since it contains as a submatrix the $N \times N$ identity. It follows from Carathéodory's Theorem that there exists also a satisfiable system of the form

$$A''x'' = b, \ x'' \geq 0, \tag{6.5}$$

where A'' is a $N \times N$ matrix made of N linearly independent columns of A'. The variables x'_j corresponding to columns that do not occur in A'' can be set to 0. Since A'' has full rank, (6.5) has in fact a unique solution which is given by Cramer's rule: $x''_j = \det(A''_j)/\det(A'')$ for $j = 1, \ldots, N$, where A''_j results from A'' by replacing its jth column by b. Since all except perhaps the n first columns of A'' contain exactly one 1 and are otherwise 0, this matrix can be brought by a permutation σ of its rows into the form

$$B = \begin{pmatrix} C & 0 \\ D & I \end{pmatrix},$$

where C is a square matrix of dimension at most n. The entries of C are integers with absolute value at most a, hence the determinant of B is an integer of absolute value smaller than $(na)^n$.

Consider a variable x''_j corresponding to an original variable x_i (therefore $j \le n$). The matrix A''_j can be brought by the same permutation of rows σ into the form

$$B_j = \begin{pmatrix} C_j & 0 \\ D_j & I \end{pmatrix},$$

where B_j is obtained from B by replacing its jth column vector by the vector $(b_{\sigma(1)}, \ldots, b_{\sigma(N)})$. Since the entries of b are also integers bounded in absolute value by a the same bound as before applies.

Finally, if the entries of b are arbitrary real numbers, expanding $\det(C_j)$ along its jth column proves equation 6.3 with $I = \{\sigma(j) \mid j \le n\}$.   □

Theorem 6   *Let A be an $m \times n$ integer matrix and b an m-vector and let $a \in \mathbb{N}$ be a bound for the absolute value of the entries of A and the components of b. Then, if $Ax = b$ has a solution in $\mathbb{N}^n$, it also has one whose components are bounded in absolute value by $n(ma)^{2m+1}$.*

Proof.   Consider $M = (ma)^m$ and a solution $x \in \mathbb{N}^n$ of $Ax = b$ which is minimal with respect to sum of components. If all the components of x are bounded by M, we are done. Otherwise, we can assume

(i) the components $x_1, \ldots, x_k$ of x are greater than M.

Denote by $v_1, \ldots, v_k$ the first k columns of A and by V the $m \times k$ matrix having these columns.

If there exist integers $\alpha_1, \ldots, \alpha_k$ between 0 and M, not all zero, such that $V\alpha = 0$ for $\alpha = (\alpha_1, \ldots, \alpha_k)$, then the vector $x' = (x_1 - \alpha_1, \ldots, x_k - \alpha_k, x_{k+1}, \ldots, x_n)$ is also a solution of $Ax = b$ contradicting the minimality of x. Therefore we have

(ii) there is no vector of integers $\alpha = (\alpha_1, \ldots, \alpha_k)$ with its components between 0 and M and not all zero, such that $V\alpha = 0$.

In the rest of this proof we show that (i) and (ii) lead to a contradiction. First we claim

(iii) there is a vector $h \in \{0, \pm 1, \ldots, \pm M\}^m$ such that $h^T V \geq 1$.

Suppose (iii) does not hold. Then there are no real solutions to $h^T V \geq 1$. If there were real solutions, Theorem 5 would ensure the existence of a rational solution z having numerators and denominator bounded in absolute value by $(ma)^m = M$. Multiplying both sides of the inequation $h^T V \geq 1$ by the common denominator of the coordinates of z we would obtain an integer solution to $h^T V \geq 1$ bounded in absolute value by M. But this contradicts our assumption that (iii) does not hold.

Now, from the absence of real solutions to $h^T V \geq 1$ and Corollary 4 applied with $b = (0, \ldots, 0)$ and $c = (1, \ldots, 1)$ one deduces that the set

$$\{\sum_{i=1}^k \alpha_i \mid V\alpha = 0, \alpha \geq 0\}$$

is either empty or unbounded. But this latter set is nonempty since $\alpha_j = 0$ for $j = 1, \ldots, k$ satisfies the requirement and therefore it must be unbounded. This implies the existence of $\alpha_1, \ldots, \alpha_k \in \mathbb{R}$ nonnegative and not all zero satisfying $V\alpha = 0$.

Let $j \leq k$ satisfy $\alpha_j > 0$. We can suppose that, in fact, $\alpha_j = 1$ and consider the system

$$V'z = w,$$

where V' is the matrix obtained by deleting the jth column in V and w is minus this column. We know that this system has a nonnegative solution. If r denotes the rank of V', by Carathéodory's Theorem we know that there exists an $m \times r$ submatrix V'' of V' such that the system

$$V''z = w$$

has a nonnegative solution. By deleting some redundant rows in V'' we obtain an $r \times r$ system whose only solution is nonnegative. Again, Cramer's rule together with the bound given in Lemma 2 yield a nonnegative integer solution α with components bounded by $(ra)^r \leq (ma)^m = M$ (note that the denominator given by Cramer's rule now becomes α_j). In this way we get a nonnegative integer vector α such that $V\alpha = 0$ whose components are bounded by M. But this is in contradiction with (ii) and therefore (iii) is proved.

We finally consider an h as given by (iii) and multiply on the left both sides of $Ax = b$ by h^T. Splitting the first sum in two parts we obtain

$$\sum_{j=1}^k h^T \cdot v_j x_j = h^T \cdot b - \sum_{j=k+1}^n h^T \cdot v_j x_j,$$

where, we recall, v_j denotes the jth column of A. Therefore

$$\sum_{j=1}^k x_j \leq maM + (n-k)amM^2 \leq nmaM^2$$

from which the statement follows.   □

Corollary 5 *Let A an $m \times n$ integer matrix and b a vector in $\mathbb{Z}^m$ be an instance of the* IPF. *If the system $Ax \geq b$ has a solution, then it has a solution whose coordinates are bounded in absolute value by $2(m + 2n)(ma)^{2m+1}$ where a is a bound for the absolute value of the entries of A and the components of b.*

Proof.   Let (A, b) be as in the statement. Then we have, as in Corollary 3 (ii), that $Ax \geq b$ has an integer solution if and only if $A'x' = b$ has a solution $x' = (x_1, x_2, x_3) \in \mathbb{N}^{m+n+n}$, where $A' = [-I, A, -A]$.

Now, let a be a bound for the absolute value of the entries of A and b. Then this is also a bound for the entries of A' and Theorem 6 yields that if a solution x' exists, there is one bounded in absolute value by $(m + 2n)(ma)^{2m+1}$.   □

Corollary 6 *The integer programming feasibility problem belongs to* NP.   □

We have thus also completed the proof of Theorem 4.

We close this section noting that the choice of the ground ring along with its definition of height and cost can drastically affect the complexity of a given problem. In our case, the same problem —the linear programming feasibility problem— shows this variance.

- It is NP-complete over the integers with bit cost,

- can be solved in polynomial time over the rationals with bit cost (see Chapter 15), and

- it belongs to $\text{NP}_\mathbb{R}$ over the reals but it is conjectured to be neither in $\text{P}_\mathbb{R}$ nor $\text{NP}_\mathbb{R}$-complete.

6.4   Additional Comments and Bibliographical Remarks

Classical computation theory extensively uses binary expansions of integers as well as computable bijections between $\mathbb{N}$ and $\mathbb{N}^k$ or $\mathbb{N}^\infty$. In fact, although classical models of computation were designed to operate on the simplest possible elements (finite words from a finite alphabet), interest was primarily on arithmetic problems. The constructions of Section 6.1 are thus quite classical.

A standard reference for the theory of linear inequalities and linear programming is the book of Schrijver [1986]. The exposition in Section 6.2, with the exception of Proposition 2 and its corollary, closely follows this reference.

Integer programming feasibility is one of the 21 NP-complete problems in [Karp 1972]. Its membership in NP was first shown in [Boroch and Treybig 1976]. Our proof, and more concretely Theorem 6, is due to Papadimitriou [1981]. Theorem 5 is taken from [Cucker and Koiran 1995].

7

Algebraic Settings for the Problem "P $\neq$ NP?"

When complexity theory is studied over an arbitrary unordered field K, the classical theory is recaptured with $K = \mathbb{Z}_2$. The fundamental result that the Hilbert Nullstellensatz as a decision problem is NP-complete over K allows us to reformulate and investigate complexity questions within an algebraic framework and to develop transfer principles for complexity theory.

Here we show that over algebraically closed fields K of characteristic 0 the fundamental problem "P$\neq$NP?" has a single answer that depends on the tractability of the Hilbert Nullstellensatz over the complex numbers $\mathbb{C}$. A key component of the proof is the Witness Theorem enabling the elimination of transcendental constants in polynomial time. This chapter requires a stronger background (in algebra in particular) than the rest of the book.

7.1 Statement of Main Theorems

We consider the Hilbert Nullstellensatz in the form HN/K: given a finite set of polynomials in n variables over a field K, decide if there is a common zero over K. At first the field is taken as the complex number field $\mathbb{C}$. Relationships with other fields and with problems in number theory are developed here.

Only machines and algorithms that branch on "$h(x) = 0$?" are considered here. The symbol $\leq$ is not used. Thus the development is quite algebraic, eventually using properties of the height function of algebraic number theory. A main theme is eliminating constants. The moral is roughly: using transcendental and algebraic numbers does not help much in speeding up integer decision problems.

Let $\overline{\mathbb{Q}}$ be the algebraic closure of the rational number field $\mathbb{Q}$. The following is proved.

Theorem 1 *If* $P = NP$ *over* $\mathbb{C}$, *then* $P = NP$ *over* $\overline{\mathbb{Q}}$, *and the converse is also true.*

Remark 1 Here $\mathbb{C}$ may be replaced by any algebraically closed field containing $\overline{\mathbb{Q}}$.

Now we are going to define an invariant τ of integers (and polynomials over $\mathbb{Z}$) that describes how many arithmetic operations are necessary to build up an integer starting from 1. More precisely, a *computation of length* l of the integer m is a sequence of integers, $x_0, x_1, \ldots, x_l$, where $x_0 = 1$, $x_l = m$ and given k, $1 \leq k \leq l$, there are i, j, $0 \leq i, j < k$ such that $x_k = x_i \circ x_j$, where $\circ$ is addition, subtraction, or multiplication. We define $\tau : \mathbb{Z} \to \mathbb{N}$ by letting $\tau(m)$ be the minimum length of a computation of m.

The following is easy to check, where here and in the following log denotes $\log_2$.

Proposition 1 *For all* $m \in \mathbb{Z}^+$ *one has* $\tau(m) \leq 2 \log m$.

If m is of the form 2^{2^k}, then $\tau(m) = \log \log m + 1$. The same is essentially true even if m is any power of 2.

Open Problem. Is there a constant c such that

$$\tau(k!) \leq (\log k)^c \text{ all } k \in \mathbb{Z}^+?$$

We remark that if "factoring is hard" using inequalities, then the open problem has a negative answer.[1]

Definition 1 Given a sequence of integers a_k we say that a_k is *easy to compute* if there is a constant c such that $\tau(a_k) \leq (\log k)^c$, all $k > 2$, and hard to compute otherwise. We say that the sequence a_k is *ultimately easy to compute* if there are nonzero integers m_k such that $m_k a_k$ is easy to compute and *ultimately hard to compute* otherwise.

In Sections 7.5 and 7.6 we prove the following.

Theorem 2 *If the sequence of integers* $k!$ *is ultimately hard to compute, then* $HN/\mathbb{C}$, *the Hilbert Nullstellensatz over* $\mathbb{C}$, *is intractable and hence* $P \neq NP$ *over* $\mathbb{C}$. *Thus in that case,* $P \neq NP$ *over* $\overline{\mathbb{Q}}$.

---

[1]Here is a sketch of the proof. Suppose to the contrary that $k!$ is *easy to compute* and n is the product of primes p and q where $p < k < q$. We show how to easily factor n.

Let $x_0, x_1, \ldots, x_l = k!$ be a short computation of $k!$, $l \leq (\log k)^c$. Then we induce a short computation of $r = k! \bmod n$ using

$$(x_0 \bmod n, x_1 \bmod n, \ldots, r = x_l \bmod n).$$

By the Euclidean Algorithm, $y = \gcd(r, n)$ may be quickly computed. By our hypothesis it follows that $y = p$, and thus our assertion is proved.

Next consider the analogous situation for polynomials with integer coefficients $f \in \mathbb{Z}[t]$. A *computation* of length ℓ of f is a sequence of $u_i \in \mathbb{Z}[t]$, where $u_0 = 1$, $u_1 = t$, $u_{\ell+1} = f$, and for each k, $1 < k \le \ell + 1$ there are i, j, $0 \le i, j < k$ such that $u_k = u_i \circ u_j$, where $\circ$ is addition, subtraction, or multiplication. Define $\tau : \mathbb{Z}[t] \to \mathbb{N}$ by $\tau(f)$ is the minimum length of a computation of f.

Let $\text{Zer}(f)$ be the number of distinct integral zeros of f. The following has a certain plausibility.

HYPOTHESIS. $\text{Zer}(f) \le (\tau(f) + 1)^c$ for all nonzero $f \in \mathbb{Z}[t]$.

Here c is a universal positive constant. We do not know if the hypothesis is true or false, even, for example, with the constant $c = 1$.

Theorem 3 *If the preceding Hypothesis is true, then* NP $\neq$ P *over* $\mathbb{C}$, *and* NP $\neq$ P *over* $\overline{\mathbb{Q}}$.

7.2   Eliminating Constants: Easy Cases

In this section we begin a study of the problem of eliminating the constants of a computation without an exponential increase in the time. Our first result asserts that this can always be done if the constants lie in an algebraic extension of the given field K. The result holds for fields of any characteristic.

For the rest of the chapter, unless otherwise stated, we assume that at any computation node of a machine M, the computation performed is either addition, multiplication, subtraction, or division of two elements of L. This represents no loss of generality since any machine can be so converted with at most a multiplicative constant increase in halting time.

Definition 2 Suppose $K \subset L$ are fields and (Y, Y_0) is a decision problem over L. The *restriction of* (Y, Y_{yes}) *to* K is $(Y \cap K^{\infty}, Y_{\text{yes}} \cap K^{\infty})$. The same applies to the case where K is a ring.

Proposition 2 *Let M be a machine over a field L which is an algebraic extension of a field K. Then there is a machine M' over K and a constant $c > 0$ (depending on M) with the following property. For any decision problem (Y, Y_{yes}) over L decided by M, the restriction of (Y, Y_{yes}) to K is decided by M', and the halting time satisfies*
$$T_{M'}(y) \le c T_M(y), \text{ for all } y \in Y \cap K^{\infty}.$$

Proof.   Since M has only a finite number of constants then, by restriction, M is also a machine over a subfield of L that is a finite algebraic extension of K. Thus, our proposition follows if we assume that L is a finite algebraic extension of K, and show it for this case. So we make this assumption.

Consider L as a vector space over K of dimension q. Thus L may be represented as K^q, where the inclusion $K \subset L$ is represented as the inclusion of K in K^q as the first coordinate.

We now construct a machine M' over K that on inputs from K^∞ simulates M on these inputs with halting time increased by no more than a multiplicative constant. The state space of M' is considered as $(K^q)_\infty$ so that it also represents L_∞. An initial subroutine of M' in effect takes an input from K^∞ and writes it as the first coordinates in $(K^q)_\infty$.

Since addition and multiplication in L are represented by fixed symmetric bilinear maps over K

$$B_+ : K^q \times K^q \to K^q$$
$$B_\times : K^q \times K^q \to K^q,$$

M' can simulate the addition and multiplication nodes of M by incorporating these polynomial maps in computation nodes. Subtraction nodes of M are simulated in M' by multiplication by (-1) followed by B_+. Division of b by a is accomplished by solving the linear system $B_\times(a, y) = b$ for y by Gaussian elimination. This requires on the order of q^3 steps. In each of these simulations, constants from L that occur in M are replaced by their corresponding q-tuples over K.

Since $x = (x_1, \ldots, x_q)$ represents the zero element in L if and only if $x_i = 0$ for $i = 1, \ldots, q$, branching in M is simulated by checking if the first q coordinates of an element in the state space of M' are zero. Shifting right or left in M is simulated by shifting right or left q times in M'. Care is taken to keep track of the intended lengths of sequences in the computation. A final subroutine ensures that the appropriate finite sequence $(x_1, x_{q+1}, x_{2q+1}, \ldots, x_{mq+1})$ of the coordinates of the "final state" x in a computation is output.

Using the isomorphism between K^q and L, one can see that on inputs $y \in Y \cap K^\infty$, M' gives the same answers as M with the desired time bound where c is on the order of q^3.   □

The proof does not require M to be a decision machine, merely that the outputs of M are defined over K, that is, are in K^∞, for inputs over K.

Proposition 3 *Let R be an integral domain and K its quotient field. Let (Y, Y_{yes}) be a decision problem solved by a machine M over K in halting time T. Then M can be replaced by an equivalent machine without division, with constants only from R, and with halting time cT for some $c \in \mathbb{N}$. Thus, there is a machine over R solving the restriction of (Y, Y_{yes}) to R in time cT for some $c \in \mathbb{N}$.*

Proof.   A machine M' without division that simulates M is obtained by "doubling" the space used in the computation. An initial subroutine of M' takes an input $(x_1, x_2, \ldots, x_s)$ to $(\ldots, 0, \widehat{2s}.x_1, 1, x_2, 1, \ldots, x_s, 1, 0, \ldots)$ in the state space of M'. Note that elements x in the state space of M of the form $x = (\ldots, 0, x_1, x_2, \ldots, x_s, 0, \ldots)$ may be represented (nonuniquely) by elements in the state space of M' of the form

$$(\ldots, 0, x_1^{[n]}, x_1^{[d]}, x_2^{[n]}, x_2^{[d]}, \ldots, x_s^{[n]}, x_s^{[d]}, 0, \ldots),$$

where $x_i = x_i^{[n]}/x_i^{[d]}$ for $1 \le i \le s$. Since K is the quotient field of R, elements of K_∞ have representation in R_∞.

Computation nodes of M' perform the natural modification of the operations associated with the computation nodes of M which, as previously, are assumed to be the basic arithmetic operations over K. In particular, a computation node in M that performs a division $f(x_1, x_2) = (x_1/x_2)$ is replaced in M' by a computation node with associated map

$$g(x_1^{[n]}, x_1^{[d]}, x_2^{[n]}, x_2^{[d]}) = (x_1^{[n]} x_2^{[d]}, x_1^{[d]} x_2^{[n]}).$$

Constants from K in M are replaced in M' by pairs of constants from R. So, for example, a computation node of M with associated map $f(x_1) = kx_1$ with constant $k \in K$ is replaced in M' by a computation node with associated map $g(x_1^{[n]}, x_1^{[d]}) = (px_1^{[n]}, qx_1^{[d]})$ for some $p, q \in R$ with $k = p/q$.

Thus, if an initial input to M' comes from R^∞, all states in the subsequent computation will be in R_∞.

A branch node in M that tests if $x_1 = 0$ is replaced in M' by one that tests if $x_1^{[n]} = 0$ and $x_1^{[d]} \neq 0$.

Finally, M accepts an input x; that is, M outputs the value 1 given input x, if the first coordinate of the final state in the computation is 1 (and $\widehat{1}$ is to the left of the distinguished marker). Consequently, M' is designed to accept an input if the first and second coordinates of the final state in the computation are equal but not 0 (and $\widehat{2}$ is to the left of the distinguished marker). Similar considerations apply for rejecting an input.

The overall slowdown of M' with respect to M is linear.

Thus, M' has the requisite properties for both conclusions in the statement of the proposition. $\square$

7.3   Witness Theorem

We need an algebraic theorem, which we call the Witness Theorem, for the proof of our main results. This section is devoted to that theorem.

The first step is to extend the definition of τ to polynomials in several variables over $\mathbb{Z}$. Let $G \in \mathbb{Z}[t_1, \ldots, t_n]$. Quite similarly to the one-variable case, consider finite sequences

$$(u_0, u_1, \ldots, u_n, u_{n+1}, \ldots, u_{n+\ell} = G),$$

where $u_0 = 1, u_1 = t_1, \ldots, u_n = t_n$, and for $n < k \leq n + \ell, u_k = v \circ w$ for some $v, w \in \{u_0, u_1, \ldots, u_{k-1}\}$, and $\circ$ is $+, -$ or $\times$. Then $\tau(G)$ is the minimum such ℓ.

Definition 3 Define a *witness* $w \in \overline{\mathbb{Q}}^s$ for $f \in \overline{\mathbb{Q}}[t_1, \ldots, t_s]$ as a w satisfying the property that if $f(w) = 0$, then $f = 0$; that is, f is the zero polynomial.

In situations we encounter, f is presented so that it is not obvious if it is zero.

Theorem 4 (Witness Theorem)   *Let $F(x, t) = F(x_1, \ldots, x_r, t_1, \ldots, t_s)$ be a polynomial in $r + s = n$ variables with coefficients in $\mathbb{Z}$ and let $F_x \in \overline{\mathbb{Q}}[t_1, \ldots, t_s]$*

be defined by $F_x(t) = F(x, t)$ for each $x \in \overline{\mathbb{Q}}^r$. Suppose that N is a positive integer satisfying:

$$\log N \geq 4n\tau^2 + 4\tau, \quad \tau = \tau(F).$$

Then for $x \in \overline{\mathbb{Q}}^r$, there exists an algebraic number w_1 in $\{2^N, x_1^N, \ldots, x_r^N\}$ such that the point $w = (w_1, \ldots, w_s)$, where $w_i = w_{i-1}^N$, $i = 2, \ldots, s$ is a witness for $F_x \in \overline{\mathbb{Q}}[t_1, \ldots, t_s]$.

Our proof of the Witness Theorem depends heavily on the use of *heights* of algebraic numbers.

The height $H : \overline{\mathbb{Q}} \to \overline{\mathbb{Q}}$ is a function, in this chapter only, whose properties are summarized in the following proposition. Note that the height of Chapter 6 is approximately $\log H$.

Proposition 4

(a) $H(1) = H(0) = 1$, $H(2) = 2$, $H(w) \geq 1$, $H(-w) = H(w)$, $H\left(\frac{1}{w}\right) = H(w)$.

(b) $H(v + w) \leq 2H(v)H(w)$.

(c) $H(w^k) = H(w)^k$, $H(vw) \leq H(v)H(w)$.

(d) $H(v + w) \geq \frac{1}{2}\frac{H(v)}{H(w)}$.

(e) $H(vw) \geq \frac{H(v)}{H(w)}$ if $w \neq 0$.

A definition of H and proofs of (a) and (c) are given in [Lang 1991]. Moreover, (b) is proved in the appendix to this section. Note that (d) follows from (b) by

$$H(v) = H((v + w) - w) \leq 2H(v + w)H(w).$$

Now divide by $2H(w)$.

Similarly, we obtain (e) from (c) by

$$H(v) = H\left((vw)\frac{1}{w}\right) \leq H(vw)H(w).$$

Note that, in general,

$$H\left(\sum_{i=0}^{n} x_i\right) \leq 2^n \prod_{i=0}^{n} H(x_i).$$

All that is used in this section is the existence of a function $H : \overline{\mathbb{Q}} \to \overline{\mathbb{Q}}$ with properties (a), (b), and (c) (and hence also (d) and (e)). It is a good exercise to prove Proposition 4 for $\mathbb{Q}$ with $H(r) = \max(|p|, |q|)$ where $r = p/q$ and $\gcd(p, q) = 1$.

Remark 2 It is important to notice that the height function H is used in this chapter as a technical tool to prove the Witness Theorem and not to define the size of elements and vectors in $\overline{\mathbb{Q}}$. In this chapter computations over $\overline{\mathbb{Q}}$ are considered with the algebraic cost, that is, with $\mathrm{ht}_{\overline{\mathbb{Q}}} \equiv 1$.

If $g \in \overline{\mathbb{Q}}[t]$ is a one-variable polynomial, and $g(t) = \sum_{i=0}^{d} a_i t^i$, define $H(g) = \prod_{i=0}^{d} H(a_i)$.

Proposition 5 *For all $g \in \overline{\mathbb{Q}}[t]$ and all $w \in \overline{\mathbb{Q}}$*

$$H(g(w)) \leq 2^d H(w)^d H(g).$$

Proof.   Use Horner's rule

$$H\left(\sum_{i=0}^{d} a_i w^i\right) = H(a_0 + w(a_1 + w(a_2 + \ldots + w(a_{d-1} + wa_d))\ldots))$$

$$\leq 2^d H(a_0)H(w)H(a_1)H(w)H(a_2)\ldots$$

$$= 2^d \prod_{i=0}^{d} H(a_i)H(w)^d. \tag{7.1}$$

$\square$

If $G(x) = \sum a_\alpha x^\alpha$ is a polynomial in n variables over $\overline{\mathbb{Q}}$, let

$$H(G) = \prod_\alpha H(a_\alpha).$$

Proposition 6 *For $G \in \mathbb{Z}[t_1, \ldots, t_n]$, let $\tau = \tau(G)$. Then*

$$H(G) \leq 2^{2^{2^{2n\tau^2}}}.$$

Toward the proof we have the following lemma whose proof is simple and straightforward.

Lemma 1 *The degree of G is less than or equal to 2^τ. The number of monomials in G, indexed by α, is less than D^n, where $D = 2^\tau$.* $\square$

We now prove Proposition 6.

Proof of Proposition 6.   It goes by induction on τ. One checks it by inspection for $\tau = 1$.

Now let $G = FF'$, where $\tau(F), \tau(F') < \tau$ (the case $G = F + F'$ or $G = F - F'$, is even simpler). Write $F(x) = \sum a_\alpha x^\alpha$, $F'(x) = \sum b_\beta x^\beta$, and $G(x) = \sum c_\gamma x^\gamma$. Then

$$c_\gamma = \sum_\beta a_{\gamma - \beta} b_\beta.$$

Note that by Lemma 1, the degrees of F, F', and G are less than or equal to D and the number of terms in F, F', and G is even less than D^n. Then

$$H(c_\gamma) \leq \prod_\beta 2 H(a_{\gamma - \beta}) H(b_\beta)$$

$$\leq 2^{D^n} H(F) H(F').$$

Thus

$$H(G) \leq (2^{D^n} H(F)H(F'))^{D^n}.$$

By the induction hypothesis

$$H(G) \leq 2^{D^{2n}} \cdot 2^{D^n \cdot 2^{2n(\tau-1)^2+1}},$$

so

$$\log H(G) \leq D^{2n} + D^n \cdot 2^{2n(\tau-1)^2+1}$$
$$\leq 2^{2n\tau} + 2^{n\tau} 2^{2n(\tau-1)^2+1}$$
$$\leq 2^{2n\tau^2}$$

for $\tau \geq 2$.   □

The next proposition, although simple, with a short proof, is crucial for the Witness Theorem.

Proposition 7 *Let $g \in \overline{\mathbb{Q}}[t]$ be a nonconstant polynomial in one variable of degree d. Then for every $x \in \overline{\mathbb{Q}}$,*

$$H(g(x)) \geq \frac{H(x)}{2^d H(g)}.$$

Proof.   Write

$$g(t) = \sum_{i=0}^{d} a_i t^i, \ a_d \neq 0, \ d > 0.$$

Then

$$H(g(x)) = H\left(a_d x^d + \sum_{i=0}^{d-1} a_i x^i\right)$$
$$\geq \frac{1}{2} \frac{H(a_d x^d)}{H\left(\sum_{i=0}^{d-1} a_i x^i\right)}$$
$$\geq \frac{1}{2^d} \frac{H(x)^d}{H(a_d)H(x)^{d-1}H(a_0)\dots H(a_{d-1})}$$
$$\geq \frac{1}{2^d} \frac{H(x)}{H(g)}.$$

Here we have used Propositions 4 and 5.   □

Corollary 1   *For $g \in \overline{\mathbb{Q}}[t]$, if $H(x) > 2^d H(g)$, then $g(x) \neq 0$ unless g is zero.*

                                                                     □

For $x \in \overline{\mathbb{Q}}^n$ let

$$H(x) = \max_{1 \leq i \leq n} H(x_i).$$

For $G \in \overline{\mathbb{Q}}[t_1, \ldots, t_n]$ and $x = (x_1, \ldots, x_r) \in \overline{\mathbb{Q}}^r, r < n$, let

$$G_{x_1, \ldots, x_r}(t_{r+1}, \ldots, t_n) = G(x_1, \ldots, x_r, t_{r+1}, \ldots, t_n).$$

Proposition 8 *For any $G \in \overline{\mathbb{Q}}[t_1, \ldots, t_n]$ and $x = (x_1, \ldots, x_r) \in \overline{\mathbb{Q}}^r$ with $r < n$ we have*

$$H(G_{x_1, \ldots, x_r}) \leq H(G)(2H(x))^{D^{n+1}},$$

where degree $G \leq D$, and $H(x) = H(x_1, \ldots, x_r)$.

Proof. Let $G(t) = \sum_\alpha a_\alpha t^\alpha$. Note that $G_{x_1, \ldots, x_r} \in \overline{\mathbb{Q}}[t_{r+1}, \ldots, t_n]$ is a polynomial whose coefficients may be indexed by $(\alpha_{r+1}, \ldots, \alpha_n)$ and, for each $(\alpha_{r+1}, \ldots, \alpha_n)$, have the form

$$\sum a_\alpha x_1^{\alpha_1}, \ldots, x_r^{\alpha_r},$$

where the sum is over $\alpha = (\alpha_1, \ldots, \alpha_n)$ such that the last $n - r$ entries of α are $(\alpha_{r+1}, \ldots, \alpha_n)$. We must estimate the product of the heights of these coefficients to obtain the proposition. The estimate is similar to that used in Proposition 5.

The estimate for the height of a coefficient of $G_{x_1, \ldots, x_r}$ is

$$\leq 2^{D^r} \prod_{\alpha = (\alpha_1, \ldots, \alpha_r)} H(a_\alpha)H(x_1)^{\alpha_1} \ldots H(x_r)^{\alpha_r}$$

$$\leq 2^{D^r} \prod_{\alpha = (\alpha_1, \ldots, \alpha_r)} H(a_\alpha)H(x)^D.$$

Take the product over all the coefficients to get

$$H(G_{x_1, \ldots, x_r}) \leq 2^{D^n} H(G)H(x)^{D^{n+1}}$$

yielding the necessary estimate. $\qquad\square$

For the proof of the Witness Theorem we may assume that w_1 is one of $2^N, x_1^N, \ldots, x_r^N$ with largest height so

$$H(w_1) \geq \max(2^N, H(x_i)^N).$$

Then $H(w_1) > 1$ and $H(w_i) > H(w_{i-1})$.

Now with these x, w as in the Witness Theorem, for each $j = 1, \ldots, s$ and $\widehat{\beta} = (\widehat{\beta}_{j+1}, \ldots, \widehat{\beta}_s)$, we define a one-variable polynomial $G_{\widehat{\beta}}^j$ so that we are able to apply the one-variable lower bound of Proposition 7.

Write

$$F(x, t) = \sum_{\substack{\alpha = (\alpha_1, \ldots, \alpha_r) \\ \beta = (\beta_1, \ldots, \beta_s)}} a_{\alpha, \beta} x^\alpha t^\beta,$$

then define

$$G_{\widehat{\beta}}^j(t) = \sum_{\substack{\alpha=(\alpha_1,\ldots,\alpha_r) \\ \beta=(\beta_1,\ldots,\beta_j,\widehat{\beta}_{j+1},\ldots,\widehat{\beta}_s)}} a_{\alpha,\beta} x^\alpha w_1^{\beta_1} \ldots w_{j-1}^{\beta_{j-1}} t^{\beta_j}.$$

Lemma 2 *For each $j = 1, \ldots, s$ and $\widehat{\beta}$ as previously,*

$$H(w_j) > 2^D H(G_{\widehat{\beta}}^j).$$

Proof. Fix j. It is sufficient to prove that

$$H(w_j) > 2^D H(F_{x,w_1,\ldots,w_{j-1}}),$$

or yet by Proposition 8 that

$$H(w_j) > 2^D H(F)(2H(x_1, \ldots, x_r, w_1, \ldots, w_{j-1}))^{D^{n+1}}.$$

Now use Proposition 6. The needed estimate is:

$$H(w_j) > \begin{cases} 2^D \cdot 2^{2^{2n\tau^2}} (2H(w_{j-1}))^{D^{n+1}} & \text{if } j > 1 \\ 2^D \cdot 2^{2^{2n\tau^2}} (2\max(2, H(x)))^{D^{n+1}} & \text{if } j = 1. \end{cases}$$

Now use $D = 2^\tau$, and verify that

$$\log N > \tau + 2n\tau^2 + 2(n+1)\tau.$$

$\square$

The Witness Theorem is almost proved. Use Proposition 7 with $j = s$ in Lemma 2. We obtain

$$G_{\emptyset}^s(t) = F_{x,w_1,\ldots,w_{s-1}}(t), \ \widehat{\beta} = \emptyset$$

and

$$H(w_s) > 2^D H(G_{\emptyset}^s).$$

Therefore $F_{x,w_1,\ldots,w_{s-1}}$ is zero. So for each $\widehat{\beta}_s$,

$$\sum_{\substack{\alpha=(\alpha_1,\ldots,\alpha_n) \\ \beta=(\beta_1,\ldots,\beta_{s-1},\widehat{\beta}_s)}} a_{\alpha,\beta} x^\alpha w_1^{\beta_1} \ldots w_{s-1}^{\beta_{s-1}} = 0.$$

Continuing the same process for $s - 1, s - 2, \ldots, 1$, we obtain eventually for any $\widehat{\beta} = (\widehat{\beta}_1, \ldots, \widehat{\beta}_s)$ that

$$\sum_\alpha a_{\alpha,\widehat{\beta}} x^\alpha = 0.$$

This yields our theorem. $\square$

Appendix to Section 7.3

In this appendix we prove part (b) of Proposition 4. We could not find it in the literature.

To do so, we need to use some notions from algebraic number theory. References for these notions can be found in Section 7.9.

Definition 4 $H_K(u) = \prod_{v \in M_K} \max(1, |u|_v^{N_v})$, where M_K is the set of valuations of K. If v restricts to v_0, then define

$$N_v = [K_v : F_{v_0}],$$

the degree relative to the completions.

Definition 5 $H(u) = H_K(u)^{1/[K:Q]}$, for $u \in K$. It can be shown that this is independent of K.

Lemma 3 $\sum_{v \in M_K^\infty} N_v = [K : Q]$, where $M_K = M_K^\infty \cup M_K^*$, M_K^∞ the Archimedean valuations and M_K^* the nonArchimedean valuations. □

Now for the proof of Part (b) of Proposition 4.

Proof of Proposition 4 (b). We may write:

$$
\begin{aligned}
&H_K(x + y) \\
&= \prod_{v \in M_K^\infty} \max(1, |x + y|_v^{N_v}) \prod_{v \in M_K^*} \max(1, |x + y|_v^{N_v}) \\
&\leq \prod_{v \in M_K^\infty} 2^{N_v} \prod_{v \in M_K^\infty} (\max(1, |x|_v, |y_v|))^{N_v} \prod_{v \in M_K^*} (\max(1, |x|_v, |y|_v))^{N_v}
\end{aligned}
$$

from properties of $|\ |_v$, Archimedean and nonArchimedean, respectively. Since

$$\max(1, |x|_v, |y|_v) \leq \max(1, |x|_v) \max(1, |y|_v),$$

we have

$$H_K(x + y) \leq 2^{\sum_{v \in M_K^\infty} N_v} \prod_{v \in M_K} (\max(1, |x|_v))^{N_v} \prod_{v \in M_K} (\max(1, |y|_v))^{N_v}.$$

Using the lemma it follows that

$$H_K(x + y) \leq 2^{[K:Q]} H_K(x) H_K(y).$$

Taking roots we thus obtain

$$H(x + y) \leq 2H(x)H(y).$$

□

7.4   Elimination of Constants: General Case

The main focus of this section is the following proposition.

Proposition 9 (Elimination of Constants)   *Let $K \subset L$ be fields where $K \subset \overline{\mathbb{Q}}$. Let (Y, Y_{yes}) be a decision problem solved by a machine M over L. Then there is a machine M' over K solving the restriction of (Y, Y_{yes}) to K and a constant $c \in \mathbb{N}$ such that $T_{M'}(y) \leq T_M(y)^c$ for all $y \in Y \cap K^\infty$.*

Lemma 4   *For the proof of the preceding proposition, it is sufficient to consider the case $L = K(s_1, \ldots, s_l)$, where $s_1, \ldots, s_l$ is a transcendence base for L over K (i.e., the s_i are algebraically independent over K).*

Proof.   The machine M over L uses a finite number of constants $\eta_1, \ldots, \eta_r$ from L. Therefore M can be considered as a machine over $K(\eta_1, \ldots, \eta_r) \subset L$ by restriction.

By a standard theorem of algebra (in field theory) one may rewrite the sequence $\eta_1, \ldots, \eta_r$ as $s_1, \ldots, s_l, \mu_1, \ldots, \mu_q$, where the $s_1, \ldots, s_l$ form a transcendence basis for $K(s_1, \ldots, s_l)$ over K and the $\mu_1, \ldots, \mu_q$ are algebraic over $K(s_1, \ldots, s_l)$. Now apply Proposition 2 to obtain a machine over $K(s_1, \ldots, s_l)$ with the same values on inputs from K^∞ as M and only a constant multiple increase in time.   □

Proof of Proposition 9.   We give the proof of the Elimination of Constants Proposition where L has the form given in Lemma 4. By Proposition 3 we may suppose that each computation node of the machine M is an arithmetic node ($+$, $-$, $\times$) and that each constant in M is a polynomial in $s = (s_1, \ldots, s_l)$ over K. Let $\alpha = (\alpha_1, \ldots, \alpha_k)$ be a sequence of all the coefficients occurring in these polynomials. We may then suppose each constant in M is of the form $p(\alpha, s)$, where p is a polynomial over $\mathbb{Z}$. Let C be the sum of the $\tau(p)$ over all constants in M.

We construct a machine M' over K that given input $y = (y_1, \ldots, y_n) \in Y \cap K^\infty$ generates a computation path that simulates the computation path $\gamma_y = (\eta_0, \ldots, \eta_t, \ldots)$ traversed by y when input to M. The critical construction is to simulate the "branching structure" in γ_y, and to do this with at most a polynomial increase in time.

So suppose η_t is a branch node and f_t the associated (*modified*) *step t branching polynomial* (on the active state space S_m where $m = K_M + t + n$). That is, f_t is the (projection onto the first coordinate of the) composition of the successive (modified) computations occurring along the computation path γ_y through step t. We may consider f_t as a polynomial in y, α, and s over $\mathbb{Z}$ with $\tau(f_t) \leq t + C$. The computation path γ_t branches Yes or No according to whether or not $f_t(y, \alpha, s) = 0$.

We construct M' so that given input $y = (y_1, \ldots, y_n)$ and "time" t, M' generates elements $w_1, \ldots, w_l$ in K to replace $s_1, \ldots, s_l$ and thus obtain a machine over K. To produce w_1, let $W = 4(n + k + l)(t + C)^2 + 4(t + C)$ and repeat squaring

W times each of $2, y_1, \ldots, y_n, \alpha_1, \ldots, \alpha_k$ (not giving a unique w_1, but a set of them). Let $w_1, \ldots, w_l$ be as in the Witness Theorem. Now test if $f_t(y, \alpha, w) = 0$ successively for each one of the $n + k + 1$ choices.

If any one of these $f_t(y, \alpha, w) \neq 0$, then $f_t(y, \alpha, s) \neq 0$ and we branch accordingly. On the other hand, by the Witness Theorem, if $y \in K^\infty$ and all the $f_t(y, \alpha, w) = 0$, then $f_t(y, \alpha, s) = 0$. It is easy to check that the total increase in time is polynomial so that we have proved our proposition. $\qquad\square$

Lemma 5 *If $K \subset L$ are algebraically closed fields, then the restriction of* HN$/L$ *to K is* HN$/K$.

Proof. What we need to show here is if $f_1, \ldots, f_k \in K[x_1, \ldots, x_n]$ have a common zero $\zeta \in L^n$, then they must have a common zero in K^n. But this follows from Hilbert's Nullstellensatz since, if the f_i have no common zero in K^n, then there exist $g_i, i = 1, \ldots, k$, polynomials in n variables over K, such that $\sum g_i f_i = 1$. Evaluation at ζ gives a contradiction. Alternatively, this follows directly from the model completeness of the theory of algebraically closed fields (see Section 7.8). This proves Lemma 5. $\qquad\square$

Now we can prove the first statement of Theorem 1.

Proof of Theorem 1 ("If" Direction). Suppose $P = NP$ over $\mathbb{C}$. Then HN$/\mathbb{C} \in$ P by a machine M over $\mathbb{C}$. Then M "solves" HN$/\overline{\mathbb{Q}}$ (inputs from $\overline{\mathbb{Q}}^\infty$) by Lemma 5, but M is still a machine over $\mathbb{C}$. Now apply Proposition 9 to obtain a machine over $\overline{\mathbb{Q}}$ solving HN$/\overline{\mathbb{Q}}$ in polynomial time. By the NP-completeness of this problem over $\overline{\mathbb{Q}}$, $P = NP$ over $\overline{\mathbb{Q}}$. $\qquad\square$

7.5 Twenty Questions

Toward the proofs of Theorems 2 and 3 we introduce a decision problem we call "Twenty Questions" which is of independent interest.

Let R be a ring (integral domain) or field of characteristic 0 which we consider without order and let $\mathbb{Z}^+$ be the positive integers. Then *Twenty Questions over R* is the following problem.

Given input $(k, \mathrm{ht}_{\mathbb{Z}}(k), z) \in \mathbb{Z}^+ \times \mathbb{Z}^+ \times R$, decide if $z \in \{1, 2, \ldots, k\}$.

Here, we recall, $\mathrm{ht}_{\mathbb{Z}}(k)$ is defined to be the smallest natural number greater than or equal to $\log(1 + k)$. Strictly speaking this is not a "decision problem" in the sense of Chapter 4 (see Remark 7 there).

Even if R happens to be an ordered ring as $\mathbb{Z}$, we continue to branch only on equality tests.

Twenty Questions over any ring R can be decided within halting time $3k + 3$ by the machine in Figure A.

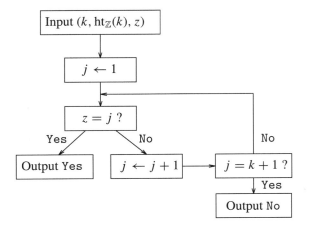

Figure A   A machine for Twenty Questions.

Can one do better? We do not know. But if $R = \mathbb{Z}$, and branching on order is permitted, then the decision (halting) time is approximately $\log k$, with the algorithm used in the parlor game called Twenty Questions.

We say that *Twenty Questions over R is tractable* if it can be decided in halting time $(\log k)^c$ over R where c is some constant (depending only on R). The next theorem shows that if Twenty Questions over $\mathbb{Z}$ is tractable, then so is the order relationship itself.

Theorem 5 *If Twenty Questions over $\mathbb{Z}$ is tractable, then on input $(x, y) \in \mathbb{Z} \times \mathbb{Z}$, one can decide if $x < y$ with halting time bounded by a polynomial in* $\max(\log |x|, \log |y|)$.

Proof.   Figure B shows a machine that solves the problem. This machine halts after visiting at most $3k + 4$ nodes where k is the first integer greater than $\max(\log |x|, \log |y|)$. Of these nodes, $2k$ are Twenty Questions for $2, 2^2, \ldots, 2^k$ twice each, hence the total halting time is bounded by

$$2 \sum_{j=1}^{k} j^c + k + 4,$$

which is less than or equal to

$$2 \left(\frac{k(k+1)}{2} \right)^c + k + 4.$$

$\square$

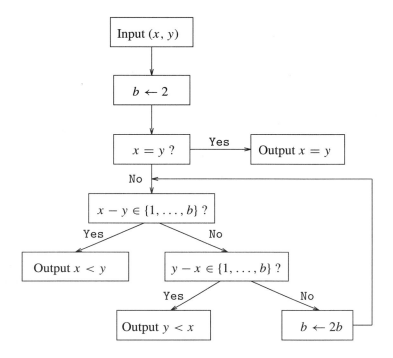

Figure B   A machine computing ≤ in $\mathbb{Z}$.

Theorem 6   *If* P = NP *over* $\mathbb{C}$, *then Twenty Questions over* $\mathbb{C}$ *is tractable.*

Proof.   The method is to embed Twenty Questions in a decision problem (Y, Y_{yes}) which is in NP over $\mathbb{C}$. Then if NP = P over $\mathbb{C}$, (Y, Y_{yes}) is in P over $\mathbb{C}$ and there is a machine M which decides Twenty Questions in halting time bounded by $(\log k)^c$, c a constant. Here M is the restriction of the machine that decides (Y, Y_{yes}) in polynomial time.
    The decision problem (Y, Y_{yes}) is described as follows.

$$Y = \mathbb{C}^\infty \qquad \text{and} \qquad Y_{\text{yes}} = \bigcup_{k \in \mathbb{Z}^+} Y_{\text{yes},k},$$

where

$$Y_{\text{yes},k} = \{(k, \text{ht}_{\mathbb{Z}}(k), z_1, \ldots, z_{\text{ht}_{\mathbb{Z}}(k)}) \mid z_1 \in \{1, \ldots, k\}\}.$$

The embedding of Twenty Questions in (Y, Y_{yes}) is simply

$$(k, \text{ht}_{\mathbb{Z}}(k), z) \rightarrow (k, \text{ht}_{\mathbb{Z}}(k), z, 1, \ldots, 1),$$

where the number of ones is $\text{ht}_{\mathbb{Z}}(k) - 1$. The proof is finished by the next lemma.

□

Lemma 6 (Y, Y_{yes}) *is in* NP *over* $\mathbb{C}$.

Proof.   The $\text{NP}_{\mathbb{C}}$ machine operates on variables

$$(u_1, u_2, z_1, \ldots, z_n, w_0, \ldots, w_{n-1}, v_0, \ldots, v_{n-1}, y_0, \ldots, y_{n-1})$$

It checks if u_2 is a positive integer by addition of 1s. It checks if the input size (given with the input by definition) is $4u_2 + 2$. If so, $n = u_2$. It checks if $w_{n-1} = 1$, $w_i(w_i - 1) = 0$, $v_i(v_i - 1) = 0$, and $y_i(y_i - 1) = 0$ for $i = 0, \ldots, n - 1$. It checks if $u_1 = \sum_{i=0}^{n-1} 2^i w_i$. It sets $a = \sum_{i=0}^{n-1} 2^i v_i$ and $b = \sum_{i=0}^{n-1} 2^i y_i$. Finally it checks if $z_1 = 1 + a$ and $u_1 = z_1 + b$. If so, it outputs Yes. Note that if the tests are verified, the ws, vs, and ys are 0 or 1 and a and b are nonnegative integers. Hence z_1 and u_1 are positive integers, $1 \leq z_1 \leq u_1$, and $u_2 = \text{ht}_{\mathbb{Z}}(u_1)$. The time required is a constant times u_2.

Finally we show that every element of $Y_{\text{yes},k}$ has a positive test. Let

$$(k, \text{ht}_{\mathbb{Z}}(k), z_1, \ldots, z_{\text{ht}_{\mathbb{Z}}(k)}) \in Y_{\text{yes},k}.$$

Suppose $\text{ht}_{\mathbb{Z}}(k) = n$. Then $z_1 = 1 + a$ for some nonnegative integer a with $\text{ht}_{\mathbb{Z}}(a) \leq n$ and $k = z_1 + b$ for some nonnegative integer b with $\text{ht}_{\mathbb{Z}}(b) \leq n$. ☐

Theorem 7   *If Twenty Questions over* $\mathbb{C}$ *is tractable, then Twenty Questions over* $\mathbb{Z}$ *is tractable.*

Proof.   It follows immediately from the elimination of constants in Sections 7.2 and 7.4. ☐

Remark 3   The result and proof of Theorems 6 and 7 are valid if $\mathbb{C}$ is replaced by any field or ring R containing $\mathbb{Z}$.

7.6   Proof of Theorems 2 and 3

Proof of Theorem 3.   Suppose that $\text{P} = \text{NP}$ over $\mathbb{C}$. Then by Theorems 6 and 7, Twenty Questions over $\mathbb{Z}$ is tractable. Thus there is a machine over $\mathbb{Z}$ deciding:

$$\text{given } (k, \text{ht}_{\mathbb{Z}}(k), z) \in \mathbb{Z}^+ \times \mathbb{Z}^+ \times \mathbb{Z}, \text{ is } z \in [1, k]?$$

in halting time $(\log k)^c$. By an extension to $\mathbb{Z}$ of the Canonical Path Theorem (see formula 2.26 and Theorem 5 of Chapter 2) for each k there is a one-variable nontrivial polynomial $F_k \in \mathbb{Z}[t]$ vanishing on the set $\{1, 2, \ldots, k\}$ with $\tau(F_k) \leq (\log k)^c$ for some constant c.

Observe that the hypothesis preceding Theorem 3 is now violated. That is, for each c' there is a k_0 such that

$$\text{Zer}(F_k) \geq k \geq ((\log k)^c + 1)^{c'} \geq (\tau(F_k) + 1)^{c'} \text{ for } k \geq k_0.$$

☐

We now prove Theorem 2.

Proof of Theorem 2. Suppose to the contrary that $P = NP$ over $\mathbb{C}$. For each k, let $F_k \in \mathbb{Z}[t]$ be as in the proof of Theorem 3. Then we know that the degree of F_k is less than or equal to $2^{\tau(F_k)}$. So there is an integer ℓ, $|\ell| \leq 2^{\tau(F_k)}$ with $F_k(\ell) \neq 0$. We may assume $|\ell|$ is minimal satisfying $F_k(\ell) \neq 0$.

By Proposition 1, $\tau(\ell) \leq 2\tau(F_k)$ so (since $\tau(F_k) \leq (\log k)^c$ for some constant c) we have $\tau(\ell) \leq 2(\log k)^c$. Thus, by evaluating F_k at ℓ, we have

$$\tau(F_k(\ell)) \leq 3(\log k)^c. \tag{7.2}$$

We now observe that $F_k(\ell)$ has $k!$ as a factor by checking the two possibilities, $\ell > k$ or $\ell \leq 0$. For if $\ell > k$, then F_k is zero on $\{1, \ldots, \ell - 1\}$ and so $F_k(t) = \prod_{i=1}^{\ell-1}(t-i)g_1(t)$, whereas if $\ell \leq 0$, then F_k is zero on $\{\ell + 1, \ldots, k\}$ and so $F_k(t) = \prod_{i=\ell+1}^{k}(t-i)g_2(t)$.

Thus, letting $m_k = F_k(\ell)/k!$ (ℓ depends on k also) we have by (7.2) that $\tau(m_k k!) \leq 3(\log k)^c$ which implies that $k!$ is ultimately easy to compute. This finishes the proof of Theorem 2. □

7.7 Main Theorem, an Algebraic Proof of the Converse

Let K be an algebraically closed field and L a field, $K \subset L$. A set $S \subset K[t_1, \ldots, t_n]$ determines an algebraic set $V_K \subset K^n$ by $x \in V_K$ if and only if $f(x) = 0$ for all $f \in S$. Moreover S also determines an algebraic set $V_L \subset L^n$ by $x \in V_L$ if and only if $f(x) = 0$ all $f \in S$.

Lemma 7 *With S and notation as previously, let V'_L be the algebraic set defined by*

$$V'_L = \{x \in (L)^n \mid f(x) = 0 \text{ all } f \in K[t_1, \ldots, t_n] \ni f \equiv 0 \text{ on } V_K\}.$$

Then $V'_L = V_L$.

Proof. Since clearly $V'_L \subset V_L$, it is sufficient to show that any $f \in K[t_1, \ldots, t_n]$ vanishing on V_K must also vanish on V_L. But by the Hilbert Nullstellensatz such an f satisfies, for some $l > 0$, $f^l \in I_K(S)$, the ideal generated by S over K. Therefore f^l also vanishes on V_L and hence f does. □

Therefore V_L is determined by V_K.

Proposition 10 *Let $S \subset K[x_1, \ldots, x_n]$ and $g_1, \ldots, g_l \in K[x_1, \ldots, x_n]$. Let V_K and V_L be defined by S. If there is a point $z \in V_L$ such that $g_i(z) \neq 0$, for all $i = 1, \ldots, l$, then there is a point $z' \in V_K$ such that $g_i(z') \neq 0$, for all $i = 1, \ldots, l$.*

Proof.   We first prove the proposition in case V_K is irreducible. Now proceed by induction on l. The case $l = 0$ is already done in the proof of Lemma 4.

By induction we suppose the assertion proven for $l - 1$ and establish it for l. Assume that $z \in V_L$ and $g_i(z) \neq 0$ for all $i = 1, \ldots, l$. By induction the set U of $z' \in V_K$ such that $g_i(z') \neq 0$, for all $i = 1, \ldots, l - 1$ is nonempty and Zariski open. If there is no $z' \in U$ such that $g_l(z') \neq 0$, then g_l is zero on U and hence zero on V_K by the irreducibility of V_K. Hence by the Nullstellensatz there is an m such that g_l^m is in the ideal $I_K(S)$ generated by S in $K[x_1, \ldots, x_n]$. Hence g_l^m is also in the ideal $I_L(S)$ generated by S in $L[x_1, \ldots, x_n]$ and g_l vanishes on V_L which is a contradiction. The general case is finished by the next lemma.   □

Lemma 8   *Let $V_K \subset K^n$ be an algebraic set with V_K the union of algebraic sets V_1 and V_2. Then*

$$V_L = V_{1,L} \cup V_{2,L}.$$

Proof.   For $i = 1, 2$, the ideals satisfy $I(V_i) \supset I(V_K)$. Thus if $x \in L_{i,L}, i = 1$ and 2, then $x \in V_L$. On the other hand, if $x \notin V_{1,L} \cup V_{2,L}$, then there exist $f_i \in I(V_{1,K})$, $i = 1, 2$ such that $f_i(x) \neq 0$. Thus $f_1 f_2(x) \neq 0$ and $f_1 f_2 \notin I(V_1) \cup I(V_2) = I(V_K)$ so $x \notin V_L$.   □

Now we refer back to Definitions 3 and 4 of Chapter 2.

A *basic quasi-algebraic formula* over a ring R is a system

$$f_1(x) = 0, \ldots, f_l(x) = 0$$
$$g_1(x) \neq 0, \ldots, g_k(x) \neq 0,$$

where the f_i and g_j are elements of $R[t_1, \ldots, t_m]$, for some $m \in \mathbb{N}$.

A basic quasi-algebraic formula over $R \subset K$, K a field, defines a *basic quasi-algebraic set* over R in K^n by

$$V = \{x \in K^m \mid f_i(x) = 0, \ i = 1, \ldots, l, \ g_j(x) \neq 0, \ j = 1, \ldots, k\}.$$

A basic quasi-algebraic formula over $\mathbb{Z}$ defines a basic quasi-algebraic set over $\mathbb{Z}$ in K^m for any field K.

A subset of K^m is *quasi-algebraic over R* if it is the union of a finite number of basic quasi-algebraic sets over R. Quasi-algebraic sets over R in K^m are closed under finite union, finite intersection, and the operation of taking complements.

Proposition 11   *Given n, m there is a finite set of basic quasi-algebraic formulas over $\mathbb{Z}$ such that: given any field K, $n \times m$ matrix A over K, and vector $b \in K^n$, the linear equation $A(x) = b$ has a solution in K^m if and only if (A, b) is in the quasi-algebraic set in $K^{n \times m + n}$ which is the union of the basic quasi-algebraic sets defined by these formulas.*

Proof.   The system $A(x) = b$ has a solution if and only if there are k columns of A such that the $(n \times k)$ matrix B determined by them has rank k and the $n \times (k + 1)$ matrix obtained by adjoining the column b also has rank k, $0 \leq k \leq m$.

This condition is expressed in terms of the determinants of the minors of A that are polynomial over $\mathbb{Z}$ in the coefficients of A.   □

Corollary 2   *Given m, n, and a vector of degrees* $d = (d_1, \ldots, d_m)$, *there is a finite set of basic quasi-algebraic formulas over* $\mathbb{Z}$ *such that for any algebraically closed field K, the system of equations*

$$f_1(x) = 0, \ldots, f_m(x) = 0, \ \deg f_i = d_i$$

has a solution in K^n *if and only if the coefficients of the* f_i *lie in the quasi-algebraic set that is the union of the basic quasi-algebraic sets determined by these formulas.*

Proof. By the effective Nullstellensatz, the system $f_1(x) = 0, \ldots, f_m(x) = 0$ has no common zero if and only if there exist $g_i, i = 1, \ldots, m$ of degree $\leq C$ such that $\sum_{i=1}^{m} f_i g_i = 1$. This is a system of linear equations in the coefficients of the f_i and the preceding proposition finishes the proof.   □

Theorem 8   *Let* $K \subset L$ *be algebraically closed fields. If* P = NP *over K, then* P = NP *over L.*

Proof.   It suffices to show that the machine M which decides Hilbert's Nullstellensatz over K in polynomial time decides it over L with the same polynomial time bounds.

Fix n, m, and d. Let $K_{n,m,d}$ be the set of corresponding inputs of HN$/K$, and $L_{n,m,d}$ for HN$/L$. Thus $f \in K_{n,m,d}$ consists of m polynomials $f_1, \ldots, f_m$ of $K[t_1, \ldots, t_n]$ with degree $f_i = d_i$. The yes subset of $K_{n,m,d}$ is denoted by $K_{n,m,d,o}$, and the yes subset of $L_{n,m,d}$ by $L_{n,m,d,o}$.

Assume M has two output nodes, yes and no, and that the time bound for inputs of $K_{n,m,d}$ is T.

Consider a yes instance y of HN$/L$ and let $\eta_{y,T}$ be the node of M in the orbit of y (i.e., computation path traversed by y) at time T.

Since $K_{n,m,d,o}$ and $L_{n,m,d,o}$ are defined by the same sets of basic quasi-algebraic formulas over $\mathbb{Z}$ and the node is determined by the basic quasi-algebraic formulas over K determined by the branch nodes in the orbit of y up to time T, Proposition 10 implies that there is a yes instance of $K_{n,m,d}$ at node $\eta_{y,T}$ at time T. Thus $\eta_{y,T}$ is the yes node.

The same argument applies to a no instance, interchanging yes and no.   □

7.8   Main Theorem, a Model-Theoretic Proof of the Converse

In this section we give an alternate proof of Theorem 8 using model-theoretic results and techniques. The reader not familiar with first-order logic can find the necessary basic definitions in Section 23.1. See also Section 7.9. Assuming $K \subset L$ are algebraically closed fields, it suffices to prove the following two lemmas.

Lemma 9 *If M is a polynomial time machine over K that outputs the value 0 or 1 when input an element of K^∞, then the same is true when K is replaced by L (and hence by any field extension of K).*

Lemma 10 *If M is a time-bounded machine over K that decides HN/K, then the set of inputs to M from L^∞ that output the value 1 is exactly the set of* yes *instances of HN/L.*

Lemmas 9 and 10 follow easily from the **Model Completeness (Strong Transfer Principle)** of the theory of algebraically closed fields.

> Suppose $K \subset L$ are algebraically closed fields and Φ is a first-order sentence in the language of fields with constants from K. Then Φ is true when interpreted in K if and only if Φ is true when interpreted in L.

To prove Lemma 9, let p be the polynomial time bound for M over K and let H be the computing endomorphism of M over K. We apply the Strong Transfer Principle to each sentence Φ_n, $n > 0$ (seen easily to be writable as a first order sentence over K):

$$\forall y \exists z_0 \ldots \exists z_{p(n)} \exists w [z_0 = (1, y) \,\&_{k=1}^{p(n)} z_k = H(z_{k-1}) \,\& z_{p(n)} = (N, w)$$
$$\& \ (O(w) = 0 \vee O(w) = 1)],$$

where $y = (y_1, \ldots, y_n \underbrace{0, \ldots, 0}_{m-n})$, $w = (w_1, \ldots, w_m)$ and $m = K_M + p(n) + n$.

The sentence Φ_n asserts that for each input to M of size n, the computation halts in time bounded by $p(n)$ with output value 0 or 1. Each sentence Φ_n is true in K; so each is true in L.

We use the same technique to prove Lemma 10. For each m, d, n let

$$f_1(y^1, x) = 0, \ldots, f_m(y^m, x) = 0$$

be the general system of m polynomial equations of degree d in n variables $x = (x_1, \ldots, x_n)$ and variable coefficients $y^i = (y_1^i, \ldots, y_l^i)$, $i = 1, \ldots, m$ (here l depends on d and n). Let $p(n)$ be a (not necessarily polynomial) time bound for M and H its computing endomorphism. We apply the Strong Transfer Principle to each sentence $\Phi_{m,d,n}$, $m, d, n > 0$:

$$\forall y^1 \ldots \forall y^m \{\exists x (\&_{i=1}^m f_i(y^i, x) = 0) \Longleftrightarrow$$
$$\exists z_0 \ldots \exists z_{p(ml)} \exists w \ [\ z_0 = (1, (y^1, \ldots, y^m)) \,\&_{k=1}^{p(ml)} z_k = H(z_{k-1})$$
$$\& \ z_{p(ml)} = (N, w) \ \& \ O(w) = 1 \ ]\}.$$

The sentence $\Phi_{m,d,n}$ asserts that for each sequence of coefficients $y^1, \ldots, y^m$ (from the given field), the system $f_1(y^1, x) = 0, \ldots, f_m(y^m, x) = 0$ has a solution (in the given field) if and only if M with input $(y^1, \ldots, y^m)$ halts with output 1. Each such sentence is true in K; therefore each is true in L.

7.9   Additional Comments and Bibliographical Remarks

This chapter is based on [Blum, Cucker, Shub, and Smale 1996a]. The part of Theorem 1 asserting P = NP over $\mathbb{C}$ implies P = NP over $\overline{\mathbb{Q}}$, is proved there for the first time. The same is true for the Witness Theorem of Section 7.3 and Proposition 9 as well. The converse in Theorem 1 is due to Michaux [1994] who gave a model-theoretic proof similar to ours. Much of the rest is from [Shub and Smale 1995]. In particular, Theorems 2 and 6 are proved in that paper. A version of Theorem 5 is used in [Shub 1993b].

The function τ is a version of standard concepts in algebraic complexity theory as in [Heintz and Morgenstern 1993]. There is also a simpler function defined without allowing multiplication in the old subject of additive chains (see [Scholz 1937] and [Knuth 1981]). Some results on τ are in [de Melo and Svaiter 1996] and in [Moreira TA].

The relationship of the open problem in Section 7.1 to factoring was first pointed out to us by Don Coppersmith. A way of computing $n!$ in $O(\log n)$ steps was shown by Shamir [1979]. He assumes though that, besides addition, subtraction, multiplication, and tests for equality, his computational model can also compute $\lfloor x/y \rfloor$ in a single step. From here he factors an integer n in $O(\log^2 n)$ steps. For related results on factoring see [Strassen 1976].

For material on heights used in Section 7.3 and its appendix, see [Lang 1991]. [Lang 1993a] provides background for the algebra and, in particular, for the field theory we use (e.g., Lemma 4 of Section 7.4). Relevant background for diophantine approximation in polynomial computations is in [Krick and Pardo 1996]. The effective Nullstellensatz used in Corollary 2 is, for example, in [Brownawell 1987].

For background in model theory see [Hodges 1993] or [Ebbinghaus, Flum, and Thomas 1994].

We wish to thank Jean-Pierre Dedieu and Teresa Krick who read an earlier version of this chapter and made several corrections.

We end this chapter by stating some relating results and posing an open problem. Bruno Poizat [1995] has pointed out the following.

Theorem 9 *If* P = NP *over an infinite field* K, *then* K *is algebraicaly closed.*

The proof is based on a result of Angus Macintyre [1971] stating that if an infinite field admits elimination of quantifiers, then it is algebraicaly closed. For if P = NP over K, then HN/K is solved by a time bounded machine over K and this implies that K admits elimination of quantifiers. An analogue of Macintyre's result for ordered and valued fields can be found in [Macintyre, McKenna, and van den Dries 1983]. Hence, we also have: if P = NP over an ordered field K, then K is real closed.

Theorem 10 (Michaux [1994]) *If* $\mathbb{C} \subset K \subset L$ *where* K, L *are fields and* K *is algebraically closed, then* P = NP *over* L *implies* P = NP *over* K.

Remark 4 It follows from Theorem 1 together with the preceding theorems that the problem P = NP ? over K reduces to the single problem P = NP ? over $\overline{\mathbb{Q}}$ in the case of fields of characteristic zero.

Open Problem. Does a similar result prevail in the case of characteristic $p \neq 0$? And for ordered fields?

Appendix A

In this appendix we introduce some basic notions and terminology of algebraic geometry. We state most of them without proof and point the reader to Section A.2 for references.

A.1 Basic Notions of Algebraic Geometry

Let k be a field.

Definition 1 An *algebraic set* $X \subseteq k^n$ is the set of common zeros of a finite set of polynomials $f_1, \ldots, f_m \in k[x_1, \ldots, x_n]$.

If $I \subseteq k[x_1, \ldots, x_n]$ is the ideal generated by $f_1, \ldots, f_m$, then X is also the set of zeros of the elements of I. One could also consider sets of zeros of infinite families of polynomials in $k[x_1, \ldots, x_n]$. The following theorem shows that there is no gain in doing so.

Theorem 1 (Hilbert's Basis Theorem) *Every ideal of $k[x_1, \ldots, x_n]$ is finitely generated.*

In more geometric terms, Theorem 1 states that every algebraic set is the intersection of a finite number of hypersurfaces (i.e., of sets defined by $f(x_1, \ldots, x_n) = 0$ for a polynomial f).

It is easy to check that the set of algebraic subsets of k^n satisfies the axioms for the closed sets in a topology. The topology they determine in k^n is called the *Zariski topology*. Denote the zeros in k^n of a family F of polynomials in $k[x_1, \ldots, x_n]$ by

$\mathcal{Z}(F)$. Also, for a set $X \subseteq k^n$ denote by $\mathcal{I}(X)$ the set of polynomials in $k[x_1, \ldots, x_n]$ that vanish on X. Then the closure of X in the Zariski topology is $\mathcal{Z}(\mathcal{I}(X))$.

Remark 1 For $k = \mathbb{R}$ or $\mathbb{C}$ the Zariski topology in k^n is coarser than the Euclidean one. Every Zariski closed set is closed in the Euclidean topology. The converse is false. The unit ball in $\mathbb{R}^n$ (or in $\mathbb{C}^n$) is not the zero set of any ideal in $\mathbb{R}[x_1, \ldots, x_n]$ (respectively, in $\mathbb{C}[x_1, \ldots, x_n]$).

Proposition 1 *Let k be a field with an infinite number of elements. If $X \subseteq k^n$ is nonempty and open for the Zariski topology, then X is dense.*

Proof. For $n = 1$ we need only prove that if a polynomial f vanishes over all the elements of k except perhaps a finite number, then f is the zero polynomial. But this is true. For $n > 1$ the result follows by induction. □

Definition 2 An algebraic set $X \subseteq k^n$ is said to be *reducible* if there exist algebraic sets $X_1, X_2 \subseteq k^n$ such that $X = X_1 \cup X_2$ and $X \neq X_i$ for $i = 1, 2$. Otherwise, we say that X is *irreducible* or, equivalently, that it is an *affine variety*.

Proposition 2 *An algebraic set $X \subseteq k^n$ is irreducible if and only if $\mathcal{I}(X)$ is a prime ideal.*

Proof. If $\mathcal{I}(X)$ is not prime, there are $f_1, f_2 \notin \mathcal{I}(X)$ such that $f_1 f_2 \in \mathcal{I}(X)$. Then $X = (X \cap \mathcal{Z}(f_1)) \cup (X \cap \mathcal{Z}(f_2))$ and $X \cap \ddagger(f_i) \neq X$ for $i = 1, 2$ so X is reducible.

Conversely, if $X = X_1 \cup X_2$ and $X \neq X_i$ for $i = 1, 2$, then $\mathcal{I}(X) \neq \mathcal{I}(X_i)$ for $i = 1, 2$. Let $f_i \in \mathcal{I}(X) - \mathcal{I}(X_i)$. Then $f_1 f_2 \in \mathcal{I}(X)$ and thus $\mathcal{I}(X)$ is not prime. □

The next result is a weak form of the Hilbert Nullstellensatz.

Theorem 2 *Let k be algebraically closed and I be an ideal of $k[x_1, \ldots, x_n]$. Then $\mathcal{Z}(I) = \emptyset$ if and only if $1 \in I$. Or equivalently, if $f_1, \ldots, f_m \in k[x_1, \ldots, x_n]$, then the f_i have no common zero if and only if there are $g_1, \ldots, g_m \in k[x_1, \ldots, x_n]$ such that $\sum_{i=1}^{m} f_i g_i = 1$.*

Now we describe the Nullstellensatz in its usual form.

An ideal I of a commutative ring R is said to be *radical* if for every $f \in R$, if f^d belongs to I for some $d \geq 1$, then $f \in I$. The set

$$\sqrt{I} = \{f \in R \mid \exists d \geq 1 f^d \in I\}$$

is a radical ideal, called the radical of I.

The strong form of the Hilbert Nullstellensatz can now be stated.

Theorem 3 *Let k be algebraically closed and I be an ideal of $k[x_1, \ldots, x_n]$. Then $\mathcal{P}(\mathcal{Z}(I)) = \sqrt{I}$.*

A.2   Additional Comments and Bibliographical Remarks

Since $\sqrt{I} = I$ for a radical ideal I, the Nullstellensatz in its strong form establishes a bijection between radical ideals of $k[x_1, \ldots, x_n]$ and algebraic sets in k^n. One can further associate with the algebraic set $X \subseteq k^n$ the quotient ring $R = k[\bar{x}]/\mathcal{P}(X)$, where $\bar{x} = (x_1, \ldots, x_n)$. It turns out that all features of the set X which are invariant under algebraic changes of coordinates can be deduced from the ring R. Therefore the study of algebraic sets reduces to the study of rings.

The preceding sketches the close relationship between algebraic geometry —which studies algebraic sets— and commutative algebra —which studies commutative rings. Some textbooks in commutative algebra are [Atiyah and Mac-Donald 1969; Zariski and Samuel 1979]. For algebraic geometry see [Hartshorne 1977; Shafarevich 1977]. The latter references assume the reader is familiar with the concepts of commutative algebra as developed in the former ones. Books simultaneously developing both subjects are [Eisenbud 1995; Kunz 1985]. For a proof of Hilbert's basis theorem see [Atiyah and MacDonald 1969; Kunz 1985]. For a proof of the Hilbert Nullstellensatz see [Kendig 1977; Kunz 1985; Zariski and Samuel 1979].

The Hilbert Nullstellensatz requires the base field to be algebraically closed. Nullstellensätze over nonalgebraically closed fields have been proved in the last decades for other classes of fields. Over $\mathbb{R}$, and more generally over any real closed field, such results can be found in [Dubois 1969; Risler 1970] and a weak form in [Krivine 1964]. We prove a special case in Chapter 19. The existence of a Nullstellensatz over the reals motivated the development of a real algebraic geometry. References for this subject are the books by Bochnak, Coste, and Roy [1987] and by Benedetti and Risler [1990].

Part II

Some Geometry of Numerical Algorithms

8
Newton's Method

We have called Newton's method the "... 'search algorithm' sine qua non of numerical analysis and scientific computation." Yet we have seen that even for a polynomial of one complex variable we cannot decide if Newton's method will converge to a root of the polynomial on a given input. In this chapter we begin a more comprehensive study of Newton's method. We introduce quantities α, β, and γ which play an important role in analyzing the complexity of algorithms that approximate the solutions of systems of equations. Our main results, Theorems 1 and 2, give the speed of convergence to a root in terms of these quantities, while other results such as Proposition 3 estimate them. In particular, Theorem 2 gives us a criterion, computable at a point x, to confirm that x is "close" to an actual zero ζ of a system of equations. Here close is defined in a strong sense and Newton's method doubles precision at each step starting with x.

Theorem 1 and its n-dimensional generalization play a crucial role in the complexity results for Bezout's Theorem. Proposition 2 is used in the next chapter for a complexity analysis of an algorithm for the fundamental theorem of algebra. Theorem 4 is used in Chapter 15 in a complexity analysis of a basic algorithm in linear programming.

For purposes of exposition the one-variable case is treated first. Then it is noted how the results extend to systems of equations $f : \mathbb{C}^n \to \mathbb{C}^n$ and even maps of Banach spaces $f : \mathbb{E} \to \mathbb{F}$.

8.1   Approximate Zeros

We begin this section by solving linear equations. Given a linear equation in one variable

$$f(x) = ax + b$$

with $a \neq 0$, we solve the equation $f(x) = 0$ by

$$x = -a^{-1}b.$$

For quadratic equations

$$f(z) = az^2 + bz + c \qquad a \neq 0,$$

we solve for the two roots, $f(z) = 0$, by the quadratic formula

$$\zeta_+ = \frac{-b + \sqrt{b^2 - 4ac}}{2a} \qquad \text{and} \qquad \zeta_- = \frac{-b - \sqrt{b^2 - 4ac}}{2a}.$$

But even these simple formulas for the roots involve square roots that are themselves only computed approximately using the elementary arithmetic operations and inequalities.

Newton's method is an iterative method designed to approximate the roots of nonlinear equations. Given an initial approximation a to a root of the equation $f(z) = 0$, Newton's method replaces a by the exact solution a' of the best linear approximation to f which is given by the tangent to the graph of f at the point $(a, f(a))$.

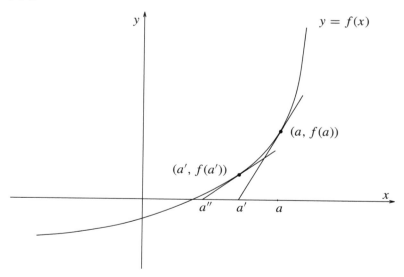

Figure A   Starting from a, two steps of Newton's method give a'', a close approximation to a zero of f.

Suppose that

$$f(z) = a_0 + a_1 z + \ldots + a_n z^n + \ldots = \sum_{j=0}^{\infty} a_i z^i$$

is an analytic function of one complex (or real) variable defined on all of $\mathbb{C}$ (or $\mathbb{R}$). Thus, for example, f may be a polynomial, the sine or cosine functions, the exponential function, or sums, products, and composition of these, and so on. Our main application for the theory developed in this chapter is to polynomials.

Newton's method is an iteration based on the map from $\mathbb{C}$ to itself,

$$N_f(z) = z - (f'(z))^{-1} f(z),$$

where $f'(z)$ is the derivative of f at z. This formula is defined as long as $(f'(z))^{-1}$ exists.

The formula for N_f is also written $N_f(z) = z - (f(z)/f'(z))$. We say $(f'(z))^{-1}$ exists in place of $f'(z) \neq 0$ because the theory we are presenting is valid in the much more general context of maps between n-dimensional or even Banach spaces. In this context the derivative $f'(z)$ is a continuous linear map that we assume has an inverse. We also write $N_f'(z)$ as we do because the formula is valid in n-dimensional or Banach spaces where linear maps do not necessarily commute. See the end of the chapter for more discussion of this point.

We recall that if $f(\zeta) = 0$ and $f'(\zeta)^{-1}$ exists, then $N_f(\zeta) = \zeta$ and in that case $N_f'(\zeta) = f'(\zeta)^{-1} f''(\zeta) f'(\zeta)^{-1} f(\zeta) = 0$. The Taylor series of N_f near ζ is then

$$N_f(z) - \zeta = c_2(z - \zeta)^2 + \text{ higher order terms},$$

where c_2 is the second derivative of Newton's map divided by two. Thus the distance from $N_f(z)$ to ζ is decreasing quadratically. We now proceed to make this more precise.

Definition 1 Say that z is an *approximate zero* of f if the sequence given by $z_0 = z$ and $z_{i+1} = N_f(z_i)$ is defined for all natural numbers i, and there is a ζ such that $f(\zeta) = 0$ with

$$|z_i - \zeta| \leq \left(\frac{1}{2}\right)^{2^i - 1} |z - \zeta|.$$

Call ζ the *associated zero*.

Given an approximate zero z of f with associated zero ζ and an upper bound M for $|z - \zeta|$, we may approximate ζ to any accuracy ϵ we wish by applying Newton's method to z and iterating k times, where k is the first integer greater than $\log(|\log \epsilon| + \log M)$; that is, $|N_f^k(z) - \zeta| < \epsilon$ for $k \geq \log(|\log \epsilon| + \log M)$. Thus locating an approximate zero of a function f gives us an effective termination procedure for an algorithm that approximates the zero to any preassigned accuracy. The two main theorems of this section, Theorems 1 and 2, give criteria for a point z to be an approximate zero of f. An important feature of these theorems is that the criteria are computable from f either at the roots of f or the point z alone, respectively.

First we need to define an auxiliary quantity. Let

$$\gamma = \gamma(f, z) = \sup_{k \geq 2} \left| \frac{f'(z)^{-1} f^{(k)}(z)}{k!} \right|^{1/k-1},$$

where we use $f^{(k)}$ to denote the kth derivative of f. This definition applies to analytic functions f. If f is analytic and $f'(z)^{-1}$ exists, then this sup exists as well since $f^{(k)}/k! = a_k$ has a geometric growth rate.

Theorem 1 *Suppose that $f(\zeta) = 0$ and that $f'(\zeta)^{-1}$ exists. If*

$$|z - \zeta| \leq \frac{3 - \sqrt{7}}{2\gamma} \quad \text{for} \quad \gamma = \gamma(f, \zeta),$$

then z is an approximate zero of f with associated zero ζ.

For the proof of this theorem we first prove two lemmas and a proposition.

Lemma 1 *For $0 \leq r < 1$,*

(a) $\displaystyle\sum_{i=0}^{\infty} r^i = \frac{1}{1 - r}$.

(b) $\displaystyle\sum_{i=1}^{\infty} i r^{i-1} = \frac{1}{(1 - r)^2}$.

Proof.   In (a) we have summed the geometric series which gives an analytic function of r. In (b) we have differentiated both sides of (a), term by term on the left. □

The following simple quadratic polynomial plays an important role in the estimates in this chapter.

$$\psi(u) = 1 - 4u + 2u^2. \tag{8.1}$$

Lemma 2 *If $|z_1 - z|\gamma(f, z) < 1 - (\sqrt{2}/2)$, then*

(a) $f'(z)^{-1} f'(z_1) = 1 + B$, where $|B| \leq \dfrac{1}{(1 - u)^2} - 1 < 1$;

(b) $|f'(z_1)^{-1} f'(z)| \leq \dfrac{(1 - u)^2}{\psi(u)}$,

where $u = |z_1 - z|\gamma(f, z)$.

Proof. (a) Recall $\gamma = \gamma(f, z)$. Then

$$f'(z)^{-1} f'(z_1) = f'(z)^{-1}(f'(z) + f''(z)(z_1 - z) + \ldots)$$
$$= f'(z)^{-1} \left(f'(z) + \sum_{k=2}^{\infty} \frac{f^{(k)}(z)(z_1 - z)^{k-1}}{(k-1)!} \right)$$
$$= 1 + B,$$

where

$$B = \sum_{k=2}^{\infty} k \frac{f'(z)^{-1} f^{(k)}(z)(z_1 - z)^{k-1}}{k!}.$$

Then $|B| \leq \sum_{k=2}^{\infty} k(\gamma |z_1 - z|)^{k-1} = (1/(1 - u)^2) - 1$ which is less than 1 since $u < 1 - (\sqrt{2}/2)$.

(b)

$$|f'(z_1)^{-1} f'(z)| = |(f'(z)^{-1} f'(z_1))^{-1}| = |(1 + B)^{-1}|$$
$$\leq \sum_{k=0}^{\infty} |B|^k \leq \frac{1}{1 - (\frac{1}{(1-u)^2} - 1)} = \frac{(1 - u)^2}{\psi(u)}$$

$\square$

Proposition 1 *Let $f(\zeta) = 0$, and let $u = |z - \zeta| \gamma(f, \zeta)$. Suppose $u < (5 - \sqrt{17})/4$. Then*

(a) $|N_f(z) - \zeta| < \dfrac{\gamma(f, \zeta)|z - \zeta|^2}{\psi(u)} = \dfrac{u|z - \zeta|}{\psi(u)}.$

(b) $|N_f^k(z) - \zeta| \leq \left(\dfrac{u}{\psi(u)} \right)^{2^k - 1} |z - \zeta|$ *for all $k \geq 0$.*

Proof. (a) Let $\gamma = \gamma(f, z)$. Then

$$f(z) = \sum_{k=1}^{\infty} \frac{f^{(k)}(\zeta)(z - \zeta)^k}{k!}$$

and

$$f'(z) = \sum_{k=1}^{\infty} \frac{f^{(k)}(\zeta)}{(k-1)!} (z - \zeta)^{k-1},$$

so

$$f'(z)(z - \zeta) - f(z) = \sum_{k=1}^{\infty} \left(\frac{1}{(k-1)!} - \frac{1}{k!} \right) f^{(k)}(\zeta)(z - \zeta)^k$$
$$= \sum_{k=1}^{\infty} (k-1) \frac{f^{(k)}(\zeta)}{k!} (z - \zeta)^k.$$

Then

$$|N_f(z) - \zeta| = |(z - \zeta) - f'(z)^{-1}(f(z))|$$

$$= |f'(z)^{-1} f'(\zeta) \sum_{k=1}^{\infty}(k-1)\frac{f'(\zeta)^{-1} f^{(k)}(\zeta)}{k!}(z - \zeta)^k|$$

$$\leq |f'(z)^{-1} f'(\zeta)| \, |z - \zeta| \sum_{k=1}^{\infty}(k-1)(\gamma|z - \zeta|)^{k-1}$$

$$\leq \frac{(1-u)^2}{\psi(u)} |z - \zeta| \left(\frac{1}{(1-u)^2} - \frac{1}{(1-u)} \right)$$

$$\leq \frac{u|z - \zeta|}{\psi(u)} \, .$$

Note that $u/\psi(u) < 1$ for $0 \leq u < (5 - \sqrt{17})/4$, using the quadratic formula. This proves (a) of Proposition 1.

Now, let us prove (b).

For $k = 0$ this is trivial. For $k \geq 1$ assume by induction that

$$|N_f^{k-1}(z) - \zeta| < \left(\frac{u}{\psi(u)} \right)^{2^{k-1}-1} |z - \zeta|$$

Then apply (a) to get

$$|N_f^k(z) - \zeta| < \frac{\gamma}{\psi(u)} \left(\left(\frac{u}{\psi(u)} \right)^{2^{k-1}-1} \right)^2 |z - \zeta|^2$$

$$< \left(\frac{u}{\psi(u)} \right)^{2^k-1} |z - \zeta|$$

and we are done. □

We can now give the proof of theorem 1.

Proof of Theorem 1. $(3 - \sqrt{7})/2$ is the first positive solution of $u/\psi(u) = 1/2$. Thus if $u < (3 - \sqrt{7})/2$, then $u/\psi(u) < 1/2$ and Proposition 1(b) finishes the proof. □

Remark 1 Proposition 1 implies that Newton's method converges if $u/\psi(u) < 1$; that is, $|z - \zeta|\gamma(f, \zeta) < (5 - \sqrt{17})/4$.

The constant is better than in Theorem 1, but z is not guaranteed to be an approximate zero.

The following result follows immediately from the preceding remark.

Corollary 1 *If ζ, ζ' are zeros of f, then they are separated by a distance that can be estimated from below by*

$$|\zeta' - \zeta| \geq \frac{5 - \sqrt{17}}{4\gamma(f, \zeta)}.$$

$\square$

Example 1 As an example of an application of Theorem 1 let us consider the problem of computing the dth roots of the unity; that is, we want to compute the roots of the polynomial

$$f(x) = x^d - 1.$$

Let $\zeta \in \mathbb{C}$ be such that $f(\zeta) = 0$. We have that

$$
\begin{aligned}
\gamma(f, \zeta) &= \sup_{k \geq 2} \left| \frac{f'(\zeta)^{-1} f^{(k)}(\zeta)}{k!} \right|^{1/(k-1)} \\
&= \sup_{k \geq 2} \left(\frac{d(d-1)\ldots(d-k+1)}{k!} \right)^{1/(k-1)} \\
&\leq \frac{d}{2}.
\end{aligned}
$$

According to Theorem 1 all points z such that

$$|z - \zeta| < \frac{3 - \sqrt{7}}{d}$$

are approximate zeros of f with associated zero ζ.

Remark 2 The invariant $\gamma(f, \zeta)$, Theorem 1, its proof, and its corollaries extend immediately to systems $f : \mathbb{C}^n \to \mathbb{C}^n$ and even to maps of Banach spaces. See Remark 7 at the end of the chapter. When we use Theorem 1 in Chapter 14 we refer to this extension.

8.2 Point Estimates for Approximate Zeros

Theorem 1 is useful if we have information about one or more of the roots of f, but we would like a criterion computable at the point z itself that guarantees that z is an approximate zero of f. To this end we define two more auxiliary quantities, the length of the Newton step

$$\beta(f, z) = |z - N_f(z)| = |f'(z)^{-1} f(z)|$$

and

$$\alpha(f, z) = \beta(f, z)\gamma(f, z)$$

In Theorem 2 we show that if $\alpha(f, z) < \alpha_0$ for some universal constant α_0, then z is an approximate zero of f. First we prove some preliminary propositions that are interesting in their own right. Proposition 2 estimates the reduction in the absolute value of f after one iterate of Newton's method. As a consequence of Theorem 4 we obtain the following result.

Theorem 2   *There is a universal constant α_0 with the following property. If $\alpha(f, z) < \alpha_0$, then z is an approximate zero of f in the sense of Definition 1. Moreover, the distance from z to the associated zero ζ is at most $2\beta(f, z)$.*

Remark 3 The invariant $\alpha(f, z)$ depends only on derivatives of f at the point z, which can be computed if f is a polynomial map. Thus Theorem 2 gives a criterion that can be used in principle and in practice to give certainty that z is indeed an approximation to a solution.

Proposition 2   *Let $z' = N_f(z)$. If $\alpha(f, z) < 1$, then*

$$\frac{|f(z')|}{|f(z)|} \leq \frac{\alpha(f, z)}{1 - \alpha(f, z)}.$$

Remark 4 This is the only result in this chapter that does not generalize to n-dimensional or Banach spaces. In the proof we use the fact that $f'(z)$ and $f^{(k)}(z)$ commute. It is not used in the rest of this chapter. However we use it in Section 9.2.

Proof.   Since $z' = N_f(z)$ one has $z' - z = -(f(z)/f'(z))$ and

$$f(z') = f(z) - f'(z)\left(\frac{f(z)}{f'(z)}\right) + \sum_{k=2}^{\infty}(-1)^k\frac{f^{(k)}(z)}{k!}\left(\frac{f(z)}{f'(z)}\right)^k$$

so

$$|f(z')| \leq |f(z)|\sum_{k=2}^{\infty}\left|\frac{f^{(k)}(z)}{k!f'(z)}\right|\left|\frac{f(z)}{f'(z)}\right|^{k-1}$$

$$\leq |f(z)|\sum_{k=2}^{\infty}\gamma^{k-1}(f, z)\beta^{k-1}(f, z)$$

$$\leq |f(z)|\frac{\alpha(f, z)}{1 - \alpha(f, z)} \qquad \text{as long as } \alpha(f, z) < 1.$$

□

The next proposition estimates α, β and γ for a point z_1 near z in terms of the values of these quantities at z.

Proposition 3   *If $u < 1 - (\sqrt{2}/2)$ and $|z_1 - z|\gamma(f, z) = u$, then*

(a) $\beta(f, z_1) \leq \dfrac{(1 - u)}{\psi(u)}((1 - u)\beta(f, z) + |z_1 - z|);$

(b) $\gamma(f, z_1) \leq \dfrac{\gamma(f, z)}{\psi(u)(1 - u)};$

(c) $\alpha(f, z_1) \leq \dfrac{(1 - u)\alpha(f, z) + u}{\psi(u)^2}.$

We use the following two lemmas to prove the proposition. The first is the kth derivative version of Lemma 1.

Lemma 3 *Let* $0 \leq r < 1$ *and* k *be a positive integer; then*

$$\sum_{\ell=0}^{\infty} \frac{(k+\ell)!}{k!\ell!} r^{\ell} = \frac{1}{(1-r)^{k+1}}.$$

Proof. By induction it is easy to see that

$$\sum_{\ell=0}^{\infty} \frac{(k+\ell)! r^{\ell}}{\ell!} = \left(\sum_{i=1}^{\infty} r^{i}\right)^{(k)} \qquad \text{and that} \qquad \left(\frac{1}{1-r}\right)^{(k)} = \frac{k!}{(1-r)^{k+1}}.$$

As in Lemma 1, $\left(\displaystyle\sum_{i=0}^{\infty} r^{i}\right)^{(k)} = (1/(1-r))^{(k)}.$ $\qquad\qquad\qquad\square$

Lemma 4 *If* $u < 1 - (\sqrt{2}/2)$ *and* $|z_1 - z|\gamma(f, z) = u$, *then*

(a) $\left|\dfrac{f'(z_1)^{-1} f^{(k)}(z_1)}{k!}\right| \leq \dfrac{1}{\psi(u)} \left(\dfrac{\gamma(f, z)}{1-u}\right)^{k-1}$ *for* $k \geq 2$;

(b) $|f'(z)^{-1} f(z_1)| \leq \beta(f, z) + \dfrac{|z_1 - z|}{1 - u}$.

Proof. **(a)** Write γ for $\gamma(f, z)$. Using the Taylor expansion of $f^{(k)}$ at z,

$$\left|\frac{f'(z_1)^{-1} f^{(k)}(z_1)}{k!}\right| \leq |f'(z_1)^{-1} f'(z)| \left|\frac{f'(z)^{-1}}{k!} \sum_{\ell=0}^{\infty} \frac{f^{k+\ell}(z)(z_1 - z)^{\ell}}{\ell!}\right|.$$

Using Lemma 2(b) and rearranging the terms in the second factor, this is

$$\leq \frac{(1-u)^2}{\psi(u)} \left|\sum_{\ell=0}^{\infty} \frac{(k+\ell)!}{k!\ell!} \frac{f'(z)^{-1} f^{(k+\ell)}(z)(z_1 - z)^{\ell}}{(k+\ell)!}\right|$$

$$\leq \frac{(1-u)^2}{\psi(u)} \sum_{\ell=0}^{\infty} \frac{(k+\ell)!}{k!\ell!} \gamma^{k+\ell-1} |z_1 - z|^{\ell} \text{ which by Lemma 3 is}$$

$$\leq \frac{(1-u)^2}{\psi(u)} \gamma^{k-1} \frac{1}{(1-u)^{k+1}}$$

$$\leq \frac{1}{\psi(u)} \left(\frac{\gamma}{1-u}\right)^{k-1};$$

(b)

$$|f'(z)^{-1} f(z_1)|$$

$$= \left| f'(z)^{-1} f(z) + (z_1 - z) + \sum_{k=2}^{\infty} \frac{f'(z)^{-1} f^{(k)}(z)}{k!} (z_1 - z)^k \right|$$

$$\leq |f'(z)^{-1} f(z)| + |z_1 - z| \left| 1 + \sum_{k=2}^{\infty} \gamma^{k-1} |z_1 - z|^{k-1} \right|$$

$$\leq \beta(f, z) + |z_1 - z| \left| 1 + \left(\frac{1}{1-u} - 1 \right) \right|$$

$$\leq \beta(f, z) + \frac{|z_1 - z|}{|1-u|}$$

$$= \beta(f, z) + \frac{|z_1 - z|}{1-u}.$$

The last step follows from the bound $u < 1$. $\qquad\square$

Proof of Proposition 3. (a)

$$\beta(f, z_1) = |f'(z_1)^{-1} f(z_1)| \leq |f'(z_1)^{-1} f'(z)| \, |f'(z)^{-1} f(z_1)|$$

$$\leq \frac{(1-u)^2}{\psi(u)} \cdot \left(\beta(f, z) + \frac{|z_1 - z|}{1-u} \right) \qquad \text{by Lemmas 2(b) and 4(b)}$$

$$= \frac{(1-u)}{\psi(u)} ((1-u)\beta(f, z) + |z_1 - z|).$$

(b) By definition

$$\gamma(f, z_1) = \sup_{k \geq 2} \left| \frac{f'(z_1)^{-1} f^{(k)}(z_1)}{k!} \right|^{1/(k-1)},$$

and by Lemma 4 (a) this is less than

$$\sup_{k \geq 2} \left(\frac{1}{\psi(u)} \right)^{1/(k-1)} \frac{\gamma(f, z)}{1-u}.$$

Since $\psi(u) < 1$ for $u < 1 - (\sqrt{2}/2)$, the supremum is achieved at $k = 2$ and we are done.

(c) Multiplying the inequalities in (a) and (b) proves (c). $\qquad\square$

Next we bound the derivative of Newton's map in terms of α.

Proposition 4 *For all analytic f and all z, $|N'_f(z)| \leq 2\alpha(f, z)$.*

Proof.

$$|N'_f(z)| = |f'(z)^{-1} f''(z) f'(z)^{-1} f(z)|$$

$$\leq 2 \left| \frac{f'(z)^{-1} f''(z)}{2} \right| |f'(z)^{-1} f(z)|$$

$$\leq 2\gamma(f, z)\beta(f, z) = 2\alpha(f, z).$$

$\qquad\square$

The next proposition states a fact about contraction maps of complete metric spaces X. For most of our applications X is a closed ball and $d(x, y) = |x - y|$. We use $B(r, z)$ to denote the closed ball of radius r around z defined by $B(r, z) = \{z' \mid d(z, z') \leq r\}$.

Definition 2 Suppose that X is a complete metric space. A map $f : X \to X$ satisfying that $d(f(x), f(y)) \leq cd(x, y)$ for all x, y in X with $c < 1$ is called a *contraction map* with *contraction constant* c.

Proposition 5 *Let $f : X \to X$ be a contraction map with contraction constant c. Then there is a unique fixed point $p \in X$, $f(p) = p$, and $f^n(x)$ converges to p as $n \to \infty$ for all x in X. Moreover, for any $x \in X$,*

$$\frac{d(x, f(x))}{1 + c} \leq d(x, p) \leq \frac{d(x, f(x))}{1 - c}.$$

Proof.   By induction it follows that $d(f^n(x), f^{n+1}(x)) \leq c^n d(x, f(x))$ for $n \geq 1$. By summing the geometric series it follows that for each $n \geq 1$,

$$d(f^n(x), f^m(x)) \leq \frac{c^n}{1 - c} d(x, f(x))$$

for all $m \geq n$. Since c^n tends to zero $\{f^n(x)\}_{n \geq 1}$ is a Cauchy sequence and, since X is complete, converges to a point p in X. The sequence $\{f^{n+1}(x)\}_{n \geq 1}$ also converges to p so by continuity of f, $f(p) = p$. Since $d(f(p), f(q)) \leq cd(p, q)$ it follows that p is the unique fixed point of f and that every orbit $f^n(x)$ converges to p as $n \to \infty$. Since $d(x, p) \leq d(x, f(x)) + d(f(x), f^2(x)) + \ldots \leq \sum_{n=0}^{\infty} c^n d(x, f(x))$, by summing the geometric sequence once again it follows that $d(x, p) \leq 1/(1 - c)d(x, f(x))$. Finally, by the triangle inequality,

$$d(x, f(x)) \leq d(x, p) + d(p, f(x))$$
$$= d(x, p) + d(f(p), f(x)) \leq (1 + c)d(x, p).$$

$\square$

Theorem 3 *If*

$$r < \frac{1 - \frac{\sqrt{2}}{2}}{\gamma(f, z)},$$

then

(a) *for all z_1 with $|z_1 - z| < r$,*

$$|N_f'(z_1)| \leq \frac{2(\alpha(f, z) + u)}{\psi(u)^2}, \quad u = r\gamma(f, z), \quad \psi \text{ as in (8.1)}.$$

(b) $N_f(B(r, z)) \subset B(r', N_f(z))$, *where* $r' = ((2(\alpha(f, z) + u))/\psi(u)^2)r$

Proof.   Part (a) follows immediately from Proposition 4 and Proposition 3(c). For Part (b) we need the following lemma.

Lemma 5   *Suppose $g : B(r, z) \rightarrow B(r, z)$ is continuously differentiable with $|g'(z_1)| \leq c$ for all $z_1 \in B(r, z)$. Then $|g(z_1) - g(z_2)| \leq c|z_1 - z_2|$ for all $z_1, z_2 \in B(r, z)$.*

Proof.   Let L be the straight-line segment connecting z_1 and z_2. So the length of L is $|z_1 - z_2|$ and $L \subset B(r, z)$. The distance $|g(z_1) - g(z_2)|$ equals the length of the straight-line segment connecting $g(z_1)$ and $g(z_2)$, which is the shortest differentiable curve joining them. Thus

$$|g(z_1) - g(z_2)| \leq \text{length } g(L) \leq \text{ length } L \cdot \max_{z' \in L} |g'(z')| \leq |z_1 - z_2| \cdot c$$

by the mean value theorem.   □

Proof of Theorem 3(b).   By Theorem 3(a) and Lemma 5,

$$|N_f(z_1) - N_f(z)| \leq \frac{2(\alpha(f, z) + u)}{\psi(u)^2}|z_1 - z|$$

$$\leq \frac{2(\alpha(f, z) + u)}{\psi(u)^2}r$$

for all z_1 in $B(r, z)$.   □

Corollary 2   *If $u < 1 - (\sqrt{2}/2)$, $c = (2(\alpha(f, z) + u))/(\psi(u)^2) < 1$ and $\alpha(f, z) + cu \leq u$, then N_f is a contraction map of the ball $B(u/(\gamma(f, z)), z)$ into itself with contraction constant c. Hence there is a unique root ζ of f in $B(u/(\gamma(f, z)), z)$ and all $z' \in B(u/(\gamma(f, z)), z)$ tend to ζ under iteration of N_f.*

Proof.   By Theorem 3(a) , c is a contraction constant on $B(u/(\gamma(f, z)), z)$. By Theorem 3(b) and the triangle inequality, if $\beta(f, z) + cu/(\gamma(f, z)) < u/(\gamma(f, z))$, then $N_f(B(u/(\gamma(f, z)), z)) \subset B(u/(\gamma(f, z)), z)$. But $\beta(f, z) + cu/(\gamma(f, z)) < u/(\gamma(f, z))$ follows from $\alpha(f, z) + cu < u$ by dividing by $\gamma(f, z)$. Now the rest of the proof follows from Proposition 5.   □

Corollary 2 gives us a good criterion in terms of α and γ for convergence of the iterates of Newton's map by a contraction map in a neighborhood of a point z. The next theorem gives a simpler criterion in terms of α and u. The three inequalities in Corollary 2 hold if α and u are small enough. Further restrictions on α and u guarantee that $B(u/(\gamma(f, z)), z)$ consists of approximate zeros.

Theorem 4 (Robust α Theorem)   *There are positive real numbers α_0 and u_0 such that: if $\alpha(f, z) < \alpha_0$, then there is a root ζ of f such that*

$$B\left(\frac{u_0}{\gamma(f, z)}, z\right) \subset B\left(\frac{3 - \sqrt{7}}{2\gamma(f, \zeta)}, \zeta\right)$$

and N_f maps $B(u_0/(\gamma(f, z)), z)$ into $B(u_0/(\gamma(f, \zeta)), \zeta)$ with contraction constant less than or equal to $1/2$.

Remark 5 It follows from Theorem 1 that $B(u_0/(\gamma(f,z)),z)$ consists of approximate zeros with associated zero ζ.

Proof.   Choose $\alpha_0 > 0$, $u_0 > 0$, $\ell_0 > 2$ to satisfy:

(i) $c_0 = \dfrac{2(\alpha_0 + u_0)}{\psi(u_0)^2} < \dfrac{1}{\ell_0}$;

(ii) $\alpha_0 + c_0 u_0 < u_0$;

(iii) $\left(\dfrac{\alpha_0}{1 - c_0} + u_0 \right) \left(\dfrac{1}{\psi(\frac{\alpha_0}{1-c_0})(1 - \frac{\alpha_0}{1-c_0})} \right) < \dfrac{3 - \sqrt{7}}{2}$;

(iv) $\dfrac{1}{\psi(\frac{\alpha_0}{1-c_0})(1 - \frac{\alpha_0}{1-c_0})} \leq \dfrac{\ell_0}{2}$.

Let ζ be the root of f given by (i), (ii), and Corollary 2. Then by Proposition 5,

(v) $|z - \zeta| \leq \frac{\beta(f,z)}{1-c_0}$.

By the triangle inequality, if $z' \in B(u_0/(\gamma(f,z)),z)$, then

$$|z' - \zeta| \leq \frac{\beta(f,z)}{1 - c_0} + \frac{u_0}{\gamma(f,z)} \ .$$

Multiplying by $\gamma(f,z)$ gives

$$|z' - \zeta| \cdot \gamma(f,z) \leq \frac{\alpha(f,z)}{1 - c_0} + u_0$$

and

$$|z' - \zeta| \cdot \gamma(f,\zeta) \leq \left(\frac{\alpha(f,z)}{1 - c_0} + u_0 \right) \frac{\gamma(f,\zeta)}{\gamma(f,z)} \ .$$

By Proposition 3(b) and (v) multiplied by $\gamma(f,z)$,

(vi) $\dfrac{\gamma(f,\zeta)}{\gamma(f,z)} \leq \dfrac{1}{\psi(\frac{\alpha_0}{1-c_0})(1 - \frac{\alpha_0}{1-c_0})}$

so

$$|z' - \zeta| \gamma(f,\zeta) < \left(\frac{\alpha(f,z)}{1 - c_0} + u_0 \right) \frac{1}{\psi(\frac{\alpha_0}{1-c_0})(1 - \frac{\alpha_0}{1-c_0})} < \frac{3 - \sqrt{7}}{2}$$

and

$$B\left(\frac{u_0}{\gamma(f,z)}, z \right) \subset B\left(\frac{3 - \sqrt{7}}{2\gamma(f,\zeta)}, \zeta \right) \ .$$

Moreover, by Corollary 2, (i), and (ii), $\zeta \in B(u_0/(\gamma(f,z)),z)$ and N_f has contraction constant less than $1/\ell_0$ on $B(u_0/(\gamma(f,z)),z)$. Hence if z_1 belongs to the ball $B(u_0/(\gamma(f,z)),z)$, then

$$|z_1 - \zeta| < \frac{2u_0}{\gamma(f,z)}$$

and

$$|N_f(z_1) - \zeta| \cdot \gamma(f,\zeta) \leq \frac{2}{\ell_0} \frac{u_0}{\gamma(f,z)} \gamma(f,\zeta) \leq \frac{2}{\ell_0} u_0 \cdot \frac{1}{\psi(\frac{\alpha_0}{1-c_0})(1 - \frac{\alpha_0}{1-c_0})} \leq u_0$$

by (iv) and (vi) and so we are done. $\square$

Remark 6 We may take $\ell_0 = 3$ and $\alpha_0 = .03$, $u_0 = .05$. This may be checked by substitution.

Theorem 2 now follows.

We close this chapter with a discussion about the level of generality of the results we have just proved.

Remark 7 We began this chapter assuming that f was an analytic function of one complex or real variable defined on all of $\mathbb{C}$ or $\mathbb{R}$. In fact, we have been careful to present our definitions, theorems, and proofs to be valid in a broader context. Now we explain the context.

We suppose that $\mathbb{E}$ and $\mathbb{F}$ are complete normed vector spaces, that is, Banach spaces, over the real or complex numbers. So $\mathbb{E}$ and $\mathbb{F}$ might be $\mathbb{R}^n$ or $\mathbb{C}^m$ or subspaces of them, or they might even be infinite-dimensional spaces such as $C^0([0, 1], \mathbb{R})$, the space of continuous functions ϕ with domain the closed unit interval $[0, 1]$ and taking real values. When dealing with elements of $\mathbb{E}$ or $\mathbb{F}$ where we have used absolute value it should be replaced by the norm so, for example, in $C^0([0, 1], \mathbb{R})$ a standard norm which makes it a complete normed vector space is $|\phi| = \sup_{x \in [0, 1]} |\phi(x)|$.

Next f is presumed to be defined and analytic on some open set $D \subset \mathbb{E}$ with values in $\mathbb{F}$. Where we have written f' it should be considered as a continuous linear operator $f' : \mathbb{E} \to \mathbb{F}$ which is the derivative of f. Then $f^{(k)}$ is the kth derivative of f and is a symmetric multilinear operator, operating on k-tuples of elements in $\mathbb{E}$ with values in $\mathbb{F}$. When the k-tuple has a vector x repeated ℓ times, $f^{(k)}x^\ell$ denotes the operator on $k - \ell$ tuples obtained by substituting x in ℓ places. In the definition of γ, $f'(z)^{-1} f^{(k)}(z)$ is a composition so that it operates on k-tuples of elements of $\mathbb{E}$ and takes values in $\mathbb{E}$. Absolute values of operators are understood to be operator norms; that is, for an operator A, its operator norm is

$$\|A\| = \sup_{x \neq 0} \frac{\|Ax\|}{\|x\|}.$$

That $f'(z)^{-1}$ exists means that $f'(z)$ has a continuous linear operator inverse. So that now

$$N_f'(z) = f'(z)^{-1} f''(z) f'(z)^{-1} f(z)$$

makes sense as a linear operator from $\mathbb{E}$ to itself and is indeed the derivative of Newton's map. That $f'(z) = 0$ means it is identically zero as a linear operator. Several places where we have written 1, such as in Lemma 2, should be read as the identity linear map.

The entire chapter now makes sense for analytic $f : \mathbb{E} \to \mathbb{F}$, where $\mathbb{E}$ and $\mathbb{F}$ are Banach spaces over the real or complex numbers. Our definitions, theorems, corollaries, lemmas, and propositions remain the same with the exception of Proposition 2 which is restricted to one dimension.

When our map f is defined on an open set $D \subset \mathbb{E}$ and not on all of $\mathbb{E}$, $f : D \to \mathbb{F}$, our theorems, corollaries, lemmas, and propositions remain valid with the additional hypothesis that the ball of radius $(1 - (\sqrt{2}/2))/\gamma(f, z)$ around the point z is contained in D. In fact it is natural to have the open ball of radius $1/\gamma(f, z)$ contained in D as the next proposition shows.

Proposition 6 *Let f be analytic at z and r be the radius of convergence of the Taylor series of f at z. Then $r \geq 1/\gamma(f, z)$.*

Proof. One has

$$r \geq \frac{1}{\limsup \left| \frac{f^{(k)}(z)}{k!} \right|^{1/k}}$$

and

$$\limsup \left| \frac{f^{(k)}(z)}{k!} \right|^{1/k} \leq \limsup |f'(z)|^{1/k} \left| \frac{f'(z)^{-1} f^{(k)}(z)}{k!} \right|^{1/k}$$

$$\leq \limsup \left| \frac{f'(z)^{-1} f^{(k)}(z)}{k!} \right|^{1/k}$$

$$\leq \limsup \left| \frac{f'(z)^{-1} f^{(k)}(z)}{k!} \right|^{1/(k-1)}$$

$$\leq \sup \left| \frac{f'(z)^{-1} f^{(k)}(z)}{k!} \right|^{1/(k-1)}$$

$$\leq \gamma(f, z).$$

□

In later chapters when we deal with multivariable polynomials or optimization problems we use our results on Newton's method in this wider context.

We end this chapter with a version of the inverse function theorem that is valid in this context and which gives an estimate of the size of the ball on which the inverse is defined in terms of γ.

If $f'(z)^{-1}$ exists, the inverse function theorem asserts that there is an inverse function f_z^{-1} defined on a ball B around $f(z)$, with the property that $f_z^{-1}(f(z)) = z$, $f(f_z^{-1}(w)) = w$ for all $w \in B$, and f_z^{-1} is differentiable. We use Theorem 4 to estimate the size of this ball.

Proposition 7 (Inverse Function Theorem) *Let $f : B(r, z_0) \to \mathbb{F}$ be analytic. Then, the open ball of radius $\alpha_0/(|f'(z_0)^{-1}|\gamma(f, z_0))$ about $f(z_0)$ is in the image of the open ball of radius $(1 - (\sqrt{2}/2))/\gamma(f, z_0)$ about z_0 and $f_{z_0}^{-1}$ exists and is differentiable on this ball.*

Proof. Let $c \in \mathbb{F}$ with $|c| \leq \alpha_0/(|f'(z_0)^{-1}|\gamma(f, z_0))$ and $f_c(z) = f(z) - c - f(z_0)$.

Then $\gamma(f_c, z_0) = \gamma(f, z_0)$ and

$$\beta(f_c, z_0) = |f'(z_0)^{-1} c| \leq |f'(z_0)^{-1}||c|.$$

Thus $\alpha(f_c, z_0) < \alpha_0$ and $N_{f_c}^k(z_0)$ converges to the unique root ζ_c of f_c in the open ball of radius $(1 - (\sqrt{2}))/\gamma(f, z_0)$ around z_0. Moreover $f(\zeta_c) = c + f(z_0)$ and $f'(\zeta_c)^{-1}$ exists by Lemma 2. The proposition follows.   □

8.3   Additional Comments and Bibliographical Remarks

Kantorovich pioneered a general modern treatment of Newton's method (see [Kantorovich and Akilov 1964]). The approach here is modeled after the paper by Smale [1986b] which emphasizes data at one point in contrast to Kantorovich hypotheses of estimates over a region.

Earlier work is [Traub and Woźniakowski 1979] and [Woźniakowski 1977]. Theorem 2 is proved in [Smale 1986b] with the constant $\alpha_0 = 0.130707$ and Kim [1988] has a one-dimensional version. The best constant for α_0 (equal to $\frac{1}{4}(13 - 3\sqrt{17}) \approx 0.157671$)) was subsequently found by Royden [1986] and Wang Xinghua with Han Danfu (see [Wang 1993]). In [Shub and Smale 1985, 1986] this last way is presented with the best constants for α_0. Curry [1989] and Chen [1994] have found generalizations.

The first version of *approximate zero* in the way it is used here is in [Smale 1981] with various developments in [Shub and Smale 1985, 1986]. Some applications and other developments of this α-theory were given in [Smale 1986a; Rheinboldt 1988; Renegar and Shub 1992; Ye 1994]. The approach to proving Theorems 3 and 4 is in [Shub and Smale 1994].

Dedieu [TAb] has a version of the separation theorem, Corollary 1. See also [Malajovich-Muñoz 1993].

This chapter covers only one approach to Newton's method, one which we find useful for complexity analysis. But there is a vast and fascinating literature on the subject of Newton's method which we have ignored here.

9

Fundamental Theorem of Algebra: Complexity Aspects

This chapter is devoted to producing a complexity analysis of a "homotopy method" for finding approximately a zero of a polynomial. The homotopy method is realized as a sequence of applications of Newton's method. This algorithm is a prototype of the main methods of numerical analysis for solving a nonlinear system of equations. The complexity analysis here (and extensions in later chapters) gives a depth of understanding of these methods which is missing in the usual treatment based on convergence proofs. In this chapter we only consider the one-variable case. We use from the previous chapter only the easily proved Proposition 2. We also use a result from the theory of Schlicht functions due to Loewner.

9.1   The Fundamental Theorem of Algebra

The fundamental theorem of algebra asserts that every nonconstant polynomial with complex numbers as coefficients has a complex zero. In other words, for every nonconstant complex polynomial f, there is a point ζ in the complex plane with $f(\zeta) = 0$.

In this section we give a proof of it. The proof gives rise to a numerical algorithm to find an approximate zero and we analyze the algorithm from a complexity perspective, using the theorems we have proved about Newton's method in the last chapter. We do not attempt to get the best constants or most efficient algorithms (even so our algorithms are quite efficient) but rather to illustrate some of the principles of a complexity theory for numerical analysis.

The main tool is the inverse function theorem for one variable which is used in the following form. Let f be a complex polynomial and $z \in \mathbb{C}$ with $f'(z) \neq 0$. Then there is a $\delta > 0$ and a complex differentiable (i.e., complex analytic) map

$$f_z^{-1} : D_\delta(f(z)) \to \mathbb{C}$$

with $f_z^{-1}(f(z)) = z$ and $f(f_z^{-1}(w)) = w$ for all $w \in D_\delta(f(z))$. Here $D_\delta(f(z))$ is the set of all w such that $|w - f(z)| < \delta$.

Moreover, if z ranges over a closed and bounded set $K \subset \mathbb{C}$ where f' is never zero, then the corresponding $\delta > 0$ can be chosen independent of $z \in K$. The proof of this last fact can be obtained by taking a convergent subsequence as in usual compactness arguments.

A complex polynomial of *degree d* has an expression

$$f(z) = \sum_{i=0}^{d} a_i z^i, \ a_d \neq 0, \ a_i \in \mathbb{C} \text{ for all } i. \tag{9.1}$$

Theorem 1 (Fundamental Theorem of Algebra) *Let f be a complex polynomial of degree d, $d \geq 1$. Then there is $\zeta \in \mathbb{C}$ with $f(\zeta) = 0$.*

First we prove two lemmas about such polynomials.

Lemma 1 *The polynomial f has at most d roots.*

Proof.   Suppose now that $f(\zeta) = 0$. Then by polynomial division we may write $f(z) = (z - \zeta)g(z)$ where g is a polynomial with degree $d - 1$. Now induction finishes the proof.   □

Lemma 2 *Let f be as in (9.1). If $f(\zeta) = 0$, then*

$$|\zeta| < 2 \max_{1 \leq k \leq d} \left(\left| \frac{a_{d-k}}{a_d} \right|^{1/k}, 1 \right).$$

Proof.   Let

$$b = \max_{1 \leq k \leq d} \left(\left| \frac{a_{d-k}}{a_d} \right|^{1/k}, 1 \right) \quad \text{and} \quad g(z) = \frac{1}{a_d b^d} f(bz).$$

Then

$$g(z) = \sum_{i=0}^{d} c_i z^i, \ c_d = 1, \ |c_i| \leq 1 \text{ all } i.$$

Thus

$$\left| \sum_{0}^{d-1} c_i z^i \right| < \sum_{k=0}^{d-1} |z|^k = \frac{|z|^d - 1}{|z| - 1}$$

and the last is less than $|z|^d$ for $|z| \geq 2$. Hence $g(z) \neq 0$ for $|z| \geq 2$ and $f(z) \neq 0$ for $|z|/b \geq 2$.   □

We recall that a map $f : \mathbb{C}^n \to \mathbb{C}^m$ is said to be *proper* when the pre-image $f^{-1}(S)$ of every closed and bounded set $S \subset \mathbb{C}^m$ is also closed and bounded. An immediate consequence of the preceding lemma is the following result.

Corollary 1 *A nonconstant polynomial $f : \mathbb{C} \to \mathbb{C}$ is proper.*   □

Definition 1 A *critical point* θ of a polynomial f is a point where the derivative f' vanishes. If f has degree $d > 0$, then f' is a polynomial of degree $(d - 1)$ and so has at most $(d - 1)$ zeros. Thus f has at most $(d - 1)$ critical points. A *critical value* of f is a point $f(\theta)$, where θ is some critical point. It follows that a polynomial of degree d has at most $(d - 1)$ critical values.

Toward the proof of Theorem 1, we may assume that 0 is not a critical value of f, since if $f'(\theta) = 0$ and $f(\theta) = 0$, θ is our desired solution.

Next we say that z_0 is a *good point* (of f) if the line segment

$$L = \{tf(z_0) \mid 0 \le t \le 1\} = L_{z_0} \tag{9.2}$$

contains no critical value of f.

Let us prove that there are good points. There are $z_1 \in \mathbb{C}$ that are not critical points of f. By the inverse function theorem, the image of a disk around z_1 contains a disk around $f(z_1)$. The image disk contains infinitely many points with infinitely many nonintersecting line segments. Not all of them can contain critical values. So there exists a good point z_0.

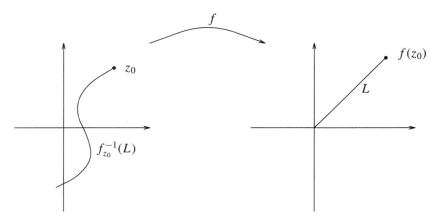

Figure A   The line L for z_0 and its pre-image by f_{z_0}.

Let L be as in (9.2) where z_0 is a good point. By the preceding corollary, $f^{-1}(L)$ is closed and bounded.

From the lemmas and the inverse function theorem, take $\delta_z > 0$ so that the inverse $f_z^{-1} : D_{\delta_z}(f(z)) \to \mathbb{C}$ is defined for each $z \in f^{-1}(L)$. Since $f^{-1}(L)$ is closed and bounded there is a $\delta > 0$ independent of z such that δ_z may be choosen bigger than δ. Let n be a positive integer with $n > |f(z_0)|/\delta$.

Define w_i, $i = 0, \ldots, n$, by

$$w_i = \frac{(n-i)f(z_0)}{n}$$

and observe that $|w_i - w_{i-1}| < \delta$, for $i = 1, \ldots, n$; thus

$$w_i \in D_\delta(w_{i-1}).$$

Now one can define inductively

$$z_i = f_{z_{i-1}}^{-1}(w_i), \quad i = 1, 2, \ldots, n.$$

Since $f(z_i) = w_i$, $f(z_n) = w_n = 0$. $\square$

Remark 1 Note that

$$f_{z_0}^{-1} : D_\delta(f(z_0)) \to \mathbb{C}$$

may be extended continuously to

$$D_\delta(f(z_0)) \cup D_\delta(f(z_1)) \to \mathbb{C}$$

by uniting it with $f_{z_1}^{-1}$. It can be shown that they agree on the overlap. In this way eventually $f_{z_0}^{-1}$ becomes defined continuously on the δ-neighborhood of L.

9.2   A Homotopy Method

We have shown the existence of zeros of a complex polynomial. We now begin our construction and analysis of an algorithm to locate them. The search problem is as follows

> Input: $(\varepsilon, a_0, \ldots, a_d)$, $\varepsilon > 0$, and $f(z) = \sum_{i=0}^{d} a_i z^i$
> Output: $x \in \mathbb{C}$ with the property $\left| \sum_{i=0}^{d} a_i x^i \right| < \varepsilon$.

Here $|f(x)| < \varepsilon$ is one possible definition of a solution x. There are many, and the literature on algorithms for this problem is rich. We describe an algorithm and associated constructions which we believe are insightful with widespread implications. This algorithm on one hand is close to and inspired by the proof of the fundamental theorem of algebra in the preceding section. On the other hand, it is a prototype of algorithms called "homotopy methods," "global Newton," and "continuation." The basic increment is one step of Newton's method for an appropriately related polynomial and initial point. We produce a complexity result in two stages, Theorem 2 in this section and Theorem 4 in Section 9.4. They exhibit well the situation described in Section 1.6.

We must be content with determining an approximation. Since Abel and Galois, it has been known that if the degree of f is greater than four, its zeros are not in general expressible as rational functions of its coefficients, and this remains true even allowing the additional power of extracting roots to our computational abilities.

We are now more precise. Given an accuracy $\varepsilon > 0$, and say $\varepsilon < 1$, $z_0 \in \mathbb{C}$, and f a nonconstant complex polynomial, let $f_t(z) = f(z) - tf(z_0)$, $0 \le t \le 1$. Compare this with Section 9.1. But in what follows we do not exclude the case that 0 is a critical value. Thus let ζ_t be a curve in $\mathbb{C}$, with $f_t(\zeta_t) = 0$, $\zeta_1 = z_0$, and $f_t'(\zeta_t) \ne 0$ for all $t \in (0, 1]$.

We describe a sequence t_i with $t_0 = 1$ and $t_0 > t_1 > \ldots > t_k$ in order to "follow" ζ_t. Let

$$x_0 = z_0 \,(= \zeta_1) \quad \text{and} \quad x_i = N_{f_i}(x_{i-1}) \quad \text{for } i \le k, \tag{9.3}$$

where $f_i = f_{t_i}$ (abuse of notation) and $N_{f_i}(x) = x - ((f_i(x))/(f_i'(x)))$ is Newton's endomorphism.

Thus one might hope that x_i is defined for all i and that x_i is a good approximation of the zero $\zeta_i = \zeta_{t_i}$ of f_i (another abuse of notation) and even that $|f(x_k)| < \varepsilon$. For $|t_i - t_{i-1}| \le \Delta$ small enough (implying that k is large) this will be the case. Our goal is to achieve this with a good bound on the "complexity" k, as a function of ε, f, z_0.

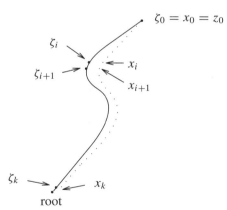

Figure B   The curve ζ_t and the dotted path of the x_i that approximate the points $\zeta_i = \zeta_{t_i}$.

The next step is to describe an invariant θ_{f_0, z_0} which gives a measure of how close our path ζ_t comes to a critical point of f. It is some kind of condition number of the pair (z_0, f), as described in Section 1.6.

Definition 2   Given $z_0 \in \mathbb{C}$ and a polynomial f, let the *wedge* $W = W_{f, z_0, \theta}$ of angle θ around $f(z_0)$ be defined by

$$W = \{w \in \mathbb{C} \mid 0 < |w| < 2|f(z_0)|, \arg \frac{w}{f(z_0)} < \theta\}.$$

Here $\arg(w/f(z_0))$ is the angle between w and $f(z_0)$ considered as vectors in $\mathbb{C}$.

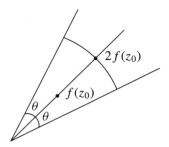

Figure C   The wedge of angle θ around $f(z_0)$.

Now we suppose that $f'(z_0) \neq 0$ so that the branch $f_{z_0}^{-1}$ of the inverse of f is well-defined on some disk containing $f(z_0)$. As we saw in the remark after the proof of the fundamental theorem of algebra, $f_{z_0}^{-1}$ may be extended to a larger domain. Let $\theta = \theta_{f,z_0}$ be the largest angle θ with the property that $f_{z_0}^{-1}$ extends to the wedge $W_{f,z_0,\theta} = W$. If, for example, there are no critical values in W, then such an extension always exists. Thus we can speak of the map $f_{z_0}^{-1} : W \to \mathbb{C}$; let $\mathcal{U} = \mathcal{U}_{f,z_0} = f_{z_0}^{-1}(W)$. We may also write $W = W_{f,z_0}$.

Note that there must be a critical value on the boundary of W. However there may be extraneous critical values inside the wedge, which are images of critical points not in $\mathcal{U}$, and perhaps far from ζ_t.

With f, z_0, ε given (as inputs say), let

$$M = M_{f,z_0} = \frac{25}{25 + \sin\theta_{f,z_0}}$$

$$t_i = M^i, \ i = 0, 1, 2, \ldots. \tag{9.4}$$

Theorem 2   *With $\{t_i\}$ as in (9.4), x_i is well-defined for all i in (9.3) and satisfies*

$$|f(x_i)| < 2M^i|f(z_0)| \qquad \text{for all } i. \tag{9.5}$$

Moreover if

$$k > \left(1 + \frac{25}{\sin\theta_{f,z_0}}\right)\left(\ln|f(z_0)| + \ln\frac{1}{\varepsilon} + 1\right), \tag{9.6}$$

then $|f(x_k)| < \varepsilon$.

Remark 2   Inequality (9.6) is an upper bound for the complexity k. Note that the bound is independent of the degree d of f. Compare this with Section 1.6. This theorem is used subsequently to obtain bounds for k depending on d and the magnitude of the coefficients of f, but not on θ_{f,z_0}.

Proof.   We first show how the last statement of the theorem is a consequence of (9.5). From (9.5) our condition to verify is:

$$\frac{1}{M^k} \geq \frac{2|f(z_0)|}{\varepsilon}.$$

Take logarithms to get

$$k \ln \frac{1}{M} > \ln |f(z_0)| + \ln \frac{1}{\varepsilon} + 1.$$

For $0 < s$,

$$\frac{1}{\ln(1+s)} \leq 1 + \frac{1}{s}$$

and the rest follows.

We prove (9.5) by induction on a slightly stronger statement. Let $f_i = f_{t_i}$ (abuse of notation) and consider the statements

(A_i) x_i is defined and $x_i \in \mathcal{U}_{f,z_0}$; $|f_i(x_i)| \leq (t_i - t_{i+1})|f(z_0)|$

for $i \in \mathbb{N}$. Here the (x_i) are as in (9.3). Note that (A_i) implies (9.5) since $|f_i(x_i)| = |f(x_i) - t_i f(z_0)| \leq (t_i - t_{i+1})|f(z_0)|$ and thus $|f(x_i)| \leq 2t_i|f(z_0)|$. (Recall $t_i = M^i$.) Moreover (A_0) is true since $f_0(x_0) = 0$. The following claim then proves Theorem 2.

Claim   *The sentences (A_i) are true for all $i \in \mathbb{N}$.*

As we have already noted (A_0) is true. We prove the claim by induction on i. We proceed in a series of lemmas assuming that (A_i) is true.

To ensure that $x_{i+1} \in \mathcal{U}$, we introduce three curves w_s, v_s, and u_s, for $s \in [0, 1]$ in the following way.

Let w_s be the arc in W_{f,z_0} consisting of two line segments joining:

$$f(x_i) \text{ to } f(\zeta_i) \quad 0 \leq s \leq \frac{1}{2},$$

$$f(\zeta_i) \text{ to } f(\zeta_{i+1}) \quad \frac{1}{2} \leq s \leq 1.$$

Next let $v_s = f_{z_0}^{-1}(w_s)$ lie in $\mathcal{U}$ "over w_s." Note that $v_0 = x_i$, $v_{1/2} = \zeta_i$, and $v_1 = \zeta_{i+1}$. Finally let $u_s = N_{f_{i+1}}(v_s)$, so that u_s is obtained by Newton's method for f_{i+1} applied to v_s for each $s \in [0, 1]$. Thus $u_0 = x_{i+1}$ and $u_1 = \zeta_{i+1}$ (the last since $f_{i+1}(\zeta_{i+1}) = 0$).

For a polynomial f and a complex number z, we defined in the previous chapter a number $\alpha = \alpha(f, z)$ about which we proved in Proposition 2 there that it satisfies the following property: if $\alpha < 1$, then $z' = N_f(z)$ is defined and

$$|f(z')| \leq \frac{\alpha}{1 - \alpha}|f(z)|. \tag{9.7}$$

The next proposition comes from Schlicht function theory, the theory of complex analytic functions which are one-to-one on the unit disk. Its proof would carry us too far afield, so we do not give it here. See the last section of this chapter for references for it.

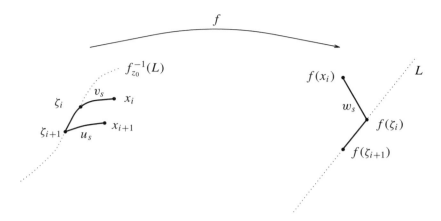

Figure D   The curves w_s, v_s, and u_s.

Proposition 1   *For every polynomial f and every $z \in \mathbb{C}$ we have*

$$\alpha(f, z) \le \frac{4|f(z)|}{r(f_z^{-1})}.$$

□

Here recall that f_z^{-1} is the branch of the inverse of the polynomial f that takes $f(z)$ into z and is defined on a disk of radius r by the implicit function theorem. The largest such r is called the radius of convergence of f_z^{-1} and denoted by $r(f_z^{-1})$.

The following lemmas use the preceding definitions, $\theta = \theta_{f,z_0}$, f_{i+1}, v_s, and M, and suppose that $s \in [0, 1]$.

Lemma 3 (Main Lemma)   *If (A_i) holds, then*

$$\alpha(f_{i+1}, v_s) \le \frac{8(1 - M)}{\sin \theta - (1 - M)}.$$

Toward the proof of the main lemma, we need some smaller lemmas.

Lemma 4   *If (A_i) holds, then*

$$|f(v_s) - f(\zeta_i)| \le (t_i - t_{i+1})|f(z_0)|.$$

Proof.   This is the same as the inequality $|w_s - f(\zeta_i)| \le (t_i - t_{i+1})|f(z_0)|$. It is sufficient to check this at the endpoints of the two segments defining w_s. This is quite immediate, using the hypothesis (A_i).   □

Recall that $f_i(v_s) = f(v_s) - f(\zeta_i) = f(v_s) - t_i f(z_0)$.

Lemma 5 *If (A_i) holds, then*

$$|f_{i+1}(v_s)| \le 2(t_i - t_{i+1})|f(z_0)|.$$

Proof.

$$
\begin{aligned}
f_{i+1}(v_s) &= f(v_s) - t_{i+1} f(z_0) \\
&= f(v_s) - t_i f(z_0) + (t_i - t_{i+1}) f(z_0).
\end{aligned}
$$

So

$$|f_{i+1}(v_s)| \le |f(v_s) - t_i f(z_0)| + (t_i - t_{i+1})|f(z_0)|.$$

Now use Lemma 4. □

Lemma 6 *As usual $\theta = \theta_{f,z_0}$. Then if (A_i) holds, one has*

$$r(f_{\zeta_i}^{-1}) \ge |f(\zeta_i)| \sin \theta.$$

Proof. The lemma follows from the diagram (Figure E) and the definitions.

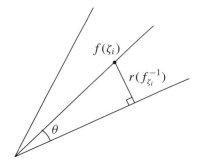

Figure E

□

Lemma 7 *If (A_i) holds, then*

$$r(f_{i+1,v_s}^{-1}) \ge |f(z_0)| M^i (\sin \theta - (1 - M)).$$

Proof. From the definitions,

$$r(f_{i+1,v_s}^{-1}) \ge r(f_{\zeta_i}^{-1}) - |f(v_s) - f(\zeta_i)|.$$

Now use Lemmas 4 and 6 noting again that $f(\zeta_i) = t_i f(z_0)$ and $t_i = M^i$. □

Now Lemmas 5 and 7 with Proposition 1 yield Lemma 3. Note that $r(f_{t,z}^{-1})$ is independent of t.

Lemma 8 *Suppose that (A_i) is true. If $\alpha = \alpha(f_{i+1}, v_s)$, $0 \le s \le 1$, then*

$$\frac{2\alpha}{1-\alpha} \le M.$$

Proof.   This is a gentle exercise using Lemma 3 and the definition of M in (9.4). We leave it to the reader.   □

Now we are ready to prove the Claim and with it, to finish the proof of Theorem 2.

Proof of the Claim.   Let us suppose that (A_i) holds. By inequality 9.7, for $s \in [0, 1]$, $|f_{i+1}(u_s)| \le \alpha/(1-\alpha)|f_{i+1}(v_s)|$, α as in Lemma 8. Now by Lemmas 5 and 8 we obtain

$$|f_{i+1}(u_s)| \le M(M^i - M^{i+1})|f(z_0)|$$
$$\le (t_{i+1} - t_{i+2})|f(z_0)|.$$

This estimate yields that $f(u_s) \in W_{f,z_0}$ for all $s \in [0, 1]$ and since $u_1 = \zeta_{i+1} \in \mathcal{U} = \mathcal{U}_{f,z_0}$, this means that $u_s \in \mathcal{U}$ for all s. In particular $u_0 = x_{i+1} \in \mathcal{U}$. Moreover for $s = 0$, the last estimate is that of A_{i+1} and with it we finish our proof.   □

9.3   Where to Begin the Homotopy

The goal of this section is to make an estimate, which yields many points z with fairly large $\theta_{f,z}$. This estimate is used in the next section to eliminate $\theta_{f,z}$ from the complexity estimate of Theorem 2. Throughout this section, suppose $f(z) = \sum_{i=0}^{d} a_i z^i$, $a_d = 1$, and $|a_i| \le 1$.
Let S_R^1 be the circle of radius R about 0 in $\mathbb{C}$.

Theorem 3   *The set of points $z \in S_R^1$ such that $\theta_{f,z} < a$ is contained in the union of $2(d-1)$ arcs of angle*

$$\frac{2}{d}(a + 2\arcsin\frac{1}{R-1}), \text{ for } R > 2.$$

To prove Theorem 3 we consider the family of curves in $\mathbb{C}$ defined by taking inverse images under the polynomial f of the family of rays in $\mathbb{C}$. If $w \in \mathbb{C}$ and $w \ne 0$, then the ray through w is $L = L_w = \{\lambda w \mid \lambda \in \mathbb{R}, \lambda > 0\}$. If $f(z) \ne 0$ and $f'(z) \ne 0$, then by the inverse function theorem f_z^{-1} is defined in a disk B around $f(z)$, where as usual f_z^{-1} takes $f(z)$ to z and $f(f_z^{-1}(z')) = z'$ for $z' \in B$; so f_z^{-1} maps $L_{f(z)} \cap B$ onto a smooth curve c_z through z. The vector $-f(z)$ is tangent to $L_{f(z)}$ so $(f_z^{-1})'(-f(z))$ is tangent to c_z at z. By the inverse function theorem $(f_z^{-1})' = (f'(z))^{-1}$ so the Newton vector $-f'(z)^{-1}f(z)$ is tangent to c_z at z. The c_z fit together into maximal connected curves $\widehat{c_z}$ that partition $\mathbb{C}$ minus the zeros and critical points of f, and $f(\widehat{c_z})$ is contained in the ray $L_{f(z)}$.

If θ is a critical point of f and $f^{(k)}$ is the first nonvanishing derivative of f at θ, then

$$f(z) = f(\theta) + (z - \theta)^k g(z), \tag{9.8}$$

where degree $g = d - k$ and $g(\theta) \neq 0$.

Recall that we may write a nonzero complex number z as $re^{i\phi}$, where r is the modulus $|z|$ of z, and ϕ which is defined up to 2π is the argument of z and is written $\arg(z)$. The representation of a complex number as $re^{i\phi}$ can be accomplished via the exponential mapping

$$e^z = 1 + z + \frac{z^2}{2!} + \ldots + \frac{z^n}{n!} + \ldots,$$

which has the property that

$$e^{z_1 + z_2} = e^{z_1} e^{z_2}.$$

In particular if $z = a + bi$ with $a, b \in \mathbb{R}$, then $r = e^a$ and $\phi = b$. Also, $\arg(z_1 z_2) = \arg(z_1) + \arg(z_2)$. If we denote by $\ln$ an inverse function to e^z, then $\arg(z) = \operatorname{Im} \ln(z)$, where $\operatorname{Im}(z)$ is the imaginary part of the complex number z. It then follows from (9.8) that

$$\arg(f(z) - f(\theta)) = k \arg(z - \theta) + \arg(g(z))$$

and that if we restrict f to a small circle $S^1(r, \theta) = \{\theta + re^{i\phi} \mid \phi \in \mathbb{R}\}$ then, taking derivatives, that

$$\frac{d(\arg f(\theta + re^{i\phi}) - f(\theta))}{d\phi} = k + \operatorname{Im} \frac{g'(\theta + re^{i\phi})}{g(\theta + re^{i\phi})} \cdot ire^{i\phi}.$$

We summarize some implications of the previous discussion in a lemma.

Lemma 9 *For f, θ as in the preceding with $f'(\theta) = 0$, $f(\theta) \neq 0$, and for r small enough, $\arg(f(\theta + re^{i\phi}) - f(\theta))$ is monotone in ϕ and f maps $S^1(r, \theta)$ k times around $f(\theta)$ intersecting both $L^+(f(\theta)) = \{\lambda f(\theta) \mid \lambda > 1\}$ and $L^-(f(\theta)) = \{\lambda f(\theta) \mid 0 < \lambda < 1\}$ k times in alternating fashion.*

Thus there are k curves $\widehat{c_z}$ that map to $L^+(f(\theta))$ and k that map to $L^-(f(\theta))$. If we orient the curves by $-f'(z)^{-1}f(z)$, then those mapping to $L^+(f(\theta))$ point into θ and those mapping to $L^-(f(\theta))$ point out of θ (see Figure F). □

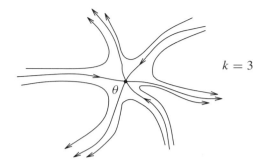

Figure F   The curves $\widehat{c}_z$ close to a critical point θ that is not a root.

It remains to consider the family of curves $\widehat{c}_z$ near roots ζ of f. First we recall that if $z_1 = x_1 + iy_1$, $z_2 = x_2 + iy_2$ where x_i, y_i are real for $i = 1, 2$, then

$$\mathrm{Re}(\bar{z}_1 z_2) = \mathrm{Re}(z_1 \bar{z}_2) = x_1 x_2 + y_1 y_2$$

and the last is the usual dot product of z_1 and z_2 considered as vectors in $\mathbb{R}^2$ which we denote in the rest of this chapter by $\langle z_1, z_2 \rangle$. The next lemma helps to understand the overall geometric picture.

Lemma 10   *One has*

$$-\mathrm{grad}\left(\frac{1}{2}|f(z)|^2\right) = -f(z)\overline{f'(z)}.$$

Proof.   For any $v \in \mathbb{C}$

$$
\begin{aligned}
\left\langle -\mathrm{grad}\left(\frac{1}{2}|f(z)|^2\right), v\right\rangle &= -\frac{1}{2}(\langle f(z), f(z)'v\rangle + \langle f'(z)v, f(z)\rangle) \\
&= -\langle f(z), f'(z)v\rangle = -f(z)\overline{f'(z)}\bar{v} \\
&= \langle -f(z)\overline{f'(z)}, v\rangle.
\end{aligned}
$$

$\square$

From Lemma 10 it follows that the Newton vector $-f'(z)^{-1}f(z)$ and $-\mathrm{grad}\left(\frac{1}{2}|f(z)|^2\right)$ differ by a positive real multiple since $|f'(z)|^2 = f'(z)\overline{f'(z)}$. Thus the curves $\widehat{c}_z$ are tangent to $-\mathrm{grad}\left(\frac{1}{2}|f(z)|^2\right)$ and orthogonal to the level curves of $|f(z)|^2$. Near a zero ζ these curves limit on ζ and their orientation points toward ζ (toward Figure G).

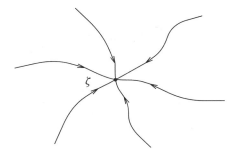

Figure G The curves $\widehat{c}_z$ close to a multiple root ζ.

Now we show that the curves $\widehat{c}_z$ cross and enter into any large circle centered at 0. This leads to the proof of Theorem 3.

Lemma 11 *Let $f(z) = z^d + a_{d-1}z^{d-1} + \ldots + a_0$ be a complex polynomial with $|a_i| \leq 1$ for $0 \leq i \leq d-1$. Let $R \geq 2$. Then for any $z \in S_R^1$,*

$$\mathrm{Re}\left(\bar{z}\frac{f(z)}{f'(z)}\right) > 0.$$

Proof. Let $w_1, \ldots, w_d$ be the roots of f; then $|w_i| < 2$ by Lemma 2. The w_i are in the half-plane defined by the tangent line to the circle at z; that is, $\mathrm{Re}(\bar{z}w_i) < \mathrm{Re}(\bar{z}z)$.

So $\mathrm{Re}(\bar{z}(z - w_i)) > 0$ for all i, $1 \leq i \leq d$ and $\mathrm{Re}(z/(z - w_i)) > 0$ for all i, $1 \leq i \leq d$ and $\mathrm{Re}\left(z\sum_{i=1}^{d} 1/(z - w_i)\right) > 0$; so

$$\mathrm{Re}\left(z\frac{f'(z)}{f(z)}\right) > 0 \text{ and } \mathrm{Re}\left(\bar{z}\frac{f(z)}{f'(z)}\right) > 0.$$

(In the proof we have used

$$\mathrm{Re}(\bar{u}v) > 0 \text{ iff } \mathrm{Re}(u\bar{v}) > 0 \text{ iff } |v|^2 \,\mathrm{Re}\left(u\frac{1}{v}\right) > 0.)$$

$\square$

From the last lemma if $z \in S_R^1$, then the inner product satisfies

$$\langle z, \frac{-f(z)}{f'(z)}\rangle = \mathrm{Re}\left(\frac{-\bar{z}f(z)}{f'(z)}\right) < 0.$$

Lemma 12 *Suppose $f(z) = z^d + a_{d-1}z^{d-1} + \ldots + a_0$ with $|a_i| \leq 1$ for $0 \leq i \leq d-1$. Let $z_1, z_2 \in S_R^1$ such that*

$$\beta \leq \left|\arg\frac{z_1}{z_2}\right|.$$

Then

$$\left| \arg \frac{f(z_1)}{f(z_2)} \right| \geq d\beta - 2 \arcsin \frac{1}{R-1}.$$

Proof.

$$\frac{f(z)}{z^d} = 1 + \frac{a_{d-1}z^{d-1} + \ldots + a_0}{z^d}$$

and for $z \in S_R^1$

$$\frac{\left| \sum_{j=0}^{d-1} a_j z^j \right|}{|z^d|} \leq \frac{\sum_{j=0}^{d-1} R^j}{R^d} = \frac{R^d - 1}{(R-1)R^d} < \frac{1}{R-1}.$$

Thus $f(z)/z^d$ is inside the circle of radius $1/(R-1)$ centered by 1 (see Figure H), and

$$\left| \arg \frac{f(z)}{z^d} \right| \leq \arcsin \frac{1}{R-1}.$$

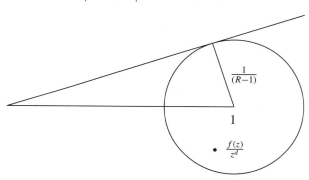

Figure H   The location of $f(z)/z^d$.

Now

$$\frac{f(z_1)}{f(z_2)} = \frac{z_1^d(f(z_1)/z_1^d)}{z_2^d(f(z_2)/z_2^d)} = \left(\frac{z_1}{z_2} \right)^d \frac{f(z_1)/z_1^d}{f(z_2)/z_2^d}$$

and the lemma follows from the fact that the argument of a product is additive and the triangle inequality.   □

Let θ be a critical point of f. Let L be the ray through $f(\theta)$, $L = \{\lambda f(\theta) \mid \lambda > 0\}$, and Σ_θ the component of $f^{-1}(L)$ that contains θ. Let

$$\Sigma = \bigcup_{\theta \mid f'(\theta)=0} \Sigma_\theta.$$

Lemma 13   *If $R \geq 2$ and f is a monic polynomial of degree d whose coefficients are bounded by 1 in absolute value, then $\Sigma \cap S_R^1$ is a set of at most $2(d-1)$ points.*

Proof.   Recall that the curves $\widehat{c}_z$ which are inverse images of rays by f are tangent to the point $(-f(z))/f'(z)$ which by Lemma 11 is transversal to S_R^1 for any $R \geq 2$. It follows that for any fixed θ_i such that $f'(\theta_i) = 0$, $\sum_{\theta_i} \cap S_R^1$ consists of at most $k_i + 1$ points where k_i is the multiplicity of θ_i *as a root of* f'. Thus

$$\sum_{\theta_i | f'(\theta_i)=0} k_i = d - 1 \quad \text{and} \quad \sum_{\theta_i | f'(\theta_i)=0} (k_i + 1) \leq 2(d - 1).$$

Equality is only obtained here if each critical point θ_i is a simple zero of f', that is, if each $k_i = 1$.   □

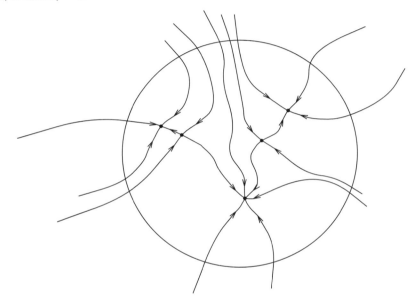

Figure I   An overall geometric picture.

Proof of Theorem 3.   Let $x_1, \ldots, x_l$ with $0 \leq l \leq 2(d-1)$ be the set of points, $\Sigma \cap S_R^1$, in Lemma 13. If for all i, $1 \leq i \leq l$,

$$\left| \arg \frac{z}{x_i} \right| \geq \frac{1}{d} \left(a + 2 \text{ arc sin } \frac{1}{R-1} \right),$$

then by Lemma 12

$$\left| \arg \frac{f(z)}{f(x_i)} \right| \geq a$$

and $\theta_{f,z} \geq a$. So if $\theta_{f,z} < a$, then there exist some x_i, $i = 1, \ldots, l$, such that

$$\left| \arg \frac{z}{x_i} \right| < \frac{1}{d} \left(a + 2 \text{ arc sin } \frac{1}{R-1} \right)$$

and this proves Theorem 3.   □

9.4   The Algorithm and Its Complexity

The method of Theorem 2 for finding roots depends on a choice of z_0. Using Theorem 3, we show how to make that choice intelligently. This leads to an algorithm with a good complexity bound for approximating a zero of a polynomial.

Now motivated by Theorem 3, let a and $R > 2$ satisfy

$$a + 2 \arcsin \frac{1}{R-1} < \frac{\pi}{4}. \qquad (9.9)$$

Then using Theorem 3, we may conclude that the set of points $z \in S_R^1$ with $\theta_{f,z} < a$ is contained in the union of $2(d-1)$ arcs, each of angle less than $\pi/2d$. Thus no two of the $4d$ evenly spaced points on S_R^1, $Re^{(2\pi j/4d)\sqrt{-1}}$, $j = 0, \ldots, 4d-1$ lie in the same one of these $2(d-1)$ arcs. So at least one-half of these points have the associated $\theta_{f,z} \geq a$.

Let us now consider M as given in (9.4) and k the smallest integer satisfying

$$k \geq \left(1 + \frac{25}{\sin a}\right)\left(d \ln R + \ln \frac{1}{\varepsilon} + 1\right). \qquad (9.10)$$

For these a, R, M, and k consider the algorithm:

      Input (ε, f), $f(z) = z^d + a_{d-1}z^d + \ldots + a_0$, $|a_i| \leq 1$
(1) Let k be as in (9.10).
(2) Choose $j \in \{0, 1, \ldots, 4d-1\}$ not previously chosen.
(3) Let $x_0 = Re^{(2\pi j\sqrt{-1})/4d}$. Let $x_i = N_{f_i}(x_{i-1})$, for $i = 1, \ldots, k$, be as in Theorem 2 (using $t_i = M^i$, etc.).
(4) If $|f(x_k)| < \varepsilon$, output x_k, else go to (2).

Note that Theorems 2 and 3 ensure that the algorithm is correct in the sense that it halts in Step (4). Thus, it performs at most $4d$ iterations of its main loop. On the other hand, Step (3) performs exactly k evaluations of the Newton's endomorphism. We can summarize this in our main theorem.

Theorem 4   *The algorithm just described returns an approximate root of f performing at most $4kd$ evaluations of the Newton's endomorphism k as in (9.10). If j in Step (2) is chosen at random, then on the average it performs only $2k$ such evaluations.*   □

Remark 3   One may choose $a = 0.628 < \pi/5$, $R = 12.745$ (that satisfies $2 \arcsin 1/(R-1) < \pi/20$). This choice yields for k the smallest integer greater than $43.56(2.55d - \ln \varepsilon + 1)$.

Remark 4   One may obtain the requisite number of arithmetic operations (and comparisons) by multiplying the number of evaluations of the Newton's endomorphism by the number of operations done at each one of these evaluations. This latter number is less than $3d$, and hence the complexity estimates of Theorem 4 get multiplied by $3d$. The total number of arithmetic operations is therefore —for the a and R of the preceding remark— bounded by

$$514d^2(2.55d + \ln \frac{1}{\varepsilon} + 1)$$

and thus within $O(d^2(d + \ln 1/\varepsilon))$.

Remark 5 The numbers $e^{(2\pi j\sqrt{-1})/4d}$ in Step (3) are not represented by rational numbers. But that is dealt with by simple approximations, and a little thought. The complexity estimates are not significantly affected.

Remark 6 The algorithm used in the first half of Theorem 4 is one in the sense of our machines over $\mathbb{R}$ defined in Part I, taking into account Remark 5. Then our result is a prototype of a polynomial time bound of a numerical analysis problem as suggested in Section 1.6.

Remark 7 The algorithm in the second part of Theorem 4 uses a random choice. In terms of machines over $\mathbb{R}$, this could be encompassed by adding an appropriate type of node, one which yields a random choice. Note the speedup factor of d.

Remark 8 Suppose one is given as input ε and $f(z) = \sum_0^d a_i z^i, a_d \neq 0$ without the normalizing conditions on the coefficients. Then one uses a preliminary procedure explicit in the proof of Lemma 2 to obtain a normalized polynomial g. In order to obtain an element x satisfying $|f(x)| < \varepsilon$ it is enough to obtain a y satisfying that

$$g(y) < \frac{\varepsilon}{a_d b^d}, \qquad \text{where} \qquad b = \max_{1 \leq i \leq d} \left(\left| \frac{a_{d-i}}{a_d} \right|^{1/i}, 1 \right).$$

Therefore, the $\ln(1/\varepsilon)$ of the complexity bound in Remark 4 now becomes $\ln(1/\varepsilon) + d \ln b + |\ln |a_d||$.

Remark 9 Note that in the implementation of this algorithm it is not necessary to store all the values x_i in Step (3). As a matter of fact, one can initialize a variable x with the value of x_0 and update this value for $i = 1$ to k.

Remark 10 The same algorithm slightly modified yields all the roots of a polynomial with a corresponding complexity analysis.

9.5   Additional Comments and Bibliographical Remarks

Proofs of the fundamental theorem of algebra are legion. The one here follows [Smale 1986b] closely but has roots in the thesis of Gauss (see [Smale 1981] for historical remarks). The statement in Remark 1 follows immediately from considerations proved in elementary complex variable books (see, e.g., Lang [1993b]).

The complexity result of Section 9.4 is a distillation of the efforts of Smale [1981], Shub and Smale [1985, 1986], and Smale [1986b]. However, the proof in Section 9.2 via Proposition 2 of Chapter 8 is an improvement over the previous accounts. Proposition 1 is derived very quickly in [Smale 1981] from a more difficult theorem of Loewner (and Loewner's theorem has four proofs in [Schober 1980]).

For Remark 10 on finding all the roots, one can see [Shub and Smale 1985; Shub and Smale 1986].

There is some history of complexity results for algorithms implementing the fundamental theorem of algebra as in [Dejon and Henrici 1969; Ostrowski 1973; Henrici 1977; Schönhage 1982]. Sharper results than Theorem 4 with different (and more complicated) algorithms are, for example, in [Renegar 1987b; Neff 1994; Neff and Reif 1996; Pan 1995, TA].

The main algorithm of this chapter is a homotopy or continuation method. This is the method of choice in recent times for solving systems of equations by numerical techniques. Here is some of the literature: [Kellog, Li, and Yorke 1976; Eaves and Scarf 1976; Smale 1976; Keller 1978; Garcia and Gould 1980; Hirsch and Smale 1979; Garcia and Zangwill 1981; Allgower and Georg 1990, 1993, 1997; Morgan 1987; Li, Sauer, and Yorke 1987]. These works mostly do not deal with complexity considerations. Later ones which do are [Renegar 1987a] and [Shub and Smale 1993a, 1993b, 1993c, 1996, 1994].

A main alternative class of methods for solving polynomial systems of equations are based on algebra (elimination theory, Gröbner bases, effective Nullstellensatz). We do not try to deal with them here.

10
Bézout's Theorem

Bézout's Theorem is the n-dimensional generalization of the Fundamental Theorem of Algebra. It "counts" the number of solutions of a system of n complex polynomial equations in n-unknowns. It is the goal of this chapter to prove Bézout's Theorem. In Chapter 16 we use Bézout's Theorem as a tool to derive geometric upper bounds on the number of connected components of semi-algebraic sets and complexity-theoretic lower bounds on some problems such as the Knapsack.

We proceed rather leisurely. First we revisit the Fundamental Theorem of Algebra and prove it again in such a way that the proof goes over to prove Bézout's Theorem in an extended geometric form. The proof proceeds by constructing a homotopy from a system with known roots. Most of the effort goes into showing that, generally speaking, pairs of systems of equations may be joined by a path that avoids the set of systems of equations with multiple roots, the discriminant variety. As in the proof of the Fundamental Theorem of Algebra, the homotopy gives rise to a path of solutions that may be followed numerically by Newton's method. We defer a complexity analysis of this approach to finding roots of systems of equations to Chapter 14.

10.1   The Fundamental Theorem of Algebra Revisited

Let $f(z) = a_d z^d + \ldots + a_0$ with $a_d \neq 0$ a complex polynomial. From the Fundamental Theorem of Algebra we know that there is a root r_1 of $f(z)$ and $a_d^{-1} f(z) = (z - r_1) g(z)$, where the degree of g is $d - 1$. Thus via induction there are d complex numbers $r_1, \ldots, r_d$ such that $f(z) = a_d(z - r_1)(z - r_2) \ldots (z - r_d)$.

There may be fewer than d distinct numbers in the list $r_1, \ldots, r_d$. The *multiplicity* of a root is the number of times it appears in the list $r_1, \ldots, r_d$ of roots. Thus another version of the Fundamental Theorem of Algebra asserts that a complex polynomial f has d roots, with the provisos that the roots are counted with multiplicity and $a_d \neq 0$.

In order to better understand the proviso $a_d \neq 0$ let us consider in some more detail the case $d = 2$, with real coefficients; that is,

$$f(z) = az^2 + bz + c \qquad a \neq 0.$$

It is well known that the two roots of this equation are

$$\zeta_+ = \frac{-b + \sqrt{b^2 - 4ac}}{2a} \quad \text{and} \quad \zeta_- = \frac{-b - \sqrt{b^2 - 4ac}}{2a}.$$

If we fix b and c with $b > 0$ and we let a tend to zero we find that ζ_+ tends to $-c/b$ and ζ_- "escapes" to infinity. One way to avoid this type of escape is to embed the space where the roots are found in a compact space where escape is impossible. The space obtained after this compactification, called projective space, contains some new points that are "at infinity." In one variable the addition of a single point is sufficient to capture the behavior of a polynomial at infinity. When dealing with systems of polynomials in several variables, the set of points at infinity in multidimensional projective space will be rich enough to reflect the direction that finite roots take in their escape to infinity.

Now let us be more precise about the way this compactification is done and the form that the Fundamental Theorem of Algebra takes in this compact space.

If $f(z, w) = a_d z^d + a_{d-1} z^{d-1} w + \ldots + a_0 w^d$ and some $a_i \neq 0$, we say that f is an *homogeneous polynomial* of z and w of degree d. We are no longer assuming $a_d \neq 0$. If (z, w) is a zero of an homogeneous polynomial, then so is $(\lambda z, \lambda w)$ for any $\lambda \in \mathbb{C}$. We express this in a lemma.

Lemma 1 *If $f(z, w)$ is an homogeneous polynomial of degree d, then*

$$f(\lambda z, \lambda w) = \lambda^d f(z, w) \text{ for all } \lambda \in \mathbb{C},$$

and hence if $f(z, w) = 0$, then $f(\lambda z, \lambda w) = 0$ for all $\lambda \in \mathbb{C}$.

We say that a (nontrivial) line $\{(\lambda z, \lambda w) \mid x \in \mathbb{C}\}$ is a *solution* of $f(z, w) = 0$ if

$$f(\lambda z, \lambda w) = 0 \text{ for all } \lambda \in \mathbb{C}.$$

The fundamental theorem of algebra now takes the following form.

Theorem 1 (The Fundamental Theorem of Algebra—Version 2.) *Let $f(z, w) = a_d z^d + a_{d-1} z^{d-1} w + \ldots + a_0 w^d$ be a nonzero complex homogeneous polynomial of degree d. Then f has d complex solutions provided the solutions are counted with multiplicity.*

Theorem 1 is intimately related to Theorem 1 of Chapter 9. Given a polynomial $f(z) = a_d z^d + a_{d-1} z^{d-1} + \ldots + a_0$ its *homogenization* is the homogeneous polynomial $\widehat{f}(z, w) = a_d z^d + a_{d-1} z^{d-1} w + \ldots + a_0 w^d$. If $f(z) = 0$, then $\widehat{f}(z, 1) = 0$, and if $\widehat{f}(z, w) = 0$, then $\widehat{f}((z/w), 1) = 0$ and $f(z/w) = 0$ as long as $w \neq 0$.

The zeros of $\widehat{f}(z, w)$ with $w = 0$ but $z \neq 0$ are called *zeros at infinity* for f. They occur only if $a_d = 0$. Thus except for zeros at infinity the zeros of f are given by intersecting the line $w = 1$ with the solution lines of $\widehat{f}$ (see Figure A).

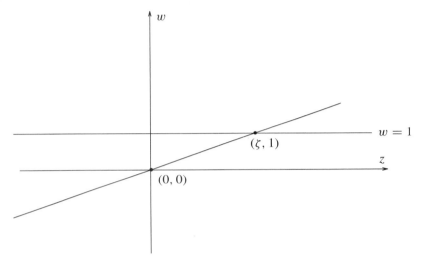

Figure A   A zero line of $\widehat{f}$, corresponding to the zero ζ of f.

If we let $\mathcal{P}_d$ be the vector space of all polynomials of degree less than or equal to d, and $\mathcal{H}_d$ the vector space of all homogeneous polynomials of degree d, then the map given by homogenization

$$\wedge : \quad \begin{matrix} \mathcal{P}_d & \to & \mathcal{H}_d \\ f & \to & \widehat{f} \end{matrix}$$

is a linear isomorphism. The inverse is obtained by setting $w = 1$. Both vector spaces $\mathcal{P}_d$ and $\mathcal{H}_d$ have dimension $(d + 1)$.

Using these ideas, it is not difficult to derive the second version of the Fundamental Theorem of Algebra from the first one in the preceding chapter. Notwithstanding, we devote the next three sections to a different proof of Theorem 1. The reason is that the ideas behind this latter proof naturally extend to polynomial systems in several variables and, eventually, they lead to an algorithm for computing approximate zeros of these systems.

Version 2 of the Fundamental Theorem of Algebra asserts that the homogeneous complex polynomial f of degree d has d solutions which are lines through zero in $\mathbb{C}^2$. Thus it is natural to define the space of lines through zero. We do this in general in the next section.

10.2   Projective Space

Definition 1 Let $\mathbb{C}^{n+1}$ be the $(n+1)$-dimensional complex Cartesian space so that $\mathbb{C}^{n+1} = \{(x_0, \ldots, x_n) \mid x_i \in \mathbb{C} \text{ for } i = 0, \ldots, n\}$. Then *n-dimensional complex projective space* $\mathbb{P}(\mathbb{C}^{n+1})$ is the set of one-dimensional vector subspaces of $\mathbb{C}^{n+1}$ (i.e., complex lines through zero) or we may say $\mathbb{P}(\mathbb{C}^{n+1})$ is the set of $x \in \mathbb{C}^{n+1}$, where we identify x with λx whenever λ is a nonzero complex number.

With this definition the map

$$\mathbb{C}^n \to \mathbb{P}(\mathbb{C}^{n+1})$$
$$x \to (1, x)$$

is an embedding of $\mathbb{C}^n$ into $\mathbb{P}(\mathbb{C}^{n+1})$ whose image is $\mathbb{P}(\mathbb{C}^{n+1})$ less the points of the form $(0, x)$ with $x \in \mathbb{C}^n - \{0\}$ (or the "points at ∞"). The addition of these points at infinity makes $\mathbb{P}(\mathbb{C}^{n+1})$ compact. "Escape to infinity" is no longer a problem. "Infinity" is in our space.

In fact, one is not actually using the coordinate structure so that if W is a finite-dimensional complex vector space, there is a naturally *associated projective space*, $\mathbb{P}(W)$. Thus, $\mathbb{P}(W)$ is defined as the set of nontrivial $w \in W$, where $v, w \in W$ are identified when $v = \lambda w$ for $\lambda \in \mathbb{C} - \{0\}$.

In general, there is a natural map $Q : W - \{0\} \to \mathbb{P}(W)$ which takes a nonzero vector v to the unique one-dimensional vector subspace in which it lies. Thus, $Q^{-1}(x)$ for $x \in \mathbb{P}(W)$ is a line through 0 minus $\{0\}$.

An *Hermitian product* on a complex vector space W is a map

$$W \times W \to \mathbb{C}$$
$$(v, w) \qquad \langle v, w \rangle$$

satisfying, for all $u, v, w \in W$, and $\lambda \in \mathbb{C}$,

1. $\langle v, w \rangle = \overline{\langle w, v \rangle}$,

2. $\langle u + v, w \rangle = \langle u, w \rangle + \langle v, w \rangle$,

3. $\langle \lambda v, w \rangle = \lambda \langle v, w \rangle$,

4. $\langle v, v \rangle \geq 0$ and $\langle v, v \rangle = 0$ implies $v = 0$.

The standard Hermitian product on $\mathbb{C}^{n+1}$ is given by

$$\langle v, w \rangle = \sum_{i=0}^{n} v_i \bar{w}_i,$$

where $v = (v_0, \ldots, v_n)$ and $w = (w_0, \ldots, w_n)$. An Hermitian product on a vector space W determines many properties of the vector space W which are related to it. We mention a few here which we use.

Two vectors $v, w \in W$ are *orthogonal* if $\langle v, w \rangle = 0$. It follows that if $v \in W$ and $v \neq 0$, then the set of vectors orthogonal to v,

$$T_v = \{w \in W \mid \langle v, w \rangle = 0\},$$

is a vector subspace of W. If W has dimension n, then T_v has dimension $n - 1$. A basis $w_1, \ldots, w_n$ of W is called orthogonal if $\langle w_i, w_j \rangle = 0$ for $i \neq j$. Any finite-dimensional vector space has an orthogonal basis.

An Hermitian product also determines a norm on W by

$$\|w\| = \langle w, w \rangle^{1/2}.$$

Once we have a norm on W we have the unit sphere $S_1 \subset W$,

$$S_1 = \{w \in W \mid \|w\| = 1\}.$$

An orthogonal basis $w_1, \ldots, w_n$ of W is called *orthonormal* if $\|w_i\| = 1$ for all $i = 1, \ldots, n$. If $v_1, \ldots, v_n \in V$ and $w_1, \ldots, w_n \in W$ are the orthonormal basis of V and W, respectively, then the unique linear map L from V to W taking v_i to w_i for $i = 1, \ldots, n$ is an isomorphism and preserves lengths; that is, $\|L(v)\| = \|v\|$ for all $v \in V$. Thus, L maps the unit sphere in V to the unit sphere in W and they are homeomorphic. Since the unit sphere S^{2n-1} in $\mathbb{C}^n$ is compact it follows that the unit sphere in any finite-dimensional vector space is compact.

We denote by ρ the restriction of Q to S_1,

$$\rho : S_1 \to \mathbb{P}(W).$$

Since ρ is continuous it follows that $\mathbb{P}(W)$ is also compact. For $x \in \mathbb{P}(W)$, $\rho^{-1}(x)$ is the unit circle in $Q^{-1}(x)$. Another way to express this is that the unit circle S^1 in $\mathbb{C}$ acts on S_1 and the quotient by this action is $\mathbb{P}(W)$.

Let L be a line through 0 in W and let $0 \neq w \in L$. The set

$$A_w = w + T_w$$

is an affine subspace of W. Since T_w has dimension $n - 1$ and w is orthogonal to T_w it follows that any line L' close to L intersects the space A_w in a unique point. If we let $\phi_w(L')$ be this point, then ϕ_w is defined in a neighborhood U_L of L in $\mathbb{P}(W)$ and is a homeomorphism there. Let $\psi_w = \phi_w^{-1}$. Then ψ_w is defined in a neighborhood V_w of w in A_w,

$$\psi_w : V_w \to \mathbb{P}(W).$$

In the language of differentiable manifolds the ψ_w are *charts* for $\mathbb{P}(W)$.

A map g defined on U_L taking values in a differentiable manifold N, $g : U_L \to N$, is *differentiable at L* if $g \circ \psi_w : V_w \to N$ is differentiable at w. In that case the derivative of $g \circ \psi_w$ is defined on T_w. Now just as $\mathbb{P}(W)$ is defined as equivalence classes of points $x \in W$ where we identify x with λx for $\lambda \in \mathbb{C}, \lambda \neq 0$, the tangent bundle of $\mathbb{P}(W)$ is defined as equivalence classes $(x, v) \in W \times W$ where $v \in T_x$ and (x, v) is equivalent to $(\lambda x, \lambda v)$ for $\lambda \in \mathbb{C}, \lambda \neq 0$. The necessity for the scaling in the vector v is evident in Figure B or from the chain rule.

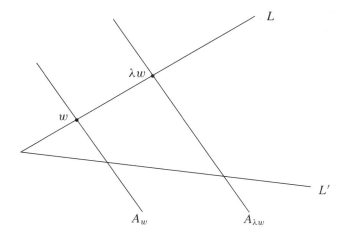

Figure B   The vector from L' to L in $A_{\lambda w}$ is λ times the vector from L' to L in A_w.

Remark 1 The definition of complex projective space we have given does not hinge on any special feature of the complex field and can be extended to vector spaces over arbitrary fields. In the case of the field of real numbers, the rest of the discussion holds as well replacing the Hermitian structure in $\mathbb{C}^{n+1}$ by the standard dot product in $\mathbb{R}^{n+1}$. We meet real projective spaces in Chapter 16.

We meet differentiable manifolds at several points in this book. We close this section with a formal definition of differentiable manifolds and maps. For a detailed treatment of such concepts see the references in Section 10.6.

Definition 2 A topological space M is an *n-dimensional differentiable manifold* if there is a set A, open sets $U_a \subset \mathbb{R}^n$ for $a \in A$, and functions $\psi_a : U_a \to M$ such that

1. $\psi_a : U_a \to M$ is a homeomorphism from U_a onto its image for all $a \in A$,

2. $\cup_{a \in A} \psi_a(U_a) = M$, and

3. $\psi_b^{-1}\psi_a$ is a smooth map on the open set in which it is defined for all $a, b \in A$.

The maps ψ_a are called *charts*. The set of charts $\{\psi_a\}$ for $a \in A$ is called an *atlas*. An atlas $\{\psi_a\}_{a \in A}$ is *maximal* if any homeomorphism $\Phi : U \to M$ with $U \subset \mathbb{R}^n$ an open subset such that $\psi_a \Phi$ and $\Phi \psi_a$ are smooth maps on the open sets on which they are defined for all $a \in A$ is already an element of $\{\psi_a\}_{a \in A}$.

Any atlas is contained in a maximal atlas, which defines a *differentiable structure* on M.

If M and N are differentiable manifolds, then $f : M \to N$ is *differentiable* if $\psi_a^{-1} f \psi_b$ is differentiable for all $\{\psi_a\}_{a \in A}$ and $\{\psi_b\}_{b \in B}$ atlases of M and N, respectively.

Let $\mathbb{R}^k \subset \mathbb{R}^n$ be the vector subspace of $\mathbb{R}^n$ consisting of those vectors whose last $(n - k)$ coordinates are 0.

Definition 3 A subset N of the n-dimensional differentiable manifold M is a *submanifold* of M if there are charts for M, $\psi_b : U \to M$ for $b \in B$, such that

1. $\psi_b : (U_b \cap \mathbb{R}^k) \subset N$ for all $b \in B$, and

2. $(\psi_b | U_b \cap \mathbb{R}^k) : U_b \cap \mathbb{R}^k \to N$ for $b \in B$ is an atlas for N.

10.3   The Varieties V, Σ', and Σ

Define $\mathcal{H}_d$ as the set of all homogeneous polynomials in $n + 1$ variables of degree d. Let the vector of degrees $(d) = (d_1, \ldots, d_k)$, fix $n \geq 2$, and let $\mathcal{H}_{(d)} = \mathcal{H}_{d_1} \times \ldots \times \mathcal{H}_{d_k}$, so $\mathcal{H}_{(d)}$ is the space of all polynomial mappings $f = (f_1, \ldots, f_k)$, $f : \mathbb{C}^{n+1} \to \mathbb{C}^k$ such that $f_i \in \mathcal{H}_{d_i}$ for all i. We may define the *solution variety*

$$V = \{(f, x) \in \mathbb{P}(\mathcal{H}_{(d)}) \times \mathbb{P}(\mathbb{C}^{n+1}) \mid f(x) = 0\}.$$

Then V is well-defined. It plays a crucial role in our proof. Here we use $f \in \mathbb{P}(\mathcal{H}_{(d)})$ and $x \in \mathbb{P}(\mathbb{C}^{n+1})$ by choosing representatives "f" $\in \mathcal{H}_{(d)}$ and "x" $\in \mathbb{C}^{n+1}$. This is legitimate essentially because our definitions are so natural.

We now establish that V is a smooth and connected submanifold of $\mathbb{P}(\mathcal{H}_{(d)}) \times \mathbb{P}(\mathbb{C}^{n+1})$ of complex codimension k whose tangent space at the point (f, x) is the vector space of (h, w), where $h \in T_f$, $w \in T_x$, and $h(x) + Df(w) = 0$.

Recall that if $f : M \to N$ is a differentiable map, then $q \in N$ is a *regular value* of f if and only if for every $p \in f^{-1}(q)$, the derivative of f at p is a surjective linear map. The derivative maps the tangent space of M at p to the tangent space of N at q. If $M = \mathbb{C}^m$ and $N = \mathbb{C}^n$, the tangent spaces are $\mathbb{C}^m$ and $\mathbb{C}^n$, respectively. For the manifolds and smooth varieties we consider, we usually try to describe the tangent space.

Proposition 1 *The solution variety V is a smooth connected subvariety of $\mathbb{P}(\mathcal{H}_{(d)}) \times \mathbb{P}(\mathbb{C}^{n+1})$ of codimension k. The tangent space to V at (f, x) is the vector space of (h, w) where $h \in T_f$, $w \in T_x$, and $h(x) + Df(x)(w) = 0$.*

Proof.   Let the evaluation map ev be defined by

$$
\begin{array}{ccc}
ev : & \mathcal{H}_{(d)} \times \mathbb{C}^{n+1} & \longrightarrow & \mathbb{C}^k \\
& (f, x) & \to & f(x).
\end{array}
$$

Then the derivative of the evaluation map at (f, x) applied to (h, w) is $D(ev_{(f,x)})(h, w) = h(x) + Df(x)w$. Taking $w = 0$ and varying h only we see that 0 is a regular value of ev. Let $V' = \{(f, x) \in \mathcal{H}_{(d)} - \{0\} \times \mathbb{C}^{n+1} - \{0\} \mid f(x) = 0\}$. Then V' is smooth by the inverse function theorem, its codimension is k, and its tangent space at (f, x) is the vector space of (h, w) such that $h(x) + Df(x)w = 0$.

We also claim that V' is connected. For $x \in \mathbb{C}^{n+1} - \{0\}$ let $V'_x = \{(f, x) \in V'\}$. The elements $f \in \mathcal{H}_{(d)} - \{0\}$ appearing in the first coordinate of V'_x form a linear subspace of $\mathcal{H}_{(d)} - \{0\}$ of codimension k. Hence V'_x is connected. Now we show that given $x, x' \neq 0$ there is a path connecting V'_x to $V'_{x'}$. Let $f \in V'_x$ and let L_t be a family of invertible linear maps $L_t : \mathbb{C}^{n+1} \to \mathbb{C}^{n+1}$ for $0 \leq t \leq 1$ such that the association $t \to L_t$ is continuous, $L_0(x) = x$, and $L_1(x) = x'$. Then $(f_t, x_t) = (f \circ L_t^{-1}, L_t x)$ is a path in V' connecting V'_x to $V'_{x'}$.

Let $\mathbb{C}^*$ be the multiplicative group of nonzero complex numbers; then $\mathbb{C}^* \times \mathbb{C}^*$ acts freely on $\mathcal{H}_{(d)} - \{0\} \times \mathbb{C}^{n+1} - \{0\}$ by $(\lambda, \mu)(f, x) = (\lambda f, \mu x)$. That the action is free means that $(\lambda, \mu)(f, x) \neq (\lambda', \mu')(f, x)$ unless $(\lambda, \mu) = (\lambda', \mu')$. This action leaves V' invariant. The equivalence relation defined by this action is the one identifying (f, x) and (g, y) if there is a (λ, μ) such that $(\lambda, \mu)(f, x) = (g, y)$. The quotient of $\mathcal{H}_{(d)} - \{0\} \times \mathbb{C}^{n+1} - \{0\}$ under this relation is $\mathbb{P}(\mathcal{H}_{(d)}) \times \mathbb{P}(\mathbb{C}^{n+1})$ and the quotient of V' is V. It follows that V is connected. To see that V is a smooth subvariety of $\mathbb{P}(\mathcal{H}_{(d)}) \times \mathbb{P}(\mathbb{C}^{n+1})$ we may first intersect V' with $S_1(\mathcal{H}_{(d)}) \times S_1(\mathbb{C}^{n+1})$, the product of the unit spheres in $\mathcal{H}_{(d)}$ and $\mathbb{C}^{n+1}$. Since V' contains the product of the lines through x and y for any $(x, y) \in V'$ it follows that V' is transversal to $S_1(\mathcal{H}_{(d)}) \times S_1(\mathbb{C}^{n+1})$ and hence $V'' = V' \cap (S_1(\mathcal{H}_{(d)}) \times S_1(\mathbb{C}^{n+1}))$ is a smooth manifold. The variety V is now the quotient of V'' by a free S^1 action. Since S^1 is compact it follows that V is a smooth manifold.

Now it follows that the codimension of V equals the codimension of V', and similarly that the tangent space is as asserted. □

Remark 2 We have taken care here to use the compact group S^1 because the quotients of smooth manifolds by free noncompact group actions, even free $\mathbb{R}$-actions, are not always smooth manifolds.

From now on we assume $k = n$. In that case $\dim V = \dim \mathbb{P}(\mathcal{H}_{(d)})$.

Our analysis concentrates on the restriction to V of the projection from $\mathbb{P}(\mathcal{H}_{(d)}) \times \mathbb{P}(\mathbb{C}^{n+1})$ onto the first factor. We denote this map by π_1:

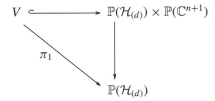

The map $\pi_1 : V \to \mathbb{P}(\mathcal{H}_{(d)})$ has the property that for each $f \in \mathbb{P}(\mathcal{H}_{(d)})$, $\pi_1^{-1}(f)$ is naturally identified with the set of zeros of f in $\mathbb{P}(\mathbb{C}^{n+1})$.

10.4 The Discriminant Variety in One Variable

We now return to the case $n = 1$. Then $\mathcal{H}_{(d)} = \mathcal{H}_d$. Version 2 of the fundamental theorem of algebra may be interpreted as saying that the fiber of $\pi_1 : V \to \mathbb{P}(\mathcal{H}_d)$

has d points everywhere as long as the points are counted with multiplicity. Now let us consider the derivative of $\pi_1 : V \to \mathbb{P}(\mathcal{H}_d)$. We are interested in the *critical variety* Σ', that is, the set $\Sigma' \subseteq V$ of points where this derivative is singular. For a nonzero tangent vector (h, v) at (f, x) in V to map to zero requires that h be zero or that $Df(x)v = 0$. That is, we simultaneously have that $f(x) = 0$ and $Df(x)(v) = 0$ where $\langle x, v \rangle = 0$. The following elementary lemma helps us identify the critical set Σ' of $\pi_1 : V \to P_{\mathcal{H}_d}$.

Lemma 2 *The pair $(f, x) \in \mathbb{P}(\mathcal{H}_d) \times \mathbb{P}(\mathbb{C}^2)$ belongs to Σ' if and only if $(\partial f/\partial z)(x) = (\partial f/\partial w)(x) = 0$ where $f(z, w)$ is an homogeneous polynomial and $x \in \mathbb{C}^2 - \{0\}$.*

Proof.   If $(f, x) \in \Sigma'$ we have seen that there exists $v \in \mathbb{C}^2 - \{0\}$ such that $\langle x, v \rangle = 0$ and $Df(x)v = 0$, but also $f(\lambda x) = 0$ for $\lambda \in \mathbb{C} - \{0\}$ so $Df(x)(x) = 0$. Thus $Df(x)$ is zero at x and $(\partial f/\partial z)(x) = (\partial f/\partial w)(x) = 0$. On the other hand, if $(\partial f/\partial z)(x) = (\partial f/\partial w)(x) = 0$, then $Df(x)$ is zero; so $Df(x)(x) = 0$ and $Df(\lambda x)(x) = \lambda^{d-1}Df(x)(x) = 0$ for all λ. So f is constant on the line through x and hence x is a zero of f; so $(f, x) \in V$. Alternatively, we can see that $f(x) = 0$ by Euler's identity

$$z\frac{\partial f}{\partial z}(x) + w\frac{\partial f}{\partial w}(x), = df(x)$$

where $x = (z, w)$. $\square$

It follows that Σ' is a subvariety of $\mathbb{P}(\mathcal{H}_d) \times \mathbb{P}(\mathbb{C}^2)$ of complex codimension 2 (it is defined by two independent equations), and hence Σ' is a subvariety of V of complex codimension 1.

Definition 4 The image $\pi_1(\Sigma') = \Sigma$ is called the *discriminant variety* $\Sigma \subset \mathbb{P}(\mathcal{H}_d)$. In other words, $f \in \Sigma$ if and only if f has a multiple root.

It is not difficult to give equations for the variety Σ itself.

Definition 5 Given two homogeneous polynomials f, g of degrees d and ℓ,

$$f(z, w) = \sum_{i=0}^{d} a_i z^i w^{d-i} \quad \text{and} \quad g(z, w) = \sum_{i=0}^{\ell} b_i z^i w^{\ell-i},$$

their *resultant* is the polynomial in their coefficients given by the determinant of the following $(\ell + d) \times (\ell + d)$ matrix (called the *Sylvester matrix*),

$$
\left.
\begin{matrix}
\ell \text{ rows} \left\{ \vphantom{\begin{matrix}a\\a\\a\end{matrix}} \right. \\
\\
d \text{ rows} \left\{ \vphantom{\begin{matrix}a\\a\\a\end{matrix}} \right. \\
\end{matrix}
\right.
\begin{pmatrix}
a_d & \cdots & a_{d-\ell+1} & \cdots & \cdots & a_0 & \cdots & & \\
& \ddots & & & & & & & \\
& & a_d & & & \cdots & a_1 & a_0 \\
b_\ell & \cdots & b_0 & \cdots & & & & \\
& \ddots & & & & & & \\
& & \ddots & & & & & \\
& & & b_\ell & \cdots & b_1 & b_0 \\
\end{pmatrix}.
$$

More precisely the entry c_{ij} in row i and column j in this matrix is given by

$$
c_{ij} = \begin{cases} a_{d-j+i} & \text{for } 1 \leq i \leq \ell \quad \text{and} \quad i \leq j \leq d+i \\ b_{i-j} & \text{for } \ell + 1 \leq i \leq \ell + d \quad \text{and} \quad i - \ell \leq j \leq i \\ 0 & \text{otherwise.} \end{cases}
$$

The *discriminant* of an homogeneous polynomial $f(z, w)$ of degree d is the homogeneous polynomial of degree $2d - 1$ in the coefficients of f defined by the resultant of $(\partial f/\partial z)$ and $(\partial f/\partial w)$.

The main property of the resultant is given in the following lemma.

Lemma 3 *The homogeneous polynomials f and g have a common zero in $\mathbb{P}(\mathbb{C}^2)$ if and only if their resultant is zero.*

Proof. The pair (r_1, r_2) is a solution of $f(z, w) = 0$ iff $(r_2 z - r_1 w)$ divides $f(z, w)$ and the same holds for g. It follows that f and g of degrees d and ℓ have a common solution if and only if there are nontrivial homogeneous polynomials k and h of degrees $\ell - 1$ and $d - 1$, respectively, such that $fk + gh = 0$. Viewing $(k, h) \rightarrow fk + gh$ as a linear map from $\mathcal{H}_{\ell-1} \times \mathcal{H}_{d-1} \rightarrow \mathcal{H}_{\ell+d-1}$ we see that the dimension of the spaces $\mathcal{H}_{\ell-1} \times \mathcal{H}_{d-1}$ and $\mathcal{H}_{\ell+d-1}$ are equal and therefore there is a nontrivial common zero of f and g if and only if the determinant of the linear map is zero. If

$$
f(z, w) = \sum_{i=0}^{d} a_i z^i w^{d-i} \quad \text{and} \quad g(z, w) = \sum_{i=0}^{\ell} b_i z^i w^{\ell-i},
$$

then this determinant is the Sylvester determinant and the lemma follows. $\square$

Proposition 2 *The discriminant variety $\Sigma \subset \mathbb{P}(\mathcal{H}_d)$ is the zero set of the discriminant polynomial.*

Proof. For any $f \in \mathbb{P}(\mathcal{H}_d)$ one has that $f \in \Sigma$ if and only if the resultant of $(\partial f/\partial z)$ and $(\partial f/\partial w)$ is zero. $\square$

Thus Σ is a subvariety of $\mathbb{P}(\mathcal{H}_d)$ and Σ also has complex codimension 1. Now it is easy to see that Σ cannot separate $\mathbb{P}(\mathcal{H}_d)$, or even any ball $B \subset \mathbb{P}(\mathcal{H}_d)$, into two connected components. For if distinct f, g lie in $B - (\Sigma \cap B)$, let L be the complex line $L = \{sf + tg \mid s, t \in \mathbb{C}\}$ considered in $\mathbb{P}(\mathcal{H}_d)$. Then $L \cap B$ is a disc in the Riemann sphere and $(L \cap B \cap \Sigma)$ has at most $2d - 1$ points. So we may connect f to g in $L \cap B$ by an arc that avoids Σ. Similarly, $\mathbb{P}(\mathcal{H}_d) - \Sigma$ is connected.

Proof of Theorem 1. First we note that if $\zeta = (\zeta_1, \zeta_2)$ is a solution of $f(z, w) = 0$, then $\zeta_2 z - \zeta_1 w$ divides $f(z, w)$. So $f(z, w)$ has at most d solutions. Now let $r_1, \ldots, r_d \in \mathbb{C}$ such that $r_i \neq r_j$ for $i \neq j$. Let $g(z, w) = (z - r_1 w) \cdots (z - r_d w)$. Then g has d solutions. We consider f and g in $\mathbb{P}(\mathcal{H}_d)$. Then $g \notin \Sigma$. If h is any other element of $\mathbb{P}(\mathcal{H}_d) - \Sigma$, h and g may be joined by an arc g_t, $0 \leq t \leq 1$, in

$\mathbb{P}(\mathcal{H}_d) - \Sigma$ such that $g_0 = g$ and $g_1 = h$. By the inverse function theorem applied to the map $\pi_1 : V - \pi_1^{-1}(\Sigma) \to \mathbb{P}(\mathcal{H}_d) - \Sigma$ we may lift the arc g_t to $(g_t, r_{it}) \subset V$ starting at $(g_0, (r_i, 1))$ for each i. Each of these d distinct arcs gives a solution of h for $t = 1$. Hence h has d solutions.

Thus if $f \in \mathbb{P}(\mathcal{H}_d) - \Sigma$, we are done. If otherwise $f \in \Sigma$, there is a sequence $g_i \in \mathbb{P}(\mathcal{H}_d) - \Sigma$, $i = 1, 2, \ldots$ such that g_i converges to f. Let ζ_i be a solution of g_i. Then since V is compact, the sequence (g_i, ζ_i) has a limit point (f, ζ) in V and $f(\zeta) = 0$. Thus f has at least one solution.

Now we show that the sum of the multiplicities of the roots of f is d. To do this we take a different definition of multiplicity, one which will go over to the proof of Bézout's Theorem. Let $\zeta_1, \ldots, \zeta_k$ be the roots of f and let U_i be disjoint open neighborhoods of the ζ_i, $i = 1, \ldots, k$. Since f is not zero in the compact set $\mathbb{P}(\mathbb{C}^2) - \bigcup_{i=1}^k U_i$, by continuity there is a ball B around f in $\mathbb{P}(\mathcal{H}_d)$ such that any $g \in B$ has all its roots in $\cup_{i=1}^k U_i$. Let $g \in B - \Sigma$ and let the multiplicity of ζ_i be the number of roots of g in U_i. The sum of the multiplicities is d since g has d roots. We now show that this definition makes sense. If g_0 and g_1 are both in $B - \Sigma$, they may be connected by an arc g_t in $B - \Sigma$ since Σ does not separate B. Now we may lift the arc g_t to d arcs in V as for the preceding homotopy (we already know that each g_t has d roots, since g_t is not in Σ). Since all the roots of g_t stay in $\cup_{i=1}^k U_i$, g_0 and g_1 have the same number of roots in each U_i. We leave the equivalence of the definitions of multiplicity as an exercise.   □

We conclude with one last point, that the multiplicity of any root is at least one in our definition. Given f, ζ_i, $i = \{1, \ldots, k\}$ as in the preceding, consider $(B \times U_i) \cap V$. This is an open set that contains (f, ζ_i), and contains points not in Σ'. Thus $\pi_1((B \times U_i) \cap V)$ contains an open set in $\mathbb{P}(\mathcal{H}_d)$ and $\pi_1((B \times U_i) \cap V)$ contains points $g \in B - \Sigma$. Such a g has a root in U_i.

Remark 3 From the inverse function theorem it follows that the roots of a polynomial vary differentiably in the neighborhood of a polynomial that is not in the discriminant variety. From the properties of the multiplicity index it follows that the set of unordered roots of a polynomial written with multiplicity vary continuously. We formalize these statements in the next corollary. Our roots lie in $\mathbb{P}(\mathbb{C}^2)$. So d unordered roots lie in $\mathbb{P}(\mathbb{C}^2)^d / S_d$ where S_d is the symmetric group that acts on $\mathbb{P}(\mathbb{C}^2)^d$ by permuting the factors. Our polynomial systems lie in $\mathbb{P}(\mathcal{H}_d)$. The topological space $\mathbb{P}(\mathbb{C}^2)^d / S_d$ is compact and is a differentiable manifold in the complement of the large diagonal Δ. As a consequence of Theorem 1 we obtain a differentiable version of the classical continuity of the roots as a function of the coefficients.

Corollary 1 *The function*

$$\sigma : \mathbb{P}(\mathcal{H}_d) \longrightarrow \mathbb{P}(\mathbb{C}^2)^d / S_d$$

which takes polynomial equations to their unordered roots is continuous in $\mathbb{P}(\mathcal{H}_d)$ and differentiable on $\mathbb{P}(\mathcal{H}_d) - \Sigma$.   □

10.5   Bézout's Theorem

Bézout's Theorem is the n-dimensional version of the Fundamental Theorem of
Algebra. It deals with n homogeneous polynomial equations in $n + 1$ unknowns.
The relationship with n polynomial equations in n unknowns is the same that we
have already seen setting one of the homogeneous variables equal to 1.

Even for two homogeneous linear equations in three unknowns we may have a
whole line of solution lines without the system of equations being identically zero.
In this case the solutions of nearby problems cannot vary uniformly continuously
with the problem. Arbitrarily close to a problem with a whole line of solutions are
problems with unique solutions that are more than $\pi/4$ apart. Since projective space
is compact and continuous functions on compact spaces are uniformly continuous,
the roots of polynomial systems cannot vary continuously. In general the solutions
of a system of n homogeneous polynomial equations $f = (f_1, f_2, \ldots, f_n)$ in $n+1$
unknowns may not be finite and may have quite complicated geometry. We denote
by $\mathcal{D}$ the *Bézout number* of f given by

$$\mathcal{D} = \prod_{i=1}^{n} d_i,$$

where d_i is the degree of f_i for $i = 1, \ldots, n$.

The definitions of Σ and Σ' of the previous section may be extended to the
n-dimensional case as follows.

The subvariety $\Sigma' \subset V$ is defined as the set of critical points of the projection
$\pi_1 : V \to \mathbb{P}(\mathcal{H}_{(d)})$ and $\Sigma = \pi_1(\Sigma')$. From Proposition 1 it follows that a point
$(f, x) \in \Sigma'$ if and only if $f(x) = 0$ and there is a $w \in T_x$, $w \neq 0$ such that
$Df(x)w = 0$; that is, x is a degenerate zero of f. Thus $f \in \Sigma$ if and only if f
has a degenerate zero. Although it is not immediately obvious that the conditions
$f(x) = 0$, $Df(x)w = 0$, $w \in T_x$, and $w \neq 0$ define an algebraic variety we may
rephrase them as $f(x) = 0$ and rank $Df(x) < n$ or yet as $f(x) = 0$ and $\det(M) = 0$
for all $n \times n$ minors M of $Df(x)$. This last is a system of polynomial equations and
we see that Σ' is an algebraic subvariety of the smooth connected variety V. The
image $\Sigma \subset \mathbb{P}(\mathcal{H}_{(d)})$ is likewise an algebraic subvariety, the discriminant variety.
(We prove that Σ is a variety in Theorem 2 of Appendix B.)

Note that the polynomial system $f_0 = (f_{01}, \ldots, f_{0n})$, where

$$f_{0i}(z_1, \ldots, z_{n+1}) = z_i^{d_i} - z_{n+1}^{d_i}$$

has $\mathcal{D}$ roots and they are all nondegenerate. Thus $f_0 \notin \Sigma$ and $\Sigma \neq \mathbb{P}(\mathcal{H}_{(d)})$.

We now state and prove Bézout's Theorem in two versions, an extended
geometric version and a more elementary version.

Theorem 2 (Bézout's Theorem)   *Suppose $f \in \mathcal{H}_{(d)} - \Sigma$. Then f has $\mathcal{D}$ zeros in*
$\mathbb{P}(\mathbb{C}^{n+1})$.

Proof.   It follows just as in the previous section that $\mathbb{P}(\mathcal{H}_{(d)}) - \Sigma$ is connected.
Now by the inverse function theorem it follows that any two systems f_1 and

$f_2 \in \mathbb{P}(\mathcal{H}_{(d)}) - \Sigma$ have the same number of zeros since they may be connected by an arc in $\mathbb{P}(\mathcal{H}_{(d)}) - \Sigma$, and this number is $\mathcal{D}$ from the preceding example. This finishes the proof of Theorem 2.   □

The next theorem extends Bézout's Theorem to the case of multiple or even infinitely many zeros. The index m is the multiplicity.

Theorem 3 (Bézout's Theorem—Extended Geometric Version) *Let* $f_i : \mathbb{C}^{n+1} \to \mathbb{C}$, $i = 1, \ldots, n$ *be homogeneous polynomials of degree* d_i, *not all identically zero, and let* $f = (f_1, \ldots, f_n)$. *Let* $Z_j = Z_j(f)$, $j = 1, \ldots, k$ *be the connected components of the set of zeros of* f *in* $\mathbb{P}(\mathbb{C}^{n+1})$. *Then we may assign an index* $m(Z_j)$ *to each* Z_j *that satisfies*

(a) $m(Z_j) \geq 1$ *for* $j = 1, \ldots, k$;

(b) $\sum_{j=1}^{k} m(Z_j) = \mathcal{D}$;

(c) $m(Z_j) = 1$ *for a nondegenerate isolated zero; and*

(d) *there are neighborhoods* U_j *of* Z_j *in* $\mathbb{P}(\mathbb{C}^{n+1})$, *for* $j = 1, \ldots, k$, *and* B *of* f *in* $\mathbb{P}(\mathcal{H}_{(d)})$ *such that if* $g = (g_1, \ldots, g_n) \in B$, *then* $Z(g) \subset \cup_{j=1}^{k} U_j$ *and* $\sum_t m(Z_t(g)) = m(Z_j)$ *where the sum is taken over all components of* $Z(g)$ *contained in* U_j.

Proof.   If $f \in \Sigma$, let $f_i \in \mathbb{P}(\mathcal{H}_{(d)}) - \Sigma$ be a sequence such that f_i converges to f. Let $(f_i, \zeta_i) \in V$ and (f, ζ) a limit point of the sequence (f_i, ζ_i). Then $f(\zeta) = 0$. So $Z(f)$ is not empty.

Now let $Z_j = Z_j(f)$, $j = 1, \ldots, k$ be the connected components of the zeros of $f \in \mathbb{P}(\mathcal{H}_{(d)})$, where f may be in Σ. Let U_j be disjoint open sets with $Z_j \subset U_j$, $j = 1, \ldots, k$. By continuity of the polynomial system f it follows that there is a ball B around f in $\mathbb{P}(\mathcal{H}_{(d)})$ such that all the zeros of any $g \in B$ are contained in $\cup_{j=1}^{k} U_j$. Suppose $g \in B - \Sigma$ and define $m(Z_j)$ to be the number of zeros of g in U_j. This is well-defined since $B - \Sigma$ is connected. Properties (b), (c), and (d) of the index m follow directly from the definition.

We now turn to the proof of (a). Let $V_j = (B \times U_j) \cap V$. Then V_j is an open set in V that contains $\{f\} \times Z_j$. Since Σ' is a proper subvariety of the connected variety V, V_j cannot be contained in Σ. Hence the projection of V_j into B contains an open set and a point $g_j \in B - \Sigma$. By definition $\pi_1^{-1}(g_j)$ has at least one point in V_j; that is, g_j has at least one solution in U_j and $m(Z_j) > 0$.   □

10.6   Additional Comments and Bibliographical Remarks

Bézout's Theorem is a classical theorem in algebraic geometry. See, for example, [Mumford 1976; Shafarevich 1977]. The approach taken here follows closely

[Shub 1993b] and [Shub and Smale 1993a, 1993b, 1993c, 1996, 1994]. The resultant of two one-variable polynomials is described in [Lang 1993a] and more generally in [Van der Waerden 1949] (old edition of "modern algebra" and "algebraic geometry"). See [Griffiths and Harris 1978] for background on projective space and complex algebraic geometry. Background in differential topology can be found in [Guillemin and Pollack 1974; Hirsch 1976; Milnor 1965].

11

Condition Numbers and the Loss of Precision of Linear Equations

The condition number of an invertible real or complex $n \times n$ matrix A is defined as

$$\kappa(A) = \|A\| \, \|A^{-1}\|,$$

where $\|A\|$ is the operator norm

$$\|A\| = \sup_{x \neq 0} \frac{\|Ax\|}{\|x\|}$$

and $\mathbb{R}^n$ or $\mathbb{C}^n$ is given the usual inner product. The condition number measures the relative error in the solution of the system of linear equations

$$Ax = v.$$

If the error in v is δv causing an error δx so that $A(x + \delta x) = v + \delta v$, then

$$\frac{\|\delta x\|}{\|x\|} \leq \kappa(A) \frac{\|\delta v\|}{\|v\|}. \tag{11.1}$$

Here is the argument

$$\frac{\|\delta x\|}{\|x\|} = \frac{\|\delta x\|}{\|v\|} \frac{\|v\|}{\|x\|} = \frac{\|A^{-1}\delta v\|}{\|v\|} \frac{\|Ax\|}{\|x\|} \leq \|A^{-1}\| \|A\| \frac{\|\delta v\|}{\|v\|} = \kappa(A) \frac{\|\delta v\|}{\|v\|}.$$

Moreover, it is easily seen that there are vectors v and δv such that $\|\delta x\|/\|x\| = \kappa(A)\|\delta v\|/\|v\|$, where $x = A^{-1}v$ and $\delta x = A^{-1}(v + \delta v) - x$. So $\kappa(A)$ is a sharp worst case estimate of the relative error in x, as a function of the relative error in v.

Condition numbers play a wide role in numerical linear algebra. For example, they are used to estimate the relative error in x, not only due to the relative error in v, but as a function of the relative error in A as well. They also are used to estimate the relative error in x due to rounding errors in Gaussian elimination, or after k steps of the conjugate gradient algorithm for $k < n$.

If we take log to base 10 or 2, of (11.1) then we see that if v has t digits or bits of accuracy, then x has $t - \log \kappa(A)$ digits or bits of accuracy. If the number of digits or bits of accuracy is the input precision, then $\log \kappa(A)$ measures the *loss of precision*.

If the loss of precision were to exceed the machine precision of current computers for most instances, algorithms for the solution of linear equations would not be very meaningful from a computational perspective.

The main results of this chapter, Theorems 3 and 5, estimate the average loss of precision for random matrices whose entries are independently distributed with respect to the standard normal distribution.

Main Theorems.   *The average loss of precision is at most*

$$5/2 \log_b n + \frac{1}{\ln b}$$

for real $n \times n$ matrices and

$$2 \log_b n + \frac{1}{2 \ln b}$$

for complex $n \times n$ matrices.

Here b is the base of the log defining the loss of precision. Thus, in general, even moderate input precision will be sufficient to give meaningful output for quite large problems.

It is immediate to check that the condition number of any nonzero one-by-one matrix is one and therefore the loss of precision is zero. Consequently, in the rest of the chapter we focus on $n \times n$ matrices for $n \geq 2$.

11.1   The Eckart–Young Theorem

We let Σ be the subset of matrices of determinant 0. Thus Σ corresponds to the set of ill-posed problems, problems that may not have solutions or for which solutions do not vary continuously. As we have remarked in Chapter 10, even if we homogenize our system of n linear equations in n unknowns so that solutions always exist, the solutions do not vary continuously near Σ. Let

$$d_F(A, \Sigma) = \min_{M \in \Sigma} \|A - M\|_F.$$

Here $\|A\|_F$ denotes the Frobenius or Hilbert–Schmidt norm of A,

$$\|A\|_F = \sqrt{\text{trace}(AA^*)} = \sqrt{\sum_{i,j=1}^{n} |a_{ij}|^2},$$

where a_{ij} are the entries of A and A^* is the adjoint of A, that is, the transpose A^T of A for real matrices and the transpose conjugate $\overline{A}^T$ of A for complex matrices. The norm $\|A\|_F$ arises from the Hermitian product $\langle A, B \rangle = \text{trace}(AB^*) = \sum_{i,j=1}^{n} a_{ij} \bar{b}_{ij}$.

Theorem 1 (Eckart–Young) *Let A be an invertible matrix; then*

$$d_F(A, \Sigma) = \frac{1}{\|A^{-1}\|}.$$

This is the first of a series of theorems we prove relating the condition of a problem to its distance to the ill-posed problems.

Our proof of the Eckart–Young Theorem exploits the invariance of $\|A\|_F$ under composition with orthogonal matrices.

Lemma 1 *Let A be a real $n \times n$ matrix and B an orthogonal $n \times n$ matrix. Then*

$$\|BA\|_F = \|AB\|_F = \|A\|_F.$$

Proof. Note that multiplication by B on the right preserves the length squared of the rows of A and on the left preserves the length squared of the columns of A; thus the Frobenius norm of A is preserved. $\qquad\square$

Now we turn to the proof of the Eckart–Young theorem.

Proof of Theorem 1. Let $v \in \mathbb{R}^n$ such that $\|v\| = 1$ and $\|A^{-1}(v)\| = \|A^{-1}\|$. Let $u = (1/\|A^{-1}\|)A^{-1}(v)$. Then $\|u\| = 1$ and $\|Au\| = 1/\|A^{-1}\|$. Composing by an orthogonal transformation B on the right so that $B(e_1) = u$ we have that Au is the first column of AB. Changing the first column of AB to be zero we see $d_F(AB, \Sigma) \leq 1/\|A^{-1}\|$. By Lemma 1 we have that $d_F(A, \Sigma) = d_F(AB, \Sigma)$ so, $d_F(A, \Sigma) \leq 1/\|A^{-1}\|$. To prove the opposite inequality let $M \in \Sigma$ and $d_F(A, \Sigma) = \|A - M\|_F$. Let $M(v) = 0$ for $\|v\| = 1$. Let B be an orthogonal matrix mapping e_1 to v. Then the first column of MB is zero and therefore $\|AB(e_1)\| \leq \|AB - MB\|_F = d_F(A, \Sigma)$. So

$$\|A^{-1}\| \geq \frac{\|B(e_1)\|}{\|AB(e_1)\|} \geq \frac{1}{d_F(A, \Sigma)}$$

and we are done. $\qquad\square$

Remark 1 The proof for complex invertible $n \times n$ matrices is the same except we use unitary matrices instead of orthogonal matrices. Lemma 1 holds with B unitary and A an $n \times n$ complex matrix.

11.2   Probabilities and Integrals

In this section we introduce some basic concepts of probability theory. All the
probabilities we consider are given by point density functions, as follows.

Definition 1 Let M be $\mathbb{R}^m$, $\mathbb{C}^m$, the unit sphere $S^{m-1} \subset \mathbb{R}^m$, or even a differentiable
manifold with a volume form $d\omega$. Let $f : M \to \mathbb{R}_+$ be defined almost everywhere
on M, say in the complement of a finite union of lower-dimensional submanifolds
of M, and suppose that f is sufficiently regular to be integrable with $\int_M f\,d\omega = 1$.
Then f is a *point density* function and we define for $A \subset M$ (a domain on which
we may integrate) the *probability measure* of A to be

$$m(A) = \int_A f\,d\omega.$$

We have written a single integral sign to represent the multiple integral of f
instead of repeating the integral sign as many times as the dimension of M as is
frequently done in calculus texts. We use the repeated integral sign to represent
repeated or iterated integration that is performed one variable at time. Of course,
it is an important theorem that in Euclidean space the multiple integral equals the
repeated integral. We call this theorem and some of its generalizations *Fubini's
Theorem*. Fubini's Theorem depends on the fact that the area of a rectangle is the
product of the area of its sides. Sometimes when we try to decompose a space
into factors this simple product formula for volumes is violated and the repeated
integral formula needs extra factors. We begin to confront this situation in the next
section.

Now let us see some examples of probability measures given by point density
functions.

Example 1 The uniform distribution on the unit cube $[0, 1]^n$ is given by

$$m(A) = \int_A 1\,dx.$$

Example 2 Analogously, the uniform distribution on the unit sphere S^{n-1} is given
by

$$m(A) = \frac{1}{\text{Vol}\,(S^{n-1})} \int_A 1\,d\Theta,$$

where $d\Theta$ is the volume on the sphere.

Example 3 The standard Gaussian probability distribution on the real line is given
by

$$m(A) = \frac{1}{(2\pi)^{1/2}} \int_A e^{-(x^2/2)}\,dx, \qquad A \subset \mathbb{R}.$$

Example 4 In $\mathbb{R}^k$, $x_1, \ldots, x_k$ are independently distributed with respect to the standard normal distribution if

$$m(A) = \frac{1}{(2\pi)^{k/2}} \int \int \cdots \int_A \prod_{i=1}^{k} e^{-(x_i^2/2)} dx_1 \cdots dx_k, \qquad A \subset \mathbb{R}^k.$$

One sees quite immediately from this formula that if A is a k-dimensional rectangular region, then the integral is the product of the one-dimensional integrals on the sides. Otherwise said, the probability of A is the product of the probabilities of the A_i where the A_i are the k one-dimensional sides of A, or yet that the A_i are independent.

Example 5 By Fubini's Theorem the preceding example is the same as the standard Gaussian on $\mathbb{R}^k$,

$$m(A) = \frac{1}{(2\pi)^{k/2}} \int_A e^{-(\|x\|^2/2)} dx_1 \cdots dx_k, \qquad A \subset \mathbb{R}^k.$$

Example 6 For $n \times n$ real matrices with entries m_{ij}, the standard Gaussian may be written

$$m(A) = \frac{1}{(2\pi)^{n^2/2}} \int_A e^{-\|M\|_F^2/2} dm_{11} \cdots dm_{nn}.$$

Example 7 The dimension of $\mathbb{C}^n$ as a real vector space is $2n$. Reflecting this fact, Examples 3 to 6 for the complex numbers become:

$$(3c) \qquad m(A) = \frac{1}{2\pi} \int_A e^{-(|z|^2/2)} dz \qquad A \subset \mathbb{C}.$$

$$(4c) \qquad m(A) = \frac{1}{(2\pi)^k} \int \int \cdots \int_A \prod_{i=1}^{k} e^{-(|z_i|^2/2)} dz_1 \cdots dz_k.$$

$$(5c) \qquad m(A) = \frac{1}{(2\pi)^k} \int_A e^{-(\|z\|^2/2)} dz_1 \cdots dz_k.$$

$$(6c) \qquad m(A) = \frac{1}{(2\pi)^{n^2}} \int_A e^{-(\|M\|_F^2/2)} dm_{11} \cdots dm_{nn}.$$

11.3 Some Integration Formulas

Besides Fubini's Theorem for integration in $\mathbb{R}^n$ we use various extensions of it. As we have mentioned in the last section, the repeated integral formula sometimes needs extra factors to adjust for the fact that the product of the volumes in the factors is not exactly the volume in the the product space. We illustrate this first with polar coordinates and their n-dimensional generalization spherical coordinates.

Example 8 Integration in polar coordinates in $\mathbb{R}^2$ has the following familiar formula,

$$\int_{\mathbb{R}^2} g(x, y) dx dy = \int_{\theta \in S^1} \int_{r \in \mathbb{R}_+} g(r\theta) r \, dr \, d\theta,$$

where $d\theta$ is the arc length on S^1. The ray $\{r\theta\}_{r \in \mathbb{R}_+}$ is orthogonal to the circle of radius r centered at zero, but the infinitesimal volume element at $(r\theta)$ is $r dr d\theta$ since the circle of radius r has length $2\pi r$ instead of 2π. This explains the factor of r on the right-hand side.

Spherical coordinates in $\mathbb{R}^n$ are the n-dimensional generalization of polar coordinates,

$$\int_{\mathbb{R}^n} g(x_1, \ldots, x_n) \, dx_1 \cdots dx_n = \int_{\Theta \in S^{n-1}} \int_{r \in \mathbb{R}_+} g(r\Theta) r^{n-1} dr d\Theta,$$

where $d\Theta$ is the volume on the unit sphere S^{n-1}.

For both polar and spherical coordinates we have assumed that $g : \mathbb{R}^n \to \mathbb{R}$, and that the left-hand integrals exist.

Example 9 If $S^{n-1} \subset \mathbb{R}^n$ is the unit sphere, $B^k \subset \mathbb{R}^k$ is the unit ball, $\mathbb{R}^k \subset \mathbb{R}^n$, and $\pi : S^{n-1} \to B^k$ is the restriction of the orthogonal projection from $\mathbb{R}^n$ to $\mathbb{R}^k$, then

$$\int_{S^{n-1}} g(\Theta) d\Theta = \int_{x \in B^k} \int_{y \in \pi^{-1}(x)} g(y)(1 - \|x\|^2)^{-1/2} dy dx, \tag{11.2}$$

where g is integrable.

We sketch why this formula is true.

For $x \in B^k$, $\pi^{-1}(x)$ is an $(n - k - 1)$ sphere of radius $(1 - \|x\|^2)^{1/2}$. To compare the volume in S^{n-1} with the product of the volume in this $(n - k - 1)$ sphere and the volume in B^k, we take the orthogonal space to the tangent space of the $(n - k - 1)$ sphere in S^{n-1} and multiply by the inverse of the Jacobian determinant of π restricted to this orthogonal space. We call the Jacobian determinant of π restricted to the orthogonal space of the tangent space of the $(n - k - 1)$ spheres the *normal Jacobian* of π, and denote it $\mathrm{NJ}(\pi)$. The integral formula (11.2) now follows from Lemma 2.

Lemma 2 *The normal Jacobian of* $\pi : S^{n-1} \to B^k$ *at* $x \in S^{n-1}$ *is* $(1 - \|\pi(x)\|^2)^{1/2}$.

Proof. Let $\|\pi(x)\| = r$. Let S_r^{k-1} be the sphere of radius r about 0 in $\mathbb{R}^k$. Then $\pi^{-1}(S_r^{k-1}) = S_r^{k-1} \times S_{\sqrt{1-r^2}}^{n-k-1}$ which leaves only one normal direction to $\pi^{-1}(\pi(x))$, namely, the one that maps to the ray in D^k. Thus the problem reduces to $S^1 \subset \mathbb{R}^2$ and the norm of the projection is easily seen to be $\sqrt{1 - r^2}$. $\qquad\square$

Formula (11.2) and its treatment via normal Jacobian is a special case of the co-area formula which we encounter in Chapter 13.

We next apply integration in spherical coordinates to a special case of the right-hand integral in equation (11.2), and derive a formula that is useful for the proof of the Main Theorems. Recall that

$$\text{Vol}(S^{n-1}) = \frac{2\pi^{n/2}}{\Gamma(\frac{n}{2})},$$

where Γ is the gamma function

$$\Gamma(x) = \int_0^\infty e^{-t} t^{x-1} dt$$

and that

$$\int_0^1 t^p (1-t^2)^q dt = \frac{1}{2} \frac{\Gamma(\frac{p+1}{2})\Gamma(q+1)}{\Gamma(\frac{p+1}{2}+q+1)}.$$

The gamma function satisfies the functional equation $\Gamma(x+1) = x\Gamma(x)$ and $\Gamma(1) = 1$. Therefore, $\Gamma(n) = (n-1)!$ for any positive integer n. References for the gamma function and its properties are given in Section 11.7.

Lemma 3 Let $B^{m-\ell} \subset \mathbb{R}^{m-\ell}$ be the unit ball and $k \geq 0$; then

$$\frac{1}{\text{Vol}(S^{m-1})} \int_{B^{m-\ell}} \frac{\text{Vol}(S^k_{(1-\|x\|^2)^{1/2}})}{(1-\|x\|^2)^{1/2}} dx_1 \ldots dx_{m-\ell} = \frac{\pi^{(k-\ell+1)/2}\Gamma(\frac{m}{2})}{\Gamma(\frac{m+k-\ell+1}{2})}$$

Proof. The integral on the left equals

$$\frac{\text{Vol}(S^k)}{\text{Vol}(S^{m-1})} \int_{B^{m-\ell}} (1-\|x\|^2)^{(k-1)/2} dx_1 \ldots dx_{m-\ell},$$

which using spherical coordinates equals

$$\frac{\text{Vol}(S^k)\text{Vol}(S^{m-\ell-1})}{\text{Vol}(S^{m-1})} \int_0^1 (1-r^2)^{(k-1)/2} \cdot r^{m-\ell-1} dr.$$

Substituting the expression preceding the lemma gives the right-hand side. □

Definition 2 Suppose we are given a probability distribution m via a point density function f on $X = \mathbb{R}^n$, $\mathbb{C}^n$ or S^{n-1}. For example, $f(x) = 1/(2\pi)^{n/2} e^{-\|x\|^2/2}$ for the standard Gaussian on $\mathbb{R}^n$ or $f(x) = 1/\text{Vol}(S^{n-1})$ for the uniform probability distribution on S^{n-1}. If $\phi : X \to \mathbb{R}$ and ϕf is integrable, the *expected or average value* of ϕ with respect to the distribution given by f is

$$\int_X \phi f \, dx.$$

The next proposition shows that if $\phi : \mathbb{R}^n \to \mathbb{R}$ is homogeneous of degree 0, (i.e., $\phi(\lambda x) = \phi(x)$ for $\lambda \neq 0$, $\lambda \in \mathbb{R}$), then the average value of ϕ with respect to the standard Gaussian on $\mathbb{R}^n$ is the same as the average value of ϕ with respect to the uniform distribution on S^{n-1}.

Proposition 1 *Let $\phi(x)$ be homogeneous of degree 0, then*

$$\frac{1}{(2\pi)^{n/2}} \int_{\mathbb{R}^n} \phi(x) e^{-\|x\|^2/2} dx = \frac{1}{\mathrm{Vol}\,(S^{n-1})} \int_{S^{n-1}} \phi(\theta) d\theta,$$

assuming the integral exists.

Proof. By integration in spherical coordinates

$$\frac{1}{(2\pi)^{n/2}} \int_{\mathbb{R}^n} \phi(x) e^{-(\|x\|^2/2)} dx = \frac{1}{(2\pi)^{n/2}} \int_{S^{n-1}} \int_{r=0}^{\infty} \phi(r\theta) r^{n-1} e^{-(r^2/2)} dr\, d\theta$$

$$= \frac{1}{(2\pi)^{n/2}} \int_{S^{n-1}} \phi(\theta) d\theta \int_{r=0}^{\infty} r^{n-1} e^{-(r^2/2)} dr.$$

Now substitute $u = r^2/2$ to obtain

$$\frac{1}{(2\pi)^{n/2}} \int_{S^{n-1}} \phi(\theta) d\theta\; 2^{(n-2)/2} \int_0^{\infty} u^{(n-2)/2} e^{-u} du$$

$$= \frac{\Gamma(\frac{n}{2})}{2\pi^{n/2}} \int_{S^{n-1}} \phi(\theta) d\theta$$

$$= \frac{1}{\mathrm{Vol}\,(S^{n-1})} \int_{S^{n-1}} \phi(\theta) d\theta.$$

$\square$

We end this section with an estimate on average values.

Definition 3 Let X be $\mathbb{R}^n$, $\mathbb{C}^n$, or S^n, with a probability distribution m given by a point density function f (or even more generally, a probability space with no atoms, i.e., no points of positive probability). Let $S : X \to \mathbb{R}_+$ be a real valued non-negative integrable function and let $g : (0, 1) \to \mathbb{R}$ be decreasing and integrable. We say that $S(x) \leq g(y)$ *with probability* $1 - y$ if $m\{x\,|\,S(x) \leq g(y)\} \geq 1 - y$ for all $0 < y < 1$.

Proposition 2 *With the previous notation, suppose that $S(x) < g(y)$ with probability $1 - y$. Then the average $E(S)$ satisfies*

$$E(S) = \int_X S(x) f(x) dx \leq \int_0^1 g(y) dy.$$

Proof. Let $y_i \in (0, 1)$ for $-\infty < i < \infty$ be a decreasing sequence with 0 and 1 as limit points. Let $\cdots \supset M_i \supset M_{i-1} \supset \cdots$ be chosen so that $m(M_i) = 1 - y_i$, and $S(x) \le g(y_i)$ for all $x \in M_i$. Then

$$\int S(x)f(x)dx = \sum_{i \in \mathbb{Z}} \int_{M_i - M_{i-1}} S(x)f(x)dx$$
$$\le \sum_{i \in \mathbb{Z}} g(y_i)m(M_i - M_{i-1})$$
$$= \sum_{i \in \mathbb{Z}} g(y_i)(y_{i-1} - y_i),$$

which converges to the Riemann integral of g between 0 and 1.   □

11.4   A Linear Algebra Estimate

The next lemma relates the norm of the inverse of an $n \times n$ matrix to a geometric condition on the columns of the matrix.

For an $n \times n$ matrix A, let $\widehat{A}_i$ refer to A with the ith column removed and, for an $n \times (n - 1)$ matrix A and $b \in \mathbb{R}^n$, let $(A, b)_i$ refer to the $n \times n$ matrix obtained from A by inserting b as the ith column. Also, let us denote by A_i the ith column of A.

Lemma 4   *Let A be an $n \times n$ matrix. If $\|A^{-1}\| = x$, then there is an $i \in \{1, 2, \ldots, n\}$ and a vector $B_i \in \mathbb{R}^n$ with $\|B_i\| \le \sqrt{n}/x$ and such that $A_i + B_i$ is in the space spanned by A_j, $j \ne i$, $j \in \{1, \ldots, n\}$.*

Proof.   We assume that $\|A^{-1}\|$ exists so that the columns of A are independent. That $\|A^{-1}\| = x$ means that there is a unit vector $b \in \mathbb{R}^n$ such that $v = A^{-1}b$ and $\|v\| = x$. Choose i such that $|v_i| \ge x/\sqrt{n}$. By Cramer's rule

$$v_i = \frac{\det(\widehat{A}_i, b)_i}{\det A}.$$

Hence $\det(\widehat{A}_i, (1/v_i)b)_i = \det A$ and $\det(\widehat{A}_i, A_i - (1/v_i)b)_i = 0$. As the columns of A are independent, this implies that $A_i - (1/v_i)b$ is spanned by the other columns, $\|(1/v_i)b\| \le \sqrt{n}/x$ and we are done.   □

Remark 2   The lemma holds for real or complex $n \times n$ matrices

In the following we denote by $\mathcal{M}^{\mathbb{R}}_{(n,m)}$ the space of real $n \times m$ matrices and if $m = n$ we denote this space just by $\mathcal{M}^{\mathbb{R}}_n$. Let $S^{n^2-1} \subset \mathcal{M}^{\mathbb{R}}_n$ be the unit sphere with respect to the Frobenius norm. Also let B^{n^2-n} be the unit ball in the space $\mathcal{M}^{\mathbb{R}}_{(n,n-1)}$ and for $i = 1, \ldots, n$ the map

$$\pi_i : \mathcal{M}^{\mathbb{R}}_n \to \mathcal{M}^{\mathbb{R}}_{(n,n-1)}$$
$$A \qquad \widehat{A}_i.$$

For any $\varepsilon > 0$ we consider $N_{\varepsilon}^{\mathbb{R}} \subset S^{n^2-1}$ the set of $A \in S^{n^2-1}$ such that A is not invertible or $\|A^{-1}\| > 1/\varepsilon$. The Eckardt–Young theorem provides a geometric characterization of $N_{\varepsilon}^{\mathbb{R}}$, namely,

$$N_{\varepsilon}^{\mathbb{R}} = \{A \in S^{n^2-1} \mid d_F(A, \Sigma) < \varepsilon\}.$$

In the next section we estimate the volume of $N_{\varepsilon}^{\mathbb{R}}$. To do so, consider for $\varepsilon > 0$ and $i = 1, \ldots, n$ the set

$$W_{\varepsilon,i}^{\mathbb{R}} = \{A \in S^{n^2-1} \mid A_i \text{ is within } \sqrt{n}\varepsilon \text{ of the}$$
$$\text{space spanned by the columns of } \widehat{A}_i\}.$$

Lemma 5 *For all $\varepsilon > 0$,*

$$\mathrm{Vol}\,(N_{\varepsilon}^{\mathbb{R}}) \le \sum_{i=1}^{n} \mathrm{Vol}\,(W_{\varepsilon,i}^{\mathbb{R}}).$$

Proof.   By Lemma 4 we have that $N_{\varepsilon}^{\mathbb{R}} \subset \bigcup_{i=1}^{n} W_{\varepsilon,i}^{\mathbb{R}}$.   □

Now we turn to $W_{\varepsilon,i}^{\mathbb{R}}$.

Proposition 3 *For all $\varepsilon > 0$, $B \in \mathcal{M}_{(n,n-1)}^{\mathbb{R}}$, and $i = 1, \ldots, n$ one has*

$$\mathrm{Vol}\,(W_{\varepsilon,i}^{\mathbb{R}} \cap \pi_i^{-1} B) \le 2\sqrt{n}\varepsilon \,\mathrm{Vol}\,(S_{(1-\|B\|_F^2)^{1/2}}^{n-2}),$$

where S_r^{m-1} is the sphere of radius r in $\mathbb{R}^m$.

Proof.   If $A \in W_{\varepsilon,i}^{\mathbb{R}} \cap \pi_i^{-1} B$ then $\|A_i\| = (1-\|B\|_F^2)^{1/2}$; that is, $A_i \in S_{(1-\|B\|_F^2)^{1/2}}^{n-1}$, and A_i is within $\sqrt{n}\varepsilon$ of the codimension space spanned by the columns of B.

The volume of $W_{\varepsilon,i}^{\mathbb{R}} \cap \pi_i^{-1} B$ is thus at most the volume of the cylinder of base $S_{(1-\|B\|_F^2)^{1/2}}^{n-2}$ and height $2\sqrt{n}\varepsilon$, that is, at most $2\sqrt{n}\varepsilon \,\mathrm{Vol}\,(S_{(1-\|B\|_F^2)^{1/2}}^{n-2})$.   □

11.5   The Main Theorems for the Reals

We can now estimate the volume of $N_{\varepsilon}^{\mathbb{R}}$.

Proposition 4 *For any $\varepsilon > 0$ and any $n \ge 2$*

$$\frac{\mathrm{Vol}\,(N_{\varepsilon}^{\mathbb{R}})}{\mathrm{Vol}\,(S^{n^2-1})} \le \varepsilon n^{5/2}.$$

Proof.   By Lemma 5 it suffices to show that

$$\frac{\mathrm{Vol}\,(W_{\varepsilon,i}^{\mathbb{R}})}{\mathrm{Vol}\,(S^{n^2-1})} \le \varepsilon n^{3/2}.$$

Let $\chi(W_{\varepsilon,i}^{\mathbb{R}})$ be the characteristic function of $W_{\varepsilon,i}^{\mathbb{R}}$. Then

$$
\frac{\text{Vol}\,(W_{\varepsilon,i}^{\mathbb{R}})}{\text{Vol}\,(S^{n^2-1})}
$$

$$
= \frac{1}{\text{Vol}\,(S^{n^2-1})} \int_{S^{n^2-1}} \chi(W_{\varepsilon,i}^{\mathbb{R}})dS^{n^2-1}
$$

$$
= \frac{1}{\text{Vol}\,(S^{n^2-1})} \int_{B^{n^2-n}} \text{Vol}\,(W_{\varepsilon,i}^{\mathbb{R}} \cap \pi_i^{-1}B)(1 - \|B\|_F^2)^{-1/2} dB^{n^2-n},
$$

the last equality by formula (11.2). Now by Proposition 3

$$
\frac{\text{Vol}\,(W_{\varepsilon,i}^{\mathbb{R}})}{\text{Vol}\,(S^{n^2-1})}
$$

$$
\leq \frac{2\sqrt{n}\varepsilon}{\text{Vol}\,(S^{n^2-1})} \int_{B^{n^2-n}} \text{Vol}\,(S_{(1-\|B\|_F^2)^{1/2}}^{n-2})(1 - \|B\|_F^2)^{-1/2} dB^{n^2-n}
$$

which by Lemma 3 (with $m = n^2$, $k = n - 2$, and $\ell = n$) is at most

$$
\frac{2\sqrt{n}\varepsilon}{\pi^{1/2}} \frac{\Gamma(\frac{n^2}{2})}{\Gamma(\frac{n^2-1}{2})}
$$

which by the next lemma is bounded by

$$
\frac{2^{1/2}}{\pi^{1/2}} \varepsilon n^{3/2} \leq \varepsilon n^{3/2}.
$$

$\square$

Lemma 6 *For any $x > 0$ one has*

$$
\frac{\Gamma(x + 1/2)}{\sqrt{x}\Gamma(x)} < 1.
$$

Proof.   It is easy to prove from the fact that $\Gamma(z + 1) = z\Gamma(z)$ that $(\Gamma(x + n + 1/2))/(\sqrt{x + n}\Gamma(x + n)) \to 1$ as $n \to \infty$ for any $x \in \mathbb{R}$. On the other hand, for each $n \geq 0$ one has that

$$
\frac{\Gamma(x + n + 1 + 1/2)}{\sqrt{x + n + 1}\Gamma(x + n + 1)} = \frac{(x + n + 1/2)\Gamma(x + n + 1/2)}{\sqrt{x + n + 1}\sqrt{x + n}\sqrt{x + n}\Gamma(x + n)}
$$

$$
> \frac{\Gamma(x + n + 1/2)}{\sqrt{x + n}\Gamma(x + n)},
$$

so the preceding sequence is monotone increasing and hence each term is less than 1.

$\square$

Theorem 2 *The probability that $\kappa(A) > 1/\varepsilon$ for the standard Gaussian probability distribution on $\mathcal{M}_n^{\mathbb{R}}$ is less than or equal to $n^{5/2}\varepsilon$.*

We first prove a simple lemma. Define a second condition number μ of $n \times n$ matrices by

$$\mu(A) = \|A^{-1}\| \, \|A\|_F.$$

Lemma 7 $\kappa(A) \leq \mu(A)$.

Proof. We have to prove $\|A\|_F \geq \|A\|$. For $v \in \mathbb{R}^n$

$$\|A(v)\|^2 = \sum_{i=1}^n \|A_i \cdot v\|^2,$$

where A_i is the ith row of A. By Cauchy–Schwartz this is

$$\leq \sum_{i=1}^n \|A_i\|^2 \|v\|^2$$
$$\leq \|A\|_F^2 \|v\|^2.$$

So $\|A\| \leq \|A\|_F$.   □

Proof of Theorem 2. By Lemma 7 it suffices to prove the theorem for μ. Since μ is homogeneous of degree 0 it suffices by Proposition 1 to prove the theorem for the uniform distribution on the sphere. And since $N_\varepsilon^{\mathbb{R}}$ is the set of $A \in S^{n^2-1}$, where $\mu(A) \geq 1/\varepsilon$, Proposition 4 finishes the proof.   □

We now can estimate the average loss of precision.

Theorem 3 *The average loss of precision for $A \in \mathcal{M}_n^{\mathbb{R}}$ with the standard Gaussian probability distribution*

$$\frac{1}{(2\pi)^{n^2/2}} \int_{\mathcal{M}_n^{\mathbb{R}}} \log_b \kappa(A) e^{-(\|A\|_F^2/2)} dm_{11} \cdots dm_{nn}$$

is less than or equal to $(5/2)\log_b n + (1/\ln b)$ where $b \in \mathbb{N}$ is the base of the logarithm defining the loss of precision.

Proof. By Theorem 2, the probability that $\log_b \kappa(A) \leq \log_b(n^{5/2}/y)$ is greater than or equal to $1 - y$. Thus by Proposition 2 the average of $\log_b \kappa(A)$ is less than or equal to

$$\int_0^1 \log_b(\frac{n^{5/2}}{y}) = 5/2 \log_b n + \frac{1}{\ln b}.$$

     □

11.6   The Main Theorems for the Complex Numbers

Let us now consider complex matrices. As previously, $\mathcal{M}_n^{\mathbb{C}}$ denotes the space of complex $n \times n$ matrices, S^{2n^2-1} is the unit sphere in it, and we consider $N_\varepsilon^{\mathbb{C}} = \{A \in S^{2n^2-1} \mid \|A^{-1}\| \geq 1/\varepsilon\}$ and $W_{\varepsilon,i}^{\mathbb{C}} = \{A \in S^{2n^2-1} \mid A_i \text{ is within } \sqrt{n}\varepsilon \text{ of the}$ space spanned by the columns of $\widehat{A}_i\}$.

We also have

$$N_\varepsilon^{\mathbb{C}} \subset \bigcup_{i=1}^n W_{\varepsilon,i}^{\mathbb{C}} \qquad \text{and} \qquad \text{Vol}\,(N_\varepsilon^{\mathbb{C}}) \leq \sum_{i=1}^n \text{Vol}\,(W_{\varepsilon,i}^{\mathbb{C}}).$$

Let us also consider, for $i = 1, \ldots, n$, the map

$$\pi_i : \mathcal{M}_n^{\mathbb{C}} \to \mathcal{M}_{(n,n-1)}^{\mathbb{C}}$$

$$A \qquad \widehat{A}_i.$$

Proposition 5   *For all $\varepsilon > 0$, $B \in \mathcal{M}_{(n,n-1)}^{\mathbb{C}}$, and $i = 1, \ldots, n$ one has*

$$\text{Vol}\,(W_{\varepsilon,i}^{\mathbb{C}} \cap \pi_i^{-1}B) \leq \pi n \varepsilon^2 \text{Vol}\,(S_{(1-\|B\|_F^2)^{1/2}}^{2n-3}).$$

Proof.   The proof is the same as Proposition 3 except that for $A \in \pi_i^{-1}(B)$, A_i lies in an orthogonal disc neighborhood of radius $\sqrt{n}\varepsilon$ of the ball of radius $(1-\|B\|_F^2)^{1/2}$ in $\mathbb{C}^{n-1}$. The volume of the intersection with the sphere is at most the volume of the cylinder which is $\pi n \varepsilon^2 \text{Vol}\, S_{(1-\|B\|_F^2)^{1/2}}^{2n-3}$.   □

As for the reals we estimate the volume of $N_\varepsilon^{\mathbb{C}}$.

Proposition 6   *One has*

$$\frac{\text{Vol}\,(N_\varepsilon^{\mathbb{C}})}{\text{Vol}\,(S^{2n^2-1})} \leq n^4 \varepsilon^2.$$

Proof.   By Lemma 5 it suffices to show that

$$\frac{\text{Vol}\,(W_{\varepsilon,i}^{\mathbb{C}})}{\text{Vol}\,(S^{2n^2-1})} \leq n^3 \varepsilon^2.$$

Let $\chi(W_{\varepsilon,i}^{\mathbb{C}})$ be the characteristic function of $W_{\varepsilon,i}^{\mathbb{C}}$. Then

$$\frac{\text{Vol}\,(W_{\varepsilon,i}^{\mathbb{C}})}{\text{Vol}\,(S^{2n^2-1})}$$

$$= \frac{1}{\text{Vol}\,(S^{2n^2-1})} \int_{S^{2n^2-1}} \chi(W_{\varepsilon,i}^{\mathbb{C}}) dS^{2n^2-1}$$

$$= \frac{1}{\text{Vol}\,(S^{2n^2-1})} \int_{B^{2n^2-2n}} \text{Vol}\,(W_{\varepsilon,i}^{\mathbb{C}} \cap \pi_i^{-1}(B))(1 - \|B\|_F^2)^{-1/2} dB^{2n^2-2n},$$

the last by formula (11.2). Now by Proposition 5

$$\frac{\text{Vol}\,(W^{\mathbb{C}}_{\varepsilon,i})}{\text{Vol}\,(S^{2n^2-1})}$$

$$\leq \frac{\pi n \varepsilon^2}{\text{Vol}\,(S^{2n^2-1})} \int_{B^{2n^2-2n}} \text{Vol}\,(S^{2n-3}_{(1-\|B\|^2_F)^{1/2}})(1 - \|B\|^2_F)^{-1/2} d\,B^{2n^2-2n}$$

which by Lemma 3 is

$$\leq n\varepsilon^2 \frac{\Gamma(n^2)}{\Gamma(n^2-1)} \leq n^3 \varepsilon^2.$$

$\square$

The next two theorems are the analogues of Theorems 2 and 3 for the complex case. Their proofs are analogous using Proposition 6 in place of Proposition 4.

Theorem 4 *The probability that $\mu(A) > 1/\varepsilon$ or that $\kappa(A) > 1/\varepsilon$ for either the standard Gaussian probability distribution on $\mathcal{M}^{\mathbb{C}}_n$ or the uniform distribution on S^{2n^2-1} (the unit sphere in $\mathcal{M}^{\mathbb{C}}_n$) is less than or equal to $n^4 \varepsilon^2$.*

Theorem 5 *The average loss of precision for $A \in \mathcal{M}^{\mathbb{C}}_n$ with the standard Gaussian probability distribution is less than or equal to $2\log_b n + (1/2\ln b)$ where $b \in \mathbb{N}$ is the base of the logarithm defining the loss of precision.*

11.7   Additional Comments and Bibliographical Remarks

The study of the accuracy of numerical solutions of linear equations has a long history. The subject is treated by von Neumann and coauthors in [von Neumann 1963] and by Wilkinson [1965].

The questions as to what the average loss of precision is for linear systems was first raised in [Smale 1985]. The Main Theorem was first proved by Kostlan [1988]. The approach we have taken here was suggested in a personal letter by Jim Renegar in 1985. In particular, Lemma 4 is taken from that letter. The results we have presented are not sharp. The best results have been proven by Edelman [1988, 1989]. Indeed, up to a small additive constant the average loss of precision is $\log n$ in both the real and complex cases. Edelman's arguments are considerably more sophisticated than the ones we have used here. In the case of complex matrices Edelman [1992] proves the following.

Theorem 6

$$\frac{\text{Vol}\,N^{\mathbb{C}}_\varepsilon}{\text{Vol}\,S^{2n^2-1}} = 1 - (1 - \min(1, n\varepsilon^2))^{n^2-1}.$$

$\square$

So one can see that Proposition 6 and Theorem 4 can be improved as to the power of n.

The tractability of evaluating rational expressions is treated in a general context in [Blum and Shub 1986]. For the gamma function see [Abramowitz and Stegun 1964]. For the Eckart–Young Theorem see [Eckart and Young 1936] and [Golub and Van Loan 1989], the latter also being a good reference for the conjugate gradient algorithm mentioned at the beginning of the chapter as well as for other topics in numerical linear algebra.

12
The Condition Number for Nonlinear Problems

The goal of this chapter is to describe a measure of condition for problems that search for a solution of nonlinear systems of equations. Eventually a condition number μ is defined as a bound on the infinitesimal error of a solution caused by an infinitesimal error in the defining system of equations. With our emphasis on polynomial systems, we impose a norm on the space of such systems that reflects an important computational invariant, the distance between the zeros. To avoid the distortion caused by very large zeros, the analysis and metrics are defined in a projective space setting. The result is a unitarily invariant theory.

Finally a theorem, "the condition number theorem," is proved which is an exact formula relating the condition number to the reciprocal of a distance to a set of ill-posed problems.

12.1 Unitary Invariance

We need a way of measuring distances in the spaces of polynomials introduced in Chapter 10. We require our distance function to satisfy a natural invariance property which is lacking in the computational mathematics tradition.

The complex numbers are emphasized in the development and later we see the same considerations apply to the real case. In this analysis the unitary group plays the role of a symmetry group, and a *unitarily invariant* complexity theory is the goal. Inspiration comes from theoretical physics where unitary invariance is central. The unitary group $U(k)$ acts on $\mathbb{C}^k$ as the group of linear automorphisms (given by $k \times k$ matrices) that preserve the Hermitian product $\langle \, , \, \rangle$ on $\mathbb{C}^k$. More precisely,

$$\langle \sigma v, \sigma w \rangle = \langle v, w \rangle, \text{ for all } v, w \in \mathbb{C}^k, \ \sigma \in U(k).$$

This implies immediately that $\|\sigma w\| = \|w\|$ and $\|x - y\| = \|\sigma x - \sigma y\|$.

Let $\mathcal{H}_d$ be the linear space of homogeneous polynomials of degree d in k variables. Then the action of $U(k)$ on $\mathbb{C}^k$ induces an action of $U(k)$ on $\mathcal{H}_d$. For $f \in \mathcal{H}_d$ let $\sigma f \in \mathcal{H}_d$ be defined by $\sigma f(x) = f(\sigma^{-1}x)$, for all $x \in \mathbb{C}^k$. The problem in front of us is to endow $\mathcal{H}_d$ with an Hermitian product which is invariant under this action of $U(k)$.

Let us illustrate the situation for the case of nonhomogeneous complex polynomials of one variable. Let $\mathcal{P}_d$ be the space of all $f : \mathbb{C} \to \mathbb{C}$ with $f(x) = \sum_{i=0}^{d} a_i x^i$. Traditionally in most of mathematics one puts a norm on f by the formula

$$\|f\|^2 = \sum_{i=0}^{d} |a_i|^2. \tag{12.1}$$

On the other hand, in this case we use

$$\|f\|^2 = \sum_{i=0}^{d} |a_i|^2 \binom{d}{i}^{-1}, \tag{12.2}$$

where the weighting factor of the reciprocal of the binomial coefficient leads to unitary invariance.

Let us be more precise and more general. For $f, g \in \mathcal{H}_d$, let

$$\langle f, g \rangle = \sum_{\alpha} a_\alpha \bar{b}_\alpha \binom{d}{\alpha_1, \ldots, \alpha_k}^{-1},$$

where $f(x) = \sum a_\alpha x^\alpha$, $g(x) = \sum b_\alpha x^\alpha$, α is a multi-index, $\alpha = (\alpha_1, \ldots, \alpha_k)$, and x^α is the product $x_1^{\alpha_1} \cdot \ldots \cdot x_k^{\alpha_k}$. Moreover,

$$\binom{d}{\alpha_1, \ldots, \alpha_k} = \frac{d!}{\alpha_1! \cdot \ldots \cdot \alpha_k!}$$

is the "multinomial coefficient."

The main result of this section states the key property of this Hermitian product.

Theorem 1 *This Hermitian product on $\mathcal{H}_d$ is unitarily invariant. In other words,*

$$\langle \sigma f, \sigma g \rangle = \langle f, g \rangle \text{ for all } f, g \in \mathcal{H}_d, \ \sigma \in U(k).$$

Remark 1 It can be shown that, moreover, the Hermitian product of Theorem 1 is the unique unitarily invariant Hermitian product up to a constant scalar factor.

Towards the proof of Theorem 1 we need some preliminary results.

Let V, W be Hermitian vector spaces with orthonormal bases $v_1, \ldots, v_n$ and $w_1, \ldots, w_m$, respectively. Let $L_{v_i w_j}$ denote the linear map that maps v_i to w_j and v_k to 0 for $k \neq i$. Then the following Hermitian products on $\mathcal{L}(V, W)$, the space of linear maps from V to W, are the same.

(1) If

$$L = \sum_{\substack{i=1,\ldots,n \\ j=1,\ldots,m}} a_{j,i} L_{v_i w_j}$$

$$M = \sum_{\substack{i=1,\ldots,n \\ j=1,\ldots,m}} b_{j,i} L_{v_i w_j},$$

then

$$\langle L, M \rangle = \sum_{\substack{i=1,\ldots,n \\ j=1,\ldots,m}} a_{j,i} \overline{b_{j,i}}.$$

This Hermitian product is defined by declaring the $L_{v_i w_j}$ an orthonormal basis of $\mathcal{L}(V, W)$.

(2) If $L, M \in \mathcal{L}(V, W)$, write L, M as matrices A, B with respect to the bases $v_1, \ldots, v_n$ and $w_1, \ldots, w_m$,

$$A = (a_{ji})_{\substack{i=1,\ldots,n \\ j=1,\ldots,m}} \qquad \text{and} \qquad B = (b_{ji})_{\substack{i=1,\ldots,n \\ j=1,\ldots,m}}$$

and let

$$\langle L, M \rangle = \sum_{\substack{i=1,\ldots,n \\ j=1,\ldots,m}} a_{j,i} \overline{b_{j,i}}.$$

(3) With L, M, A, B as in (2) define

$$\langle L, M \rangle = \text{trace}(AB^*).$$

The proof of the equality of these three definitions is straightforward.

Proposition 1 *Let $U_1 : W \to W$ and $U_2 : V \to V$ be unitary transformations, and $\langle \ , \ \rangle$ the preceding Hermitian product. Then the linear maps $L \to U_1 L$ and $L \to L U_2$ for $L \in \mathcal{L}(V, W)$ preserve the Hermitian product on $\mathcal{L}(V, W)$.*

Proof. Let $L, M \in \mathcal{L}(V, W)$ and let $L' = U_1 L$ and $M' = U_1 M$. Suppose that with respect to the orthonormal basis $v_1, \ldots, v_n$ and $w_1, \ldots, w_m$ the maps L, M, and U_1 are represented as matrices by A, B, and H_1. Then, using that $\text{trace}(PQ) = \text{trace}(QP)$ and that $H_1^* H_1 = Id_W$ since H_1 is Hermitian,

$$\langle L', M' \rangle = \text{trace}(H_1 A (H_1 B)^*)$$
$$= \text{trace}(H_1 A B^* H_1^*)$$
$$= \text{trace}(AB^*)$$
$$= \langle L, M \rangle.$$

The other case is done analogously. □

We now apply Proposition 1 to the space of d-multilinear functions $\mathcal{L}_d(\mathbb{C}^k, \mathbb{C})$ for $d \in \mathbb{N}$. Recall that

$$f : \underbrace{\mathbb{C}^k \times \ldots \times \mathbb{C}^k}_{d \text{ times}} \to \mathbb{C}$$

is *multilinear* if it is linear as a function of each variable separately.

Thus, $\mathcal{L}_1(\mathbb{C}^k, \mathbb{C}) = \mathcal{L}(\mathbb{C}^k, \mathbb{C})$ and $\mathcal{L}(\mathbb{C}^k, \mathcal{L}_{d-1}(\mathbb{C}^k, \mathbb{C}))$ may be identified with $\mathcal{L}_d(\mathbb{C}^k, \mathbb{C})$.

Let $v = (v_1, \ldots, v_d) \in \mathbb{C}^k \times \ldots \times \mathbb{C}^k$. Define $X_{i,j}(v) = v_{i,j}$, the ith coordinate of v_j for $1 \le i \le k$ and $1 \le j \le d$. Then $\prod_{j=1}^{d} X_{i_j, j}$ is a multilinear map from $\mathbb{C}^k \times \ldots \times \mathbb{C}^k$ to $\mathbb{C}$ and as the $(i_1, \ldots, i_d)$ runs through the k^d possible choices, the $\prod_{j=1}^{d} X_{i_j, j}$ form a basis of $\mathcal{L}_d(\mathbb{C}^k, \mathbb{C})$. We let $\langle \, , \, \rangle$ be the Hermitian product on $\mathcal{L}_d(\mathbb{C}^k, \mathbb{C})$ that makes this basis orthonormal. Now if $U_j : \mathbb{C}^k \to \mathbb{C}^k$ is a unitary transformation for $j = 1, \ldots, d$, then

$$U : \underbrace{\mathbb{C}^k \times \ldots \times \mathbb{C}^k}_{d \text{ times}} \to \underbrace{\mathbb{C}^k \times \ldots \times \mathbb{C}^k}_{d \text{ times}}.$$

$U = (U_1, \ldots, U_d)$ is a unitary transformation of $(\mathbb{C}^k)^d$ that induces a linear endomorphism on $\mathcal{L}_d(\mathbb{C}^k, \mathbb{C})$ by $L \to LU$.

Proposition 2 *Let $U_j : \mathbb{C}^k \to \mathbb{C}^k$ be unitary for $j = 1, \ldots, d$ and $U = (U_1, \ldots, U_d)$. Then $L \to LU$ is a unitary map from $\mathcal{L}_d(\mathbb{C}^k, \mathbb{C})$ to itself.*

Proof. Apply Proposition 1, induction on d, and identify $\mathcal{L}_d(\mathbb{C}^k, \mathbb{C})$ with $\mathcal{L}(\mathbb{C}^k, \mathcal{L}_{d-1}(\mathbb{C}^k, \mathbb{C}))$. □

A multilinear map $L \in \mathcal{L}_d(\mathbb{C}^k, \mathbb{C})$ is called *symmetric* if $L(v_1, \ldots, v_d) = L(v_{\sigma(1)}, \ldots, v_{\sigma(d)})$ for any permutation σ of $\{1, \ldots, d\}$ and all $(v_1, \ldots, v_d) \in (\mathbb{C}^k)^d$. We let $\mathcal{S}_d(\mathbb{C}^k, \mathbb{C}) \subset \mathcal{L}_d(\mathbb{C}^k, \mathbb{C})$ be the linear subspace of symmetric multilinear maps.

A symmetric multilinear map $L \in \mathcal{S}_d(\mathbb{C}^k, \mathbb{C})$ defines an homogeneous polynomial function $\rho(L) = f$,

$$f : \mathbb{C}^k \to \mathbb{C}$$

by $f(v) = L(v, \ldots, v)$. For $d = 2$ this is the well-known correspondence between symmetric bilinear maps and quadratic forms. The next proposition generalizes this correspondence to arbitrary d.

Proposition 3 *The map*

$$\rho : \mathcal{S}_d(\mathbb{C}^k, \mathbb{C}) \to \mathcal{H}_d$$

is a linear isomorphism.

Proof.   Let

$$T : \mathcal{H}_d \to \mathcal{S}_d(\mathbb{C}^k, \mathbb{C})$$

be given by $T(f) = D^d f(0)$. The map T is linear and injective; $T \circ \rho = Id_{\mathcal{S}_d(\mathbb{C}^k, \mathbb{C})}$ and $\rho \circ T = Id_{\mathcal{H}_d}$ by Taylor's formula.   □

If U is a unitary transformation of $\mathbb{C}^k$, then U acts on $(\mathbb{C}^k)^d$ by

$$(v_1, \ldots, v_d) \to (U(v_1), \ldots, U(v_d))$$

and U induces a unitary transformation of $\mathcal{S}_d(\mathbb{C}^k, \mathbb{C})$ by $U(L)(v_1, \ldots, v_d) = L(U(v_1), \ldots, U(v_d))$. The following lemma is straightforward.

Lemma 1   *The actions of the unitary group on $\mathcal{S}_d(\mathbb{C}^k, \mathbb{C})$ and on $\mathcal{H}_d$ commute with the identification given by ρ; that is, if $f = \rho(L)$, then $\rho(U(L)) = f \circ U$.*   □

Lemma 2   *The isomorphism*

$$\rho : \mathcal{S}_d(\mathbb{C}^k, \mathbb{C}) \to \mathcal{H}_d$$

is a unitary isomorphism.

Proof.   Given a monomial $X^\alpha \in \mathcal{H}_d$ with $\alpha = (\alpha_1, \ldots, \alpha_k)$ and $\sum \alpha_i = d$, then

$$\rho^{-1}(X^\alpha) = \binom{d}{\alpha_1, \ldots, \alpha_k}^{-1} \sum \prod X_{i_j, j},$$

where for each $\ell \leq k$ there are α_ℓ ocurrences of ℓ in $\{i_1, \ldots, i_k\}$ and there are

$$\binom{d}{\alpha_1, \ldots, \alpha_k}$$

monomials in the sum. It follows that $\rho^{-1}(X^\alpha)$ and $\rho^{-1}(X^\beta)$ are orthogonal for $\alpha \neq \beta$ and

$$\langle \rho^{-1}(X^\alpha), \rho^{-1}(X^\alpha) \rangle = \binom{d}{\alpha_1, \ldots, \alpha_k}^{-1}.$$

Thus, ρ is a unitary isomorphism.   □

We can now prove Theorem 1.

Proof of Theorem 1.   By Proposition 2, the unitary group of $\mathbb{C}^k$ acts by unitary transformation on $\mathcal{S}_d(\mathbb{C}^k, \mathbb{C})$. Lemmas 1 and 2 translate this fact to $\mathcal{H}_d$.   □

Recall that if $\mathcal{P}_d$ is the space of all polynomials of degree d or less in $k-1$ variables, there is a natural isomorphism $\wedge : \mathcal{P}_d \to \mathcal{H}_d$ given by homogenization. From this isomorphism we obtain actions of $U(k)$ on $\mathcal{P}_d$ and an Hermitian product on $\mathcal{P}_d$ which again is unitarily invariant. Thus the norm in (12.2) (for $k=2$) is unitarily invariant whereas the one preceding it is not.

Now we consider the space $\mathcal{H}_{(d)}$ of homogeneous polynomial systems $f_1, \ldots, f_n$ in $n+1$ variables with $d = (d_1, \ldots, d_n)$ and degree $f_i = d_i$ introduced in Chapter 10. We may write $\mathcal{H}_{(d)} = \prod_{i=1}^n \mathcal{H}_{d_i}$ where $\mathcal{H}_{d_i}$ is the space of polynomials in $n+1$ variables of degree d_i. A unitarily invariant Hermitian product $\langle \ , \ \rangle_i$ and norm $\| \ \|_i$ has been defined for each $\mathcal{H}_{d_i}$. Define for $f, g \in \mathcal{H}_{(d)}$,

$$\langle f, g \rangle = \sum_{i=1}^n \langle f_i, g_i \rangle_i;$$

thus $\|f\|^2 = \sum \|f_i\|_i^2$.

If $f : \mathbb{C}^{n+1} \to \mathbb{C}^n$ is an element of $\mathcal{H}_{(d)}$ and $\sigma \in U(n+1)$, let $\sigma f \in \mathcal{H}_{(d)}$ be defined by $(\sigma f)(x) = f(\sigma^{-1} x)$. In this way $U(n+1)$ acts on $\mathcal{H}_{(d)}$ and the Hermitian product and norm on $\mathcal{H}_{(d)}$ are unitarily invariant.

12.2 Hermitian Structures and Distances in Projective Space

The action of $U(n+1)$ on $\mathcal{H}_{(d)}$ is linear:

$$(af + bg) \circ \sigma^{-1} = a(f \circ \sigma^{-1}) + b(g \circ \sigma^{-1})$$

for all $f, g \in \mathcal{H}_{(d)}$ and $a, b \in \mathbb{C}$.

Since a linear map takes lines to lines, a linear map on a vector space W induces a natural map on the projective space $\mathbb{P}(W)$. In this way $U(n+1)$ has a natural action on $\mathbb{P}(\mathbb{C}^{n+1})$ and $\mathbb{P}(\mathcal{H}_{(d)})$ so that each element $\sigma \in U(n+1)$ may be considered as an invertible map

$$\sigma : \mathbb{P}(\mathbb{C}^{n+1}) \to \mathbb{P}(\mathbb{C}^{n+1})$$
$$\sigma : \mathbb{P}(\mathcal{H}_{(d)}) \to \mathbb{P}(\mathcal{H}_{(d)})$$

having natural properties under composition, and so on. Moreover, $U(n+1)$ acts componentwise on the product $\mathbb{P}(\mathcal{H}_{(d)}) \times \mathbb{P}(\mathbb{C}^{n+1})$ as well and on the solution variety V.

Lemma 3 *The solution variety $V \subset \mathbb{P}(\mathcal{H}_{(d)}) \times \mathbb{P}(\mathbb{C}^{n+1})$ is invariant under $U(n+1)$.*

Proof.   Let $(f, \zeta) \in V$. Then $f(\zeta) = 0$, $\sigma(f, \zeta) = (f\sigma^{-1}, \sigma(\zeta))$, and the last is in V since $(f\sigma^{-1})(\sigma(\zeta)) = f(\zeta) = 0$.   □

Having defined invariant Hermitian products on the vector spaces $\mathbb{C}^{n+1}$ and $\mathcal{H}_{(d)}$, we now proceed to see how they induce invariant Hermitian products on the tangent spaces of $\mathbb{P}(\mathbb{C}^{n+1})$, $\mathbb{P}(\mathcal{H}_{(d)})$, and V, and distances on these spaces that are also invariant under the action of $U(n+1)$. First we discuss distances in differentiable manifolds.

If M is a differentiable manifold of dimension n, the tangent space of M at a point p, T_pM, is an n-dimensional vector space. A *Finsler* structure on M is an assignment of a norm to each of these vector spaces in such a fashion that the norm is a continuous function on the tangent bundle of M,

$$TM = \bigcup_{p \in M} T_pM.$$

Once we have a Finsler structure we may define the length of a continuously differentiable path

$$\phi : [0, 1] \to M$$

by

$$\ell(\phi) = \int_0^1 \|\phi'(t)\| dt.$$

Finally, the distance between two points p_1, p_2 in M, $d(p_1, p_2)$, is defined as

$$\inf\{\ell(\phi) \mid \phi : [0, 1] \to M, \ \phi(0) = p_1, \ \phi(1) = p_2\}.$$

If T_pM is a complex vector space one way to get a norm on T_pM is via an Hermitian product $\langle \, , \, \rangle_p$ on T_pM. The norm is then $\langle v, v \rangle^{1/2}$ for any $v \in T_pM$.

If W is a finite-dimensional complex vector space with an Hermitian product $\langle \, , \, \rangle$, then we may define an Hermitian product $\langle \, , \, \rangle_p$ on $T_p\mathbb{P}(W)$ by

$$\langle v, w \rangle_p = \frac{\langle v, w \rangle}{\langle p, p \rangle}$$

for $v, w \in T_p\mathbb{P}(W)$. The corresponding norm is

$$\|v\|_p = \frac{\|v\|}{\|p\|}.$$

Since

$$\langle \lambda v, \lambda w \rangle_{\lambda p} = \frac{\langle \lambda v, \lambda w \rangle}{\langle \lambda p, \lambda p \rangle} = \frac{\langle v, w \rangle}{\langle p, p \rangle}$$

we see that the Hermitian products of equivalent pairs of vectors are equal and thus our Hermitian product is defined on the tangent space to $\mathbb{P}(W)$ at p. We denote the corresponding metric on $\mathbb{P}(W)$ by d_R, and we call this distance function the *Riemannian distance* on $\mathbb{P}(W)$.

Remark 2 If $p \in W$, we now have two Hermitian products and norms on T_p, $\langle\,,\,\rangle$ and $\|\;\|$ which are the restriction of the Hermitian product and norm on W to T_p, and $\langle\,,\,\rangle_p$ and $\|\;\|_p$ which are the Hermitian product and norm of T_p considered as the tangent space to $\mathbb{P}(W)$ at p.

An assignment of an Hermitian product to each tangent space of a manifold M in a continuous fashion is called an *Hermitian structure* on M. If N is a submanifold of M and $T_p N$ is a complex vector subspace of $T_p M$ for all $p \in N$, then N inherits an Hermitian structure by restricting the Hermitian product at a point $p \in N$ to the tangent space of N at p.

In this way, the solution variety V inherits an Hermitian structure from $\mathbb{P}(\mathcal{H}_{(d)}) \times \mathbb{P}(\mathbb{C}^{n+1})$. The Hermitian structure in $\mathbb{P}(\mathcal{H}_{(d)}) \times \mathbb{P}(\mathbb{C}^{n+1})$ is the product structure; if $v_i \in T_p \mathbb{P}(\mathcal{H}_{(d)})$ for $i = 1, 2$, and $w_i \in T_q \mathbb{P}(\mathbb{C}^{n+1})$ for $i = 1, 2$, then $(v_i, w_i) \in T_{(p,q)}(\mathbb{P}(\mathcal{H}_{(d)}) \times \mathbb{P}(\mathbb{C}^{n+1}))$ for $i = 1, 2$ and $\langle(v_1, w_1), (v_2, w_2)\rangle = \langle v_1, v_2\rangle + \langle w_1, w_2\rangle$.

If M is a differentiable manifold with a Finsler structure and $f : M \to M$ is a differentiable map, then f is called an *isometry* if $\|Df(x)v\| = \|v\|$ for all $x \in M$ and $v \in T_x M$. If M has an Hermitian structure and $\langle Df(x)v, Df(x)w\rangle = \langle v, w\rangle$ for all $x \in M$ and $v, w \in T_x$, then we say that f preserves the Hermitian structure. A map that preserves an Hermitian structure is of course an isometry for the Finsler it induces and also preserves distances between points in M.

Theorem 2 *If W is a complex vector space with Hermitian product $\langle\,,\,\rangle$ and $\sigma : W \to W$ is an Hermitian linear map (i.e. $\langle \sigma w_1, \sigma w_2\rangle = \langle w_1, w_2\rangle$ for all $w_1, w_2 \in W$), then σ preserves the Hermitian structure on $\mathbb{P}(W)$ and hence is an isometry.*

Proof.   If $p \in W$ and $w_1, w_2 \in T_p$, then

$$\langle w_1, w_2\rangle_p = \frac{\langle w_1, w_2\rangle}{\langle p, p\rangle} = \frac{\langle \sigma w_1, \sigma w_2\rangle}{\langle \sigma p, \sigma p\rangle} = \langle \sigma w_1, \sigma w_2\rangle_{\sigma p}.$$

$\square$

Corollary 1 *The group $U(n + 1)$ preserves the Hermitian structure and hence acts by isometries on $\mathbb{P}(\mathbb{C}^{n+1})$, $\mathbb{P}(\mathcal{H}_{(d)})$, and V.*

Proof.   We know that if $\sigma \in U(n + 1)$, then σ preserves the Hermitian product on $\mathbb{C}^{n+1}$ by definition and on $\mathcal{H}_{(d)}$ by Theorem 1. Thus, σ preserves the Hermitian structures on $\mathbb{P}(\mathbb{C}^{n+1})$, $\mathbb{P}(\mathcal{H}_{(d)})$, their product, and on V. $\square$

We have already seen in Chapter 10 that if W is a complex vector space of dimension n with an Hermitian product, then there is a natural map from the unit sphere S_1 in W to $\mathbb{P}(W)$,

$$\rho : S_1 \to \mathbb{P}(W),$$

which takes a nonzero vector in S_1 to the one-dimensional space in which it lies. The fibers of ρ are orbits of the S^1 action on S_1 given by multiplication by the unit modulus complex numbers. The unit sphere S_1 is homeomorphic to the unit sphere S^{2n-1} in $\mathbb{C}^n$. We compare distances in S_1 and $\mathbb{P}(W)$. Since S_1 has odd dimension the tangent space at a point cannot be a complex vector space and cannot have an Hermitian product. For real vector spaces and manifolds the relevant concepts are inner product and Riemannian metric.

If W is an n-dimensional real vector space, an *inner product* on W is a function

$$\langle\,,\,\rangle : W \times W \to \mathbb{R}$$

satisfying for all $x, y, z \in W$, and $a \in \mathbb{R}$,

$$\langle x, y \rangle = \langle y, x \rangle$$
$$\langle x, y + z \rangle = \langle x, y \rangle + \langle x, z \rangle$$
$$\langle ax, y \rangle = a\langle x, y \rangle$$
$$\langle x, x \rangle \geq 0 \quad \text{and} \quad \langle x, x \rangle = 0 \text{ iff } x = 0.$$

The standard inner product on $\mathbb{R}^n$ is given by the dot product

$$\langle x, y \rangle = x \cdot y = \sum_{i=1}^{n} x_i y_i \text{ for } x, y \in \mathbb{R}^n.$$

An inner product defines a norm of a vector v by $\|v\| = \langle v, v \rangle^{1/2}$.

If W is a complex vector space of dimension n with an Hermitian product $\langle\,,\,\rangle_{\mathbb{C}}$ we may define an inner product $\langle\,,\,\rangle_{\mathbb{R}}$ on W considered as a $2n$-dimensional real vector space by

$$\langle x, y \rangle_{\mathbb{R}} = \text{Re}\,\langle x, y \rangle_{\mathbb{C}},$$

where $\text{Re}\,(\zeta)$ is the real part of a complex number ζ. Since $\langle x, x \rangle_{\mathbb{C}}$ is real, $\langle\,,\,\rangle_{\mathbb{C}}$ and $\langle\,,\,\rangle_{\mathbb{R}}$ define the same norm on W. It is easy to check that the standard Hermitian product on $\mathbb{C}^n$ gives rise to the standard inner product on $\mathbb{R}^{2n}$ when $\mathbb{C}^n$ is considered a real vector space. In fact it is sufficient to check this fact for $\mathbb{C}$ and $\mathbb{R}^2$, which we now do:

$$\text{Re}\,\langle a_1 + b_1 i, a_2 + b_2 i \rangle_{\mathbb{C}} = \text{Re}\,((a_1 + b_1 i)(a_2 - b_2 i)) = a_1 a_2 + b_2 b_2.$$

An inner product in a real vector space allows us to define orthogonality of vectors and angles. This is done as for complex vector spaces and the concepts of orthogonal and orthonormal basis are analogously defined. Also, for any $w \in W$ the space

$$T_w = \{v \in W \mid \langle w, v \rangle = 0\}$$

is an $(n-1)$-dimensional subspace of W. The tangent space to the unit sphere $S_1 \subset W$ at w is T_w, which inherits an inner product and norm from W by restriction.

The assignment of an inner product in a continuous fashion to the tangent space of a differentiable manifold M is called a *Riemannian metric* on M. A Riemannian metric has an associated Finsler structure and a distance on M. Thus an inner product on W defines a Riemannian metric on S_1.

An inner product on W allows us to define the angle between a pair w_1, w_2 of nonzero vectors in W

$$\measuredangle(w_1, w_2) = \arccos \frac{\langle w_1, w_2 \rangle}{\|w_1\| \|w_2\|}.$$

If $w_1, \ldots, w_n$ is an orthonormal basis of W, define the linear map $A : W \to \mathbb{R}^n$ by $A(w_i) = e_i$, the ith element of the standard orthonormal basis of $\mathbb{R}^n$, $e_i = (0, \ldots, 0, 1, 0, \ldots, 0)$, the one being in the ith coordinate. Then $\langle A(w_1), A(w_2) \rangle = \langle w_1, w_2 \rangle$ for all $w_1, w_2 \in W$ so A preserves lengths and angles and A maps the unit sphere S_1 in W to the unit sphere S^{n-1} in $\mathbb{R}^n$. It follows that the distance between two points w_1 and w_2 in S_1 is the angle between them.

We now can describe, for a complex vector space W, the Riemannian distance d_R on $\mathbb{P}(W)$ induced by an Hermitian product. Given two one-dimensional subspaces $\ell_1, \ell_2 \subset W$ let w_1 be a unit vector in ℓ_1 and w_2 the closest point in ℓ_2 to w_1. Let $v_2 = w_2/\|w_2\|$.

Proposition 4 *The distance between ℓ_1 and ℓ_2 in $\mathbb{P}(W)$ equals the angle between w_1 and w_2; that is, $d_R(\ell_1, \ell_2) = \theta$, where θ is the length of the arc of the circle between w_1 and v_2 in S_1.*

To prove the proposition we need a few elementary lemmas and observations.

Lemma 4 *For all $w \in W$ Re $\langle iw, w \rangle = 0$.*

Proof. $\langle iw, w \rangle = i \langle w, w \rangle$ so its real part is zero. $\square$

Lemma 5 *For all $w \in W$ the vector iw is tangent to the fiber of $\rho : S_1 \to \mathbb{P}(W)$ at any point $w \in S_1$.*

Proof.   The fiber of ρ through $w \in S_1$ is $e^{i\theta} w$. Differentiating at $\theta = 0$ proves the lemma. $\square$

If $w \in S_1$ we have used T_w in two contexts to define the orthogonal complement to w. We now distinguish them. Let

$$T_w^{\mathbb{C}} = \{ v \in W \mid \langle v, w \rangle = 0 \}$$

and

$$T_w^{\mathbb{R}} = \{ v \in W \mid \text{Re } \langle v, w \rangle = 0 \}.$$

Lemma 6 *The derivative of ρ at $w \in S_1$ is the orthogonal projection from $T_w^{\mathbb{R}}$ to $T_w^{\mathbb{C}}$ with nullspace the tangent space to the fiber of ρ at w.*

Proof.   This is evident using $T_w^{\mathbb{R}}$ and $T_w^{\mathbb{C}}$ as charts for S_1 and $\mathbb{P}(W)$ and differentiating at w. $\square$

Lemma 7 Re $\langle w_1, i w_2 \rangle = 0$.

Proof. Since w_2 is the closest point in ℓ_2 to w_1, the vector $w_1 - w_2$ is orthogonal in the inner product to ℓ_2 considered as a real two-dimensional space and hence to $i w_2$. It follows that Re $\langle w_1 - w_2, i w_2 \rangle = 0$ and since Re $\langle w_2, i w_2 \rangle = 0$ the lemma follows. □

Lemma 8 *For all $a, b, c, d \in \mathbb{R}$, Re $\langle i(a w_1 + b w_2), (c w_1 + d w_2) \rangle = 0$.*

Proof. This follows from Lemmas 4 and 7. □

Lemma 9 *Let A be the arc from w_2 to v_2 in S_1. Then the tangent vectors to A are orthogonal to the fibers of ρ for the inner product.*

Proof. The tangents to A lie in the real plane spanned by w_1 and w_2, whereas the tangents to the fibers lie in i times this plane. Now Lemma 8 finishes the argument. □

Lemma 10 *The length of $\rho(A)$ in $\mathbb{P}(W)$ equals the length of A in S_1.*

Proof. This follows from Lemmas 6 and 9 and the chain rule. □

Proof of Proposition 4. Since the length of $\rho(A)$ is θ we see that $d_R(\ell_1, \ell_2) \leq \theta$. Now we have to see that the distance is not less than θ. Suppose that we have a differentiable arc $\phi : [0, 1] \rightarrow \mathbb{P}(W)$ such that $\phi(0) = \ell_1, \phi(1) = \ell_2$, and $\ell(\phi) < \theta$. Then we may find a differentiable arc $\psi : [0, 1] \rightarrow S_1$ such that $\psi(0) = w_1$, $\rho\psi = \phi$, and ψ is orthogonal to the fibers of ρ. It follows that $\psi(1) \in \rho^{-1}(\ell_2)$ and that $d(w_1, \psi(1)) \leq \ell(\psi) = \ell(\phi) < \theta$. But v_2 is the closest point to w_1 in $\rho^{-1}(\ell_2)$ and $d(w_1, v_2) = \theta$. So we have reached a contradiction and can conclude that $d_R(\ell_1, \ell_2) = \theta$. □

12.3 The Condition Number (Nonlinear)

We begin by considering the implicit function theorem, which goes as follows (over $\mathbb{R}$). Let $F : \mathbb{R}^k \times \mathbb{R}^m \rightarrow \mathbb{R}^m$ be a C^1 map (F could also be defined locally or over $\mathbb{C}$). Suppose that $F(a_0, y_0) = 0$ and that the partial derivative (with respect to y) matrix $(\partial F / \partial y)(a_0, y_0)$ is nonsingular. Then there exists a C^1 map $G : U(a_0) \rightarrow \mathbb{R}^m$ such that $F(a, G(a)) = 0$ for all a in an open set $U(a_0) \subset \mathbb{R}^k$, and $G(a_0) = y_0$.

The matrix (of partial derivatives) $DG(a_0) : \mathbb{R}^k \rightarrow \mathbb{R}^m$ is called the *condition matrix* at (a_0, y_0). The *condition number* $\mu = \mu(a_0, y_0)$ is defined as the operator norm $\|DG(a_0)\|$ of this matrix. Thinking of a_0 as input and y_0 as output, μ is a bound on the infinitesimal output error in terms of the infinitesimal input error. The following example makes this clearer.

Example 1 Let $\mathcal{P}_d = \{f \mid f(x) = \sum_{i=0}^{d} a_i x^i, \ a_i \in \mathbb{R}\}$ so that $\mathcal{P}_d \simeq \mathbb{R}^{d+1}$ is the space of all polynomials of degree $\leq d$. Define the *evaluation map*

$$ev : \mathcal{P}_d \times \mathbb{R} \to \mathbb{R}$$
$$(f, x) \ \to \ f(x).$$

Then $ev(f, \zeta) = 0$ if and only if ζ is a zero of f. Moreover $\mu(f, \zeta)$ bounds the infinitesimal change in the solution ζ of $f(\zeta) = 0$ as a function of an infinitesimal change in the coefficients of f. Here a norm on $\mathcal{P}_d$ is required for this definition. For the moment we choose that defined by (12.1). This definition of condition number of f and ζ is quite classical.

Proposition 5 *In the preceding example*

$$\mu(f, \zeta) = \frac{\left(\sum_{i=0}^{d} |\zeta^i|^2\right)^{1/2}}{|f'(\zeta)|}.$$

Proof.   The equation $\dot{\zeta} = DG(f)\dot{f}$ is now defined by

$$\dot{f}(\zeta) + f'(\zeta)\dot{\zeta} = 0 \quad \text{or} \quad \dot{\zeta} = -\frac{\dot{f}(\zeta)}{f'(\zeta)};$$

so

$$\mu(f, \zeta) = \|DG(f)\| = \frac{1}{|f'(\zeta)|} \max_{\substack{\dot{f} \in \mathcal{P}_d \\ \|\dot{f}\|=1}} |\dot{f}(\zeta)|.$$

Since the evaluation of a polynomial $\dot{f}$ at a point ζ can be thought of as the dot product of the vector of coefficients $(a_0, \ldots, a_d)$ with $(1, \zeta, \zeta^2, \ldots, \zeta^d)$, it follows from Cauchy–Schwarz that the maximum is attained at an $\dot{f}$ whose vector of coefficients is $\dot{f} = (1, \zeta, \zeta^2, \ldots, \zeta^d) / \left(\sum_{i=0}^{d} |\zeta^i|^2\right)^{1/2}$ and evaluation at this $\dot{f}$ yields the expression in the statement.   □

This example immediately goes over to complex polynomials. We present a special case in the following corollary.

Corollary 2 *Let $f(z) = z^d - 1$ and ζ be a dth root of the unity. Then $\mu(f, \zeta) = (d + 1)^{1/2}/d$.*   □

The fact that $\mu(f, \zeta)$ in this case is less than 1 reflects the fact that the dth root increases precision near 1.

Let us return to the C^1 map $F : \mathbb{R}^k \times \mathbb{R}^m \to \mathbb{R}^m$ and the implicitly defined function G. We may differentiate the equation $F(a, G(a)) = 0$ described earlier as a function of a. This yields

$$\frac{\partial F}{\partial a}(a_0, y_0) + \frac{\partial F}{\partial y}(a_0, y_0)DG(a_0) = 0. \tag{12.3}$$

From this we obtain the explicit expression for the condition matrix

$$DG(a_0) = -\left(\frac{\partial F}{\partial y}(a_0, y_0)\right)^{-1}\frac{\partial F}{\partial a}(a_0, y_0). \qquad (12.4)$$

Thus we see that although $G : U(a_0) \to \mathbb{R}^m$ is given only implicitly, its derivative, the condition matrix, and the condition number as well, are given quite explicitly.

Now suppose that 0 is a regular value of $F : \mathbb{R}^k \times \mathbb{R}^m \to \mathbb{R}^m$. Then

$$V = \{(a, y) \in \mathbb{R}^k \times \mathbb{R}^m \mid F(a, y) = 0\}$$

is a submanifold of dimension k. Let $\pi_1 : V \to \mathbb{R}^k$ be the restriction of the projection onto the first factor.

Proposition 6 *The derivative $D\pi_1(a, y) : T_{(a,y)}(V) \to \mathbb{R}^k$, defined on the tangent space is singular if and only if $(\partial F/\partial y)(a, y)$ is singular.*

Proof.   By differentiating the equation $F(a, y) = 0$ we obtain

$$\frac{\partial F}{\partial a}(a, y)\dot{a} + \frac{\partial F}{\partial y}(a, y)\dot{y} = 0, \quad (\dot{a}, \dot{y}) \in \mathbb{R}^k \times \mathbb{R}^m. \qquad (12.5)$$

Thus $(\dot{a}, \dot{y}) \in T_{(a,y)}(V)$ if and only if $(\dot{a}, \dot{y})$ satisfies (12.5). Then $D\pi_1(a, y)$ is singular at (a, y) if and only if there is a nontrivial tangent vector $(\dot{a}, \dot{y})$ with $\dot{a} = 0$. The rest follows.   □

Suppose $D\pi_1(a_0, y_0)$ is nonsingular and that $g : U \to V$ is an inverse to π_1 defined and differentiable on $U \subset \mathbb{R}^k$, with $a_0 \in U$ and $g(a_0) = (a_0, y_0)$. Then $\pi_2 g(a) = G(a)$ with G the previously defined implicit function.

The situation permits an extension to Riemannian manifolds.

Let X be a manifold of "inputs," Y a manifold of "outputs," and $V \subset X \times Y$ the manifold of "solutions" to some computational problem. Algorithms attempt to invert, or approximate the inverse of, the restriction of the projection $\pi_1 : V \to X$. Suppose that V and X have the same dimension as some kind of local uniqueness hypothesis. If $(a, y) \in V$ is a solution (input a) with $D\pi_1(a, y)$ nonsingular, then the condition matrix,

$$DG(a) : T_a(X) \to T_y(Y),$$

is defined.

In this case we may write as before the condition number $\mu(a, y) = \|DG(a)\|$ using the operator norm. If $D\pi_1(a, y)$ is singular, then we say that (a, y) is *ill-conditioned*. Let $\Sigma' \subset V$ be the set of ill-conditioned solutions. We have

$$\Sigma' = \{(a, y) \in V \mid \text{rank } D\pi_1(a, y) \leq k - 1\}.$$

Let $\Sigma = \pi_1(\Sigma') \subset X$. Then Σ is the set of ill-conditioned inputs.

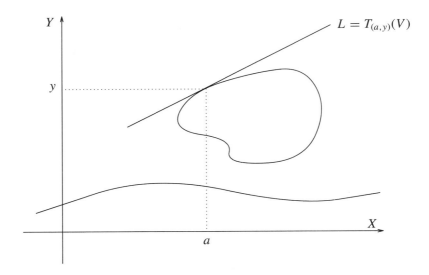

Figure A A simplified situation where $X = Y = \mathbb{R}$. One notes that the ratio of the error δy produced by an error δa is proportional to the slope of L.

We recognize as a special case, the situation of Chapter 10. There is an aspect of universality in the preceding treatment of the condition number.

In fact we now specialize to the case of Chapter 10 and the preceding section of this chapter. Thus let $X = \mathbb{P}(\mathcal{H}_{(d)})$, $Y = \mathbb{P}(\mathbb{C}^{n+1})$, and

$$V = \{(f, \zeta) \in \mathbb{P}(\mathcal{H}_{(d)}) \times \mathbb{P}(\mathbb{C}^{n+1}) \mid f(\zeta) = 0\}.$$

If $(f_0, \zeta_0) \notin \Sigma'$, let G be as in the implicit function theorem, defined in a neighborhood of f_0 with $G(f_0) = \zeta_0$. The condition number $\mu(f_0, \zeta_0)$ is defined by $\|DG(f_0)\|$.

Proposition 7

(a) $\mu : V - \Sigma' \to \mathbb{R}$ is unitarily invariant; that is, for all $\sigma \in U(n+1)$,
$\mu(\sigma(f, \zeta)) = \mu(f, \zeta)$.

(b) Σ' is unitarily invariant.

(c) $\mu(f, \zeta) = \|f\| \|Df(\zeta)|_{T_\zeta}^{-1} \Delta(\|\zeta\|^{d_i - 1})\|$.

Here $\Delta(\|\zeta\|^{d_i - 1})$ is the diagonal matrix $\mathrm{Diag}(\|\zeta\|^{d_1 - 1}, \ldots, \|\zeta\|^{d_n - 1})$. One is using the operator norm on $Df(\zeta)|_{T_\zeta}^{-1} : \mathbb{C}^n \to T_\zeta$, where $T_\zeta = \{v \in \mathbb{C}^{n+1} \mid \langle v, \zeta \rangle = 0\}$ has the induced norm, $\| \ \|$.

For the proof we use two lemmas.

Lemma 11 *Let $\sigma \in U(n+1)$ and $f : \mathbb{C}^{n+1} \to \mathbb{C}$ be a polynomial. Then*

$$D(f \circ \sigma^{-1})(\sigma x)(\sigma w) = Df(x)(w).$$

Proof. The lemma follows from the chain rule,

$$D(f \circ \sigma^{-1})\sigma(x) = Df(x)\sigma^{-1}.$$

Apply this to σw. □

Lemma 12 *The condition number $\mu(f, \zeta)$ is the maximum of $\|\dot{\zeta}\|/\|\zeta\|$ for $\|\dot{f}\|/\|f\| = 1$, where*

$$Df(\zeta)(\dot{\zeta}) + \dot{f}(\zeta) = 0, \quad \dot{\zeta} \in T_{\zeta}, \quad \dot{f} \in T_f. \tag{12.6}$$

Proof. By Proposition 1 of Chapter 10 the tangent space to V is defined by (12.6) and the rest follows as in (12.3) and (12.4) using the operator norm with respect to the norms $\| \ \|_{\zeta}$ and $\| \ \|_f$ in T_{ζ} and T_f, respectively. □

Proof of Proposition 7. We first prove Part (a). At $\sigma(f, \zeta)$ (12.6) reads as

$$D(f \circ \sigma^{-1})(\sigma \zeta)(\sigma \dot{\zeta}) + (\dot{f}\sigma^{-1})(\sigma \zeta) = 0.$$

Then by Lemma 11, $\mu(f\sigma^{-1}, \sigma\zeta) = \mu(f, \sigma)$. By the same token, if $(f, \zeta) \in V - \Sigma'$, then so does $\sigma(f, \zeta)$ proving (b).

To prove (c), note that the right-hand side scales properly relative to $\|\zeta\|$ and $\|f\|$. Then we are able to assume that $\|\zeta\| = \|f\| = 1$. Next there is a unitary transformation $\sigma \in U(n+1)$ which takes ζ to $e_0 = (1, 0, \ldots, 0)$. By unitary invariance, it is thus sufficient to prove (c) in the case that $\zeta = e_0$. From (12.6)

$$\mu(f, e_0) = \max_{\|\dot{f}\|=1} \|Df(e_0)|_{T_{e_0}}^{-1} \dot{f}(e_0)\|.$$

Since $f(e_0) = 0$ and $\dot{f} \in T_f$ the maximum is achieved at a polynomial $\dot{f}$ with $a^i_{(\alpha_0, \alpha_1, \ldots, \alpha_n)} = 0$ if $(\alpha_0, \alpha_1, \ldots, \alpha_n) \neq (d_i, 0, \ldots, 0)$ for $i = 1, \ldots, n$. Then $\dot{f}(e_0)$ is the vector whose ith coordinate is $a^i_{(d_i, 0, \ldots, 0)}$ and therefore $\|\dot{f}\| = \|\dot{f}(e_0)\|$ yielding (c) which in this case is just

$$\mu(f, e_0) = \|Df(e_0)|_{T_{e_0}}^{-1}\|.$$

□

Now we reconsider the condition number of the example $f(x) = x^d - 1$ from the unitarily invariant point of view. Let $\widehat{f}(x_0, x_1) = x_1^d - x_0^d$ be the homogenized version of f.

Proposition 8 *For each zero w of $\widehat{f}$,*

$$\mu(\widehat{f}, w) = \frac{2^{(d-1)/2}}{d}.$$

Proof.   By the previous proposition

$$\mu(\widehat{f}, w) = \|\widehat{f}\| \, \|D\widehat{f}(w)|_{T_w}^{-1}\| \, \|w\|^{d-1}.$$

By unitary invariance we may take $w = (1, 1)$. Then $\|w\| = \sqrt{2}$, $\|\widehat{f}\| = \sqrt{2}$, and one finishes the calculation using the following lemma. □

Lemma 13   *For* $w = (1, 1)$,

$$\|D\widehat{f}(w)|_{T_w}^{-1}\| = \frac{1}{d\sqrt{2}}.$$

Proof.   Note that

$$T_{(1,1)} = \{(u_0, u_1) \in \mathbb{C}^2 \mid u_0 + u_1 = 0\}$$

and

$$D\widehat{f}(1, 1)(u_0, u_1) = du_0 - du_1 = 2du_0$$

and $\|(u_0, u_1)\| = |u_0^2 + u_1^2|^{1/2} = \sqrt{2}|u_0|$. The lemma follows. □

Now we may compare Corollary 2 and Proposition 8. Note that the unitarily invariant version gives exponential increase in the condition number as d increases whereas the usual condition number is bounded and decreases to zero. Thus this sequence of polynomials from our point of view is badly conditioned. We would argue that this reflects the reality of having the roots becoming very close together as d increases. More typically polynomials have roots scattered over $\mathbb{C}$ instead of all being on a single circle. This shows the importance of our choice of a unitarily invariant Hermitian product.

12.4   The Condition Number Theorem

A very general theme in numerical analysis is a relationship between the condition number of a problem and the reciprocal of the distance to the set of ill-posed problems. This section is devoted to proving a very precise statement of this relationship for the condition number $\mu : V - \Sigma' \to \mathbb{R}$ of Proposition 7.

Recall that $V \subset \mathbb{P}(\mathcal{H}_{(d)}) \times \mathbb{P}(\mathbb{C}^{n+1})$ is represented as the set of pairs (f, ζ) with $f(\zeta) = 0$. Then the restriction $\pi_2 : V \to \mathbb{P}(\mathbb{C}^{n+1})$ of the projection onto the second factor has very nice properties (one could say that it is a $U(n + 1)$ equivariant fiber bundle map). The unitary group acts on both the source V and target $\mathbb{P}(\mathbb{C}^{n+1})$ of π_2 as we have seen.

Lemma 14   *For all* $(f, \zeta) \in V$, *and* $\sigma \in U(n + 1)$,

$$\pi_2\sigma(f, \zeta) = \sigma\pi_2(f, \zeta).$$

Proof.   The proof is simply given by

$$\pi_2\sigma(f, \zeta) = \pi_2(f\sigma^{-1}, \sigma\zeta) = \sigma\zeta = \sigma\pi_2(f, \zeta).$$

□

One says about the property of π_2 in Lemma 14, that π_2 commutes with the actions of $U(n + 1)$ or that π_2 is *equivariant*.

The inverse images $\pi_2^{-1}(\zeta) = V_\zeta$ of points under π_2, or *fibers*, have the structure of a projective space. In fact

$$V_\zeta = \{f \in \mathbb{P}(\mathcal{H}_{(d)}) \mid f(\zeta) = 0\}$$

is a projective subspace of $\mathbb{P}(\mathcal{H}_{(d)})$, being defined by the linear equation $f(\zeta) = 0$ (ζ is fixed).

Next we define a real-valued function ρ on V which represents a distance —the *fiber distance*— to the discriminant variety $\Sigma' \subset V$ defined in Chapter 10. Let

$$\rho(f, \zeta) = \text{ distance in } V_\zeta \text{ from } (f, \zeta) \text{ to } \Sigma' \cap V_\zeta.$$

Here distance is measured in the *projective metric d_P* on $\mathbb{P}(\mathcal{H}_{(d)})$ defined by $d_P = \sin d_R$.

Lemma 15 *The function $\rho : V \to \mathbb{R}$ is invariant under the action of $U(n + 1)$ on V. That is, $\rho(\sigma(f, \zeta)) = \rho(f, \zeta)$ for all $\sigma \in U(n + 1)$.*

Proof. The distance d_P in $\mathbb{P}(\mathcal{H}_{(d)})$ is defined by norms and hence is unitarily invariant. Moreover Σ' is unitarily invariant by Proposition 7. Therefore, from the definition of ρ, we have the statement.   □

To obtain the most elegant expression of our condition number theorem we modify slightly the condition number μ of Proposition 7 as follows. Recall (same proposition)

$$\mu(f, \zeta) = \|f\| \|Df(\zeta)|_{T_\zeta}^{-1} \Delta(\|\zeta\|)^{d_i - 1}\|.$$

Define a modification μ_{norm}, the *normalized condition number*, of μ by

$$\mu_{\text{norm}}(f, \zeta) = \|f\| \|Df(\zeta)|_{T_\zeta}^{-1} \Delta(\|\zeta\|^{d_i - 1} d_i^{1/2})\|.$$

Note that the introduction of the $\sqrt{d_i}$ factors is the only change. Moreover,

$$\frac{\mu_{\text{norm}}}{\sqrt{D}} \leq \mu \leq \mu_{\text{norm}},$$

where $D = \max_{1 \leq i \leq n} d_i$.

Lemma 16 *The function $\mu_{\text{norm}} : V \to \mathbb{R}$ is invariant under the action of $U(n+1)$ on V. That is, $\mu_{\text{norm}}(\sigma(f, \zeta)) = \mu_{\text{norm}}(f, \zeta)$ for all $\sigma \in U(n + 1)$.*

Proof. First we note that μ_{norm} scales correctly so it is indeed defined on V. Next we note that if $\sigma \in U(n + 1)$ and $\sigma(\zeta_1) = \zeta_2$, then $\sigma(T_{\zeta_1}) = T_{\zeta_2}$. Thus by the chain rule

$$(D(f \circ \sigma^{-1})\sigma(\zeta)|_{T_{\sigma(\zeta)}})^{-1} = \sigma \circ (Df(\zeta)|_{T_\zeta})^{-1}$$

and so, since σ preserves lengths in $\mathbb{C}^{n+1}$,

$$\|(D(f \circ \sigma^{-1})\sigma(\zeta)|_{T_{\sigma(\zeta)}})^{-1}\Delta(\|\sigma(\zeta)\|^{d_i-1}d_i^{1/2})\|$$
$$= \|(Df(\zeta)|_{T_\zeta})^{-1}\Delta(\|\sigma(\zeta)\|^{d_i-1}d_i^{1/2})\|.$$

By Theorem 1,

$$\|f \circ \sigma^{-1}\| = \|f\|. \tag{12.7}$$

Multiplying equations (12.7) and (12.7) gives

$$\mu_{\text{norm}}(f \circ \sigma^{-1}, \sigma(\zeta)) = \mu_{\text{norm}}(f, \zeta).$$

$\square$

Theorem 3 (Condition Number Theorem) *For all $(f, \zeta) \in V - \Sigma'$,*

$$\mu_{\text{norm}}(f, \zeta) = \frac{1}{\rho(f, \zeta)}.$$

One may also define the condition number μ_{norm}, and the fiber distance ρ for a polynomial system $f \in \mathbb{P}(\mathcal{H}_{(d)})$,

$$\mu_{\text{norm}}(f) = \max_{\zeta|f(\zeta)=0} \mu_{\text{norm}}(f, \zeta)$$
$$\rho(f) = \min_{\zeta|f(\zeta)=0} \rho(f, \zeta).$$

Thus, for example, $\mu_{\text{norm}}(f)$ measures the conditioning of f in terms of its worst conditioned zero.

Corollary of the Condition Number Theorem.

$$\mu_{\text{norm}}(f) = \frac{1}{\rho(f)}.$$

$\square$

Corollary 3 $\mu_{\text{norm}} \geq 1$, and hence $\mu \geq 1/\sqrt{D}$.

Proof.   This follows from the definitions since $\rho \leq 1$. $\square$

Proof of the Condition Number Theorem.   For the proof of the condition number theorem, it is sufficient to verify that

$$\mu_{\text{norm}}(f, e_0) = \frac{1}{\rho(f, e_0)},$$

where $e_0 = (1, 0, \ldots, 0) \in \mathbb{P}(\mathbb{C}^{n+1})$. This follows from the unitary invariance of μ_{norm} and ρ shown in Proposition 7 and Lemma 16. One also uses the fact that for any $\zeta \in \mathbb{P}(\mathbb{C}^{n+1})$, there is some $\sigma \in U(n+1)$ with $\sigma\zeta = e_0$. Thus $\rho(f, \zeta) = \rho(f\sigma^{-1}, e_0)$ and $\mu_{\text{norm}}(f, \zeta) = \mu_{\text{norm}}(f\sigma^{-1}, e_0)$.

For $f \in \mathcal{H}_{(d)}$, let $f = (f_1, \ldots, f_n)$ and

$$f_i(x) = a_i x_0^{d_i} + x_0^{d_i-1} \Sigma a_{ij} x_j + \ldots. \tag{12.8}$$

Let $\widehat{V}_{e_0} = \{f \in \mathcal{H}_{(d)} \mid f(e_0) = 0\}$. If $f \in \widehat{V}_{e_0}$, then the a_i are all zero and conversely. Let

$$L_{(d)} = \{f \in \mathcal{H}_{(d)} \mid f_i(x) = x_0^{d_i-1} \Sigma a_{ij} x_j\},$$

$J_{(d)}$ be the orthogonal complement of $L_{(d)}$ in $\widehat{V}_{e_0}$, and $\pi : \widehat{V}_{e_0} \to L_{(d)}$ the projection.

Note that

$$T_{e_0} = \{u \in \mathbb{C}^{n+1} \mid \langle u, e_0 \rangle = 0\} = \{(0, u_1, \ldots, u_n) \in \mathbb{C}^{n+1}\}$$

can be identified with $\mathbb{C}^n$. In this way $Df(e_0)$ may be regarded as a linear map from $\mathbb{C}^n$ to $\mathbb{C}^n$ and as a matrix. This is the matrix (a_{ij}) in (12.8).

Let $\mathcal{M}_n^{\mathbb{C}}$ be the space of complex $n \times n$ matrices endowed with the Frobenius norm $\|M\|_F^2 = \sum |M_{ij}|^2$ and its associated distance d_F.

The norm $\|\ \|$ on $\mathcal{H}_{(d)}$ induces a norm on $L_{(d)}$ and a distance d on $\mathcal{H}_{(d)}$ and $L_{(d)}$. The following is a consequence of the definition of our norm on $\mathcal{H}_{(d)}$ and the definition of Σ'. Recall that $\Delta(a_i)$ denotes the diagonal matrix having a_i at the ith entry on its diagonal.

Lemma 17 *The map* $A : L_{(d)} \to \mathcal{M}_n^{\mathbb{C}}$ *sending* f *to* $\Delta(d_i^{-1/2})Df(e_0)$ *is a norm preserving linear isomorphism. The image of* $\Sigma' \cap L_{(d)}$ *is the set* S *of singular matrices.* □

Now suppose that $\|f\| = 1$. Then

$$\mu_{\text{norm}}(f, e_0) = \|Df(e_0)^{-1}\Delta(d_i^{1/2})\|.$$

By the theorem of Eckart and Young of the previous chapter,

$$\mu_{\text{norm}}(f, e_0) = \frac{1}{d_F(Df(e_0)\Delta(d_i^{-1/2}), S)}$$

in $\mathcal{M}_n^{\mathbb{C}}$. From Lemma 17 this is equal to

$$\frac{1}{d(\pi(f), A^{-1}(S))}$$

in $L_{(d)}$. And, since $L_{(d)} \bigoplus J_{(d)}$ is an orthogonal decomposition of $\widehat{V}_{e_0}$ and $\Sigma' = \pi^{-1}(S)$,

$$d(\pi(f), A^{-1}(S)) = d(f, \Sigma' \cap \widehat{V}_{e_0})$$

and

$$\mu_{\mathrm{norm}}(f, e_0) = \frac{1}{d(f, \Sigma' \cap \widehat{V}_{e_0})} \quad \text{in } \widehat{V}_{e_0}.$$

Finally, since $\|f\| = 1$, the distance between f and the closest point in $\Sigma' \cap \widehat{V}_{e_0}$ is the sin of the angle between them and this is, by Proposition 4 and the definition of d_P, their d_P distance.

This finishes our proof of the condition number theorem.   □

12.5   Additional Comments and Bibliographical Remarks

The Hermitian product on the space of polynomials in Section 12.1 in the one-variable case, with its unitary invariance was introduced in [Weyl 1932].

However, the development of Section 12.1 does not seem to be so generally known. Eric Kostlan first pointed this out to us, and in [Kostlan 1993], the sketch of our proof of unitary invariance can be found.

The same inner product is referred to as the Bombieri inner product or norm in [Beauzamy and Dégot 1995] (where one can see further references). See also [Brockett 1973], [Reznick 1992], and [Stein and Weiss 1971] for further use of this inner product.

The condition number of Section 12.3 owes much of its development to Wilkinson [1963]. In particular there one may see the one-variable polynomial case and the definition for linear systems. See [Demmel 1987a, 1987b] for some earlier condition number theorems and their history. The condition number theorem here is in [Shub and Smale 1993a], also [Shub and Smale 1993c] and [Shub and Smale 1996]. One may see these last three papers for general background for this chapter.

Recently Dedieu [1996] has put the condition number theorem into a general setting. See also [Dedieu 1997, TAa].

13

The Condition Number in $\mathbb{P}(\mathcal{H}_{(d)})$

In this chapter we study the condition numbers $\mu(f)$ and $\mu_{\text{norm}}(f)$ as functions on $\mathbb{P}(\mathcal{H}_{(d)})$ in greater depth. Our main theorem is proven in Section 13.6.

Theorem 1 *Let $n > 1$. The probability that $\mu_{\text{norm}}(f) > 1/\varepsilon$ for $f \in \mathbb{P}(\mathcal{H}_{(d)})$ and $\varepsilon > 0$ is less than or equal to*

$$\varepsilon^4 n^3 (n + 1)(N - 1)(N - 2)\mathcal{D},$$

where $N = \dim \mathcal{H}_{(d)}$ and $\mathcal{D}$ is the Bézout number.

We first prove Theorem 1 for linear systems in Section 5 and then apply the result on linear systems to the general case. All this requires that we develop some more integral formulas for the solution variety V. After a review of integration on manifolds in Section 13.1 we apply the co-area formula of geometric integration theory in Section 13.2 to deduce Thorem 3 there.

A particular instance of Theorem 3 is developed in Section 13.3. Let $\mathcal{H}_{(d)}^{\mathbb{R}}$ be the real subspace of $\mathcal{H}_{(d)}$ consisting of systems of equations with real coefficients.

Theorem 2 *The average number of real roots of a polynomial system $f \in \mathbb{P}(\mathcal{H}_{(d)}^{\mathbb{R}})$ is $\mathcal{D}^{1/2}$ where $\mathcal{D}$ is the Bézout number.*

13.1 Integration on Manifolds

In this section we give a whirlwind tour of integration of real-valued functions defined on differentiable manifolds. We are extremely brief, omitting most details.

References for the subject can be found in Section 13.7. Our goal is to deal with spaces slightly more general than spheres and Euclidean space. Our main tools are Theorem 4 (the co-area formula) and Theorem 5. These are applied in natural settings, where the main point is that the relevant volume forms are unitarily invariant.

We recall that a *volume form* ω on an n-dimensional real vector space V is the absolute value of an alternating n-multilinear map $v : V^n \to \mathbb{R}$; that is, $\omega(v_1, \ldots, v_n) = |v(v_1, \ldots, v_n)|$, where v is n-multilinear and, for any permutation σ of $\{1, \ldots, n\}$, $v(v_{\sigma(1)}, \ldots, v_{\sigma(n)}) = \text{sign}(\sigma)v(v_1, \ldots, v_n)$. A well-known example of an alternating n-multilinear map is the determinant.

A straightforward calculation proves the following proposition.

Proposition 1 *Let $v_1, \ldots, v_n$ be a basis of V and $w_i = \sum_{j=1}^{n} a_{ij}v_j$ for $i = 1, \ldots, n$, where $A = (a_{ij})$ is an $n \times n$ matrix of real numbers. Then*

(a) $v(w_1, \ldots, w_n) = \det(A)v(v_1, \ldots, v_n)$;

(b) $\omega(w_1, \ldots, w_n) = |\det(A)|\omega(v_1, \ldots, v_n)$.

$\square$

Thus, a volume form is determined by its value on a basis of V. If V has an inner product and $e_1, \ldots, e_n$ is an orthonormal basis of V, the volume form $de_1 \ldots de_n$ which assigns the number 1 to $(e_1, \ldots, e_n)$ is the *natural volume form with respect to the inner product*. For $\mathbb{R}^n$ with the usual inner product the natural volume form is $dx_1 \ldots dx_n$, and $dx_1 \ldots dx_n(v_1, \ldots, v_n)$ is the volume of the parallelepiped spanned by the vectors $v_1, \ldots, v_n$ (see Figure A).

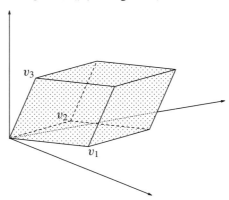

Figure A The parallelepiped spanned by v_1, v_2, v_3 in $\mathbb{R}^3$.

If V and U are now two vector spaces with $\dim V = \dim U = n$, alternating n-forms v and u with corresponding volume forms ω and μ, and $A : V \to U$ is linear, then we may define

$$\det A = \frac{u(A(v_1), \ldots, A(v_n))}{v(v_1, \ldots, v_n)}$$

for any basis $v_1, \ldots, v_n$ of V. Also,

$$|\det A| = \frac{\mu(A(v_1), \ldots, A(v_n))}{\omega(v_1, \ldots, v_n)}.$$

So, in particular, if V and U have inner products, ω and μ are equal to one on an orthonormal basis, and A is expressed as a square matrix with respect to the orthonormal basis of V and U, then $\det A$ is the determinant of this matrix.

If $x \in S^{n-1}$, the unit sphere in $\mathbb{R}^n$, the tangent space $T_x S^{n-1}$ inherits an inner product from $\mathbb{R}^n$ and has a natural volume form associated with it which is easily seen to be equal to $|\det(v_1, \ldots, v_{n-1}, x)|$, where $v_1, \ldots, v_{n-1}$ are tangent to S^{n-1} at x. An analogous statement is true for the real projective space $\mathbb{P}(\mathbb{R}^n)$ which is just S^{n-1} with x and $-x$ (and hence $v \in T_x S^{n-1}$ and $-v \in T_{-x} S^{n-1}$) identified.

For complex projective space $\mathbb{P}(\mathbb{C}^{n+1})$, the tangent space $T_x \mathbb{P}(\mathbb{C}^{n+1})$ has an inner product defined in Chapter 12 and an associated volume form $\omega(x)$ defined by

$$\omega(x)(v_1, \ldots, v_{2n}) = \frac{|\det(v_1, \ldots, v_{2n}, x, ix)|}{\|x\|^{2n+2}},$$

where $v_1, \ldots, v_{2n}$ are $2n$ real vectors in $T_x \mathbb{P}(\mathbb{C}^{n+1})$ considered as a real vector space.

Similarly for any $(n+1)$-dimensional complex vector space V with an Hermitian product,

$$\frac{|\det(v_1, \ldots, v_{2n}, f, if)|}{\|f\|^{2n+2}}$$

is the natural volume form on $T_f \mathbb{P}(V)$.

A *volume form on a differentiable manifold M* is a smoothly varying volume form on the tangent spaces $T_m M$ for $m \in M$.

Thus, for example, $dx_1 \ldots dx_n$ is a volume form on any open set in $\mathbb{R}^n$. Analogously, $\omega(x)$ is a volume form on $\mathbb{P}(\mathbb{C}^{n+1})$. If M has a smooth Riemannian metric, then the associated volume form on the tangent spaces to M are a volume form on M.

In general, if we are given a Riemannian manifold M, then unless otherwise stated the volume form on M (which we write dM when no confusion is possible) is the associated volume form to the Riemannian metric. If $V \subset M$ is a submanifold, then V inherits a Riemannian metric and an associated volume form from M.

Thus we have so far defined volume forms on open sets of $\mathbb{R}^n$, spheres S^{n-1}, the projective spaces $\mathbb{P}(\mathcal{H}_{(d)})$, $\mathbb{P}(\mathbb{C}^{n+1})$, $\mathbb{P}(\mathbb{R}^n)$, and submanifolds of Riemannian manifolds such as the solution variety $V \subset \mathbb{P}(\mathcal{H}_{(d)}) \times \mathbb{P}(\mathbb{C}^{n+1})$ and $V_x = \pi_2^{-1}(x)$, where $\pi_2 : V \to \mathbb{P}(\mathbb{C}^{n+1})$ is the projection on the second factor.

Now we define the *integral $\int_M f d\omega$ of a function f, $f : M \to \mathbb{R}$, with respect to the volume form ω on M.* If M is not compact, we assume that f is zero in the complement of a compact set.

We begin with an open set $U \subset \mathbb{R}^n$. By the previous proposition a volume form ω on U is given by a nonnegative function $\Phi : U \to \mathbb{R}$ such that $\omega = \Phi dx_1 \ldots dx_n$.

If $f : U \to \mathbb{R}$, then we define

$$\int_U f\, d\omega = \int_U f\Phi\, dx_1 \dots dx_n,$$

the right-hand side being the standard Riemann integral of $f\Phi$ on U. To reduce the definition of $\int_M f\, d\omega$ to the one we have already made for open sets in $\mathbb{R}^n$ we use charts and partitions of unity. Once again we are being very brief for the uninitiated and recommend the references given in Section 13.7.

Given a differentiable n-dimensional manifold M, a chart for M is a diffeomorphism $\psi : U \to M$ for $U \subset \mathbb{R}^n$; that is, ψ is smooth with a smooth inverse. A compact differentiable manifold M has a finite cover by charts, $\psi_i(U_i)$ for $i = 1, \dots, m$, where $\psi_i : U_i \to M$ are charts of M. Given a finite open cover W_i, $i = 1, \dots, m$ of M there is a *partition of unity* subordinate to the cover W_i, that is, functions $\Phi_i : W_i \to \mathbb{R}_+$ for $i = 1, \dots, m$ such that Φ_i is smooth, Φ_i is zero in the complement of a compact set, and $\sum_{i=1}^m \Phi_i(x) = 1$ for all $x \in M$.

Now given $f : M \to \mathbb{R}$, a volume form ω on M, charts $\psi_i : U_i \to M$ for $i = 1, \dots, m$ defining a cover $\psi_i(U_i)$ of M, and a partition of unity $\Phi_i : \psi_i(U_i) \to \mathbb{R}_+$ for $i = 1, \dots, m$ subordinate to the cover $\psi_i(U_i)$ we define

$$\int_M f\, d\omega = \sum_{i=1}^m \int_{U_i} \Phi_i f \psi_i\, d(\psi_i^* \omega),$$

where $\psi_i^* \omega(v_1, \dots, v_n) = \omega(D\psi_i(v_1), \dots, D\psi_i(v_n))$.

Using Proposition 1 and the change of variables formula in Euclidean space it is not difficult to prove that $\int_M f\, d\omega$ does not depend on the partition of unity or the charts chosen, but only on f and ω. Thus we have defined $\int_M f\, d\omega$.

13.2 A General Integral Formula

Let M, N be real compact Riemannian manifolds and V a compact submanifold of the product $M \times N$ with dim $V = $ dim M. Suppose that the restriction $\pi_2 : V \to N$ of the projection $M \times N \to N$ is regular; that is, every $z \in N$ is a regular value. Let $V_z = \pi_2^{-1}(z)$. Then V_z is a Riemannian submanifold of $M \times N$. Let x be a regular value of $\pi_1 : V \to M$. Then the number of points in $\pi_1^{-1}(x)$ is finite. Given an open subset U of V, we let $\#_U(x)$ be the number of points in $\pi_1^{-1}(x) \cap U$. If $U = V$, we simply write $\#(x)$. Since dimension $V = $ dim M we are in the context of the implicit function theorem discussed in Section 12.3 for Riemannian manifolds. We suppose that Σ', the critical set of π_1, is lower-dimensional and hence the same is true for its image by π_1, Σ. Thus, Σ and Σ' have zero measure and when integrating we may ignore them. The main goal of this section is to prove the following theorem.

Theorem 3 *Let U be an open subset of V. Then*

$$\int_{x \in \pi_1(U)} \#_U(x)\, dM = \int_{z \in N} \int_{(a,z) \in V_z \cap U} \det(DG(a)DG^*(a))^{-1/2}\, dV_z\, dN,$$

where $DG(a)$ is the condition matrix of V at the point (a, z).

For the proof of Theorem 3 we require the co-area formula which is a generalization of formula (11.2) and some lemmas from linear algebra. We begin with the statement of the co-area formula. For a proof, see the references in Section 13.7.

Suppose $F : X \to Y$ is a surjective map from a Riemannian manifold X to a Riemannian manifold Y, and suppose that the derivative $DF(x) : T_x(X) \to T_{F(x)}(Y)$ is surjective for almost all $x \in X$. The horizontal space H_x of $T_x(X)$ is defined as the orthogonal complement to $\ker DF(x)$. The horizontal derivative of F at x is the restriction of $DF(x)$ to H_x. The *Normal Jacobian* $NJ(F(x))$ is the absolute value of the determinant of the horizontal derivative, defined almost everywhere on X. For the map $\pi_1 : V \to M$ one has $NJ(\pi_1(a, x)) = |\det D\pi_1(a, x)|$. This follows from the fact that $\dim V = \dim M$ and that for almost all $x \in V$, x is a regular value of π_1, and thus $\ker D\pi_1(a, z) = 0$.

The analogue of (11.2) now takes the following form.

Theorem 4 (Co-Area Formula) *Let $F : X \to Y$ be a map of Riemannian manifolds satisfying the preceding surjectivity conditions. Then for $\phi : X \to \mathbb{R}$,*

$$\int_X \phi(x) dX = \int_{z \in Y} \int_{x \in F^{-1}(z)} \phi(x) \frac{1}{NJ(F(x))} \, dF^{-1}(z) \, dY,$$

where the integral on the left is assumed to exist. $\square$

Now we turn to the linear algebra lemmas.

Let H_1 and H_2 be finite-dimensional real vector spaces with inner product and $A : H_1 \to H_2$ be linear with graph $\Gamma(A)$; that is, $\Gamma(A) = \{(x, A(x)) \mid x \in H_1\}$. Then $\Gamma(A)$ inherits the inner product structure of the product $H_1 \times H_2$ and $\dim H_1 = \dim \Gamma(A)$. Let $\pi_i : \Gamma(A) \to H_i$ be the restriction of the projection, $i = 1, 2$. The map π_1 is a linear isomorphism from $\Gamma(A)$ to H_1, so $\det \pi_1$ makes sense. As in the previous section, the determinant of π_1 may be expressed as the determinant of any matrix representation of π_1 in orthonormal bases of $\Gamma(A)$ and H_1. The same applies to its inverse $x \to (x, A(x))$.

Recall that A^* denotes the adjoint of A, that is, its transpose A^T if A is real and its transpose conjugate $\overline{A}^T$ if A is complex.

Lemma 1 *Let V and W be vector spaces and $B : V \to W$ be an injective linear map. Then there exists an orthogonal linear map $O : V \to W$ such that $O(B^*B)^{1/2} = B$.*

Proof. Let v be any vector in V. Then,

$$\langle B^*Bv, v \rangle = \langle Bv, Bv \rangle.$$

Since B^*B is a symmetric matrix, it diagonalizes, and thus we may consider its square root. Then

$$\langle B^*Bv, v \rangle = \langle (B^*B)^{1/2}v, (B^*B)^{1/2}v \rangle.$$

Therefore, for all $v \in V$, $\langle (B^*B)^{1/2}v, (B^*B)^{1/2}v \rangle = \langle Bv, Bv \rangle$ from which the conclusion follows. $\qquad\square$

Lemma 2

$$|\det \pi_1| = \frac{1}{\det(I + A^*A)^{1/2}}.$$

Proof. Let $\begin{pmatrix} I \\ A \end{pmatrix} : H_1 \to \Gamma(A) \subset H_1 \times H_2$ be the map $x \to (x, Ax)$. By the previous lemma, there is an orthogonal map $B : H_1 \to \Gamma(A)$ such that

$$B\left(\left(\begin{pmatrix} I \\ A \end{pmatrix}^* \begin{pmatrix} I \\ A \end{pmatrix} \right)^{1/2} \right) = \begin{pmatrix} I \\ A \end{pmatrix}$$

and hence $\left| \det \begin{pmatrix} I \\ A \end{pmatrix} \right| = \det(I + A^*A)^{1/2}$. Also

$$\det \pi_1 = \frac{1}{\det \begin{pmatrix} I \\ A \end{pmatrix}}$$

since $\pi_1 \circ \begin{pmatrix} I \\ A \end{pmatrix} = I$. These two equalities give the proof. $\qquad\square$

Lemma 3 *If A is surjective, then*

(a) *The orthogonal complement of* $\ker \pi_2$ *is*

$$W = \{(A^*(AA^*)^{-1}v_2, v_2) \mid v_2 \in H_2\};$$

(b) $|\det \pi_1| \cdot \dfrac{1}{|\det(\pi_2|W)|} = \det(AA^*)^{-(1/2)}.$

Note that if A is surjective, then A^* is injective. In that case, since the image of A^* is orthogonal to $\ker A$, AA^* is invertible.

Proof.

(a) Since $\ker \pi_2 = \{(v, 0) \mid v \in \ker A\}$ it follows that $\ker \pi_2$ and W are orthogonal. Their dimensions add to $\dim H_1 + \dim H_2$. So W is the orthogonal complement of $\ker \pi_2$.

(b) Using Lemma 2 applied to $\pi_2|W$ we get

$$\frac{1}{|\det(\pi_2|W)|} = \det(I + (A^*(AA^*)^{-1})^*(A^*(AA^*)^{-1}))^{1/2}$$

$$= \det(I + (AA^*)^{-1})^{1/2}.$$

Note that A^*A and AA^* have the same nonzero eigenvalues. Hence $\det(I + A^*A) = \det(I + AA^*)$. Thus, applying Lemma 2 now to π_1 and multiplying we get

$$
\begin{aligned}
\frac{|\det\pi_1|}{\det(\pi_2|W)} &= \det(I + AA^*)^{-1/2}\det(I + (AA^*)^{-1})^{1/2} \\
&= \det(AA^*(I + (AA^*)^{-1}))^{-1/2}\det(I + (AA^*)^{-1})^{1/2} \\
&= \det(AA^*)^{-1/2}\det(I + (AA^*)^{-1})^{-1/2}\det(I + (AA^*)^{-1})^{1/2} \\
&= \det(AA^*)^{-1/2}.
\end{aligned}
$$

$\square$

We are now ready for the proof of Theorem 3.

Proof of Theorem 3. By the co-area formula with $X = V$, $Y = M$, $\phi = |\det(D\pi_1)|$, and $F = \pi_1$,

$$
\int_{(a,z)\in U} |\det(D\pi_1(a, z))| dV = \int_{a\in\pi_1(U)} \int_{\pi_1^{-1}(a)\cap U} 1 \, d\pi_1^{-1}(a) \, dM
$$
$$
= \int_{a\in\pi_1(U)} \#_U(a) dM.
$$

By the co-area formula again with $\phi = |\det(D\pi_1)|$ but projecting by π_2,

$$
\int_{(a,z)\in U} |\det(D\pi_1(a, z))| dV
$$
$$
= \int_{z\in N} \int_{(a,z)\in V_z\cap U} |\det(D\pi_1(a, z))| \frac{1}{\mathrm{NJ}(\pi_2(a, z))} \, dV_z \, dN.
$$

Since the tangent space to V at (a, z) is the graph of $DG(a)$ we apply Lemma 3 with $H_1 = T_a(M)$, $H_2 = T_z(N)$, and $A = DG(a)$ to conclude that

$$
|\det D\pi_1(a, z)| \frac{1}{\mathrm{NJ}(\pi_2(a, z))} = \det(DG(a)DG(a)^*)^{-1/2}.
$$

Putting the three equalities together, we are done. $\square$

The complex version of Theorem 3 is also true. That is, we assume that $V \subset M \times N$, where V, M, N are compact complex manifolds, that the complex dimension of V equals the complex dimension of M, and that the rest of the regularity conditions on π_1, π_2, $D\pi_1$, $D\pi_2$, and Σ' are satisfied.

Theorem 5

$$
\int_{x\in\pi_1(U)} \#_U(x) dM = \int_{z\in N} \int_{(a,z)\in V_z\cap U} \det(DG(a)DG(a)^*)^{-1} \, dV_z \, dN,
$$

where $(a, z) \in V_z \cap U$ and $DG(a)$ is the complex condition matrix of V at (a, z).

Proof. The proof follows immediately from Theorem 3 and the fact that if $A : \mathbb{C}^n \to \mathbb{C}^n$ is a complex linear map and $A_\mathbb{R} : \mathbb{R}^{2n} \to \mathbb{R}^{2n}$ the real linear map it defines, then $|\det A_\mathbb{R}| = |\det A|^2$. $\square$

13.3   Integration in V and the Average Number of Real Roots

In this section we first develop some more abstract integral formulas given the presence of symmetries and then apply them to prove our theorem on the average number of real roots.

Suppose that we are in the context of the last section, where V, M, and N are compact Riemannian manifolds and $\pi_1 : V \to M$, $\pi_2 : V \to N$ are the restrictions of π_1 and π_2 to V. A transformation $\sigma : X \to X$ of a Riemannian manifold X is called an *isometry* if the inner products are preserved by the derivative of σ; that is, $\langle D\sigma_x(v), D\sigma_x(w)\rangle_{\sigma_x} = \langle v, w\rangle_x$ for $x \in X$ and $v, w \in T_x X$.

Lemma 4   *Suppose that $\sigma_1 : V \to V$ and $\sigma_2 : N \to N$ are isometries such that $\pi_2\sigma_1 = \sigma_2\pi_2$. Then*

$$\mathrm{NJ}(\pi_2\sigma_1(x)) = \mathrm{NJ}(\pi_2(x)).$$

Proof.   Since $\pi_2\sigma_1 = \sigma_2\pi_2$ the isometry σ_1 maps $\pi_2^{-1}(x)$ to $\pi_2^{-1}(\sigma_2(x))$. Hence $D\sigma_1$ maps $\ker D\pi_2(x)$ to $\ker D\pi_2(\sigma_1(x))$, and since $D\sigma_1$ preserves inner products it maps the horizontal space H_x to $H_{\sigma_1(x)}$. Now apply the chain rule

$$D\sigma_2^{-1}(\sigma_2(x))(D\pi_2|H_{\sigma_1(x)})(D\sigma_1(x)|H_x) = (D\pi_2|H_x).$$

Since $D\sigma_2^{-1}(\sigma_2(x))$ and $D\sigma_1(x)|H_x$ are isometries $|\det(D\sigma_2)^{-1}(\sigma_2(x))| = 1 = |\det(D\sigma_1(x)|H_x)|$. Hence $|\det(D\pi_2|H_{\sigma_1}(x))| = |\det(D\pi_2|H_x)|$.   □

The next lemma is straightforward.

Lemma 5   *If $\sigma_1 : V \to V$ and $\sigma_3 : M \to M$ are isometries such that $\pi_1\sigma_1 = \sigma_3\pi_1$, then*

$$|\det D\pi_1(\sigma_1(x))| = |\det D\pi_1(x)|.$$

□

We now return to our standard example of the solution variety

$$V \subset \mathbb{P}(\mathcal{H}_{(d)}) \times \mathbb{P}(\mathbb{C}^{n+1}), \quad (d) = (d_1, \ldots, d_n).$$

Let $U \subset V$ and $U_z = U \cap V_z$ for $z \in \mathbb{P}(\mathbb{C}^{n+1})$. We say that U is *unitarily invariant* if for every unitary transformation $B : \mathbb{C}^{n+1} \to \mathbb{C}^{n+1}$ one has that $B(U)_z = U_{B(z)}$, where, we recall, $B(f, z) = (f \circ B^{-1}, Bz)$. Let $Tf(e_0) : T_{e_0} \to \mathbb{C}^n$ be defined by $Tf(e_0) = Df(e_0)|T_{e_0}$ for $\|f\| = 1$.

Proposition 2   *Let $U \subset V$ be unitarily invariant. Then*

$$\int_{\mathbb{P}(\mathcal{H}_{(d)})} \#_U \, d\mathbb{P}(\mathcal{H}_{(d)}) = \mathrm{Vol}\,(\mathbb{P}(\mathbb{C}^{n+1})) \int_{V_{e_0} \cap U} \det(Tf(e_0)^* Tf(e_0)) dV_{e_0},$$

where the representative for $f \in \mathbb{P}(\mathcal{H}_{(d)})$ is assumed to have norm 1.

Proof.   As in the proof of Theorem 3

$$\int_{\mathbb{P}(\mathcal{H}_{(d)})} \#_U \, d\mathbb{P}(\mathcal{H}_{(d)}) = \int_{\mathbb{P}(\mathbb{C}^{n+1})} \int_{V_z \cap U} |\det D\pi_1| \frac{1}{\mathrm{NJ}(\pi_2)} \, dV_z \, d\mathbb{P}(\mathbb{C}^{n+1}).$$

Let $B : \mathbb{C}^{n+1} \to \mathbb{C}^{n+1}$ be a unitary transformation with $B(z) = e_0$. Then the induced map $B : V_z \cap U \to V_{e_0} \cap U$ is an isometry. Thus, from Lemmas 4 and 5 it follows that

$$\int_{V_z \cap U} |\det D\pi_1| \frac{1}{\mathrm{NJ}(\pi_2)} dV_z = \int_{V_{e_0} \cap U} |\det D\pi_1| \frac{1}{\mathrm{NJ}(\pi_2)} dV_{e_0}$$

for all z. Since

$$|\det D\pi_1| \frac{1}{\mathrm{NJ}(\pi_2)} = \det(DG(f)DG(f)^*)^{-1},$$

where $DG(f)$ is considered as a complex linear map, we have that

$$\int_{\mathbb{P}(\mathcal{H}_{(d)})} \#_U \, d\mathbb{P}(\mathcal{H}_{(d)})$$

$$= \int_{\mathbb{P}(\mathbb{C}^{n+1})} \int_{V_{e_0} \cap U} \det(DG(f)DG(f)^*)^{-1} \, dV_{e_0} \, d\mathbb{P}(\mathbb{C}^{n+1})$$

$$= \mathrm{Vol}\,(\mathbb{P}(\mathbb{C}^{n+1})) \int_{V_{e_0} \cap U} \det(DG(f)DG(f)^*)^{-1} dV_{e_0}.$$

It remains to show that $\det(DG(f)DG(f)^*)^{-1} = \det(Tf(e_0)^* Tf(e_0))$.

Let $\widehat{K}_{e_0} \subset \mathcal{H}_{(d)}$ be the space of polynomial systems f such that $f_i = \sum a^i_{(\alpha_0,\dots,\alpha_n)} x^\alpha$ with $a^i_{(\alpha_0,\dots,\alpha_n)} = 0$ if $(\alpha_0,\dots,\alpha_n) \neq (d_1, 0, \dots, 0)$. Then $\widehat{K}_{e_0}$ is orthogonal to $\widehat{V}_{e_0}$ and $\widehat{K}_{e_0} \oplus \widehat{V}_{e_0} = \mathcal{H}_{(d)}$. Since e_0 is a root of all the polynomial systems in $\widehat{V}_{e_0}$, $DG|\widehat{V}_{e_0} = 0$. As in the proof of Proposition 7 of Chapter 12, one has that $DG|\widehat{K}_{e_0}$ can be identified with $Tf(e_0)^{-1}$ and the result follows.   □

Remark 1 The equation

$$\det(DG(f)DG(f)^*)^{-1} = \det(Tf(\zeta)^* Tf(\zeta))$$

holds for every pair $(f, \zeta) \in V$ with $\|f\| = \|\zeta\| = 1$. This follows from the validity of the equation for $\zeta = e_0$, since both expressions are unitarily invariant.

Proposition 2 has a real version. We let $\mathcal{H}^{\mathbb{R}}_{(d)} \subset \mathcal{H}_{(d)}$ be those systems of polynomials all of whose coefficients are real. The set $\mathcal{H}^{\mathbb{R}}_{(d)}$ inherits the Riemannian structure from $\mathcal{H}_{(d)}$. Let $V^{\mathbb{R}} \subset \mathbb{P}(\mathcal{H}^{\mathbb{R}}_{(d)}) \times \mathbb{P}(\mathbb{R}^{n+1})$ be the algebraic set defined by $\{(f, z) \mid f(z) = 0\}$. Then $V^{\mathbb{R}}$ is a smooth compact manifold. The proof is the same as in the complex case. We let $\pi_1 : V^{\mathbb{R}} \to \mathbb{P}(\mathcal{H}^{\mathbb{R}}_{(d)})$ and

$\pi_2 : V^{\mathbb{R}} \to \mathbb{P}(\mathbb{R}^{n+1})$ be the restrictions of π_1 and π_2 to $V^{\mathbb{R}}$. By the inverse function theorem dim $V^{\mathbb{R}} = \dim \mathbb{P}(\mathcal{H}_{(d)}^{\mathbb{R}})$. The set of critical points Σ' of π_1 is still lower-dimensional and has measure 0. Now orthogonal matrices act on the solution variety $V^{\mathbb{R}}$ by isometries. If B is orthogonal, $B(f, z) = (f \circ B^{-1}, Bz)$. Let $V_z = \pi_2^{-1}(z)$.

Proposition 3   *Let $U \subset V^{\mathbb{R}}$ be orthogonally invariant. Then*

$$\int_{\mathbb{P}(\mathcal{H}_{(d)}^{\mathbb{R}})} \#_U \, d\mathbb{P}(\mathcal{H}_{(d)}^{\mathbb{R}}) = \text{Vol}\,(\mathbb{P}(\mathbb{R}^{n+1})) \int_{V_{e_0}^{\mathbb{R}} \cap U} (\text{DET})^{1/2} dV_{e_0}^{\mathbb{R}},$$

where $\text{DET} = \det(Tf(e_0)^* Tf(e_0))$ *and the representative for* $f \in \mathbb{P}(\mathcal{H}_{(d)}^{\mathbb{R}})$ *is assumed to have norm 1.*

Proof.   It is the same as Proposition 2 using Theorem 3.   □

For f in $\mathbb{P}(\mathcal{H}_{(d)}^{\mathbb{R}})$ let $\#(f)$ denote the number of real roots of f. Then

$$A_{(d)} = \frac{1}{\text{Vol}\,(\mathbb{P}(\mathcal{H}_{(d)}^{\mathbb{R}}))} \int_{\mathbb{P}(\mathcal{H}_{(d)}^{\mathbb{R}})} \#(f) \, d\mathbb{P}(\mathcal{H}_{(d)}^{\mathbb{R}})$$

is by definition the average number of real roots. The main result of this section is the evaluation of this average value, which is given in Theorem 2. We restate it here.

Theorem 2 $A_{(d)} = \mathcal{D}^{1/2}$ *where $\mathcal{D}$ is the Bézout number.*

Proof.   It follows in a straightforward manner from the next lemma applied to the case where $(d) = (1, 1, \ldots, 1)$. Then clearly $A_{(d)} = 1$ and $\mathcal{D} = 1$; so $C(n)$ is identically 1, and $A_{(d)} = \mathcal{D}^{1/2}$.   □

Thus it is sufficient to prove Lemma 6.

Lemma 6 $A_{(d)} = \mathcal{D}^{1/2} C(n)$, *where C is a function of n.*

Proof.   By Proposition 3

$$\int_{\mathbb{P}(\mathcal{H}_{(d)}^{\mathbb{R}})} \#(f) d\mathbb{P}(\mathcal{H}_{(d)}^{\mathbb{R}})$$

$$= \text{Vol}\,(\mathbb{P}(\mathbb{R}^{n+1})) \int_{V_{e_0}^{\mathbb{R}}} (\text{DET})^{1/2} dV_{e_0}^{\mathbb{R}}$$

$$= \frac{\text{Vol}\,(\mathbb{P}(\mathbb{R}^{n+1}))}{2} \int_{S_{e_0}} (\text{DET})^{1/2} dS_{e_0}.$$

Here S_{e_0} is the unit sphere in $\widehat{V}_{e_0}^{\mathbb{R}} = \{f \in \mathcal{H}_{(d)}^{\mathbb{R}} \mid f(e_0) = 0\}$. The space $V_{e_0}^{\mathbb{R}} = \mathbb{P}(\widehat{V}_{e_0}^{\mathbb{R}})$ is obtained from S_{e_0} by identifying the vector x and $-x$. So $\text{Vol}\,(V_{e_0}^{\mathbb{R}}) = \frac{1}{2}\text{Vol}\,(S_{e_0})$, which gives rise to the $\frac{1}{2}$ in the previous equality.

Now

$$A_{(d)} = \frac{1}{\mathrm{Vol}\,(\mathbb{P}(\mathcal{H}_{(d)}^{\mathbb{R}}))} \int_{\mathbb{P}(\mathcal{H}_{(d)}^{\mathbb{R}})} \#(f)d\mathbb{P}(\mathcal{H}_{(d)}^{\mathbb{R}})$$

$$= \frac{\mathrm{Vol}\,(\mathbb{P}(\mathbb{R}^{n+1}))}{2\mathrm{Vol}\,(\mathbb{P}(\mathcal{H}_{(d)}^{\mathbb{R}}))} \int_{S_{e_0}} (\mathrm{DET})^{1/2} dS_{e_0}$$

$$= \frac{\Gamma(\frac{N}{2})}{2\pi^{N/2}} \mathrm{Vol}\,(\mathbb{P}(\mathbb{R}^{n+1})) \int_{S_{e_0}} (\mathrm{DET})^{1/2} dS_{e_0},$$

where $N = \dim \mathcal{H}_{(d)}^{\mathbb{R}}$. The next lemma completes the proof of Lemma 6 and hence of Theorem 2.   □

Lemma 7

$$\int_{S_{e_0}} (\mathrm{DET})^{1/2} dS_{e_0} = \mathcal{D}^{1/2} \frac{\pi^{N/2}}{\Gamma(\frac{N}{2})} H(n),$$

where $H(n)$ depends only on n.

Proof. As in Section 12.4 let $L_{(d)}^{\mathbb{R}}$ be the linear subspace of $f \in \widehat{V}_{e_0}^{\mathbb{R}}$ of the form $f_i(x) = x_0^{d_i-1} \sum_{j=1}^n a_{ij}x_j$ and $\pi : \widehat{V}_{e_0}^{\mathbb{R}} \to L_{(d)}^{\mathbb{R}}$ be the orthogonal projection. The kernel of π, $J_{(d)}^{\mathbb{R}}$, consists of polynomial systems with no terms of the form $x_0^{d_i}, x_0^{d_i-1} \Sigma a_{ij}x_j$ in the ith coordinate. It follows that $Dg(e_0)$ is identically zero for any $g \in J_{(d)}^{\mathbb{R}}$ and hence that $Tf(e_0) = T(\pi(f))(e_0)$ for any $f \in \widehat{V}_{e_0}^{\mathbb{R}}$. By equality 11.2

$$\int_{S_{e_0}} (\mathrm{DET})^{1/2} dS_{e_0}$$

$$= \int_{\substack{f \in L_{(d)}^{\mathbb{R}} \\ \|f\| \le 1}} \int_{\pi^{-1}(f)} (\mathrm{DET})^{1/2} (1 - \|f\|^2)^{-1/2} d\pi^{-1}(f) dL_{(d)}^{\mathbb{R}}$$

$$= \mathrm{Vol}\,(S^{N-n^2-n-1}) \int_{\substack{f \in L_{(d)}^{\mathbb{R}} \\ \|f\| \le 1}} (\mathrm{DET})^{1/2} (1 - \|f\|^2)^{(N-n^2-n-2)/2} dL_{(d)}^{\mathbb{R}}.$$

Now we use the change of variables given by Lemma 17 of Chapter 12,

$$f \to \Delta(d_i^{-1/2})Tf(e_0),$$

which is a linear isometry from $L_{(d)}^{\mathbb{R}}$ to the space of $n \times n$ real matrices $\mathcal{M}_n^{\mathbb{R}}$. It follows that

$$\int_{\substack{f \in L_{(d)}^{\mathbb{R}} \\ \|f\| \le 1}} (\mathrm{DET})^{1/2} (1 - \|f\|^2)^{(N-n^2-n-2)/2} dL_{(d)}^{\mathbb{R}}$$

$$= \int_{\substack{M \in \mathcal{M}_n^{\mathbb{R}} \\ \|M\| \le 1}} \det((\Delta(d_i^{1/2})M)^* \Delta(d_i^{1/2})M)^{1/2} (1 - \|M\|^2)^{(N-n^2-n-2)/2} d\mathcal{M}_n^{\mathbb{R}}$$

or yet since $\det\Delta^2 = \mathcal{D}$,

$$= \mathcal{D}^{1/2} \int_{\substack{M\in\mathcal{M}^{\mathbb{R}}_n \\ \|M\|\leq 1}} \det(M^*M)^{1/2}(1 - \|M\|^2)^{(N-n^2-n-2)/2} d\mathcal{M}^{\mathbb{R}}_n.$$

Now we use spherical coordinates to see that this last is

$$\mathcal{D}^{1/2} \int_0^1 (1 - r^2)^{(N-n^2-n-2)/2} \int_{\substack{M\in\mathcal{M}^{\mathbb{R}}_n \\ \|M\|=r}} \det(M^*M)^{1/2} \, dS_r^{n^2-1} \, dr.$$

But

$$\int_{\substack{M\in\mathcal{M}^{\mathbb{R}}_n \\ \|M\|=r}} \det(M^*M)^{1/2} dS_r^{n^2-1} = r^{n^2+n-1} \int_{\substack{M\in\mathcal{M}^{\mathbb{R}}_n \\ \|M\|=1}} \det(M^*M)^{1/2} dS^{n^2-1},$$

where r^{n^2-1} scales the volume element and r^n scales $\det(M^*M)^{1/2}$. Here $S_r^{n^2-1}$ is the sphere of radius r in $\mathcal{M}^{\mathbb{R}}_n$ and S^{n^2-1} is the unit sphere. Also,

$$\int_0^1 r^{n^2+n-1}(1 - r^2)^{(N-n^2-n-2)/2} dr = \frac{1}{2} \frac{\Gamma(\frac{N-n^2-n}{2})\Gamma(\frac{n^2+n}{2})}{\Gamma(\frac{N}{2})}.$$

It follows that

$$\int_{S_{e_0}} (\mathrm{DET})^{1/2} dS_{e_0} = \mathcal{D}^{1/2} \mathrm{Vol}\,(S^{N-n^2-n-1}) \frac{1}{2} \frac{\Gamma(\frac{N-n^2-n}{2})}{\Gamma(\frac{N}{2})} h_1(n).$$

Substituting $\mathrm{Vol}\,(S^{N-n^2-n-1}) = \dfrac{2\pi^{\frac{N-n^2-n}{2}}}{\Gamma(\frac{N-n^2-n}{2})}$ proves the lemma. $\square$

13.4   The Condition Number in $\mathcal{M}^{\mathbb{C}}_{(n,n+1)}$

Let $\mathcal{M}^{\mathbb{C}}_{(n,n+1)}$ be the space of $n \times (n + 1)$ complex matrices. If $M \in \mathcal{M}^{\mathbb{C}}_{(n,n+1)}$ has rank n, then we let $M^{\dagger}$ be the right inverse of M with the smallest operator norm. Thus, $M^{\dagger}$ is characterized by the following properties.

(1) $MM^{\dagger}$ is the identity matrix;

(2) $M^{\dagger}(\mathbb{C}^n)$ is orthogonal to $\ker M$.

The matrix $M^{\dagger}$ is called the *Moore–Penrose inverse* of M. It can be expressed in terms of M by $M^{\dagger} = M^*(MM^*)^{-1}$. If M has rank n, we define the condition number of M as

$$\kappa(M) = \|M\|\|M^{\dagger}\|.$$

Just as in Chapter 11, the condition number serves to estimate the relative error in the solution x of the linear equation $Mx = b$ as a function of the relative error in the right-hand side b; that is to say:
if $Mx = b$, then there is a solution of $M(x + \delta x) = b + \delta b$ with

$$\frac{\|\delta x\|}{\|x\|} \le \kappa(M) \frac{\|\delta b\|}{\|b\|}.$$

Now we give another interpretation to $\kappa(M)$. The space $\mathcal{M}^{\mathbb{C}}_{(n,n+1)}$ coincides with $\mathcal{H}_{\underbrace{(1, \ldots, 1)}_{n+1}}$. Let

$$V \subset \mathbb{P}(\mathcal{M}^{\mathbb{C}}_{(n,n+1)}) \times \mathbb{P}(\mathbb{C}^{n+1})$$

be our usual solution variety $V = \{(M, x) \mid M(x) = 0\}$ together with the restrictions of the projections

$$\pi_1 : V \to \mathbb{P}(\mathcal{M}^{\mathbb{C}}_{(n,n+1)})$$

and

$$\pi_2 : V \to \mathbb{P}(\mathbb{C}^{n+1}).$$

In this context as in Chapter 12 we let $G : U \to \mathbb{P}(\mathbb{C}^{n+1})$ be defined as in the implicit function theorem where U is a neighborhood of M in $\mathcal{M}^{\mathbb{C}}_{(n,n+1)}$. Then $N(G(N)) = 0$ for all $N \in U$ and the condition matrix $DG(M)$ is identified with $(M|T_x)^{-1}$, where $M(x) = 0$, $\|M\|_F = 1 = \|x\|$. Here we recall that $\| \ \|_F$ is the Frobenius norm on $\mathcal{M}^{\mathbb{C}}_{(n,n+1)}$, $\|M\|_F = \langle M, M \rangle_F^{1/2} = \mathrm{trace}(M^*M)^{1/2}$. For $d = (1, \ldots, 1)$ the Frobenius norm and unitarily invariant norm on $\mathcal{H}_{(d)} = \mathcal{M}^{\mathbb{C}}_{(n,n+1)}$ are the same. We have the condition number $\mu(M) = \|(M|T_x)^{-1}\|$ and $(M|T_x)^{-1} = M^\dagger$, thus $\mu(M) = \|M^\dagger\|$. More generally, $\mu(M) = \|M\|_F \|M^\dagger\|$ for any M where $M(x) = 0$ and $\|x\| = 1$.

Lemma 8

$$\kappa(M) \le \mu(M).$$

Proof. The proof is the same as Lemma 7 in Chapter 11.   □

The next sections are devoted to studying the probability distribution of $\mu(M)$ on the unit sphere $S \subset \mathcal{M}^{\mathbb{C}}_{(n,n+1)}$ whose real dimension is $2n^2 + 2n - 1$. Since $\mu(M)$ is unitarily invariant as in Section 12.1 the probability distribution of $\mu(M)$ on S is the same as on $\mathbb{P}(\mathcal{M}^{\mathbb{C}}_{(n,n+1)})$. It is also the same as the probability distribution of μ on $\mathcal{M}^{\mathbb{C}}_{(n,n+1)}$ with the probability distribution given by the point density function

$$\frac{1}{(2\pi)^{n^2+n}} e^{-\|M\|_F^2/2}.$$

We proceed as in Sections 11.4 and 11.6. We end this section by characterizing $\mu(M)$.
Let $\Sigma_x = \{N \in \mathcal{M}^{\mathbb{C}}_{(n,n+1)} \mid N(x) = 0 \text{ and } \exists y \ne 0 \text{ with } \langle y, x \rangle = 0 \text{ and } N(y) = 0\}$.

Proposition 4 *If* $M \in V_x - \Sigma_x$, *then*

(a) $d_F(M, \Sigma_x) = \frac{1}{\mu(M)}$.

(b) $\displaystyle\min_{\substack{\langle y,x \rangle = 0 \\ \|y\| = 1}} \frac{\|M(y)\|}{\|M\|_F} = \frac{1}{\mu(M)}$.

Proof.   Part (a) is a special case of the condition number theorem.
Part (b) follows directly from the definition.   □

Remark 2 Part (a) may be rewritten as

$$\min_{N \in \Sigma_x} \frac{\|N - M\|_F}{\|M\|_F} = \frac{1}{\mu(M)}.$$

13.5   A Linear Algebra Estimate and the Distribution of the Condition Number

In this section, we prove a proposition which relates the norm of the Moore–Penrose inverse of a matrix in $\mathcal{M}^{\mathbb{C}}_{(n,n+1)}$ in terms of a geometric condition on the columns of the matrix. We use the results to derive probability information about the distribution of μ in $\mathcal{M}^{\mathbb{C}}_{(n,n+1)}$.

If $n = 1$ and $M \in \mathcal{M}^{\mathbb{C}}_{(1,2)}$ is not trivial, then $\mu(M) = 1$, as a short computation shows. So in the rest of the chapter we assume $n > 1$.

In Section 11.4, we showed that if the norm of this inverse of an $n \times n$ matrix A is x, then we can alter one column of A by a perturbation of size $\sqrt{n}/x$ and make A singular. In Proposition 5 we show that if $M \in \mathcal{M}^{\mathbb{C}}_{(n,n+1)}$, $M(u) = 0$, and the norm of the Moore–Penrose inverse of M is x, then we can alter two columns of M by a perturbation of size $\sqrt{2n}/x$ so that $M(u)$ is still equal to 0 and M restricted to T_u, the orthogonal complement to u, is singular.

We begin with several lemmas.

Lemma 9   *Let* $u, v \in \mathbb{C}^{n+1}$ *such that* $\langle u, v \rangle = 0$ *and* $\|u\| = \|v\| = 1$. *Let M be the $n+1$ by 2 matrix* $(u \, v)$. *For every pair* (i, j) *with* $1 \leq i < j \leq n+1$ *we consider the 2×2 minor of M*, M_{ij}, *given by its ith and jth rows. Then* $\sum_{i<j} |\det M_{ij}|^2 = 1$.

Proof.

$$|\det M_{ij}|^2 = (u_i v_j - u_j v_i)(\overline{u_i v_j} - \overline{u_j v_i}).$$

Then

$$\sum_{i<j} |\det(M_{ij})|^2 = \frac{1}{2} \sum_{i,j} (u_i v_j - u_j v_i)(\overline{u_i v_j} - \overline{u_j v_i})$$

$$= \frac{1}{2} \sum_{i,j} ((|u_i|^2 |v_j|^2 + |u_j|^2 |v_i|^2) - (\overline{u_i v_j} u_j v_i + u_i v_j \overline{u_j v_i}))$$

$$= \frac{1}{2} \sum_j \left(\left(\sum_i |u_i|^2 \right) |v_j|^2 + \left(\sum_i |v_i|^2 \right) |u_j|^2 \right) - (0 + 0)$$

$$= \frac{1}{2} \left(\sum_j |v_j|^2 + |u_j|^2 \right) = 1.$$

$\square$

Lemma 10 *Let* $a_i, b_i \geq 0$ $i = 1, \ldots, k$, *where*

$$\sum_{i=1}^{k} a_i > 0, \quad \sum_{i=1}^{k} b_i > 0.$$

Then

$$\min_{\substack{i=1,\ldots,k \\ b_i \neq 0}} \frac{a_i}{b_i} \leq \frac{\sum_{i=1}^{k} a_i}{\sum_{i=1}^{k} b_i}.$$

Proof. If $x, y, w, z > 0$, and $x/y \leq w/z$, then $x/y \leq (w+x)/(z+y)$, and the lemma follows by induction. $\square$

Lemma 11 *Let* $u, v \in \mathbb{C}^{n+1}$ *be such that* $\langle u, v \rangle = 0$ *and* $\|u\| = 1 = \|v\|$. *Let* M *be the* $n + 1$ *by* 2 *matrix* (uv). *For every pair* (i, j) *with* $1 \leq i < j \leq n+1$ *we consider the* 2×2 *minor of* M, M_{ij}, *given by its* ith *and* jth *rows. Then there is an* (i, j) *such that* $\|M_{i,j}^{-1}\|_F \leq \sqrt{2n}$.

Proof.

$$\|M_{i,j}^{-1}\|_F^2 = \frac{|v_i|^2 + |v_j|^2 + |u_i|^2 + |u_j|^2|}{|\det M_{i,j}|^2}.$$

By the previous lemma there is an i_0, j_0 such that

$$\frac{|v_{i_0}|^2 + |v_{j_0}|^2 + |u_{i_0}|^2 + |u_{j_0}|^2}{|\det M_{i_0,j_0}|^2} \leq \frac{\sum_{\substack{i,j=1 \\ i \neq j}}^{n+1} |v_i|^2 + |v_j|^2 + |u_i|^2 + |u_j|^2|}{\sum_{\substack{i,j=1 \\ i \neq j}}^{n+1} |\det M_{ij}|^2}$$

$$\leq \frac{n(|v|^2 + |u|^2)}{1} = 2n$$

and thus

$$\|M_{i_0,j_0}^{-1}\|_F \leq \sqrt{2n}.$$

$\square$

For $M \in \mathcal{M}^{\mathbb{C}}_{(n,n+1)}$ let $\widehat{M}_{ij} \in \mathcal{M}^{\mathbb{C}}_{(n,n-1)}$ denote M with the ith and jth columns deleted. For $M \in \mathcal{M}^{\mathbb{C}}_{(n,n-1)}$ and $u_i, u_j \in \mathbb{C}^n$ let (M, u_i, u_j) denote the element of $\mathcal{M}^{\mathbb{C}}_{(n,n+1)}$ obtained from M by inserting u_i as the ith column and u_j as the jth column. Thus $(\widehat{M}_{i,j}, A_i, A_j)$ is M with the ith column replaced by A_i and the jth column replaced by A_j.

Proposition 5 *Let* $M \in \mathcal{M}^{\mathbb{C}}_{(n,n+1)}$ *have rank* n. *Suppose* $M(x) = 0$, $\|y\| = 1$, $\langle x, y \rangle = 0$ *and* $\|M(y)\| = \varepsilon$. *Then there is a matrix* $N \in \mathcal{M}^{\mathbb{C}}_{(n,n+1)}$ *and* i, j *satisfying* $1 \le i, j \le n + 1$ *with* $i \ne j$ *such that*

(1) $\widehat{N}_{ij} = \widehat{M}_{ij};$

(2) $\|N - M\|_F \le \sqrt{2n}\varepsilon;$

(3) $N(x) = N(y) = 0;$

(4) N *has rank* $n - 1$.

Proof.　We may assume $\|x\| = 1$. Let $M(y) = w$. By the previous lemma choose i, j such that $\|B^{-1}\|_F \le \sqrt{2n}$, where

$$B = \begin{pmatrix} x_i & y_i \\ x_j & y_j \end{pmatrix}.$$

Let $(A_i \ A_j) = -(0, w)B^{-1}$, and let $N = (\widehat{M}_{ij}, M_i + A_i, M_j + A_j)$. Then

$$\|(A_i A_j)\|_F \le \|(0, w)\|_F \|B^{-1}\|_F \le \varepsilon\sqrt{2n};$$

so $\|N - M\|_F \le \varepsilon\sqrt{2n}$. Since

$$(A_i A_j)B = -(0, w)$$
$$N(x) = M(x) = 0 \text{ and}$$
$$N(y) = M(y) - w = 0.$$

It remains to prove (4).

Since A_i and A_j are both multiples of w, $N(z) = M(z) + \lambda w$ for some λ depending on z. If $N(z) = 0$, then $M(z)$ is a multiple of w and z is an element of the space spanned by x and y. Since M has rank n it follows that N has rank $n - 1$.　　□

Recall that $S = S^{2n^2+2n-1} \subset \mathcal{M}^{\mathbb{C}}_{(n,n+1)}$ is the unit sphere. Let $N_\varepsilon = \{M \in S \mid \mu(M) > 1/\varepsilon\}$. We use Proposition 5 to estimate $\mathrm{Vol}\,(N_\varepsilon)/\mathrm{Vol}\,(S)$ in Theorem 6. Once again we follow Section 11.4 with some modifications. We begin with some lemmas. Let

$$N'_\varepsilon = \{M \in N_\varepsilon \mid \mathrm{rank}\, M_{ij} = n - 1 \text{ for all } i, j$$
$$\text{satisfying } 1 \le i, j \le n + 1, \text{ and } i \ne j\}$$

and

$$\pi_{ij} : \mathcal{M}^{\mathbb{C}}_{(n,n+1)} \to \mathcal{M}^{\mathbb{C}}_{(n,n-1)}$$
$$M \to \widehat{M}_{ij}.$$

Lemma 12   Vol $(N'_\varepsilon) =$ Vol (N_ε).

Proof.   The set of matrices of rank smaller than $n-1$ has zero volume in $\mathcal{M}^{\mathbb{C}}_{(n,n-1)}$ and the same is true for $(\pi_{ij})^{-1}$ of this set by the co-area formula. Taking the union over i, j still gives zero volume.

Given $M \in S$ let $E(\widehat{M}_{i,j})$ be the subspace of $\mathbb{C}^{n+1}$ generated by the columns of $\widehat{M}_{ij}$. Given a subspace $V \subset \mathbb{C}^{n+1}$ and a vector $x \in \mathbb{C}^{n+1}$ let $d(x, V)$ be the distance from x to V, that is, the norm of the orthogonal projection of x on the orthogonal complement to V. Let

$$W_{\varepsilon,i,s} = \{M \in S \mid d(M_i, E(\widehat{M}_{ij}))^2 + d(M_j, E(\widehat{M}_{i,j}))^2 < 2n\varepsilon^2\}.$$

$\qquad\qquad\qquad\qquad\qquad\qquad\qquad\qquad\qquad\qquad\qquad\qquad\qquad\qquad$ $\square$

Lemma 13

(a)

$$N'_\varepsilon \subset \bigcup_{\substack{i,j=1 \\ i \neq j}}^{n+1} W_{\varepsilon,i,j} \text{ and so}$$

(b)

$$\text{Vol}(N_\varepsilon) \leq \sum_{\substack{i,j=1 \\ i \neq j}}^{n+1} \text{Vol}(W_{\varepsilon,i,j}).$$

Proof.   (a) By Proposition 4(b) of Section 4: if $\mu(M) \geq 1/\varepsilon$, then there are x such that $M(x) = 0$ and $y \in \mathbb{C}^{n+1}$, $\|y\| = 1 < x$, $y >= 0$, and $\|M(y)\| \leq \varepsilon$. Now apply Proposition 5. There are i, j, and N such that $\widehat{N}_{ij} = \widehat{M}_{ij}$ and rank $N =$ rank $\widehat{N}_{ij} = n - 1$. Thus N_i and N_j are in $E(\widehat{N}_{ij})$ which equals $E(\widehat{M}_{ij})$. Moreover, $d(M_i, E(\widehat{M}_{ij}))^2 + d(M_j, E(\widehat{M}_{ij}))^2 \leq \|M - N\|_F^2 \leq 2n\varepsilon^2$.

Part (b) follows from (a) and the previous lemma. $\qquad\qquad\qquad\qquad\qquad\qquad$ $\square$

Note that if V is a codimension 1 subspace of $\mathbb{C}^n$ and $A_1, A_2 \in \mathbb{C}^n$, then $d(A_1, V)^2 + d(A_2, V)^2 \leq 2n\varepsilon^2$ if and only if $d((A_1, A_2), V \times V)^2 \leq 2n\varepsilon^2$, where $(A_1, A_2) \in \mathbb{C}^{2n}$ and $V \times V$ is the complex codimension 2 subspace of $\mathbb{C}^{2n}$ formed by producting V with itself.

Let B^{2n^2-2n} be the unit ball in $\mathcal{M}^{\mathbb{C}}_{(n,n-1)}$.

Proposition 6   *Let* $A \in B^{2n^2-2n}$. *Then*

$$\text{Vol}(W_{\varepsilon,i,j} \cap \pi_{ij}^{-1}(A)) \leq 4n^2\varepsilon^4 \text{Vol } B_1^4 \text{Vol } S_{(1-\|A\|^2)^{1/2}}^{4n-5}.$$

Here the volume on the left is $(4n - 1)$-*dimensional volume and the volumes on the right are* 4- *and* $(4n - 5)$-*dimensional volumes, respectively.*

Proof.   If $M \in \pi_{ij}^{-1}(A) \cap W_{\varepsilon,i,j}$, then

$$\|M_i\|^2 + \|M_j\|^2 = (1 - \|A\|^2)^{1/2};$$

so $(M_i, M_j) \in S_{(1-\|A\|^2)^{1/2}}^{4n-1}$ and (M_i, M_j) is within $\sqrt{2n}\varepsilon$ of the codimension 2 subspace of $\mathbb{C}^{2n}$ given by $E(A) \times E(A)$, where $E(A)$ is the subspace of $\mathbb{C}^n$ generated by the columns of A. Thus the $(2n - 1)$-dimensional volume of $W_{\varepsilon,i,j} \cap \pi_{ij}^{-1}(A)$ is at most the volume of the cylinder of base $S_{(1-\|A\|^2)^{1/2}}^{4n-5}$ and side $B_{\sqrt{2n}\varepsilon}^4$ where B_r^4 is the ball of radius r in $\mathbb{C}^2$ or $\mathbb{R}^4$. The volume of $B_{\sqrt{2n}\varepsilon}^4$ is $4n^2\varepsilon^4 \mathrm{Vol}\, B_1^4$. So the volume of the cylinder is $4n^2\varepsilon^4 \mathrm{Vol}\, B_1^4 \mathrm{Vol}\, S_{(1-\|A\|^2)^{1/2}}^{4n-5}$ and we are done.   □

We are now ready for Theorem 1 in the linear case. We state the volume estimate for the sphere instead of the projective space. As we have already remarked, the probabilities and ratio of volumes are the same for the sphere and projective space.

Theorem 6

$$\frac{\mathrm{Vol}\,(N_\varepsilon)}{\mathrm{Vol}\,(S)} \leq \frac{n^3(n+1)\Gamma(n^2+n)}{\Gamma(n^2+n-2)}\varepsilon^4.$$

Proof.   It is sufficient to prove that

$$\frac{\mathrm{Vol}\,(W_{\varepsilon,i,j})}{\mathrm{Vol}\,(S)} \leq \frac{2n^2\Gamma(n^2+n)}{\Gamma(n^2+n-2)}\varepsilon^4,$$

by Lemma 12(b).

Let $\chi(W_{\varepsilon,i,j})$ be the characteristic function of $W_{\varepsilon,i,j}$. Then

$$\mathrm{Vol}\,(W_{\varepsilon,i,j})$$

$$= \int_S \chi(W_{\varepsilon,i,j})dS \quad \text{which by (11.2)}$$

$$= \int_{M \in B_1^{2n^2-2n}} \int_{\pi_{ij}^{-1}(M)} \chi(W_{\varepsilon,i,j})(1 - \|M\|^2)^{-1/2}\,d\pi_{ij}^{-1}(M)\,dB_1^{2n^2-2n}$$

$$= \int_{M \in B_1^{2n^2-2n}} \mathrm{Vol}\,(W_{\varepsilon,i,j} \cap \pi_{i,j}^{-1}(M))(1 - \|M\|^2)^{-1/2}dB_1^{2n^2-2n}$$

$$\leq 4n^2\varepsilon^4 \mathrm{Vol}\,(B_1^4)\mathrm{Vol}\,(S_1^{4n-5}) \int_{M \in B_1^{2n^2-2n}} (1 - \|M\|^2)^{4n-6}dB_1^{2n^2-2n}.$$

But by spherical coordinates,

$$\int_{M \in B_1^{2n^2-2n}} (1 - \|M\|^2)^{4n-6}dB_1^{2n^2-2n}$$

$$\leq \mathrm{Vol}\,(S^{2n^2-2n-1}) \int_0^1 r^{2n^2-2n-1}(1-r^2)^{2n-3}dr$$

$$\leq \mathrm{Vol}\,(S^{2n^2-2n-1})\frac{\Gamma(n^2-n)\Gamma(2n-2)}{2\Gamma(n^2+n-2)}.$$

Now substituting $\mathrm{Vol}\,(S^m) = 2\pi^{(m+1)/2}/\Gamma((m+1)/2)$, and $\mathrm{Vol}\,(B_1^4) = \pi^2/2$, and dividing by $\mathrm{Vol}\,(S)$ we are done.   □

13.6   The Distribution of the Normalized Condition Number for $\mathbb{P}(\mathcal{H}_{(d)})$

In this section we prove our theorem on the distribution of the normalized condition number in $\mathbb{P}(\mathcal{H}_{(d)})$. We have already dealt with the linear case in the last section.

Let $U_\varepsilon \subset V \subset \mathbb{P}(\mathcal{H}_{(d)}) \times \mathbb{P}(\mathbb{C}^{n+1})$ be the set $\{(f, z) \in V \mid \mu_{\text{norm}}(f, z) > 1/\varepsilon\}$ and $N_\varepsilon = \pi_1(U_\varepsilon)$. Recall that $\mu_{\text{norm}}(f) > 1/\varepsilon$ if and only if $f \in N_\varepsilon$. So the probability that $\mu_{\text{norm}}(f) > 1/\varepsilon$ for f in $\mathbb{P}(\mathcal{H}_{(d)})$ is equal to $\text{Vol}(N_\varepsilon)/\text{Vol}(\mathbb{P}(\mathcal{H}_{(d)}))$.

Proposition 7

$$\text{Vol}(N_\varepsilon) \leq \text{Vol}(\mathbb{P}(\mathbb{C}^{n+1})) \int_{V_{e_0} \cap U_\varepsilon} \text{DET} \, dV_{e_0}$$

with equality if $\mathcal{H}_{(d)} = \mathcal{M}_{(n,n+1)}^{\mathbb{C}}$.

Proof.   Let $\chi(N_\varepsilon)$ be the characteristic function of N_ε. Then $\chi(N_\varepsilon) \leq \#_{U_\varepsilon}$ with equality for $\mathcal{M}_{(n,n+1)}^{\mathbb{C}}$. Now

$$\text{Vol}(N_\varepsilon) = \int_{\mathbb{P}(\mathcal{H}_{(d)})} \chi(N_\varepsilon) d\mathbb{P}(\mathcal{H}_{(d)}) \leq \int_{\mathbb{P}(\mathcal{H}_{(d)})} \#_{U_\varepsilon} d\mathbb{P}(\mathcal{H}_{(d)})$$

and Proposition 2 finishes the proof. $\square$

Corollary 1 *Let S^{2n^2-1} be the unit sphere in $\mathcal{M}_n^{\mathbb{C}}$, the space of $n \times n$ complex matrices. For $\varepsilon > 0$ let $N_\varepsilon \subset \mathcal{M}_n^{\mathbb{C}}$ be the matrices with condition number $> 1/\varepsilon$. Then*

$$\int_{N_\varepsilon \cap S^{2n^2-1}} \det(M^*M) dS^{2n^2-1} \leq \frac{\varepsilon^4 n^3 (n+1) \Gamma(n^2+n) \text{Vol}(S^{2n^2+2n-1})}{\Gamma(n^2+n-2)\text{Vol}(\mathbb{P}(\mathbb{C}^{n+1}))}.$$

Proof.   Let $\rho : S^{2n^2-1} \to \mathbb{P}(\mathcal{M}_n^{\mathbb{C}})$ be the natural projection. Identify $\widehat{V}_{e_0} = \{M \in \mathcal{M}^{\mathbb{C}}(n, n+1) \mid M(e_0) = 0\}$ with $\mathcal{M}_n^{\mathbb{C}}$ by $M \to M|T_{e_0}$. Then norms and condition numbers correspond. Apply Proposition 7 to the projective space V_{e_0}, lift to the sphere in $\widehat{V}_e$, and use the identification to conclude that

$$\text{Vol}(\mathbb{P}(\mathbb{C}^{n+1})) \int_{S^{2n^2-1} \cap N_\varepsilon} \det(M^*M) dS^{2n^2-1} = \text{Vol}(N_\varepsilon).$$

By Theorem 6

$$\text{Vol}(N_\varepsilon) \leq \frac{n^3(n+1)\Gamma(n^2+n)}{\Gamma(n^2+n-2)} \varepsilon^4 \text{Vol}(S^{2n^2+2n-1}).$$

Dividing by $\text{Vol}\,\mathbb{P}(\mathbb{C}^{n+1})$ finishes the proof. $\square$

We now prove Theorem 1 on the normalized condition number. We closely follow the proof of the theorem on the average number of real roots, especially Lemma 7. Let us recall its statement.

Theorem 1   *Let $n > 1$. The probability that $\mu_{\mathrm{norm}}(f) > 1/\varepsilon$ for $f \in \mathbb{P}(\mathcal{H}_{(d)})$ and $\varepsilon > 0$ is less than or equal to $\varepsilon^4 n^3 (n+1)(N-1)(N-2)\mathcal{D}$, where $N = \dim \mathcal{H}_{(d)}$ and $\mathcal{D}$ is the Bézout number.*

Proof.   The probability equals $\mathrm{Vol}\,(N_\varepsilon)/\mathrm{Vol}\,(\mathbb{P}(\mathcal{H}_{(d)}))$ and

$$\mathrm{Vol}\,(N_\varepsilon) \le \mathrm{Vol}\,(\mathbb{P}(\mathbb{C}^{n+1})) \int_{V_{e_0} \cap U_\varepsilon} \det(Tf(e_0)^* Tf(e_0)) dV_{e_0}. \tag{13.1}$$

Let $\pi : \widehat{V}_{e_0} \to L_{(d)}$ be the orthogonal projection. Then $Tf(e_0) = T(\pi(f)e_0)$ for any $f \in \widehat{V}_{e_0}$. Now lift to the unit sphere $S_{e_0} \subset \widehat{V}_{e_0}$. Recall that the natural action of the group S^1 of unit complex numbers on S_{e_0} consists of isometries and that the fibers of the natural map $\rho : S_{e_0} \to \mathbb{P}(\widehat{V}_{e_0})$ are the S^1 orbits. Let $\widetilde{U}_\varepsilon = \rho^{-1}(U_\varepsilon)$. Recall that $\mathrm{DET} = \det(Tf(e_0)^* Tf(e_0))$. Then, since DET is invariant under the S^1 action and the length of the S^1 orbits is 2π, it follows that

$$\int_{V_{e_0} \cap U_\varepsilon} \mathrm{DET}\, dV_{e_0} = \frac{1}{2\pi} \int_{S_{e_0} \cap \widetilde{U}_\varepsilon} \mathrm{DET}\, dS_{e_0} \tag{13.2}$$

and

$$\frac{1}{2\pi} \int_{S_{e_0} \cap \widetilde{U}_\varepsilon} \mathrm{DET}\, dS_{e_0}$$

$$= \frac{1}{2\pi} \int_{\substack{f \in L_{(d)} \cap \widetilde{U}_\varepsilon \\ \|f\| \le 1}} \int_{\pi^{-1}(f)} \mathrm{DET}\,(1 - \|f\|^2)^{-1/2}\, d\pi^{-1}(f)\, dL_{(d)}$$

$$= \frac{\mathrm{Vol}\,(S^{2N - 2n^2 - 2n - 1})}{2\pi} \int_{\substack{f \in L_{(d)} \cap \widetilde{U}_\varepsilon \\ \|f\| \le 1}} \mathrm{DET}\,(1 - \|f\|^2)^{N - n^2 - n - 1}\, dL_{(d)}.$$

Now we use the change of variables given by Lemma 17 of Chapter 12, stating that

$$f \to \Delta(d_i^{-1/2}) Tf(e_0)$$

is a linear isometry from $L_{(d)}$ to $\mathcal{M}_n^{\mathbb{C}}$.
    It follows that

$$\int_{\substack{f \in L_{(d)} \cap \widetilde{U}_\varepsilon \\ \|f\| \le 1}} \mathrm{DET}\,(1 - \|f\|^2)^{N - n^2 - n - 1}\, dL_{(d)}$$

$$= \int_{\substack{M \in \mathcal{M}_n^{\mathbb{C}} \cap N_\varepsilon \\ \|M\| \le 1}} \det((\Delta(d_i^{1/2})M)^* \Delta(d_i^{1/2})M)(1 - \|M\|^2)^{N - n^2 - n - 1}\, d\mathcal{M}_n^{\mathbb{C}}$$

$$= \mathcal{D} \int_{\substack{M \in \mathcal{M}_n^{\mathbb{C}} \cap N_\varepsilon \\ \|M\| \le 1}} \det(M^* M)(1 - \|M\|^2)^{N - n^2 - n - 1}\, d\mathcal{M}_n^{\mathbb{C}}. \tag{13.3}$$

Now using spherical coordinates this last is

$$\mathcal{D} \int_0^1 (1 - r^2)^{N-n^2-n-1} \int_{\substack{\|M\|=r \\ M \in \mathcal{M}_n^{\mathbb{C}} \cap N_\varepsilon}} \det(M^*M) d\mathcal{M}_n^{\mathbb{C}}.$$

But now

$$\int_{\substack{\|M\|=r \\ M \in \mathcal{M}_n^{\mathbb{C}} \cap N_\varepsilon}} \det(M^*M) d\mathcal{M}_n^{\mathbb{C}}$$

$$\leq r^{2n^2+2n-1} \int_{N \in S_1^{2n^2+2n-1} \cap N_{1/r}\varepsilon} \det(N^*N) dS_1^{2n^2+2n-1}$$

$$\leq \frac{r^{2n^2+2n-5}\varepsilon^4 n^3(n+1)\Gamma(n^2+n)\text{Vol}\,(S^{2n^2+2n-1})}{\Gamma(n^2+n-2)\text{Vol}\,(\mathbb{P}(\mathbb{C}^{n+1}))}. \tag{13.4}$$

Putting the relations (13.1) through (13.4) together and dividing by $\text{Vol}\,(\mathbb{P}(\mathcal{H}_{(d)}))$ gives that

$$\frac{\text{Vol}\,N_\varepsilon}{\text{Vol}\,\mathbb{P}(\mathcal{H}_{(d)})}$$

is at most

$$\frac{\varepsilon^4 n^3(n+1)\Gamma(n^2+n)\mathcal{D}\text{Vol}\,(S^{2n^2+2n-1})\text{Vol}\,(S^{2N-2n^2-2n-1})}{2\pi\,\Gamma(n^2+n-2)\text{Vol}\,\mathbb{P}(\mathcal{H}_{(d)})}$$

times

$$\int_0^1 (1-r^2)^{N-n^2-n-1} r^{2n^2+2n-5} dr.$$

Substituting

$$\text{Vol}\,S^{m-1} = \frac{2\pi^{m/2}}{\Gamma(\frac{m}{2})}, \qquad \text{Vol}\,(\mathbb{P}(\mathcal{H}_{(d)})) = \frac{\pi^{N-1}}{\Gamma(N)}$$

and

$$\int_0^1 (1-r^2)^{N-n^2-n-1} r^{2n^2+2n-5} dr = \frac{1}{2}\frac{\Gamma(N-n^2-n)\Gamma(n^2+n-2)}{\Gamma(N-2)}$$

we get

$$\frac{\text{Vol}\,N_\varepsilon}{\text{Vol}\,\mathbb{P}(\mathcal{H}_{(d)})} \leq \frac{\varepsilon^4 n^3(n+1)\Gamma(N)\mathcal{D}}{\Gamma(N-2)}.$$

$\square$

The case $n = 1$ in Theorem 1 is stated in the next theorem.

Theorem 7 *Let* $n = 1$ *and* $d > 1$. *The probability that* $\mu_{\mathrm{norm}}(f) > 1/\varepsilon$ *for* $f \in \mathbb{P}(\mathcal{H}_d)$ *and* $0 < \varepsilon < 1$ *is less than or equal to*

$$d(1 - (1 - \varepsilon^2)^{d-1}(1 + (d-1)\varepsilon^2)).$$

Proof.   The proof is the same except that one can evaluate the integral

$$\mathcal{D} \int_0^1 (1 - r^2)^{N-n^2-n-1} \int_{\substack{\|M\|=r \\ M \in \mathcal{M}_n^{\mathbb{C}} \cap N_\varepsilon}} \det(M^*M) \, d\mathcal{M}_n^{\mathbb{C}} \, dr$$

as

$$2\pi d \int_0^\varepsilon (1 - r^2)^{d-2} r^3 dr$$

which equals $(\pi/((d-1)d))(1 - (1 - \varepsilon^2)^{d-1}(1 + (d-1)\varepsilon^2))$. $\square$

13.7   Additional Comments and Bibliographical Remarks

Theorems 1 and 2 were first proven in [Shub and Smale 1993b]. Eric Kostlan [1993] proved Theorem 2 earlier in the case that all the d_is are the same. In this paper Kostlan has a similar result for underdetermined systems, which gives average volumes of real varieties. The proof of Theorem 1 that we have presented here is new. First we proved Theorem 6, which is Theorem 1 in the linear case. The proof mimics the proof of Proposition 6 of Chapter 11 using Proposition 5 (which was proven for us by Jim Renegar). We deduced Corollary 1 from Proposition 7. In [Shub and Smale 1993b] the integral in Corollary 1 was directly evaluated using [Edelman 1995]. That evaluation led to a slightly better estimate for Theorem 1; the power of n is reduced from 4 to 3. Although the result here is worse, the proof is easier.

The decomposition $A = OP$ for a linear operator A into orthogonal O and positive semidefinite P which is implied by Lemma 1 is called the polar decomposition of A. For more on the number of real zeros of random polynomials see [Edelman and Kostlan 1995]. For a proof of the co-area formula see [Federer 1969; Morgan 1988]. For more on the Moore–Penrose inverse (and other generalized inverses of linear maps) see [Campbell and Meyer 1979].

In Chapter 12 we have proved a condition number theorem (Theorem 3 there) showing that the condition number $\mu(f, \zeta)$ of an homogeneous polynomial system f at a root ζ is the reciprocal of the distance from f to the discriminant variety Σ along the fiber V_ζ of systems that have ζ as a root. If we do not measure the distance in V_ζ, but simply consider the distance from f to Σ, the distance may be much smaller. We illustrate this with a simple example. The polynomial $x^2 - \varepsilon x$ has derivative ε at 0. If we require that 0 remain a root, then the closest polynomial with a double root is x^2. This polynomial has distance $\varepsilon/2$ from $x^2 - \varepsilon x$ in the metric on $\mathcal{H}_2$. But $x^2 - \varepsilon x + \varepsilon^2/4$ has a double root and is at distance $\varepsilon^2/4$ from $x^2 - \varepsilon x$. Thus the distance from f to Σ is comparable to the reciprocal of the condition number squared. The singularity at 0 of the function x^2 taking $\mathbb{R}$ to $\mathbb{R}$ is called a *fold singularity* in the singularity literature.

The variety $\Sigma \subset \mathbb{P}(\mathcal{H}_{(d)})$ is $\pi_1(\Sigma')$, where $\pi_1 : V \to \mathbb{P}(\mathcal{H}_{(d)})$ and Σ' is the set of critical points of π_1. Almost all of the points in Σ' are a higher-dimensional version of the fold singularity. This is reflected in Theorem 1. Given a smooth manifold $M \subset \mathbb{P}(\mathcal{H}_{(d)})$ of real codimension 2 the volume of the set of points within distance ε of M is on the order of $\pi \varepsilon^2 \text{Vol}\,(M)$. Theorem 1 proves that the set of f with condition number greater than $1/\varepsilon$ has volume of the order of ε^4. This suggests that on the average in some sense, if the condition number of f is greater than $1/\varepsilon$, then f is on the order of ε^2 away from Σ as is the case for the fold singularity. On the other hand it is not always true that the reciprocal of the distance from f to Σ is on the order of μ_f^2. Consideration of the family of cubics $x^3 - 3\varepsilon x w^2$ as $\varepsilon \to 0$ in $\mathbb{P}(\mathcal{H}_3)$ shows that the exponent lies between 1 and 2. Chatelin and Fraysse [1993] and Chatelin, Fraysse and Braconnier [1995] compute this exponent for certain singularities.

14
Complexity and the Condition Number

The focus of this chapter is on the result that the complexity of a continuation algorithm can be bounded essentially by the square of the condition number of the homotopy.

14.1 Newton's Method in Projective Space

The goal of this section is to describe how, in projective space, Newton's method is realized by iterating a map $N_f : \mathbb{P}(\mathbb{C}^{n+1}) \to \mathbb{P}(\mathbb{C}^{n+1})$ (defined almost everywhere) with the good properties of Newton's method in $\mathbb{R}^n$ or $\mathbb{C}^n$. In particular, we prove an approximate zero theorem in this setting which runs exactly (amazingly) as the one of Theorem 1 of Chapter 8.

Let $(f, z) \in \mathcal{H}_{(d)} \times \mathbb{C}^{n+1}$ represent an element of $\mathbb{P}(\mathcal{H}_{(d)}) \times \mathbb{P}(\mathbb{C}^{n+1})$. Recall that

$$T_z = \{w \in \mathbb{C}^{n+1} \mid \langle w, z \rangle = 0\}$$

represents the tangent space of $\mathbb{P}(\mathbb{C}^{n+1})$ at z. Also A_z is the affine subspace of $\mathbb{C}^{n+1}$ given by $A_z = z + T_z$. Then the derivative of f at z is a map $Df(z) : \mathbb{C}^{n+1} \to \mathbb{C}^n$ and we can consider $Df(z)|_{T_z}^{-1}$, the inverse of the restriction of $Df(z)$ to $T_z \subset \mathbb{C}^{n+1}$.

Define

$$N_f(z) = z - Df(z)|_{T_z}^{-1} f(z).$$

Also a *line* in $\mathbb{C}^{n+1}$ means a complex line and is always supposed to pass through zero.

Then N_f takes the line through z into a line through $N_f(z)$ as seen by (easily checked)

$$\lambda z - Df(\lambda z)|_{T_{\lambda z}}^{-1} f(\lambda z) = \lambda(z - Df(z)|_{T_z}^{-1} f(z)), \ \lambda \in \mathbb{C}.$$

Denote by Λ the subset of $z \in \mathbb{P}(\mathbb{C}^{n+1})$ with $Df(z)|_{T_z}$ singular as a linear map from T_z to $\mathbb{C}^n$. We obtain the Newton's map (projective)

$$N_f : \mathbb{P}(\mathbb{C}^{n+1}) - \Lambda \to \mathbb{P}(\mathbb{C}^{n+1}).$$

Note that N_f depends only on f as an element of $\mathbb{P}(\mathcal{H}_{(d)})$ and that for $\zeta \notin \Lambda$, $N_f(\zeta) = \zeta$ if and only if $f(\zeta) = 0$.

Define a new function d_T on $\mathbb{P}(\mathbb{C}^{n+1})$ by

$$d_T(x, y) = \tan d_R(x, y), \qquad x, y \in \mathbb{P}(\mathbb{C}^{n+1}),$$

where $d_R(x, y)$ is the Riemannian distance function defined in Chapter 12. It is easily checked that d_T is characterized by

$$d_T(x, y) = \frac{\|x - y\|}{\|y\|}$$

choosing the representatives of x, y such that $x \in A_y$ by Proposition 4 of Chapter 12.

The function d_T permits the elegant assertion of Theorem 1 but d_T is not quite a distance function. It does not satisfy the triangle inequality.

Definition 1 We say that $z \in \mathbb{P}(\mathbb{C}^{n+1})$ is an *approximate zero* of $f \in \mathbb{P}(\mathcal{H}_{(d)})$ with associated actual zero $\zeta \in \mathbb{P}(\mathbb{C}^{n+1})$ $(f(\zeta) = 0)$ provided that the point

$$z_i = N_f(z_{i-1}), \ z_0 = z$$

is defined for all $i = 1, 2, \ldots$ and

$$d_T(\zeta, z_k) \le \left(\frac{1}{2}\right)^{2^k - 1} d_T(\zeta, z), \ k = 1, 2, \ldots .$$

In particular

$$d_T(\zeta, z_1) \le \frac{1}{2} d_T(\zeta, z).$$

Recall the solution variety $V = \{(f, \zeta) \in \mathbb{P}(\mathcal{H}_{(d)}) \times \mathbb{P}(\mathbb{C}^{n+1}) \mid f(\zeta) = 0\}$. For $(f, \zeta) \in V$, one defines a projective version γ_0 of the invariant γ of Chapter 8 by

$$\gamma_0(f, \zeta) = \|\zeta\| \max_{k>1} \left\| Df(\zeta)|_{T_\zeta}^{-1} \frac{D^k f(\zeta)}{k!} \right\|^{1/(k-1)}.$$

The definition is independent of the choice $(f, \zeta) \in \mathcal{H}_{(d)} \times \mathbb{C}^{n+1}$ used in the formula. Here $D^k f(\zeta)$ is the kth derivative of $f : \mathbb{C}^{n+1} \to \mathbb{C}^n$, considered as a k-linear map (see Section 12.1).

Theorem 1 (Approximate Zero Theorem, Projective Case) *Let* $(f, \zeta) \in V \subset \mathbb{P}(\mathcal{H}_{(d)}) \times \mathbb{P}(\mathbb{C}^{n+1})$, $z \in \mathbb{P}(\mathbb{C}^{n+1})$. *If* $d_T(z, \zeta)\gamma_0(f, \zeta) \leq (3 - \sqrt{7})/2$, *then* z *is an approximate zero of* f *corresponding to* ζ.

Remark 1 By Theorem 2 we could replace $\gamma_0(f, \zeta)$ in Theorem 1 by $(D^{3/2}/2)\mu_{\mathrm{norm}}(f, \zeta)$ or yet by $(D^2\mu(f, \zeta))/2$.

The proof uses Lemma 1 which in turn needs the subsequent lemmas.

Lemma 1 (Main Lemma) *Let*

$$\widehat{u} = d_T(z, \zeta)\gamma_0(f, \zeta) < \frac{3 - \sqrt{7}}{2}$$

and $\psi(t) = 1 - 4t + 2t^2$ *be as in* (8.1). *Let* $z' = N_f(z)$. *Then*

$$d_T(z', \zeta) \leq \frac{\widehat{u}}{\psi(\widehat{u})}d_T(z, \zeta).$$

In $\mathbb{C}^{n+1}$ choose representations z, ζ with the property $\zeta \in A_z$. Let $\pi : A_z \to A_\zeta$ be defined by radial projection. That is to say, if $z' \in A_z$, let L be the complex line through z' (and the origin) and $\pi(z')$ be the point $L \cap A_\zeta$.

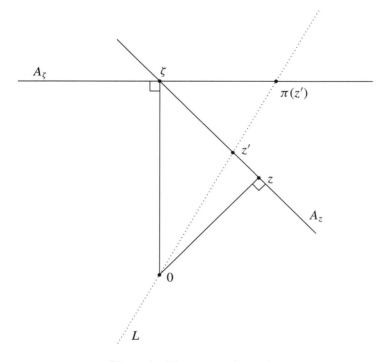

Figure A The map $\pi : A_z \to A_\zeta$.

Lemma 2 *Suppose that* $x, z' \in A_z$ *and* $\|z' - \zeta\| \leq \|z - \zeta\|$. *Then*

(1) $\pi(x) = \dfrac{\|\zeta\|^2}{\langle x, \zeta \rangle} x$

(2) $\dfrac{\|\pi(z') - \zeta\|}{\|z' - \zeta\|} \leq \dfrac{\|\pi(z) - \zeta\|}{\|z - \zeta\|}$

(3) $\dfrac{\|\pi(z) - \zeta\|}{\|z - \zeta\|} = \dfrac{\|\zeta\|}{\|z\|}$

(4) $\|\pi(z') - \zeta\| \leq \dfrac{\|\zeta\|}{\|z\|} \|z' - \zeta\|$.

Proof.   (1) $\langle \|\zeta\|^2/\langle x, \zeta \rangle x - \zeta, \zeta \rangle = 0$ so $\|\zeta\|^2/\langle x, \zeta \rangle x \in A_\zeta$.
  (2)   If U is a unitary transformation of $\mathbb{C}^{n+1}$ leaving ζ fixed, then

$$\frac{\left\| \frac{\|\zeta\|^2}{\langle x, \zeta \rangle} x - \zeta \right\|}{\|x - \zeta\|} = \frac{\left\| \frac{\|\zeta\|^2}{\langle x, \zeta \rangle} U(x) - \zeta \right\|}{\|U(x) - \zeta\|}.$$

Thus, by Part (1) it suffices to prove the statement when x is on the line segment joining ζ and $\pi(z)$. Now, elementary planar geometry finishes the argument.
  (3)   This again is elementary geometry.
  (4)   Apply Parts (2) and (3).   □

Lemma 3 *Let* $x \in A_\zeta$, $w \in T_\zeta$, *and* $u \in T_x$. *Then*

(1) *there is a unique* $\lambda \in \mathbb{C}$ *such that* $w + \lambda \zeta \in T_x$.

(2) $u = u_1 + \lambda \zeta$, *where* $u_1 \in T_\zeta$ *and* $\lambda = -\dfrac{\langle u_1, x - \zeta \rangle}{\|\zeta\|^2}$.

(3) *If* $u = u_1 + u_2$, *where* $u_1 \in T_\zeta$ *and* $u_2 = \lambda \zeta$, *then* $\dfrac{\|u_2\|}{\|u_1\|} \leq \dfrac{\|x - \zeta\|}{\|\zeta\|}$.

Proof.   (1)   If $w + \lambda \zeta$ and $w + \gamma \zeta$ are both in T_x, then $(\lambda - \gamma)\langle \zeta, x \rangle = 0$. But $\langle \zeta, x - \zeta \rangle = 0$ so $\langle \zeta, x \rangle \neq 0$ and $\lambda = \gamma$.
  (2) u may be written as $u_1 + \lambda \zeta$ where u_1 and λ are unique. So by (1) we need only verify that

$$u_1 - \frac{\langle u_1, x - \zeta \rangle}{\|\zeta\|^2} \zeta$$

is in T_x. But

$$\left\langle u_1 - \frac{\langle u_1, x - \zeta \rangle}{\|\zeta\|^2} \zeta, x \right\rangle = \langle u_1, x \rangle - \frac{\langle u_1, x - \zeta \rangle}{\|\zeta\|^2} (\langle \zeta, x - \zeta \rangle + \langle \zeta, \zeta \rangle)$$

$$= \langle u_1, x - \zeta \rangle - \frac{\langle u_1, x - \zeta \rangle}{\|\zeta\|^2} (\|\zeta\|^2)$$

$$= 0.$$

(3) By the Cauchy–Schwarz inequality

$$\|u_2\| \leq \|u_1\| \frac{\|x - \zeta\|}{\|\zeta\|}.$$

$\square$

Lemma 4 *With f, ζ, z as previously so that $\zeta \in A_z$,*

$$\|Df(\zeta)|_{T_z}^{-1} Df(\zeta)|_{T_\zeta}\| \leq \frac{\|\zeta\|}{\|z\|}.$$

Proof. Let $v \in T_\zeta$ and $u \in T_z$ such that

$$Df(\zeta)|_{T_z}^{-1} Df(\zeta)|_{T_\zeta}(v) = u$$

and consider points $u_1 \in T_\zeta$ and $u_2 = \lambda_u \zeta$ such that $u = u_1 + u_2$.

Since $Df(\zeta)u = Df(\zeta)v$ and $Df(\zeta)(\zeta) = 0$ (by the homogeneity of f), we deduce that $Df(\zeta)u_1 = Df(\zeta)v$ and since $Df(\zeta)$ is bijective on T_ζ, that $u_1 = v$. We then have

$$\frac{\|Df(\zeta)|_{T_z}^{-1} Df(\zeta)|_{T_\zeta}(v)\|}{\|v\|} = \frac{\|u\|}{\|v\|} = \frac{\|u\|}{\|u_1\|}.$$

Since by (3) of the preceding lemma applied to $x = \pi(z)$,

$$\|u_2\| \leq \|u_1\| \cdot \frac{\|\pi(z) - \zeta\|}{\|\zeta\|}$$

we have

$$\frac{\|u\|}{\|u_1\|} \leq \left(1 + \left(\frac{\|\pi(z) - \zeta\|}{\|\zeta\|}\right)^2\right)^{1/2}$$

$$= \left(\frac{\|\zeta\|^2 + \|\pi(z) - \zeta\|^2}{\|\zeta\|^2}\right)^{1/2}$$

which by the Pythagorean theorem is

$$\left(\frac{\|\pi(z)\|^2}{\|\zeta\|^2}\right)^{1/2} = \frac{\|\pi(z)\|}{\|\zeta\|} = \frac{\|\zeta\|}{\|z\|}.$$

The last is true by similar triangles. We have thus showed that

$$\frac{\|Df(\zeta)|_{T_z}^{-1} Df(\zeta)|_{T_\zeta}(v)\|}{\|v\|} \leq \frac{\|\zeta\|}{\|z\|}.$$

Since this is true for all $v \in T_\zeta$ the lemma follows. $\square$

Lemma 5 *For* $(f, \zeta) \in V$, $z \in \mathbb{P}(\mathbb{C}^{n+1})$

$$\gamma(f|_{A_z}, \zeta) \le \frac{1}{\|z\|} \gamma_0(f, \zeta), \ \zeta \in A_z.$$

Proof.   Let $\widehat{f} = f|_{A_z}$. Then $Df(\zeta)|_{T_z} = D\widehat{f}(\zeta)$ and

$$\gamma(\widehat{f}, \zeta) = \max_{k>1} \left\| Df(\zeta)|_{T_z}^{-1} Df(\zeta)|_{T_\zeta} Df(\zeta)|_{T_\zeta}^{-1} \frac{D^k \widehat{f}(\zeta)}{k!} \right\|^{1/(k-1)}.$$

By Lemma 4, this quantity is bounded by

$$\frac{\|\zeta\|}{\|z\|} \max_{k>1} \left\| Df(\zeta)|_{T_\zeta}^{-1} \frac{D^k \widehat{f}(\zeta)}{k!} \right\|^{1/(k-1)} \le \frac{1}{\|z\|} \gamma_0(f, \zeta).$$

$\square$

Lemma 6   *Let* $z, \zeta, f, \widehat{u}$ *be as in Lemma 1 with* $\zeta \in A_z$. *Define* $u = \|z - \zeta\| \gamma(f|_{A_z}, \zeta)$ *as in Theorem 1 of Chapter 8 and suppose* $\widehat{u} \le (3 - \sqrt{7})/2$. *Then* $u \le \widehat{u}$ *and* $u/\psi(u) \le \widehat{u}/\psi(\widehat{u})$.

Proof.   By Lemma 5

$$u \le \frac{\|z - \zeta\|}{\|z\|} \gamma_0(f, \zeta) = \widehat{u}.$$

Moreover, the monotonicity property of ψ yields the last part. $\square$

Proof of Lemma 1.   One has

$$
\begin{aligned}
d_T(z', \zeta) \quad &= \quad \frac{\|\pi(z') - \zeta\|}{\|\zeta\|} &&\text{by definition} \\[2mm]
&\le \quad \frac{\|z' - \zeta\|}{\|z\|} &&\text{by Lemma 2} \\[2mm]
&\le \quad \frac{u}{\psi(u)} \frac{\|z - \zeta\|}{\|z\|} &&\text{by Theorem 1 of Chapter 8.}
\end{aligned}
$$

By definition of $d_T(z, \zeta)$, then

$$d_T(z', \zeta) \le \frac{u}{\psi(u)} d_T(z, \zeta).$$

Now apply the previous lemma to finish the proof. $\square$

From Lemma 1, one obtains Theorem 1 by exactly the same process used in proving Theorem 1 of Chapter 8 from Proposition 1 there.

14.2 The Higher Derivative Estimate

The goal of this section is to estimate the important invariant γ_0 of the previous section in terms of the condition number μ.

Recall that in projective space we have two versions, μ and μ_{norm}, of the condition number defined on $V \subset \mathbb{P}(\mathcal{H}_{(d)}) \times \mathbb{P}(\mathbb{C}^{n+1})$. If we write $D = \max_{1 \le i \le n} d_i$ they are related by

$$\frac{\mu_{\text{norm}}(f, \zeta)}{D^{1/2}} \le \mu(f, \zeta) \le \mu_{\text{norm}}(f, \zeta), \ (f, \zeta) \in V.$$

The quantity μ has a direct motivation as a condition number literally, whereas in some situations, as in the condition number theorem, μ_{norm} gives a more precise expression.

Theorem 2 *For all $(f, \zeta) \in V$*

$$\gamma_0(f, \zeta) \le \frac{D^{3/2}}{2} \mu_{\text{norm}}(f, \zeta).$$

The main step in the proof is the following proposition.

Proposition 1 *Let $f \in \mathcal{H}_d$ so that $f : \mathbb{C}^{n+1} \to \mathbb{C}$ is an homogeneous polynomial of degree d. Then for all $x, w_1, \ldots, w_k \in \mathbb{C}^{n+1}$, the kth derivative of f satisfies*

$$|D^k f(x)(w_1, \ldots, w_k)| \le d(d-1)\ldots(d-k+1)\|f\|\|x\|^{d-k}\|w_1\| \ldots \|w_k\|.$$

The case of $k = 0$ is dealt with by the following lemma.

Lemma 7 *Let f be as in Proposition 1. Then $|f(x)| \le \|f\|\|x\|^d$ for all $x \in \mathbb{C}^{n+1}$.*

Proof. By homogeneity, we may suppose $\|x\| = 1$. Let $e_0 = (1, 0, \ldots, 0) \in \mathbb{C}^{n+1}$ and σ be a unitary automorphism of $\mathbb{C}^{n+1}$ taking x to e_0 ($\sigma x = e_0$). Then

$$|f(x)| = |f\sigma^{-1}\sigma x| = |g(e_0)|,$$

where $g = f\sigma^{-1}$. By the unitary invariance of our norm on $\mathcal{H}_d$, seen in Chapter 12, $\|f\| = \|g\|$. If $g(z) = \sum_\alpha b_\alpha z^\alpha$,

$$|g(e_0)| = |b_{(d,0,\ldots,0)}| \le \|g\| = \|f\|.$$

□

For fixed $w \in \mathbb{C}^{n+1}$, consider the derivative as a map $\mathcal{H}_d \to \mathcal{H}_{d-1}$ as follows. It takes f into $Df(w)$, and $Df(w)$ is interpreted as a polynomial of degree $d - 1$ in x, $x \to Df(x)(w)$.

Lemma 8 *For any $g \in \mathcal{H}_d$ and any $w \in \mathbb{C}^{n+1}$*

$$\|Dg(w)\| \leq d\|g\|\|w\|.$$

Proof. By homogeneity we may suppose $\|w\| = 1$. Moreover, by the argument of Lemma 7 and Lemma 11 of Chapter 12, we may even suppose that $w = e_0$, $e_0 = (1, 0, \ldots, 0)$.

If $f(z) = \sum a_\alpha z^\alpha$,

$$Df(x)(e_0) = \sum_{\alpha \mid \alpha_0 \neq 0} \alpha_0 a_\alpha x_0^{\alpha_0 - 1} x_1^{\alpha_1} \ldots x_n^{\alpha_n}.$$

Then

$$\|Df(e_0)\|^2 = \sum_{\alpha \mid \alpha_0 \neq 0} (\alpha_0)^2 |a_\alpha|^2 \left(\frac{(\alpha_0 - 1)!\alpha_1! \ldots \alpha_n!}{(d - 1)!} \right)$$

$$= d \sum_{\alpha \mid \alpha_0 \neq 0} \alpha_0 |a_\alpha|^2 \left(\frac{\alpha_0! \ldots \alpha_n!}{d!} \right)$$

$$\leq d^2 \sum_\alpha |a_\alpha|^2 \left(\frac{\alpha_0! \ldots \alpha_n!}{d!} \right) = d^2 \|f\|.$$

□

Lemma 9 *For $f \in \mathcal{H}_d$, $w_1, \ldots, w_k \in \mathbb{C}^{n+1}$,*

$$\|(D^k f)(w_1, \ldots, w_k)\| \leq \frac{d!}{k!} \|f\|\|w_1\| \ldots \|w_k\|,$$

where $g = (D^k f)(w_1, \ldots, w_k)$ is to be understood as an element of $\mathcal{H}_{d-k}$.

Proof. The case $k = 0$ is trivial and we proceed by induction. Write

$$\|D^{k+1} f(w_1, \ldots, w_{k+1})\| = \|Dg(w_{k+1})\|$$

which is less than $(d - k)\|g\|\|w_{k+1}\|$ by Lemma 8. But

$$\|g\| \leq d(d - 1) \ldots (d - k + 1)\|f\|\|w_1\| \ldots \|w_k\|$$

by the induction hypothesis, so

$$\|D^{k+1} f(w_1, \ldots, w_{k+1})\| \leq \frac{d!}{(k - 1)!} \|f\|\|w_1\| \ldots \|w_{k+1}\|$$

and we are finished.

□

Using Lemma 7, we obtain Proposition 1 immediately from Lemma 9.

Lemma 10 *Let $d \geq k \geq 2$ be positive integers and*

$$A_k = \left(\frac{1}{d^{1/2}} \frac{d(d-1)\ldots(d-k+1)}{k!} \right)^{1/(k-1)}.$$

Then $\max_{k>1} A_k$ is achieved at $k = 2$.

Proof. It is sufficient to show that $A_{k+1} < A_k$ for $k \geq 2$. This amounts to

$$\frac{d}{d^{1/2}} \frac{(d-1)\ldots(d-k)}{(k+1)!} < \left(\frac{d(d-1)\ldots(d-k+1)}{d^{1/2}k!} \right)^{k/(k-1)}$$

or yet

$$\frac{(d-1)\ldots(d-k)}{(k+1)!} < \left(\frac{(d-1)\ldots(d-k+1)}{k!} \right)^{1+(1/(k-1))}$$

or yet

$$\frac{d-k}{k+1} < \left(\frac{(d-1)\ldots(d-k+1)}{k!} \right)^{1/(k-1)}.$$

The last is clear. □

Lemma 11 *Let $f \in \mathcal{H}_d$ and k be an integer greater than 1. Then*

$$\left(\frac{\|D^k f(x)(w_1, \ldots, w_k)\|}{d^{1/2}\|f\|\|x\|^{d-k}k!\|w_1\|\ldots\|w_k\|} \right)^{1/(k-1)} \leq \frac{d^{1/2}(d-1)}{2}.$$

Proof. It is a consequence of Proposition 1 and Lemma 10. □

Lemma 12 *Let $f : \mathbb{C}^{n+1} \to \mathbb{C}^n$ belong to $\mathcal{H}_{(d)}$ and $x \in \mathbb{C}^{n+1}$. Then*

$$\left(\frac{\|\Delta(\|x\|^{d_i-k}d_i^{1/2})^{-1}D^k f(x)\|}{k!\|f\|} \right)^{1/(k-1)} \leq \frac{D^{3/2}}{2}.$$

Proof. By definition

$$\left(\frac{\|\Delta(\|x\|^{d_i-k}d_i^{1/2})^{-1}D^k f(x)\|}{\|f\|k!} \right)^{1/(k-1)} = \left(\sum_{i=1}^{n} \left(\frac{\|D^k f_i(x)\|}{\|x\|^{d_i-k}k!d_i^{1/2}\|f\|} \right)^2 \right)^{1/(2(k-1))}$$

and by the previous lemma this is less than or equal to

$$\left(\sum_{i=1}^{n} \left(\left(\frac{d_i^{1/2}(d_i-1)}{2} \right)^{k-1} \frac{\|f_i\|}{\|f\|} \right)^2 \right)^{1/(2(k-1))}$$

which is less than $D^{3/2}/2$. □

Now we can prove Theorem 2.

Proof of Theorem 2. One has

$$\gamma_0(f, \zeta) = \|\zeta\| \max_{k>1} \left\| Df(\zeta)|_{T_\zeta}^{-1} \frac{D^k f(\zeta)}{k!} \right\|^{1/(k-1)}.$$

We can bound the second expression by

$$\max_k \left\| (Df(\zeta)|_{T_\zeta}^{-1} \Delta(\|\zeta\|^{d_i-1} d_i^{1/2}) \|f\|) \left(\Delta \left(\frac{\|\zeta\|^{-(d_i-k)}}{d_i^{1/2} \|f\|} \right) \frac{D^k f(\zeta)}{k!} \right) \right\|^{1/(k-1)}$$

which by by Lemma 12 is at most

$$\max_k \mu_{\text{norm}}(f, \zeta)^{1/(k-1)} \frac{D^{3/2}}{2} \leq \mu_{\text{norm}}(f, \zeta) \frac{D^{3/2}}{2}$$

since $\mu_{\text{norm}}(f, \zeta) \geq 1$ by the condition number theorem. $\qquad\square$

The next result asserts how well zeros of a polynomial system must be separated in terms of the condition number of the system.

Theorem 3 *Let* $f : \mathbb{C}^{n+1} \to \mathbb{C}^n$ *represent an homogeneous system of polynomials,* $f \in \mathbb{P}(\mathcal{H}_{(d)})$. *If* ζ, ζ' *are two zeros of* f, *then*

$$d_R(\zeta, \zeta') \geq \frac{3 - \sqrt{7}}{D^2 \mu(f)},$$

where, we recall, d_R *is the Riemannian distance on the projective space* $\mathbb{P}(\mathbb{C}^{n+1})$, $D = \max(d_i)$, *and* $\mu(f) = \max_{\zeta | f(\zeta) = 0} \mu(f, \zeta)$ *is the condition number of* f.

Proof. If $d_R(\zeta, \zeta') < (3 - \sqrt{7})/D^2 \mu(f)$, then there is an $x \in \mathbb{P}(\mathbb{C}^{n+1})$ with

$$d_R(\zeta, x) < \frac{3 - \sqrt{7}}{2D^2 \mu(f)}$$

and

$$d_R(\zeta', x) < \frac{3 - \sqrt{7}}{2D^2 \mu(f)}$$

and hence

$$d_T(\zeta, x) < \frac{3 - \sqrt{7}}{D^2 \mu(f)}$$

and

$$d_T(\zeta', x) < \frac{3 - \sqrt{7}}{D^2 \mu(f)}.$$

But then, by Theorem 1, x converges to both ζ and ζ' under projective Newton iteration which is a contradiction. $\qquad\square$

14.3   Complexity of Homotopy Methods

Here we prove a theorem that estimates the number of projective Newton steps sufficient to follow a curve of zeros of a homotopy $f_t \in \mathbb{P}(\mathcal{H}_{(d)})$, $0 \leq t \leq 1$.

Example 1 (Linear Homotopy) Let ζ^* be a known zero of $f^* \in \mathbb{P}(\mathcal{H}_{(d)})$, and $f \in \mathbb{P}(\mathcal{H}_{(d)})$ be given. Define $f_t = tf + (1-t)f^*$. If f_t in $\mathcal{H}_{(d)}$ is not in Σ for any $t \in [0, 1]$, the implicit function theorem defines ζ_t such that $f_t(\zeta_t) = 0$ each t and $\zeta_0 = \zeta^*$. Then ζ_1 is a solution of the problem $f(\zeta) = 0$.

More generally consider any curve $F : I \rightarrow V - \Sigma'$, $I = [0, 1]$,

$$V = \{(f, \zeta) \in \mathbb{P}(\mathcal{H}_{(d)}) \times \mathbb{P}(\mathbb{C}^{n+1}) \mid f(\zeta) = 0\}$$

and Σ' the critical set in V. Write

$$F(t) = (f_t, \zeta_t).$$

(Thus $f_t(\zeta_t) = 0$.)

By a *partition* of I is meant an increasing sequence of t_i with $t_0 = 0$ and $t_k = 1$. By abuse of notation we write, under the preceding conditions,

$$(f_i, \zeta_i) = (f_{t_i}, \zeta_{t_i}), \ i = 0, \ldots, k.$$

Now define inductively

$$x_0 = \zeta_0 \text{ and } x_i = N_{f_i}(x_{i-1}), \ i = 1, \ldots, k, \tag{14.1}$$

where $N_{f_i} : \mathbb{P}(\mathbb{C}^{n+1}) \rightarrow \mathbb{P}(\mathbb{C}^{n+1})$ is the map of the projective Newton's method of the previous sections. One would hope that x_i is defined for each i and that x_i is an approximation of ζ_i. It is not difficult to show that is the case if $t_i - t_{i-1} < \Delta$ is small enough and hence k large. Our task is to show that there exists such a partition with a good bound on k. Since k is proportional to the number of Newton iterations, k is a basic complexity invariant.

Given a curve $F : I \rightarrow V - \Sigma'$ as above, let L_f denote the length of $\{f_t \mid 0 \leq t \leq 1\}$ and

$$\mu(F) = \max_{t \in I} \mu(f_t, \zeta_t).$$

Recall that $D = \max_{1 \leq i \leq n} d_i$, for $f \in \mathbb{P}(\mathcal{H}_{(d)})$.

Theorem 4   *Given any curve $F : I \rightarrow V - \Sigma'$, there is a partition $t_0 = 0, \ldots, t_k = 1$ of I with the x_i of (14.1) approximate zeros of f_i relative to ζ_i, respectively, and with*

$$k = \lceil 4cD^2\mu(F)^2 L_f \rceil, \ c = \frac{2}{3 - \sqrt{7}}.$$

Thus $4cD^2\mu(F)^2 L_f$ is a complexity bound on the problem of following the path of solutions ζ_t of the equations $f_t(\zeta_t) = 0$.

Remark 2 In the case of the linear homotopy, $L_f \leq \pi/2$, and we show that $t_i - t_{i-1} = \Delta = 1/k$.

Remark 3 The key complexity ingredient by Theorem 4 is $\mu(F)^2$, the condition number of the path squared. By the condition number theorem of Chapter 12, this shows that the number of Newton steps is mediated by the distance of $F(t)$ to Σ' along the fibers of π_2.

Proposition 2 *Let L_ζ denote the length of the curve ζ_t. Then*

$$L_\zeta \leq \mu(F) L_f.$$

Proof. We may write $G(f_t) = \zeta_t$, where G is defined by the implicit function theorem, using the fact that for all $t \in [0, 1], (f_t, \zeta_t) \notin \Sigma'$. Then $DG(f_t)(df_t/dt) = d\zeta_t/dt$ and $\|d\zeta_t/dt\| \leq \mu(f_t, \zeta_t)\|df_t/dt\|$. This is just the definition of condition number as in Chapter 12. So

$$L_\zeta = \int_0^1 \left\| \frac{d\zeta_t}{dt} \right\| dt \leq \mu(F) \int_0^1 \left\| \frac{df_t}{dt} \right\| dt = L_f.$$

$\square$

Theorem 4 follows immediately from Proposition 2 and the following.

Proposition 3 *Given $F : I \to V - \Sigma'$, there is a partition $t_0 = 0, \ldots, t_k = 1$ of I such that the x_i of (14.1) are approximate zeros of f_i relative to ζ_i, each i, and with $k = \lceil 4cD^2\mu(F)L_\zeta \rceil$, $c = 2/(3 - \sqrt{7})$.*

Thus it remains to prove Proposition 3. We now proceed to do so.

Proof of Proposition 3. According to Theorem 1 that follows from the assertion:

$$d_T(x_i, \zeta_i)\frac{D^2\mu(F)}{2} < \frac{1}{2}\left(\frac{3 - \sqrt{7}}{2}\right), \quad i = 0, 1, \ldots, k. \tag{14.2}$$

Of course (14.2) is true for $i = 0$ since $x_0 = \zeta_0$. We prove it in general by induction.

Since the Riemannian distance $d_R(\zeta_i, \zeta_{i-1})$ is defined by arclength there exists a partition $t_0 = 0, \ldots, t_k = 1$ such that, for each i, $d_R(\zeta_i, \zeta_{i-1}) \leq L_\zeta/k$. The bound L_ζ/k for the range of k under consideration is sufficiently small so that

$$d_T(\zeta_i, \zeta_{i-1}) = \tan d_R(\zeta_i, \zeta_{i-1}) \leq 2d_R(\zeta_i, \zeta_{i-1}) \leq \frac{2L_\zeta}{k}.$$

Thus by the hypothesis of Proposition 3

$$d_T(\zeta_i, \zeta_{i-1}) \leq \frac{3 - \sqrt{7}}{4} \frac{1}{D^2\mu(F)}.$$

By the higher derivative estimate of the previous section, it follows that

$$d_T(\zeta_i, \zeta_{i-1})\frac{D^2\mu(F)}{2} < \frac{1}{4}\left(\frac{3 - \sqrt{7}}{2}\right). \tag{14.3}$$

Now add $(14.2)_{i-1}$ and (14.3) and use the fact that $\tan(x+y) \le 7/6(\tan x + \tan y)$ in our range to obtain

$$d_T(x_{i-1}, \zeta_i)\gamma_0(f_i, \zeta_i) < \frac{7}{8}\frac{3-\sqrt{7}}{2}.$$

Apply the approximate zero theorem. Then

$$d_T(x_i, \zeta_i)\frac{D^2\mu(F)}{2} < \frac{7}{8}\left(\frac{1}{2}\left(\frac{3-\sqrt{7}}{2}\right)\right).$$

This finishes the proof by induction. $\square$

Remark 4 In the proof of Proposition 3 we have actually obtained

$$d_T(x_i, \zeta)\gamma(f_i, \zeta_i) < \frac{7}{8}(\frac{1}{2}(\frac{3-\sqrt{7}}{2})).$$

Thus even if we compute x_i with a small error relative to $D^2\mu$ we still obtain

$$d_T(x_i, \zeta)\gamma(f_i, \zeta_i) < \frac{1}{2}(\frac{3-\sqrt{7}}{2})$$

and we may proceed with the induction. This observation can help us limit the precision of the computation we do and to tolerate small errors. In the next chapter we exploit a similar observation.

Since $\alpha(f|A_z) \le \alpha(f, z)$ it follows that if $\alpha(f, z) < \alpha_0$, then z is an approximate zero of $f|A_z$ with respect to the usual Newton's method for $f|A_z$.

14.4   Additional Comments and Bibliographical Remarks

The general background for this chapter is the series of five papers by Shub and Smale [1993a, 1993b, 1993c, 1996, 1994]. In particular, Theorem 4 is the main theorem of [Shub and Smale 1993a] but the proof (and the constant) here is greatly improved. It should be especially noted, moreover, that [Shub and Smale 1996] and [Shub and Smale 1994] have more advanced complexity results than we can do here. [Renegar 1987a] is an important predecessor to those papers and also deals with the sparse case.

Theorem 1 here is new, but compare this to [Shub and Smale 1996].

The Projective Newton Method is proposed in [Shub 1993b].

Theorem 3 has a closely related version in [Dedieu TAb] where there is also a good discussion.

The higher derivative estimate of Theorem 2 is in [Shub and Smale 1993a].

15

Linear Programming

The aim of this chapter is to prove that the linear programming feasibility problem over $\mathbb{Q}$ can be solved in polynomial time. As a preliminary step, in Section 15.1 we show that inputs for rational machines can be supposed to be given by pairs of integers without substantially altering the complexity of the considered problem. In Section 15.2 we define an auxiliary problem which is a modification of the linear programming optimization problem and exhibit an algorithm to solve it with a homotopy method. The proof of correctness of this algorithm relies on the α theorem for Newton's method proved in Chapter 8. In the two sections following 15.2 we first reduce the LPF over $\mathbb{Q}$ to this auxiliary problem and then prove that the algorithm resulting from this reduction is polynomial time. The proof relies on an efficient method for solving linear systems over $\mathbb{Q}$ which is developed in Section 15.5.

It is still an open problem as to whether the linear programming feasibility problem is in P over $\mathbb{R}$. There is a notion of condition of a problem instance of LPF/$\mathbb{R}$ in terms of the distance of the problem instance to the set of ill-posed instances. Taking this condition number into account in the input size, LPF/$\mathbb{R}$ is in P, but these points are not touched on here.

15.1 Machines over $\mathbb{Q}$

In the rest of this chapter we consider rational machines with the bit input size and cost defined in Chapter 6. Recall that the bit input size is the product of the dimension of the input with the maximum height of the rational numbers that are the components of the input, and the cost is the product of the maximum height of a rational number appearing in the computation and the halting time. Recall also that the height of a rational number q/r with q and r in $\mathbb{Z}$ and relatively prime is defined to be $\max\{\mathrm{ht}_{\mathbb{Z}}(q), \mathrm{ht}_{\mathbb{Z}}(r)\}$.

The main result in this section exhibits an algorithm that computes a relatively prime numerator and denominator for a given rational with polynomial cost. We begin with a simple lemma. For a real number x let $\lfloor x \rfloor$ denote the integer part of x, that is, the largest integer smaller than or equal to x. The fractional part of x is $x - \lfloor x \rfloor$.

Lemma 1   *Let k be a subfield of $\mathbb{R}$. There is a machine over k that, given $x \in k$, $x \geq 1$, computes the integer and fractional parts of x in $O(\log(1 + x))$ arithmetic operations. If $k = \mathbb{Q}$, the machine works with polynomial bit cost.*

Proof.   The following algorithm

$$
\begin{array}{l}
\textbf{input } x \\
s := 1 \\
\textbf{while } s \leq x \textbf{ do} \\
\quad s := 2 * s \\
\textbf{end do} \\
s := s/2 \\
i := s \\
\textbf{while } x - i \geq 1 \textbf{ do} \\
\quad s := s/2 \\
\quad \textbf{if } x - i \geq s \textbf{ then } i := i + s \textbf{ end if} \\
\textbf{end do} \\
\textbf{output } (i, x - i)
\end{array}
$$

computes the integer part of x performing $O(\log(1 + x))$ arithmetic operations.

If $k = \mathbb{Q}$, this number of operations is linear in the size of the input, which is $\mathrm{ht}_{\mathbb{Q}}(x)$, and since all the numbers involved in the computation have height at most $1 + \mathrm{ht}_{\mathbb{Q}}(x)$ the result follows. $\square$

Proposition 1   *There is a machine over $\mathbb{Q}$ that, with input $x \in \mathbb{Q}$, computes $q, r \in \mathbb{Z}$ relatively prime, $q > 0$, such that $x = r/q$. The computation is done with polynomial bit cost.*

Proof.   Let $x \in \mathbb{Q}$. It is sufficient to consider the case $0 < x < 1$. Suppose $q, r \in \mathbb{N}$ are relatively prime and $x = r/q$. The Euclidean algorithm for the greatest common divisor applied to q and $r_0 = r$ produces a sequence

$$
q = d_1 r_0 + r_1
$$
$$
r_0 = d_2 r_1 + r_2
$$
$$
\vdots
$$

$$r_{s-2} = d_s r_{s-1} + 1$$

with $s \le \mathrm{ht}_{\mathbb{Z}}(q)$ and $\mathrm{ht}_{\mathbb{Z}}(d_i)$, $\mathrm{ht}_{\mathbb{Z}}(r_i) \le \mathrm{ht}_{\mathbb{Z}}(q)$ for $i = 1, \ldots, s$. Dividing the first equation by r_0 we obtain

$$\frac{1}{x} = d_1 + a_1,$$

where $a_1 = (r_1/r_0) < 1$ is the fractional part of $1/x$ and d_1 its integer part. In the same way, we have

$$\frac{1}{a_{j-1}} = \frac{r_{j-2}}{r_{j-1}} = d_j + a_j,$$

where $a_j = r_j/r_{j-1}$ is the fractional part of $1/a_{j-1}$ for $j = 2, \ldots, s-1$.

The preceding considerations lead to an algorithm to compute q and r. By applying Lemma 1 to the inverse of each element in the sequence $x, a_1, \ldots, a_s$ we obtain the next element. Then one can recover q and r from $x, a_1, \ldots, a_s$ in the following way. Inverting a_s one obtains r_{s-1} and then, for j from $s - 1$ to 1, one computes r_{j-1} from a_j and the previously computed r_j. Finally, from $r_0 = r$ and x one computes q. The bounds on s and on the heights of a_i ensure that the whole procedure has polynomial cost. □

As in Chapter 6, the map $\phi : \mathbb{Q} \to \mathbb{Z}^2$ that sends x to (q, r) as in Proposition 1 allows us to consider rational inputs and outputs for integer machines. On the other hand, the restriction to integer inputs and outputs for rational machines is provided by the ring inclusion $i : \mathbb{Z} \to \mathbb{Q}$. Let us denote by $\phi^* : \mathbb{Q}^\infty \to (\mathbb{Z}^2)^\infty$ and $i^* : \mathbb{Z}^\infty \to \mathbb{Q}^\infty$ the maps induced by ϕ and i, respectively. It is straightforward to prove the following result.

Corollary 1

(a) *The decision problem $X \subseteq \mathbb{Q}^\infty$ is in P (respectively, NP) over $\mathbb{Q}$ if and only if the decision problem $\phi^*(X) \subseteq (\mathbb{Z}^2)^\infty$ is in P (respectively, NP) over $\mathbb{Z}$.*

(b) *The decision problem $X \subseteq \mathbb{Z}^\infty$ is in P (respectively, NP) over $\mathbb{Z}$ if and only if the decision problem $i^*(X) \subseteq \mathbb{Q}^\infty$ is in P (respectively, NP) over $\mathbb{Q}$.*

□

The preceding corollary and Corollary 1 of Chapter 6 show that the question "P $\ne$ NP?" has the same answer over $\mathbb{Z}$ or $\mathbb{Q}$ as for Turing machines. Indeed the codings are so natural that classically these have all been considered the same problem.

15.2 The Barrier Method in Linear Programming

Linear programming concerns the optimization of linear functions subject to linear constraints. For many years the standard procedure for solving such problems was the simplex method which marches through the vertices of the polyhedron determined by the inequalities until it finds the optimum. The simplex method sometimes encounters difficulties in that it has to go through exponentially many vertices, although in practice it seems to work rapidly. Some explanations have been offered for this phenomenon, based on average performance considerations. The 1980s saw a revival of interior point methods for linear programming, some of which have been proven to be polynomial time for rational linear programming. In this chapter we carry out the main part of the analysis for one interior point method, the barrier method, which is a path following method. To do so we apply the α theory developed in Chapter 8. In linear programming applications today, a mixture of simplex and interior point methods is used.

Given a vector $c \in \mathbb{R}^n - \{0\}$, an $m \times n$ real matrix A, and a vector b in $\mathbb{R}^m$, the real linear programming optimization problem LPO with input (A, b, c) is

$$\text{minimize } c \cdot x$$
$$\text{subject to } Ax \geq b.$$

Let us denote by Int the open set $\{x \in \mathbb{R}^n \mid Ax > b\}$. In this section we assume that Int is nonempty and bounded. Thus A has rank n.

In this case, an approach to solving this problem, known as the *barrier method*, considers the function $h_t : \mathbb{R}^n \to \mathbb{R}$ defined by

$$h_t(x) = c \cdot x - t \sum_{i=1}^{m} \ln(A_i x - b_i),$$

where A_i are the rows of A. For fixed t we now show that this function has a unique minimum ζ_t in Int. As $t \to 0$, ζ_t tends to the solution of the LPO. The idea is that it is easier to solve the minimization problem,

$$\text{minimize } h_t$$
$$\text{subject to } x \in \text{Int},$$

where there is no boundary to worry about.

Proposition 2 *For $t > 0$ and $x \in$ Int,*

(a) $\text{grad } h_t(x) = c - t \sum_{i=1}^{m} \dfrac{A_i}{A_i x - b_i};$

(b) $D^2 h_t(x)(u, v) = t \sum_{i=1}^{m} \dfrac{(A_i \cdot u)(A_i \cdot v)}{(A_i x - b_i)^2}.$

Proof. The proof is straightforward calculus. □

Given a vector $x \in$ Int we let $L_x : \mathbb{R}^n \to \mathbb{R}^m$ be the linear map whose ith coordinate is given by

$$(L_x(y))_i = \frac{A_i y}{A_i x - b_i}.$$

Also, let $e \in \mathbb{R}^m$ be the vector whose components are all equal to one,

$$e = (1, \ldots, 1)^T.$$

Then we may rewrite Proposition 2 as follows.

Proposition 3 *For $t > 0$ and $x \in$ Int,*

(a) $\operatorname{grad} h_t(x) = c - t L_x^T e$;

(b) $D^2 h_t(x)(u, v) = t u^T L_x^T L_x v = t L_x u \cdot L_x v$.

$\square$

Proposition 4 *For $t > 0$ the function h_t is strictly convex on* Int; *that is, $D^2 h_t$ is positive definite on* Int.

Proof. Since Int is nonempty and bounded $\ker A = 0$, for if $Ax > b$, then $A(x + y) > b$ for all $y \in \ker A$. Now by the definition of L_x, $\ker A = \ker L_x$ for all $x \in$ Int. Thus $\ker L_x = 0$ from which it follows that $L_x v \cdot L_x v > 0$ for $v \neq 0$ and the second derivative $D^2 h_t$ is positive definite for $t > 0$. $\square$

Proposition 4 implies that h_t has a unique minimum on Int which we denote by ζ_t. We write L_t for L_{ζ_t}. For further ease of notation we let $f_t = \operatorname{grad} h_t$.

Proposition 5 *For any $t > 0$ the function h_t has a unique minimum $\zeta_t \in$ Int which is the unique zero of f_t. Moreover ζ_t varies smoothly with t. Any limit point x of ζ_t is a solution of the* LPO.

Proof. Implicitly differentiating $f_t(\zeta_t) = 0$ gives

$$\frac{d\zeta_t}{dt} = \frac{1}{t} Df_t(\zeta_t)^{-1} L_t^T e.$$

Now $(1/t) L_t^T e = (1/t^2) c$ since $f_t(\zeta_t) = 0$. Since c is not equal to 0 and since $Df_t(\zeta_t) = D^2 h_t(\zeta_t)$ is nondegenerate it follows that $d\zeta_t/dt \neq 0$. The implicit function theorem now proves the smoothness of ζ_t. For convergence we use the following proposition. $\square$

Let $\widehat{L}_t : \mathbb{R}^n \to \mathbb{R}^m$ be defined by $(\widehat{L}_t(x))_i = (A_i x - b_i)/(A_i \zeta_t - b_i)$. The function $\widehat{L}_t$ maps the polyhedron $P = \{x \in \mathbb{R}^n \mid Ax \geq b\}$ into $\mathbb{R}_+^m$. Moreover $\widehat{L}_t$ is an affine map and its linear part is L_t.

Proposition 6 *Let x^* be a minimum point for the* LPO. *Then*

$$c \cdot (\zeta_t - x^*) \leq tm.$$

Proof. Since $\widehat{L}_t$ maps the polyhedron $P = \{x \in \mathbb{R}^n \mid Ax \geq b\}$ into $\mathbb{R}_+^m$,

$$
\begin{aligned}
c \cdot (\zeta_t - x^*) &\leq c \cdot (\zeta_t - x^*) + t \widehat{L}_t(x^*) \cdot e \\
&= t L_t^T e \cdot (\zeta_t - x^*) + t \widehat{L}_t(x^*) \cdot e \\
&= t L_t^T e \cdot (\zeta_t - x^*) + t L_t(x^* - \zeta_t) \cdot e + t \widehat{L}_t(\zeta_t) \cdot e \\
&= t \widehat{L}_t(\zeta_t) \cdot e \\
&= tm.
\end{aligned}
$$

$\square$

The situation now is similar to the one we have already confronted in Chapter 9 for the fundamental theorem of algebra. We have proved the existence of a path ζ_t converging to an optimum.

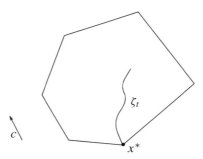

Figure A   A polyhedron $Ax \geq b$, objective function c, minimum point x^*, and the path of points ζ_t approaching x^*.

Now we attempt to numerically approximate this path in order to compute approximate solutions to the linear programming problem. In fact, let us consider the following auxiliary problem that we denote by ALPO.

> Given (A, b, c) as in LPO with Int $\neq \emptyset$ and bounded, $\zeta_1 \in$ Int that minimizes $h_1(x)$, and $\varepsilon > 0$, find $x \in \mathbb{R}^n$ such that $c \cdot (x - x^*) < \varepsilon$, where $x^* \in \mathbb{R}^n$ is a minimum of the LPO.

Note that in ALPO we suppose that we know that Int $\neq \emptyset$ as well as an exact solution ζ_1 minimizing $h_1(x)$; that is, $A\zeta_1 > b$ and grad $h_1(\zeta_1) = 0$ which are easily verifiable. These facts may make ALPO seem artificial. However, there are methods of associating an ALPO problem instance with an LPO problem instance so that the solution of the ALPO also solves the LPO. The ALPO is central in the next section where we use it to solve the rational linear programming problem.

In the algorithm that follows we begin by assuming that we know the zero ζ_1 of f_1, and use Newton's method to follow the solution. Our goal is to produce a sequence of t_i decreasing geometrically to zero with t_0 equal to 1, and good approximations x_{t_i} to ζ_{t_i}. The values of the objective function $c \cdot \zeta_{t_i}$ then tend geometrically to zero by Proposition 6. An important point is that the ratio of t_i to t_{i+1} only depends on the number of equations m and is $1 - (a/\sqrt{m})$ for a universal constant a.

Let $u \leq u_0$, $u < 1$, and $a < \alpha_0$ as in the robust α theorem (Theorem 4 of Chapter 8) further satisfy

$$\frac{1}{1 - 2a} \left(\frac{u}{2} + a \right) \leq 0.9u.$$

The numbers a and u are constants that may be chosen to be 0.01 and 0.05, respectively. Let $\lceil x \rceil$ denote the first integer greater than or equal to x, for $x \in \mathbb{R}$. Our algorithm is the following.

Algorithm 1.

Input $(A, b, c, \zeta_1, \varepsilon)$ as previously

(1) Let $M := \left\lceil \frac{(|\log \varepsilon| + \log 2m)}{a} \sqrt{m} \right\rceil$

(2) Let $x_0 := \zeta_1$ and $t_0 := 1$

(3) Let $t_{i+1} := t_i \left(1 - \frac{a}{\sqrt{m}}\right)$ and

$x_{i+1} := N_{f_{t_{i+1}}}(x_i)$ for $i = 0, \ldots, M - 1$

(4) Output x_M.

Theorem 1 *In algorithm 1, $x_i = N_{f_{t_i}}(x_{i-1})$ is defined for all i and*

$$c \cdot (x_M - x^*) < \varepsilon,$$

where x^* is a minimum of the LPO for the problem instance (A, b, c).

As in Chapter 9 let us define the *input size* of an instance of ALPO to be

$$mn + m + n + |\ln \varepsilon|.$$

Then, since the number of iterations M is polynomially bounded in the preceding quantity, one has the following corollary.

Corollary 2 *Algorithm 1 is a polynomial time algorithm for the ALPO.* $\square$

The next proposition and lemma prove the theorem.

Proposition 7 *Let $u \leq u_0$, $u < 1$, as in the Robust α Theorem and $a < \alpha_0$ further satisfy $1/(1 - 2a)((u/2) + a) \leq 0.9u$. Let L_t, ζ_t be as in the preceding, $0 < t \leq 1$. If $\|L_t x_t - L_t \zeta_t\| \leq u$, then $x_{t'} = N_{f_{t'}}(x_t)$ is an approximate zero for $f_{t'}$ with associated zero $\zeta_{t'}$ for all t', $(1 - (a/\sqrt{m}))t \leq t' \leq t$ and moreover $\|L_{t'} x_{t'} - L_{t'} \zeta_{t'}\| \leq 0.9u$.*

Remark 1 We have used 0.9 instead of 1 to leave room for computational errors. We use this room later.

Lemma 2 *For $z \in \text{Int}$ and $0 < t \leq 1$,*

(a) $c \cdot z = te \cdot L_t(z)$;

(b) $|c \cdot z - c \cdot \zeta_t| \leq t\sqrt{m}\|L_t(z) - L_t(\zeta_t)\|$.

Proof. (a) $c = tL_t^T(e)$; so $c \cdot z = tL_t^T(e) \cdot z = te \cdot L_t(z)$.

(b)

$$|c \cdot z - c \cdot \zeta_t| = |te \cdot L_t(z - \zeta_t)|$$
$$\leq t\|e\| \|L_t(z - \zeta_t)\| \quad \text{by Cauchy–Schwarz}$$
$$\leq t\sqrt{m}\|L_t(z) - L_t(\zeta_t)\|.$$

$\square$

Proof of Theorem 1.   By Proposition 7 and induction, x_i is defined for all $i > 0$ and x_i is an approximate zero for h_{t_i} with associated zero ζ_{t_i}. Moreover, by Lemma 2 and Proposition 7,

$$|c \cdot x_i - c \cdot \zeta_{t_i}| \leq t_i \sqrt{m} \|L_{t_i}(z) - L_{t_i}(\zeta_t)\| \leq t_i \sqrt{m} u < t_i \sqrt{m}.$$

By Proposition 6

$$|c \cdot \zeta_{t_i} - cx^*| < t_i m.$$

So

$$|c \cdot x_i - cx^*| \leq t_i (m + \sqrt{m}) < 2t_i m.$$

It remains to see that

$$2\left(1 - \frac{a}{\sqrt{m}}\right)^M m < \varepsilon,$$

or that

$$M \log\left(1 - \frac{a}{\sqrt{m}}\right) < \log \varepsilon - \log 2m,$$

or, since $\log(1 - (1/\sqrt{m}) < 0$, that

$$M > \frac{\log \varepsilon - \log 2m}{\log(1 - \frac{a}{\sqrt{m}})} = \frac{|\log \varepsilon| + \log 2m}{|\log(1 - \frac{a}{\sqrt{m}})|}.$$

But by hypothesis

$$M > \frac{|\log \varepsilon| + \log 2m}{\frac{a}{\sqrt{m}}}$$

so we need only note that $|\log(1 - y)| > y$ for $0 \leq y < 1$ and we are done.   □

To prove Proposition 7 we analyze Newton's method in an inner product on $\mathbb{R}^n$ that depends on t. In fact the inner product is given by the second derivative of the objective function h_t at the minimum ζ_t. This is a familiar theme in the calculus of variations. Although it may not seem immediately natural to the problems of linear programming it brings significant simplicity to the proof.

We use the previously defined linear map $L_t : \mathbb{R}^n \to \mathbb{R}^m$ to define an inner product on $\mathbb{R}^n$,

$$\langle u, v \rangle_t = \langle L_t(u), L_t(v) \rangle$$

for all $u, v \in \mathbb{R}^n$ where $\langle \ , \ \rangle$ is the standard inner product in $\mathbb{R}^m$. We write $\|v\|_t$ for the corresponding norm which equals $\|L_t(v)\|$ in $\mathbb{R}^m$.

Let $H_x = D \operatorname{grad}\left(\sum_{i=1}^m \ln(A_i x - b_i)\right) = L_x^T L_x$ be the Hessian of

$$\sum_{i=1}^m \ln(A_i x - b_i)$$

and $H_t = H_{\zeta_t}$. Note that $t H_x = D(\operatorname{grad} h_t)(x)$.

The next two propositions serve to prove Proposition 7. We defer their proofs until we have proven Proposition 7.

Proposition 8 *Let $\mathbb{R}^n$ have inner product $\langle \ , \ \rangle_t$ and norm $\| \ \|_t$. Then, for all $0 < t' \leq 1$,*

(a) $\gamma(f_{t'}, \zeta_t) \leq 1$;

(b) $\beta(f_{t'}, \zeta_t) = \|(t' H_{t'})^{-1}(\zeta_t) f_{t'}(\zeta_t)\|_t \leq \frac{|t-t'|}{|t'|} \sqrt{m}$.

Let $V_t = L_t(\mathbb{R}^n) \subset \mathbb{R}^m$. The vector space V_t inherits an inner product and norm from the standard inner product and norm on $\mathbb{R}^m$; $L_{t'} L_t^{-1} : V_t \to V_{t'}$ is a linear isomorphism and has an associated operator norm which we use in the next proposition.

Proposition 9 $\|L_{t'} L_t^{-1}\| \leq 1/(1 - \|\zeta_{t'} - \zeta_t\|_t)$ *as long as* $\|\zeta_{t'} - \zeta_t\|_t < 1$.

Now we prove Proposition 7.

Proof of Proposition 7. Choose $u \leq u_0$, $u < 1$, and $a < \alpha_0$ in the robust α theorem of Section 8.2 to further guarantee that

$$\frac{1}{1 - 2a} \left(\frac{u}{2} + a\right) \leq 0.9u.$$

Consider x_t such that $\|L_t x_t - L_t \zeta_t\| \leq u$ and t' satisfying

$$\left(1 - \frac{a}{\sqrt{m}}\right) t \leq t' \leq t.$$

Then $|(t - t')/t| \leq a/\sqrt{m}$ and, by Proposition 8,

$$\gamma(f_{t'}, \zeta_t) \leq 1$$

and

$$\beta(f_{t'}, \zeta_t) \leq \left|\frac{t - t'}{t}\right| \sqrt{m} \leq \frac{a\sqrt{m}}{\sqrt{m}} = a,$$

where $\beta(f_{t'}, \zeta_t)$ is considered for the distance given by the $\| \ \|_t$ norm. We conclude that $\alpha(f_{t'}, \zeta_t) \leq a$ and we may apply the robust α theorem. We do so again for the distance given by the $\| \ \|_t$ norm. A first consequence of the robust α theorem is that ζ_t is an approximate zero of $f_{t'}$ with associated zero $\zeta_{t'}$. Thus,

$$\|\zeta_t - \zeta_{t'}\|_t \leq \|\zeta_t - N_{f_{t'}}(\zeta_t)\|_t + \|\zeta_{t'} - N_{f_{t'}}(\zeta_t)\|_t$$
$$\leq 2\beta(f_{t'}, \zeta_t) \leq 2a.$$

Now, because of Proposition 9 we have that

$$\|L_{t'} L_t^{-1}\| \leq \frac{1}{1 - \|\zeta_t - \zeta_{t'}\|_t} \leq \frac{1}{1 - 2a}.$$

On the other hand, since by hypothesis $\|x_t - \zeta_t\|_t \leq u$,

$$\|x_{t'} - \zeta_{t'}\|_t \leq \|x_{t'} - N_{f_{t'}}(\zeta_t)\|_t + \|N_{f_{t'}}(\zeta_t) - \zeta_{t'}\|_t \leq \frac{u}{2} + a,$$

the first bound resulting again from the robust α theorem and the fact that $\|x_t - \zeta_t\|_t \leq u$, and the second from the bound on $\beta(f_{t'}, \zeta_t)$ previously computed. Now

$$
\begin{aligned}
\|x_{t'} - \zeta_{t'}\|_{t'} &= \|L_{t'}(x_{t'}) - L_{t'}(\zeta_{t'})\| \\
&= \|L_{t'}L_t^{-1}(L_t(x_{t'}) - L_t(\zeta_{t'}))\| \\
&\leq \|L_{t'}L_t^{-1}\| \, \|x_{t'} - \zeta_{t'}\|_t \\
&\leq \frac{1}{1 - 2a}\left(\frac{u}{2} + a\right) \leq 0.9u.
\end{aligned}
$$

$\square$

We now turn to the proofs of Propositions 8 and 9. Proposition 9 is easier and we do it first.

Proof of Proposition 9.   Note that the isomorphism $L_{t'}L_t^{-1}$ is the restriction to V_t of the linear map from $\mathbb{R}^m$ to itself represented by a diagonal matrix having

$$
\frac{A_i\zeta_t - b_i}{A_i\zeta_{t'} - b_i}
$$

as the ith entry on its diagonal.

Since

$$
\left|\frac{A_i\zeta_t - b_i}{A_i\zeta_{t'} - b_i}\right| = \frac{1}{1 - \left|\frac{A_i(\zeta_t - \zeta_{t'})}{A_i(\zeta_t) - b_i}\right|} \leq \frac{1}{1 - \|\zeta_t - \zeta_{t'}\|_t},
$$

we deduce that $\|L_{t'}L_t^{-1}\| \leq 1/(1 - \|\zeta_t - \zeta_{t'}\|_t)$. $\square$

Now we turn to the proof of Proposition 8. First we prove some lemmas.

Lemma 3   *Let $w \in \mathbb{R}^m$ and $t \in (0, 1)$. Then*

$$
\|H_t^{-1}(L_t^T(w))\|_t \leq \|w\|.
$$

Proof.   For any $v \in \mathbb{R}^n$

$$
\begin{aligned}
\langle H_t^{-1}(L_t^T(w)), v\rangle_t &= \langle L_t H_t^{-1}(L_t^T(w)), L_t(v)\rangle \\
&= \langle w, L_t(v)\rangle \\
&\leq \|w\| \, \|L_t v\| \qquad \text{by Cauchy–Schwarz} \\
&= \|w\| \, \|v\|_t.
\end{aligned}
$$

Take $v = H_t^{-1}(L_t^T(w))$ to conclude. $\square$

Lemma 4   *Let $u_1, \ldots, u_k \in \mathbb{R}^m$ with $\|u_j\| = 1$, $j = 1, \ldots, k$. Let $w = (w_1, \ldots, w_m)^T \in \mathbb{R}^m$ be defined by $w_i = \prod_{j=1}^k u_{ji}$, where u_{ji} is the ith component of u_j. Then $\|w\| \leq 1$.*

Proof.

$$
1 = \prod_{j=1}^k \sum_{i=1}^m u_{ji}^2 \geq \sum_{i=1}^m \prod_{j=1}^k u_{ji}^2 = \sum_{i=1}^m w_i^2.
$$

$\square$

Recall that given a vector $v = (v_1, \ldots, v_\ell) \in \mathbb{R}^\ell$ we denote by $\Delta(v)$ or by $\Delta(v_i)$ the $\ell \times \ell$ diagonal matrix whose ith diagonal entry is v_i.

Lemma 5

(a) $D(f_t)(x) = t\,H_x$

(b) $D^k(f_t)(x)(u_1, \ldots, u_k) = t(-1)^{k+1}k!L_x^T(w),$

where the ith component of w is w_i,

$$w_i = \prod_{j=1}^{k} L_x(u_j)_i \text{ for } i = 1, \ldots, m,$$

and $L_x(u_j)_i$ is the ith component of $L_x(u_j)$.

Proof. By induction the kth derivative of L_x with respect to x applied to $(u_1, \ldots, u_k)$ is $(-1)^{k+1}k!\Delta\left(\prod_{j=1}^{k}(L_x(u_j))_i\right)L_x$. Now by Proposition 3(a),

$$D^k(f_t)(x) = t(-1)^{k+1}k!L_x^T\Delta\left(\prod_{j=1}^{k}(L_x(u_j))_i\right)e$$

$$= t(-1)^{k+1}k!L_x^T(w).$$

$\square$

Proof of Proposition 8.

(a) $\gamma(f_{t'}, \zeta_t) = \gamma(f_t, \zeta_t)$ for all $t' > 0$.

For $k \geq 2$ let $u_1, \ldots, u_k \in \mathbb{R}^n$ with $\|u_i\|_t = \|L_t(u_i)\| = 1$ for all $i = 1, \ldots, k$. Then $\|w\| \leq 1$, where $w = (w_1, \ldots, w_m)$ and where $w_i = \prod_{j=1}^{k}(L_t(u_j))_i$ by Lemma 4. By Lemma 5

$$\frac{\|D(f_t)^{-1}(\zeta_t)D^k(f_t)(\zeta_t)\|_t}{k!} = \|H_t^{-1}L_t^T(w)\|_t$$

$$\leq \|w\| \leq 1 \qquad \text{by Lemma 3.}$$

This proves that $\gamma(f_t, \zeta_t) \leq 1$.

(b) $f_{t'}(\zeta_t) = c - t'L_t^T e$ by Proposition 3(a).

But $c = tL_t^T e$ since $f_{t'}(\zeta_t) = 0$, so $f_{t'}(\zeta_t) = (t - t')L_t^T e$. Now

$$\|D(f_{t'})^{-1}(\zeta_t)f_{t'}(\zeta_t)\|_t = \frac{|t - t'|}{|t'|}\|H_t^{-1}L_t^T e\|_t \qquad \text{which by Lemma 2 is}$$

$$\leq \frac{|t - t'|}{|t|}\|e\| = \frac{|t - t'|}{|t'|}\sqrt{m}.$$

$\square$

15.3   Linear Programming over $\mathbb{Q}$

In this and the next section we restrict our attention to the case of matrices with rational coefficients. Our goal is to provide a polynomial cost algorithm for the linear programming feasibility problem in this case. Thus the problem we want to solve is the following LPF over $\mathbb{Q}$.

> Given an $m \times n$ matrix A with rational entries a_{ij} and $b \in \mathbb{Q}^m$ determine if there is an $x \in \mathbb{Q}^n$ such that $Ax \geq b$.

The polynomial bound for the complexity of this problem is, as we described in Section 1.2, for the cost that measures the height of the rational numbers during the computation. We recall that the input (A, b) has size $(mn + m)L$ where L is the maximal height of the elements in A and b.

In order to solve our problem we first note that we may clear the denominators in A and b and assume, without loss of generality, that A is an integer matrix and b an integer vector. This follows from Section 15.1.

Let L be a bound for the height of the entries of A and b. Then the absolute value of the entries of A and b is bounded by $\mathbf{L} = 2^L$. We deduce that, according to Theorem 5 of Chapter 6, there exists an $x \in \mathbb{Q}^n$ such that $Ax \geq b$ if and only if there exists a rational solution of $Ax \geq b$ whose coordinates are bounded in absolute value by $K = n^n \mathbf{L}^n$.

Thus we may reduce LPF over $\mathbb{Q}$ to the case where the polyhedron is bounded and given by an integer matrix but it may not have interior yet. We now add a variable to produce a bounded polyhedron with interior, a cost vector which will help solve the feasibility problem, and a solution to the corresponding non-linear minimization problem.

Proposition 10 *Let A, b be as above and let L be a bound for the height of their entries. Then there exists $c, s, u_i, \ell_i \in \mathbb{Q}$, $i = 1, \ldots, n$, computable with polynomial cost in L, n, and m and bounded in height by $R = O(mn(L + \log n))$ such that $\ell_i \leq -K$, $u_i \geq K$ and the polyhedron $P \subset \mathbb{R}^n \times \mathbb{R}$ defined by the inequalities*

$$
\begin{aligned}
Ax + y &\geq b \\
\ell_i \leq\; &x_i \leq u_i \\
0 \leq\; &y \leq 2s
\end{aligned}
$$

has a nonempty interior. Moreover the point $(0, s)$ minimizes

$$
\begin{aligned}
h(x, y) = cy &- \sum_{i=1}^{m} \ln(A_i x + y - b_i) - \sum_{i=1}^{n} \ln(x_i - \ell_i) \\
&- \sum_{i=1}^{n} \ln(u_i - x_i) - \ln(2s - y) - \ln(y)
\end{aligned}
$$

on P.

Proof. Let us consider the gradient of h at a point $(0, s)$. Taking the derivative with respect to x_j for $j = 1, \ldots, n$ we get

$$\frac{\partial h}{\partial x_j}(0, s) = \sum_{i=1}^{m} \frac{a_{ij}}{s - b_i} + \frac{1}{\ell_j} + \frac{1}{u_j}.$$

Let $\mathbf{L} = 2^L$, $K = n^n \mathbf{L}^n$ and define $s = 2mKL$. Then $s - b_i \geq s - \mathbf{L} \geq 0$ and therefore

$$\left| \frac{a_{ij}}{s - b_i} \right| \leq \frac{\mathbf{L}}{s - b_i} \leq \frac{\mathbf{L}}{s - \mathbf{L}} = \frac{\mathbf{L}}{2mK\mathbf{L} - \mathbf{L}} = \frac{1}{2mK - 1}.$$

Let $\alpha_j = \sum_{i=1}^{m} a_{ij}/(s - b_i)$. We have that

$$|\alpha_j| \leq \frac{m}{2mK - 1}.$$

We now continue according to the sign of α_j.

(i) If $\alpha_j > 0$, set $\ell_j = -K$. Then

$$\alpha_j - \frac{1}{K} + \frac{1}{u_j} = 0 \Rightarrow u_j = \frac{K}{1 - K\alpha_j}.$$

Thus, to see that $u_j \geq K$, it is enough to see that $0 < K\alpha_j < 1$. That $0 < K\alpha_j$ is clear. The other inequality follows since $K\alpha_j \leq Km/(2Km-1)$ which is smaller than 1 if $Km > 1$.

(ii) If $\alpha_j < 0$, a similar reasoning holds for $u_j = K$ and $\ell_j = -K/(1 + K\alpha_j)$.

(iii) If $\alpha_j = 0$, we set $\ell_j = -K$ and $u_j = K$.

Thus we have found ℓ_j, u_j, and s such that the partial derivatives $(\partial h/\partial x_j)(0, s)$ vanish for $j = 1, \ldots, n$. Taking the derivative of h now with respect to y we obtain

$$\frac{\partial h}{\partial y}(0, s) = c - \sum_{i=1}^{m} \frac{1}{s - b_i} - \frac{1}{s} + \frac{1}{r - s} = c - \sum_{i=1}^{m} \frac{1}{s - b_i}.$$

Now, by requiring that $(\partial h/\partial y)(0, s) = 0$ we obtain

$$c = \sum_{i=1}^{m} \frac{1}{s - b_i}.$$

For this c the point $(0, s)$ minimizes h. The rest of the statement is easily checked.

$\square$

Thus, with any pair (A, b) we can associate a polyhedron $P \subset \mathbb{R}^{n+1}$ whose interior is bounded and nonempty and a vector $(0, \dots, 0, c) \in \mathbb{Q}^{n+1}$ defining an objective function

$$(x, y) \to (0, \dots, 0, c) \cdot (x, y) = cy.$$

Let $y^* \in \mathbb{Q}$ be the y-coordinate of a point in P minimizing this objective function. Note that the last inequality defining P forces $y^* \geq 0$. The next proposition shows a very simple relation between y^* and the feasibility problem for (A, b).

Proposition 11   *The system $Ax \geq b$ has a feasible point if and only if $y^* = 0$.*

Proof.   The "only if" direction is trivial. For the "if" direction, if there is an $x \in \mathbb{R}^n$ such that $Ax \geq b$ then we can assume that $|x_i| \leq K$ for $i = 1, \dots, n$. It is then immediate that $(x, 0) \in P$ and thus that y^* is 0. $\square$

Figure B   Two different situations for a pair (A, b). In the first one, the set given by $Ax \geq 0$ is nonempty (it is the thick segment in the x-axis) whereas in the second it is empty. In both cases, the point $(0, s)$ is in the interior of P and minimizes h.

Let us denote by A^* and b^* the matrix and vector defining P in the statement of Proposition 10 and by c^* the vector $(0, \dots, 0, c)$ in $\mathbb{Q}^{n+1}$. Note that all the entries in A^*, b^*, or c^* are bounded in height by R.

To solve the linear programming feasibility problem for (A, b) we are thus led to determine whether y^* is 0. Note that we are in the situation of the preceding section for P, h, and $\zeta_1 = (0, s)$ since we know that the the interior of P is nonempty and bounded and that $(0, s)$ minimizes h in the interior of P. Therefore we can use Algorithm 1 with input $(A^*, b^*, c^*, \zeta_1, \varepsilon^*)$ to approximate cy^* up to ε^*. The next lemma shows how small ε^* needs to be in order to decide whether $y^* = 0$ from the approximation.

Proposition 12   *If $y^* \neq 0$, then $y^* > \widehat{\varepsilon} = 1/(2(n+1)^{n+1}2^{R(n+1)})$.*

Proof.   The value of y^* is a coordinate in the rational solution of a system of $n + 1$ linear equations in $n + 1$ variables with rational coefficients. This system arises by choosing $n + 1$ independent equations from the system $A^*(x, y) = b^*$. Now these equations are either an original equation of the form $A_i x + y = b_i$

whose coefficients are integers of height bounded by L, or equations of the form $z = d$, where z is a variable and d is a rational bounded in height by R. In this latter case, we can replace this equation by an equivalent one with integer coefficients bounded in height by R.

Now a simple application of Theorem 5 of Chapter 6 bounds the absolute value of the denominator of the solution y^* by $(n + 1)^{n+1}2^{R(n+1)}$. □

Taking $\varepsilon^* = c\hat{\varepsilon}$ and since A^* has now $m + 2(n + 1)$ rows and $n + 1$ columns one deduces the following corollary.

Corollary 3 *For (A, b) as above, let M be*

$$\left\lceil \frac{|\log c| + (n + 1)(R + \log(n + 1)) + \log 2(m + 2(n + 1))}{a} \sqrt{m + 2(n + 1)} \right\rceil .$$

Then M iterations in Algorithm 1 suffice to decide whether $y^* = 0$. □

15.4 Polynomial Cost

In the preceding section we have seen how to associate with the pair (A, b) a 5-tuple $(A^*, b^*, c^*, \zeta_1, \varepsilon^*)$ whose size is polynomially bounded in the size of (A, b) and such that a solution of the LPF for (A, b) can be obtained with a single run of Algorithm 1 with input $(A^*, b^*, c^*, \zeta_1, \varepsilon^*)$ that performs a polynomial number of arithmetic operations.

This is, however, not enough to solve the LPF for (A, b) with polynomial cost since we have not necessarily polynomially bounded the heights of the intermediate results in the execution of Algorithm 1. There are essentially two places where this objective can be spoiled.

The sequence x_{N+1} may come too close to the boundary of the polyhedron P. Since $x_{N+1} = x_N - (Df_{t_{N+1}}(x_N))^{-1} f_{t_{N+1}}(x_N)$ and the $n \times n$ matrix representing $Df_{t_{N+1}}(x_N)$ is given by $t_{N+1}L_{x_N}^{\mathsf{T}} L_{x_N}$, a good bound for the height of the entries in the inverse of the matrix for $Df_{t_{N+1}}(x_N)$ will depend on a good bound for the height of the entries of L_{x_N}. These entries are quotients of the entries of A by numbers of the form $A_i x_N - b_i$ for $i = 1, \ldots, m$. Thus, we need to give a good bound for the latter. A priori, however, the sequence of points x_N could get too close to the boundary of the polyhedron P and make these quotients very large. We show in Proposition 13 that this is not the case and that the distance from x_N to the boundary of P is bounded by a geometric series in N.

On the other hand, the sequence of points x_N for $N \leq M$ could grow exponentially in height even without growing in absolute value. We show that we can slightly modify Algorithm 1 to "chop" each x_N to a polynomial number of bits without undermining the behavior of the algorithm.

Let us begin with the first problem.

Lemma 6 *There exists a universal constant μ, $0 < \mu < 1$ such that for $0 \le N \le M$ and $1 \le i \le m + 2(n + 1)$,*

$$|A_i^* \zeta_{t_N} - b_i^*| \ge \mu^N C,$$

where $C = |A_i^* \zeta_{t_0} - b_i^*| \ge 1/(\mathbf{L}(2^R + 1))$.

Proof.   Let a and u be as in Proposition 7. For $N = 0$ the result is trivial. For $N > 0$ we have seen in the proof of Proposition 7 that

$$\|\zeta_{t_{N-1}} - \zeta_{t_N}\|_{t_{N-1}} \le 2a.$$

Therefore for every $i \le m + 2(n + 1)$ we have that

$$
\begin{aligned}
|A_i^* \zeta_{t_N} - b_i^*| &\ge |A_i^* \zeta_{t_{N-1}} - b_i^*| - |A_i^* \zeta_{t_N} - A_i^* \zeta_{t_{N-1}}| \\
&\ge (1 - \|\zeta_{t_N} - \zeta_{t_{N-1}}\|_{t_{N-1}})|A_i^* \zeta_{t_{N-1}} - b_i^*| \\
&\ge (1 - 2a)|A_i^* \zeta_{t_{N-1}} - b_i^*|.
\end{aligned}
$$

The first part of the result follows for $\mu = 1 - 2a$. The bound on C is trivial since $\zeta_{t_0} = \zeta_1 = (0, s)$. $\square$

Lemma 7 *Suppose that $\|x - \zeta_t\|_t \le u < 1$. Then, for all $i \le m + 2(n + 1)$,*

$$|A_i^* x - b_i^*| \ge (1 - u)|A_i^* \zeta_t - b_i^*|.$$

Proof.

$$
\begin{aligned}
|A_i^* x - b_i^*| &\ge |A_i^* \zeta_t - b_i^*| - |A_i^* x - A_i^* \zeta_t| \\
&\ge |A_i^* \zeta_t - b_i^*| - \|x - \zeta_t\|_t |A_i^* \zeta_t - b_i^*| \\
&\ge (1 - u)|A_i^* \zeta_t - b_i^*|.
\end{aligned}
$$

$\square$

From these two lemmas we deduce the following result.

Proposition 13 *There exists a universal constant μ, $0 < \mu < 1$ such that for $1 \le i \le m + 2(n + 1)$ and $0 \le N \le M$ if $\|x - \zeta_{t_N}\|_{t_N} \le u < 1$, then*

$$|A_i^* x - b_i^*| \ge (1 - u)\mu^N C,$$

where $C = |A_i^* \zeta_{t_0} - b_i^*| \ge 1/(\mathbf{L}(2^R + 1))$. $\square$

Let us proceed now to our second problem. We first note that, since the sequence of the x_N is moving inside the polyhedron P, the coordinates of its points are bounded in absolute value by $\mathbf{R} = 2^R$.

Proposition 14 *For $0 \le N \le M$, $\|L_{t_N}\| \le \delta_N = (\mathbf{L}\sqrt{n + 1})/\mu^N C$.*

Proof. Because of Lemma 6 for all i,

$$|A_i^* \zeta_{t_N} - b_i^*| \geq \mu^N C$$

and therefore the coefficients of L_{t_N} are bounded in absolute value by

$$\frac{\mathbf{L}}{\mu^N C}$$

from which the statement follows.

□

Corollary 4 *For $0 \leq N \leq M$ and any $x, \widehat{x} \in \mathbb{R}^{n+1}$, if $\|x - \widehat{x}\| \leq u/10\delta_N$, then $\|x - \widehat{x}\|_{t_N} \leq u/10$.* □

We can now modify Algorithm 1 to avoid a possible exponential increase in the precision needed to write down x_N. The idea, depicted in Figure C, is to compute a sequence of points close to the Newton iterates which are themselves approximate zeros and have small height.

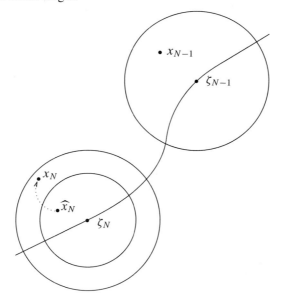

Figure C The two neighborhoods, the image $\widehat{x}_{t_N}$ of $x_{t_{N-1}}$ by $N_{f_{t_N}}$, and the result x_{t_N} of chopping $\widehat{x}_{t_N}$.

We first compute from x_{N-1} the point $\widehat{x}_N$ defined by

$$\widehat{x}_N = N_{f_{t_N}}(x_{N-1}).$$

Because of Proposition 7, this point is within $0.9u$ of ζ_{t_N} in the $\| \|_{t_N}$ norm. Therefore we can replace it by a new point x_N satisfying that $\|x_N - \widehat{x}_N\|_{t_N} \leq u/10$ and ensure that $\|x_N - \zeta_N\|_{t_N} \leq u$. Because of Corollary 4, this condition is satisfied by any x_N satisfying that $\|x_N - \widehat{x}_N\| \leq u/10\delta_N$. The next proposition states how to round off $\widehat{x}_N$ in order to ensure this bound while keeping x_N polynomially bounded in height.

Proposition 15 *For $0 \leq N \leq M$ let x_N be the point obtained by rounding off the coordinates of $\widehat{x}_N$ after the first $R + |\log(Cu/10)| + N|\log \mu| + L + \log(n + 1)$ bits of its binary expansion. Then*

$$\|x_N - \zeta_{t_N}\|_{t_N} \leq u.$$

Proof. By taking $\widehat{x}_N$ as in the statement one can ensure that the distance between each coordinate of $\widehat{x}_N$ and the same coordinate of x_N is at most

$$\frac{u\mu^N C}{10L(n + 1)} = \frac{u}{10\delta_N \sqrt{n + 1}}.$$

Therefore

$$\|x_N - \widehat{x}_N\| \leq \sqrt{n + 1}\frac{u}{10\delta_N \sqrt{n + 1}} \leq \frac{u}{10\delta_N}$$

and the preceding discussion concludes the argument. ☐

We show in Section 15.5 that one can compute inverses of rational matrices in polynomial time over $\mathbb{Q}$. As a consequence Algorithm 1 runs in polynomial time over $\mathbb{Q}$ and we can close this section stating the main result we have proved here.

Theorem 2 *The problem* LPF *is in* P *over* $\mathbb{Q}$. ☐

15.5 On the Cost of Inverting a Matrix

In this section we show how to invert an $n \times n$ invertible rational matrix $A = (a_{ij})$ in polynomial cost in the height of A, that is, $n^2(\max \mathrm{ht}_{\mathbb{Q}}(a_{ij}))$. Along the way we compute the determinant of A in polynomial cost. The computation of the determinant of A could be considered the crux of the matter for then by Cramer's rule one proves the assertion. But the computation of the determinant by its defining formula involves evaluating $n!$ terms, which takes too long, and by Gaussian elimination we may introduce intermediate results whose heights are too large.

We take another approach, which does not involve Cramer's rule, but uses another basic and well-known way of computing the inverse. Now let A be an invertible $n \times n$ matrix over any field k of characteristic 0. We begin with a basic fact about symmetric functions due to Newton.

For $k = 0, \ldots, n$, let the kth *elementary symmetric polynomial* $\sigma_k(x)$, $x = (x_1, \ldots, x_n)$, be defined by

$$\prod_{i=1}^n (t - x_i) = \sum_{k=0}^n (-1)^k \sigma_k(x_1, \ldots, x_n)t^{n-k}.$$

and $N_k(x_1, \ldots, x_n) = \sum_{i=1}^{n} x_i^k$, where $N_0(x_1, \ldots, x_n) = n$.

Proposition 16 (Newton's Identities) *For* $1 \leq k \leq n$,

$$N_k - N_{k-1}\sigma_1 + \ldots + (-1)^{k-1}N_1\sigma_{k-1} + (-1)^k k\sigma_k = 0.$$

Proof. One can proceed by comparison of the coefficients in the polynomial

$$N_k - N_{k-1}\sigma_1 + \ldots + (-1)^{k-1}N_1\sigma_{k-1} + (-1)^k \sigma_k$$
$$= (x_1^k + \ldots + x_n^k) - (x_1^{k-1} + \ldots + x_n^{k-1})(x_1 + \ldots + x_n) + \ldots +$$
$$(-1)^{k-1}(x_1 + \ldots + x_n)(x_1 \cdots x_{k-1} + \ldots + x_{n-k+2} \cdots x_n) +$$
$$(-1)^k k(x_1 \cdots x_k + \ldots + x_{n-k+1} \cdots x_n).$$

All the monomials in this expression are of type

$$x_i^\ell x_{j_1} \cdots x_{j_{k-\ell}}$$

for some ℓ, $1 \leq \ell \leq k$, with $i \neq j_s$ for $1 \leq s \leq k - \ell$.

For $\ell \geq 2$, the monomial $x_i^\ell x_{j_1} \cdots x_{j_{k-\ell}}$ appears only in the two consecutive terms $N_\ell \sigma_{k-\ell}$ and $N_{\ell-1}\sigma_{k-\ell+1}$ with opposite sign and exactly once in each. Therefore they cancel.

The remaining monomials in $N_1\sigma_{k-1}$ have the form $x_i x_{j_1} \cdots x_{j_{k-1}}$ for $i \neq j_1, \ldots, j_{k-1}$ and appear exactly k times there. They cancel with their ocurrences in $k\sigma_k$.   □

Corollary 5 *The symmetric polynomials* $\sigma_1, \ldots, \sigma_n$ *and* $N_1, \ldots, N_n$ *satisfy the linear equation*

$$M \begin{pmatrix} \sigma_1 \\ \vdots \\ \sigma_n \end{pmatrix} = \begin{pmatrix} N_1 \\ \vdots \\ N_n \end{pmatrix},$$

where M *is the lower triangular invertible matrix*

$$\begin{pmatrix} 1 & 0 & & \cdots & & 0 \\ N_1 & -2 & & & & \\ N_2 & -N_1 & 3 & & & \vdots \\ & & & \ddots & & \\ N_{n-2} & -N_{n-3} & & (-1)^n(n-1) & 0 \\ N_{n-1} & -N_{n-2} & N_{n-3} & \cdots & (-1)^n N_1 & (-1)^{n+1}n \end{pmatrix}.$$

□

We now proceed to the computation of the inverse of A via its characteristic polynomial. Recall that the characteristic polynomial of A is

$$\psi_A(t) = \det(t\,Id - A)$$

and that $\psi_A(A) = 0$.

The roots $\lambda_1, \ldots, \lambda_n$ of ψ_A in the algebraic closure of k are the *eigenvalues* of A and satisfy

$$\psi_A(t) = \sum_{k=0}^{n} (-1)^k \sigma_k(\lambda_1, \ldots, \lambda_n) t^{n-k}.$$

The product of the eigenvalues of A, which is the constant term of ψ_A, is the determinant of A and their sum is the trace of A. The eigenvalues of A^k are $\lambda_1^k, \ldots, \lambda_n^k$ for all $k \in \mathbb{N}$.

Proposition 17   *Let $N_k = \mathrm{trace}(A^k)$ for $k = 1, \ldots, n$. Then the coefficients $c_1, \ldots, c_n$ of the characteristic polynomial of A*

$$\psi_A(t) = t^n + c_1 t^{n-1} + \ldots + c_n$$

are given by

$$M_-^{-1} \begin{pmatrix} N_1 \\ \vdots \\ N_n \end{pmatrix} = \begin{pmatrix} c_1 \\ \vdots \\ c_n \end{pmatrix},$$

where M_- is the lower triangular invertible matrix

$$-\begin{pmatrix} 1 & 0 & \cdots & 0 \\ N_1 & 2 & & \\ \vdots & & \ddots & 0 \\ N_{n-1} & N_{n-2} & \cdots & n \end{pmatrix}.$$

Proof.   If $\lambda_1, \ldots, \lambda_n$ are the eigenvalues of A, then $\sum_{i=1}^{n} \lambda_i^k = N_k$. Now $c_k = (-1)^k \sigma_k(\lambda_1, \ldots, \lambda_n)$. So Corollary 5 finishes the proof. $\square$

Proposition 18   *Let A be an invertible $n \times n$ matrix and $\psi_A(t) = t^n + c_1 t^{n-1} + \ldots + c_n$ be its characteristic polynomial. Then*

$$A^{-1} = -\frac{1}{c_n}(A^{n-1} + c_1 A^{n-2} + \ldots + c_{n-1} Id).$$

Proof.   The proof follows from the equality $\psi_A(A) = 0$ and the fact that A is invertible. $\square$

Proposition 19 *Let* $\displaystyle\prod_{k=1}^{n}(t+k) = t^n + d_1 t^{n-1} + \ldots + d_n$. *Then*

$$M_{-}^{-1} = -\frac{1}{d_n}(M_{-}^{n-1} + d_1 M_{-}^{n-2} + \ldots + d_{n-1} Id).$$

Proof.   The matrix M_{-} is lower triangular and invertible with eigenvalues $-1, -2, \ldots, -n$. So Proposition 18 finishes the proof. $\square$

Now we are ready to compute the inverse of an $n \times n$ matrix A over k if it is invertible and to decide if this is not the case. We just perform the following steps.

(1)   Compute A^k and $N_k = \text{trace}(A^k)$ for $k = 1, \ldots, n$.

(2)   Compute $d_1, \ldots, d_n$ such that $\displaystyle\prod_{k=1}^{n}(t+k) = t^n + d_1 t^{n-1} + \ldots + d_n$.

(3)   Compute $M_{-}^{-1} = -\frac{1}{d_n}(M_{-}^{n-1} + d_1 M_{-}^{n-2} + \ldots + d_{n-1} Id)$.

(4)   Let

$$\begin{pmatrix} c_1 \\ \vdots \\ c_n \end{pmatrix} = M_{-}^{-1} \begin{pmatrix} N_1 \\ \vdots \\ N_n \end{pmatrix}.$$

(5)   If $c_n = 0$, then A is not invertible.

Otherwise compute $A^{-1} = -\dfrac{1}{c_n}(A^{n-1} + c_1 A^{n-2} + \ldots + c_{n-1} Id)$.

The next proposition is immediate.

Proposition 20   *The determinant of A and its inverse A^{-1} if it exists can be computed in $O(n^4)$ arithmetical operations.* $\square$

We now specialize our algorithm to the case of $\mathbb{Q}$ and the bit cost. Our main theorem follows from the facts that if $a, b \in \mathbb{Q}^n$ have their components bounded in height by L then $a \cdot b$ has height at most $2nL + \log n$ and that the dominant step in the algorithm previously described is the computation of the powers A^j and M_{-}^{j} for $j = 2, \ldots, n$.

Proposition 21   *Let A be an $n \times n$ matrix over $\mathbb{Q}$. The determinant of A and its inverse A^{-1} if it exists can be computed with polynomial cost.* $\square$

15.6   Additional Comments and Bibliographical Remarks

That the linear programming feasibility problem is in P was first proved by Khachijan [1979]. Karmarkar [1984] gave a proof employing interior point methods. Gonzaga [1989] proves the theorem using the barrier method. Our analysis of the barrier method using the robust α theorem is from [Renegar and Shub 1992]. For the reduction of the feasibility problem to the barrier method and the bit analysis we have largely followed Vavasis [1991]. Renegar [1995a, 1995b] proves that LPF/$\mathbb{R}$ is in P when the condition of the problem is taken into account. For the average speed of the simplex algorithm see [Borgwardt 1982; Smale 1983].

We have mentioned at the beginning of Section 15.5 that Gaussian elimination can produce large intermediate values. There are some ways to avoid this. Well-known ones are the Bareiss rule exposited in [Bareiss 1968; Loos 1982] and the method given in [Edmonds 1967].

The complexity of the algorithm for matrix inversion presented in Section 15.5 can be improved. In Proposition 20 the $O(n^4)$ upper bound comes from the crude $O(n^3)$ cost for matrix multiplication. Let $\alpha \in \mathbb{R}$ be such that matrix multiplication can be solved with $O(n^\alpha)$ arithmetic operations. Then the bound in Proposition 20 becomes $O(n^{\alpha+1})$. The best value for α to date is 2.376 and is given in [Coppersmith and Winograd 1990].

The algorithm for matrix inversion in Section 15.5 was given in [Csanky 1976]. Another algorithm can be found in [Berkowitz 1984]. The latter has a complexity slightly worse than the one we presented in Section 15.5 but it works for rings of arbitrary characteristic and computes the determinant without performing divisions. Our presentation owes much to discussions we had with Teresa Krick. A comprehensive reference for matrix computations is the book of Bini and Pan [1994].

The content of Proposition 1 is the well-known connection between continued fractions and the Euclidean algorithm.

Appendix B

In this appendix we prove a result, Theorem 2, which was used in Chapter 10 to prove Bézout's theorem.

B.1   The Main Theorem of Elimination Theory

As we saw in Chapter 10, sometimes it is technically convenient to compactify $\mathbb{C}^n$ and to work in $\mathbb{P}(\mathbb{C}^{n+1})$. We now extend to projective space the concepts given in Appendix A.

Let k be a field and let $f_1, \ldots, f_m \in k[x_0, \ldots, x_n]$ be homogeneous polynomials. Then the ideal I generated by $f_1, \ldots, f_m$ is homogeneous; that is, if $g \in I$ and $g = \sum_{i=0}^{d} g_i$ with g_i homogeneous of degree i, then $g_i \in I$ for $i = 0, \ldots, d$.

Definition 1 A *projective set* $X \subseteq \mathbb{P}(k^{n+1})$ is the set of common zeros of a finite set of homogeneous polynomials $f_1, \ldots, f_m \in k[x_0, \ldots, x_n]$. The projective set X is said to be a *projective variety* if the ideal $(f_1, \ldots, f_m)$ is prime.

The following projective form of Hilbert Nullstellensatz is used in our main result. Denote by $(x_0, \ldots, x_n)$ the ideal of $k[x_0, \ldots, x_n]$ generated by the indeterminates x_i, $i = 0, \ldots, n$.

Theorem 1 *Let k be algebraically closed and I be an homogeneous ideal of $k[x_0, \ldots, x_n]$. Then $\mathcal{Z}(I) = \{0\}$ if and only if there exist $d \in \mathbb{N}$ such that $(x_0, \ldots, x_n)^d \subseteq I$.*

Proof. The "if" direction is trivial. For the converse, let $f_1, \ldots, f_s \in k[x_0, \ldots, x_n]$ be homogeneous polynomials such that $I = (f_1, \ldots, f_s)$ and suppose $\mathcal{Z}(I) = \{0\}$. Consider $\widetilde{f}_i = f_i(1, x_1, \ldots, x_n)$, $i = 1, \ldots, s$ and $\widetilde{I} = (\widetilde{f}_1, \ldots, \widetilde{f}_s)$. Then $\mathcal{Z}(\widetilde{I}) = \emptyset$. Because of Theorem 2 of Appendix A, there are polynomials $g_1, \ldots, g_s \in k[x_1, \ldots, x_n]$ such that

$$1 = \sum_{i=1}^{s} g_i \widetilde{f}_i$$

and thus, replacing x_i by x_i/x_0,

$$1 = \sum_{i=1}^{s} g_i \left(\frac{x_1}{x_0}, \ldots, \frac{x_n}{x_0} \right) \widetilde{f}_i \left(\frac{x_1}{x_0}, \ldots, \frac{x_n}{x_0} \right).$$

From here we deduce that

$$x_0^{d_0} = \widehat{g}_i f_i,$$

where $\widehat{g}_i$ denotes the homogenization of g_i, and therefore that $x_0^{d_0} \in I$.

The same argument shows that for every $i \leq n$, $x_i^{d_i} \in I$. If d denotes $n + 1$ times the maximum of $d_0, \ldots, d_n$, then $(x_0, \ldots, x_n)^d \subseteq I$. $\qquad \square$

We can now proceed to our main result.

Theorem 2 (Main Theorem of Elimination Theory) *The projection*

$$\pi_2 : \mathbb{P}(\mathbb{C}^{n+1}) \times \mathbb{P}(\mathbb{C}^{m+1}) \to \mathbb{P}(\mathbb{C}^{m+1})$$

is Zariski closed; that is, if $Z \subseteq \mathbb{P}(\mathbb{C}^{n+1}) \times \mathbb{P}(\mathbb{C}^{m+1})$ is a projective set, then so is $\pi_2(Z)$.

Proof. The inclusion

$$\mathbb{C}^m \to \mathbb{P}(\mathbb{C}^{m+1})$$
$$(x_1, \ldots, x_n) \to (1, x_1, \ldots, x_n)$$

identifies $\mathbb{C}^n$ with a Zariski open subset of $\mathbb{P}(\mathbb{C}^{m+1})$. Therefore $\mathbb{P}(\mathbb{C}^{n+1}) \times \mathbb{C}^m$ is an open subset of $\mathbb{P}(\mathbb{C}^{n+1}) \times \mathbb{P}(\mathbb{C}^{m+1})$ with the product of their respective Zariski topologies. It is immediate to check that a subset X of $\mathbb{P}(\mathbb{C}^{n+1}) \times \mathbb{C}^m$ is closed if and only if there are polynomials $f_1, \ldots, f_s \in \mathbb{C}[x_0, \ldots, x_n, y_1, \ldots, y_m]$ homogeneous with respect to $x_0, \ldots, x_n$ such that X is the zero set of $f_1, \ldots, f_s$.

Since the statement of our theorem is local on the image, we can equivalently prove that

$$\pi_2 : \mathbb{P}(\mathbb{C}^{n+1}) \times \mathbb{C}^m \to \mathbb{C}^m$$

is closed. We prove this latter statement.

Let X be a Zariski closed set in $\mathbb{P}(\mathbb{C}^{n+1}) \times \mathbb{C}^m$. Then there are polynomials $f_1, \ldots, f_s$ as above such that X is the zero set of $f_1, \ldots, f_s$. Let d_i be the degree of f_i with respect to $x_0, \ldots, x_n$ for $i = 1, \ldots, s$. For all $a \in \mathbb{C}^m$,

$$y \notin \pi_2(X)$$
$$\Longleftrightarrow \quad f_i(x, a) \text{ have no common zeros in } \mathbb{P}(\mathbb{C}^{n+1})$$
$$\Longleftrightarrow \quad \exists d \geq 1 \text{ such that } (x_0, \ldots, x_n)^d \subseteq (f_1(x, a), \ldots, f_s(x, a)),$$

the second equivalence by Theorem 1. Therefore it suffices to prove that for each $d \geq 1$ the set

$$A_d = \{a \in \mathbb{C}^m \mid (x_0, \ldots, x_n)^d \subseteq (f_1(x, a), \ldots, f_s(x, a))\}$$

is a Zariski open set in $\mathbb{C}^m$. Let $\mathcal{H}_k$ be the vector space of homogeneous polynomials of degree k in $x_0, \ldots, x_n$ and consider for each $a \in \mathbb{C}^m$ the linear mapping

$$T^a : \mathcal{H}_{d-d_1} \oplus \ldots \oplus \mathcal{H}_{d-d_s} \longrightarrow \mathcal{H}_d$$
$$(g_1, \ldots, g_s) \qquad \longrightarrow \sum f_i(x, a) g_i(x).$$

Fix bases in these two vector spaces. In terms of these bases the map T^a can be given by an $n_d \times m_d$ matrix $(T_{i,j}^a)$ whose entries are polynomials in $a_1, \ldots, a_m$. It follows that

$$(x_0, \ldots, x_n)^d \subseteq (f_1(x, a), \ldots, f_s(x, a))$$
$$\Longleftrightarrow \quad T^a \text{ is surjective}$$
$$\Longleftrightarrow \quad \exists m_d \times m_d \text{ minor } M \text{ of } (T_{i,j}^a) \text{ s.t. } \det M \neq 0.$$

This shows that A_d is Zariski open and concludes the proof. $\square$

B.2   Additional Comments and Bibliographical Remarks

Our proof of Theorem 2 is taken from [Mumford 1976]. This reference also contains another proof relying on elimination theory instead of on the Nullstellensatz, and is thus more constructive.

Part III

Complexity Classes over the Reals

16
Deterministic Lower Bounds

The aim of this chapter is to provide a general technique for computing lower bounds for decision problems over the reals. These bounds come from upper bounds on the number of connected components of real algebraic sets. Besides being interesting in their own right, results proved here are used in later chapters.

16.1 A Geometric Upper Bound

The main result of this section, Theorem 1, bounds the number of connected components of a real algebraic set by an expression that is exponential in the dimension of the ambient space and polynomial in the degrees of the defining polynomials. In order to prove this fact we begin with a very special case, namely, a smooth compact hypersurface. Many of the technical notions used in this section have been defined in Chapters 8, 9, and 10.

Proposition 1 *If $f : \mathbb{R}^n \to \mathbb{R}$ is a polynomial of degree d, δ is a regular value of f, and the hypersurface $H_\delta = \{x \mid f(x) = \delta\}$ is compact, then H_δ has at most $(d(d-1)^{n-1})/2$ connected components.*

The proof relies on some results of differential topology. Before proving them, we give an outline of the proof for motivation.

Outline of the Proof of Proposition 1. Pick an orthogonal projection $L : \mathbb{R}^n \to V$, where V is a one-dimensional subspace of $\mathbb{R}^n$ so that $L \mid H_\delta$ is a Morse function; that is, any critical point of $L \mid H_\delta$ is nondegenerate.

If we change coordinates to make V the subspace $(0, 0, \ldots, x_n)$, and L the last coordinate function, then (Lagrange multipliers say that) the equations for the critical points are

$$f - \delta = 0, \quad \frac{\partial f}{\partial x_1} = 0, \ldots, \frac{\partial f}{\partial x_{n-1}} = 0.$$

Nondegeneracy asserts that the gradients of these n functions are independent. These polynomials have degree d for f and $d - 1$ for each partial derivative. Bézout's theorem says then that they have at most $d(d - 1)^{n-1}$ isolated roots even in $\mathbb{C}^n$. Since each connected component has a max and a min, the connected components are in number at most half the critical points.

To complete the argument along these lines requires several lemmas. First we include a proof that a projection L may be found. We begin by considering

$$g : H_\delta \to S^{n-1}$$
$$x \to \frac{\operatorname{grad} f(x)}{\|\operatorname{grad} f(x)\|}.$$

Since δ is a regular value of f, $\operatorname{grad} f(x) \neq 0$ for any x in H_δ and g is well-defined.

Lemma 1 *The following assertions hold.*

(i) *Almost all $y \in S^{n-1}$ are regular values of g.*

(ii) *If y is a regular value of g, then the projection π_y on the one-dimensional subspace spanned by y is a Morse function when restricted to H_δ.*

Proof. Part (i) is a special case of a basic result of differential topology, *Sard's Theorem*, which asserts that the measure of the set of critical values of a smooth mapping between differentiable manifolds is zero. We may see it in the case at hand very simply. First, we can cover H_δ by a finite set of charts (see Definition 2 of Chapter 10), each one being an $(n - 1)$-dimensional closed cube, and prove our statement for these cubes. Thus, in the following we assume H_δ is a closed cube.

If x is an element of the critical set of g, then the image of $Dg(x)$ is a proper subspace H of the tangent space $T_{g(x)}$ of S^{n-1} at $g(x)$. Consider the composition $\tau \ (= \tau_x)$,

$$H_\delta \xrightarrow{g} S^{n-1} \xrightarrow{\pi} H^\perp$$

of g with the projection π on the orthogonal complement $H^\perp$ of H in $T_{g(x)}$. Since π restricted to H is zero, so is $D\pi$ and one has that $D\tau(x) = (D\pi)(g(x)) \circ Dg(x)$ is zero as a linear map.

Let us consider $\varepsilon > 0$ and $y \in B_\varepsilon(x)$. By the mean value theorem one has that

$$d(\tau(x), \tau(y)) \leq \varepsilon \max_{\zeta \in B_\varepsilon(x)} \|D\tau(\zeta)\|.$$

Since the derivative of τ at x is 0 and H_δ is a closed cube, by the mean value theorem applied now to $D\tau$, we have that there is an $\varepsilon_0 > 0$ and a constant $c > 0$, which do not depend on the critical point x, such that $\max\limits_{\zeta \in B_\varepsilon(x)} \|D\tau(\zeta)\| \le c\varepsilon$ for all $0 < \varepsilon < \varepsilon_0$. Therefore

$$d(\tau(x), \tau(y)) \le c\varepsilon^2$$

and thus, $\tau(B_\varepsilon(x)) \subseteq B_{c\varepsilon^2}(g(x))$ where the last ball is in $H^\perp$. We deduce that $g(B_\varepsilon(x)) \subseteq \pi^{-1}(B_{c\varepsilon^2}(\tau(x)))$, that is, that the image by g of the ε ball around x is contained in the $c\varepsilon^2$ neighborhood of the sphere S tangent to $H^\perp$ through $g(x)$.

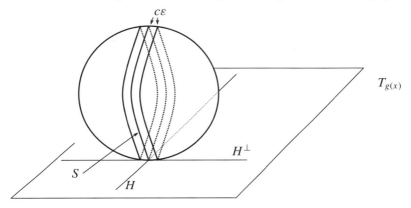

Figure A   The strip around S and the image by g of $B_\varepsilon(x)$ contained on it.

By a similar argument, there are constants $c' > 0$ and $\varepsilon_1 > 0$ such that $g(B_\varepsilon(x))$ is included in $\pi_2^{-1}(B_{c'\varepsilon}(g(x)))$ for all $0 < \varepsilon < \varepsilon_1$, where $\pi_2 : S^{n-1} \to H$. Thus, for $0 < \varepsilon < \min\{\varepsilon_0, \varepsilon_1\}$ and any critical point x of g in H_δ,

$$\text{Vol}\,(g(B_\varepsilon(x))) \le (c\varepsilon^2)^{\dim H^\perp} \cdot (c'\varepsilon)^{\dim H}$$
$$\le c_0 \varepsilon^n$$

for another constant c_0, since $\dim H^\perp + \dim H = n - 1$ and $\dim H^\perp \ge 1$.

Let L be the length of a side of the cube H_δ and $k \in \mathbb{N}$. Then H_δ can be covered by k^{n-1} subcubes of side L/k. We see that the volume of the image of the critical points of g in H_δ is at most $(c_0 L^n n^{n/2})/k$ for every $k \in \mathbb{N}$ large enough and hence is zero.

For Part (ii), let $y \in S^{n-1}$ be a regular value of g. By rotation we may assume $y = (0, \ldots, 0, 1)$. As previously computed, the critical points of the projection are then the zeros of $f - \delta = 0$, $\partial f/\partial x_1 = 0, \ldots, \partial f/\partial x_{n-1} = 0$; that is, p is a critical point if and only if $\text{grad}\, f(p)/\|\text{grad}\, f(p)\| = y$. The tangent space of H_δ at p is given by $\sum(\partial f_i/\partial x_i)v_i = 0$, thus at a critical point by $v_n = 0$.

Now by the product rule for derivatives the derivative of $\text{grad}\, f/\|\text{grad}\, f\|$ at a critical point p applied to a tangent vector v is a multiple of $(D \,\text{grad}\, f)(p)(v)$ minus a multiple of y. Since y is a regular value this linear map is an isomorphism from the tangent space of H_δ at p to the tangent space of S^{n-1} at y. But the tangent space of S^{n-1} at y is orthogonal to y and therefore $\pi_{y^\perp}(D\,\text{grad}\, f)(p)$ is an isomorphism on the tangent space to H_δ at p, where $\pi_{y^\perp}$ is the projection on the orthogonal complement of y. Once again using that the tangent space to H_δ at p is defined by $v_n = 0$ it follows that the first $n - 1$ columns of $(D\mathrm{grad}\,f)(p)$ projected into the tangent space of H_δ at p are independent vectors. Since the tangent space to H_δ at p is defined by the gradient of $f - \delta$ and the first $n - 1$ columns of $(D\,\mathrm{grad}\,f)(p)$ are the gradient vectors of $\partial f/\partial x_1, \ldots, \partial f/\partial x_{n-1}$, the gradient vectors of $f - \delta$, $\partial f/\partial x_1$, $\partial f/\partial x_2, \ldots, \partial f/\partial x_{n-1}$ are independent at a point p where $f - \delta$, $\partial f/\partial x_1$, $\partial f/\partial x_2, \ldots, \partial f/\partial x_{n-1}$ are all zero. Thus the critical points of $\pi_y | H_\delta$ are nondegenerate. $\square$

Lemma 2 *Let $g_1, \ldots, g_n : \mathbb{R}^n \to \mathbb{R}$ be polynomials with degree $g_i \le d_i$ that have k common zeros. If at each of these zeros the gradients $\mathrm{grad}\,g_i$, for $i = 1, \ldots, n$, are independent, then $k \le \prod_{i=1}^n d_i$.*

Proof.   Consider the polynomials g_i as polynomials of complex variables g_i : $\mathbb{C}^n \to \mathbb{C}$. Then the complex gradients are the same as the real gradients, hence independent. Thus, by the inverse function theorem, the roots are isolated in $\mathbb{C}^n$ and by Bézout's theorem there are at most $\prod_{i=1}^n d_i$ isolated roots. $\square$

Now we may finish the proof of Proposition 1.

Proof of Proposition 1.   Apply Lemma 2 to $f - \delta$, $\partial f/\partial x_1, \ldots, \partial f/\partial x_{n-1}$ (after rotating the regular value y in Lemma 1 to be equal to $(0, 0, \ldots, 0, 1)$). $\square$

This finishes the proof of Proposition 1. We need Sard's Theorem in a different but also elementary setting shortly. The version we use, as well as its proof, is given in the next proposition.

Proposition 2 *If $f : \mathbb{R}^n \to \mathbb{R}$ is a polynomial, then f has only finitely many critical values.*

Proof.   The critical points are zeros of $\mathrm{grad}\,f : \mathbb{R}^n \to \mathbb{R}^n$. We may complexify and consider $\mathrm{grad}\,f : \mathbb{C}^n \to \mathbb{C}^n$. The zeros of the real gradient are zeros of the complex gradient. The zeros of the complex gradient are either all of $\mathbb{C}^n$, and f is constant, or there are finitely many connected components for the homogenized equations by Bézout's theorem and f is constant on each of them. Thus there are only finitely many critical values. $\square$

Note that in the proof of the preceding proposition we needed to complexify first. Bézout's Theorem is not true over $\mathbb{R}$ as can be seen in the following example.

Example 1 Let $f_1, \ldots, f_n : \mathbb{R}^n \to \mathbb{R}$ be real polynomials with $\deg(f_i) = d_i$ such that the set of their common zeros has $\prod_{i=1}^n d_i$ isolated points. For instance, $f_i = \prod_{j=1}^{d_i}(x_i - a_{ij})$ with $a_{ij} \ne a_{ik}$ for $j \ne k$.

Then $f = f_1^2 + \cdots + f_n^2 = 0$ has degree $d = \max_i(2d_i)$. Now consider the set of polynomial equations

$$\left.\begin{cases} f = 0 \\ x_{n+1} = 0 \\ \quad\vdots \\ x_{n+1} = 0 \end{cases}\right\} n \text{ times.}$$

The zero set of this system has $\prod_{i=1}^n d_i$ connected components and the product of the degrees of the f_i is d.

Now we turn to our main theorem dealing with the more general case of a not necessarily compact or smooth set.

Theorem 1 *Let* $V \subset \mathbb{R}^n$ *be defined by* $f_1 = 0, \ldots, f_p = 0$, *where* f_i *are polynomials of degree at most* d. *Then the number of connected components of* V *is at most* $d(2d - 1)^{n-1}$.

Proof. If we denote by $B(r)$ the ball centered at the origin with radius r, it suffices to prove this estimate for $V \cap B(r)$ for a sequence of rs that go to infinity, since any connected component of V must ultimately intersect $B(r)$ for large r and contributes at least one connected component to the intersection. Consider the function

$$F_\varepsilon(x_1, \ldots, x_n) = f_1^2(x_1, \ldots, x_n) + \cdots + f_p^2(x_1, \ldots, x_n) + \varepsilon^2(x_1^2 + x_2^2 + \cdots + x_n^2).$$

For every $\varepsilon > 0$ almost every $\delta > 0$ satisfies that δ^2 is a regular value of F_ε. Let W_δ be defined by $W_\delta = \{(x_1, \ldots, x_n) \mid F_\varepsilon(x_1, \ldots, x_n) \leq \delta^2\}$ and $H_\delta = \{(x_1, \ldots, x_n) \mid F_\varepsilon(x_1, \ldots, x_n) = \delta^2\}$.

Note that $W_\delta \subseteq B(\delta/\varepsilon)$. Moreover, since δ is a regular value of F_ε, by the implicit function theorem W_δ is the closure of its interior and $H_\delta = \partial W_\delta$. See Figure B.

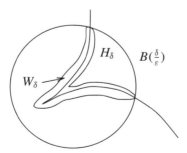

Figure B The sets W_δ and H_δ.

Lemma 3 *If* ε_n, δ_n *converge monotonically to zero,* δ_n/ε_n *converges monotonically to* r *from above, and* δ_n *is a regular value of* F_{ε_n}, *then* $W_{\delta_{n+1}} \subseteq W_{\delta_n}$ *and* $\cap_n W_{\delta_n} = V \cap B(r)$.

Proof.   To see that $W_{\delta_{n+1}} \subseteq W_{\delta_n}$ note that

$$\frac{\sum f_i^2}{\delta_{n+1}^2} + \frac{\varepsilon_{n+1}^2}{\delta_{n+1}^2} \|x\|^2 \leq 1$$

implies

$$\frac{\sum f_i^2}{\delta_n^2} + \frac{\varepsilon_n^2}{\delta_n^2} \|x\|^2 \leq 1$$

since $\delta_{n+1} \leq \delta_n$ and $\varepsilon_n/\delta_n \leq \varepsilon_{n+1}/\delta_{n+1}$.

The other part of the statement follows from the facts: (i) $\sum f_i^2 \leq \delta_n^2$ for all $n \in \mathbb{N}$ implies that $f_i = 0$ for $i = 1, \ldots, p$, and (ii) $\varepsilon_n^2 \|x\|^2 \leq \delta_n^2$ for all $n \in \mathbb{N}$ implies that $\|x\| \leq r$.   □

Now we can finish the proof of Theorem 1. For each component of W_{δ_n} there is a component of the boundary H_{δ_n}; thus W_{δ_n} has at most $d(2d-1)^{n-1}$ components. Then the intersection $\cap_{n \in \mathbb{N}} W_{\delta_n}$ also has at most $d(2d-1)^{n-1}$ connected components since for large n, W_{δ_n} is contained in an arbitrarily small neighborhood of the intersection.   □

Theorem 1 can be used to bound the number of connected components of semi-algebraic sets. We give a first bound in the next proposition that is useful in Section 16.2. A finer bound is proved in Section 16.3.

Proposition 3   *Suppose $S \subseteq \mathbb{R}^n$ is defined by*

$$\begin{cases} f_i(x) = 0, & i = 1, \ldots, p \\ f_i(x) \geq 0, & i = p+1, \ldots, p+k \\ f_i(x) > 0, & i = p+k+1, \ldots, p+\ell, \end{cases}$$

and let $d = \max\{\text{degree } f_1, \ldots, \text{degree } f_{p+\ell}, 2\}$. Then the number of connected components of S is bounded by $d(2d-1)^{n+\ell-1}$.

Proof.   Our first goal is to replace the strict inequalities by nonstrict ones. To do so, let j be the number of connected components of S and $v_1, \ldots, v_j$ be points in distinct connected components of S. Consider

$$\varepsilon = \min_{\substack{i=1,\ldots,\ell-k \\ s=1,\ldots,j}} f_{p+k+i}(v_s)$$

and S' the subset of $\mathbb{R}^n$ defined by

$$\begin{cases} f_i(x) = 0, & i = 1, \ldots, p \\ f_i(x) \geq 0, & i = p+1, \ldots, p+k \\ \widehat{f}_i(x) = f_i(x) - \varepsilon \geq 0, & i = p+k+1, \ldots, p+\ell. \end{cases}$$

Then S' has a point in each of the j connected components of S and $S' \subset S$. Thus S' has at least j connected components. Now, for each inequality

$$f_i(x) \geq 0 \quad i = p+1, \ldots, p+k$$
$$\widehat{f}_i(x) \geq 0 \quad i = p+k+1, \ldots, p+\ell,$$

add a new variable y_i and replace

$$f_i(x) \geq 0 \text{ by } f_i(x) - y_i^2 = 0$$
$$\text{and } \widehat{f}_i(x) \geq 0 \text{ by } \widehat{f}_i(x) - y_i^2 = 0.$$

Finally, let $V \subset \mathbb{R}^{n+\ell}$ be defined by the equations

$$\begin{cases} f_i(x) = 0, & i = 1, \dots, p \\ f_i(x) + y_i^2 = 0, & i = p + 1, \dots, p + k \\ \widehat{f}_i(x) + y_i^2 = 0, & i = p + k + 1, \dots, p + \ell. \end{cases}$$

Since the degree of the polynomials defining V is at most d, we apply Theorem 1 and get an upper bound of $d(2d - 1)^{n+\ell-1}$ for the number of its connected components. But S' is the projection of V on $\mathbb{R}^n$ and therefore the same bound holds for the number of connected components of S' and a fortiori for that of S. $\square$

16.2 A Complexity Lower Bound

In this section we use the upper bounds computed in the last section to get lower bounds for the depth of algebraic decision trees.

Definition 1 An *algebraic computation tree* is a finite-dimensional machine whose graph of nodes is acyclic and whose computations are arithmetic operations. That is, a computation node performs a single arithmetic operation $x \circ y$, where $\circ$ may be $+$, $-$, $\times$, or $/$, and x and y are the contents of certain machine registers or the machine constants. The number of nodes of the machine is the *size* of the tree and the length of the path from the *root* (input node) to a given node is the *depth of the node*. The largest depth of an output node is the *depth of the tree*. The ouput nodes of the tree are called *leaves*.

An *algebraic decision tree* is an algebraic computation tree in which each output node is labeled Yes or No. If T is an algebraic decision tree with input space $\mathbb{R}^n$ we say that T *accepts* $x \in \mathbb{R}^n$ if the computation of T with input x leads to a leaf labeled Yes. Otherwise we say that T *rejects* x.

Remark 1 If T is an algebraic computation tree with depth d, the number of its leaves is bounded by 2^d. This follows from the fact that each node of T has at most two next nodes.

Remark 2 In Part III, we deal mainly with decision problems and thus, our machines often compute characteristic functions. We adopt for these machines the terminology used for algebraic decision trees. That is, for a given input x, we say that machine M *accepts* x if $\Phi_M(x) = 1$ and that it *rejects* x otherwise.

Theorem 2 *Let $W \subseteq \mathbb{R}^n$ be any set and let T be an algebraic decision tree that decides W. Then the depth t of T satisfies the bounds*

(i) $t \geq \dfrac{1}{(1 + \log 3)} \log H - \dfrac{\log 3}{(1 + \log 3)} n \geq 0.38 \log H - 0.61 n,$

(ii) $t \geq \dfrac{1}{(1 + 3\log 3)} \log H - \dfrac{1 + \log 3}{1 + 3\log 3} \geq 0.21 \log H - 0.55,$

where H is the number of connected components of W.

Proof.   Let $x \in \mathbb{R}^n$ and let $\gamma_x = (v_1, \ldots, v_h)$, $h \leq t$ be the path traversed by x from the root v_1 of T to a leaf v_h. Let $\mathcal{V}_x \subseteq \mathbb{R}^n$ be the set of points in $\mathbb{R}^n$ that follow this same path. We now bound the number of connected components of $\mathcal{V}_x$ by describing it succinctly by a system of quadratic equations and inequalities.

To each computation node that is not a division node v_i add a new variable y_i and the equation $y_i = z_k \circ z_\ell$, where $z_k, z_\ell \in \{x_1, \ldots, x_n, y_1, \ldots, y_{i-1}\}$ and $\circ$ is the operation performed at v_i. If v_i is a division node, add the variable y_i and equation $y_i z_\ell = z_k$, where z_ℓ, z_k are as above and the division performed at v_i is z_k / z_ℓ.

If v_i is a branch node, add the inequality $z_k \geq 0$ if $v_{i+1} = \beta^+(v_i)$ and $z_k < 0$ otherwise, where $z_k \in \{x_1, \ldots, x_n, y_1, \ldots, y_{i-1}\}$ is the value whose sign is tested.

Let $S \subseteq \mathbb{R}^{n+h-1}$ be the semi-algebraic set defined by the system of quadratic equations and inequalities previously described. By Proposition 3 the number of connected components of S is bounded by $2(3)^{n+h-1}$ and therefore by $2(3)^{n+t-1}$. Since $\mathcal{V}_x$ is the projection onto the first n coordinates of S we deduce that the same bound holds for the number of connected components of $\mathcal{V}_x$.

The set W is the union of the $\mathcal{V}_x$ for all paths γ_x ending in a leaf labeled yes. Since there are at most 2^t such leaves, we deduce that the number of connected components of W is bounded by $2^{t+1} 3^{n+t-1}$. The lower bound for t follows, thus proving (i).

To prove (ii) let r denote the number of coordinates of $(x_1, \ldots, x_n)$ that occur in the equalities and inequalities describing $\mathcal{V}_x$. Then $r \leq n$, and the preceding reasoning shows that the number of connected components of $\mathcal{V}_x$ is bounded by $2(3)^{r+t+1}$. But $r \leq 2t$ since at each node in $\mathcal{V}_x$ at most two new coordinates can appear. Thus, the number of connected components of $\mathcal{V}_x$ is at most $2(3)^{3t+1}$ and consequently the number of connected components of W is at most $2^{t+1} 3^{3t+1}$ from which (ii) follows. $\square$

Theorem 2 yields nonuniform lower bounds since it is stated in terms of algebraic decision trees. To deduce from it lower bounds for uniform machines the following proposition is helpful.

Proposition 4   *Let M be a uniform machine with halting time bounded by a function $t : \mathbb{N} \to \mathbb{N}$. For each $n \in \mathbb{N}$ there is an algebraic computation tree $T_{M,n}$ computing the restriction of φ_M to $\mathbb{R}^n$. If M decides a set $S \subseteq \mathbb{R}^\infty$, then $T_{M,n}$ can be considered to be an algebraic decision tree and the set decided by it is $S_n = S \cap \mathbb{R}^n$.*

Proof.   We have seen in Proposition 1 of Chapter 3 that for any input size n there is a finite-dimensional machine M_n computing the restriction of φ_M to $\mathbb{R}^n$ in time at most $t(n)$. A tree T_1 is obtained from M_n by unfolding its graph up to length $t(n)$.

We now remove from T_1 all the subtrees whose leaves are not output nodes. Since M_n works in time at most $t(n)$ we know that no $x \in \mathbb{R}^n$ will reach one of these leaves. Let T_2 be the resulting tree. The tree $T_{M,n}$ is obtained from T_2 by replacing each computation node by a sequence of arithmetic operations computing the rational function associated with the node. This increases the depth of $T_{M,n}$ with respect to the depth of T_2 by only a constant factor.

If M is a decision machine, an additional step is needed. We replace each leaf by a branching node with the test $x_1 = 1$ followed by a Yes leaf corresponding to the answer $x_1 = 1$ and a No leaf corresponding to the answer $x_1 \neq 1$. $\square$

Remark 3 Theorem 2 gives a lower bound with precise constants for the depth of algebraic decision trees deciding a given set $W \subset \mathbb{R}^n$. In passing to uniform machines, this precision gets lost for two reasons: since the computation nodes of uniform machines can be arbitrary rational functions, there is a possible speedup by a constant factor; and, management introduces extra computing time that does not affect the lower bound but makes the constant multiplying of the $\log H$ meaningless.

These considerations suggest that, in the case of uniform machines, Theorem 2 will be used mainly to deduce the order of magnitude of lower bounds for the halting time. We now introduce notation that plays a role for lower bounds closely related to the "big Oh" notation introduced in Section 4.4 for upper bounds.

Definition 2 Let $f, g : \mathbb{N} \to \mathbb{R}$. We say that f is $\Omega(g)$ if there are $n_0 \in \mathbb{N}, c \in \mathbb{R}$, $c > 0$ such that for all $n \geq n_0$, $|f(n)| \geq c|g(n)|$.

Theorem 3 *Let $S \subseteq \mathbb{R}^\infty$ and let S_n be its subset of elements having size n. Then the halting time of any machine M deciding S is $\Omega(\log g(n))$ where $g(n)$ is the number of connected components of S_n.* $\square$

Remark 4 In this theorem, the constant hidden in $\Omega(\log g(n))$ depends on M. If we restrict to machines whose computation nodes only perform arithmetic operations, then this constant can be taken to be independent of the machine.

We now give some applications of Theorems 2 and 3. In each of these, we want to estimate the number of connected components of certain sets. Since these components are open subsets of $\mathbb{R}^n$ they are also path connected. We use this fact freely.

Example 2 Let us consider the problem of deciding whether a point x in $\mathbb{R}^\infty$ has two different coordinates $i, j \leq \text{size}(x)$ such that $x_i = x_j$. Thus, we want to decide the set

$$S = \{x \in \mathbb{R}^\infty \mid x_i = x_j \text{ for } i, j \leq \text{size}(x), i \neq j\}.$$

We denote by W_n the set of points having size n that do not belong to S. Then $W_n \subset \mathbb{R}^n = \{(x_1, \dots, x_n) \mid x_i \neq x_j \text{ for } i \neq j\}$.

Now, take the point $(1, 2, \ldots, n) \in W_n$ and consider points $x, y \in W_n$ given by two different permutations to its n coordinates.

There exist coordinates $i, j \leq n$ such that $x_i - x_j > 0$ and $y_i - y_j < 0$. Therefore, for any path $\gamma : [0, 1] \to W_n$ connecting x and y with $\gamma(0) = x$ and $\gamma(1) = y$, one has that the composition

$$g = (x_i - x_j) \circ \gamma : [0, 1] \to \mathbb{R}$$

sends 0 to a positive number and 1 to a negative one. By the intermediate value theorem, there is a point $c \in (0, 1)$ such that $g(c) = 0$. But then, $\gamma(c)$ has equal coordinates i and j and cannot belong to W_n. One deduces that x and y are not path connected and that the number of connected components of W_n is at least $n!$.

Since the union of the W_n is the complement of S, a direct application of Theorem 3 yields a lower bound of $\Omega(n \log n)$ for the halting time of any machine deciding S.

Example 3 In much the same manner, one shows that the problem of deciding, given $x_1, \ldots, x_n, y_1, \ldots, y_n \in \mathbb{R}$, whether $y_i = x_j$ for some $i, j \leq n$, also has a lower bound of $\Omega(n \log n)$.

We close this section with an example that has been widely studied in algebraic complexity.

Definition 3 Let $V \subset \mathbb{C}^n$ be an algebraic set defined by $f_1 = 0, \ldots, f_k = 0$, and let r be its dimension. Let L be an affine subspace of dimension $n - r$ and $\#(L \cap V)$ be the cardinality of $L \cap V$. The *degree* of V is the maximum of $\#(L \cap V)$ over all affine subspaces with $\#(L \cap V) < \infty$.

Example 4 Let $V \subseteq \mathbb{C}^n$ be as in the preceding definition and T be an algebraic decision tree that decides V. We can assume that T takes its inputs from $\mathbb{R}^{2n}$. Now let L be an affine $(n - r)$-dimensional subspace such that $\#(L \cap V) = $ degree V. With an additional $2rn$ operations one can decide whether the input belongs to L and thus to $L \cap V$. Therefore, if t is the depth of T one deduces from Theorem 2 that

$$t + 2rn \geq 0.38 \log(\text{degree } V) - 1.24n$$

and finally that

$$t \geq 0.38 \log(\text{degree } V) - (1.24 + 2r)n.$$

16.3   On the Number of Connected Components of Semi-Algebraic Sets

In this section we use Theorem 1 to bound the number of connected components of semi-algebraic sets. The bound obtained depends exponentially only on the number of variables, thus refining Proposition 3. On the way, we also obtain a bound for the number of sign conditions that can be satisfied by a set of polynomials. Both results are used in later chapters.

We begin with a particular case.

Proposition 5 *Let $S \subset \mathbb{R}^n$ be defined by*

$$\begin{cases} f_i(x) = 0, & i = 1, \ldots, p \\ f_i(x) > 0, & i = p + 1, \ldots, p + s \end{cases}$$

and let $d = \max\{\text{degree } f_1, \ldots, \text{degree } f_{p+s}\}$. Then the number of connected components of S is bounded by $(sd + 1)(2sd + 1)^n$.

Proof.   Let $W = \{x \in \mathbb{R}^n \mid f_i(x) = 0, i = 1, \ldots, p$ and $f_i(x) \neq 0, i = p + 1, \ldots, p + s\}$. Each connected component of S is a connected component of W. Therefore the number of connected components of S is at most that of W. But W is the projection on $\mathbb{R}^n$ of the algebraic set in $\mathbb{R}^{n+1}$ given by the polynomial equalities in $(x_1, \ldots, x_n, y)$,

$$f_1(x_1, \ldots, x_n) = 0, \ \ldots, \ f_p(x_1, \ldots, x_n) = 0, \ y \prod_{i=p+1}^{p+s} f_i(x_1, \ldots, x_n) - 1 = 0,$$

that has, according to Theorem 1, at most $(sd+1)(2sd+1)^n$ connected components.
□

Definition 4 A *sign condition* is any of the three elements in $\{<, =, >\}$. A polynomial $f \in \mathbb{R}[X_1, \ldots, X_n]$ *satisfies* a sign condition σ if there is an $x \in \mathbb{R}^n$ such that $f(x)\sigma 0$.

For a set $\{f_1, \ldots, f_k\}$ of polynomials in $\mathbb{R}[X_1, \ldots, X_n]$ and a tuple of sign conditions $\sigma = (\sigma_1, \ldots, \sigma_k)$, we say that *the set satisfies σ* if there is an $x \in \mathbb{R}^n$ such that $f_i(x)\sigma_i 0$ for all $i = 1, \ldots, k$. If the set of polynomials is clear from the context, we say that σ is *satisfiable*.

Proposition 6 *Let $f_i \in \mathbb{R}[X_1, \ldots, X_n]$ be a polynomial of degree d_i for $i = 1, \ldots, k$ and let $D = \sum_{i=1}^k d_i$. The number of tuples of sign conditions satisfied by the set $\{f_1, \ldots, f_k\}$ is at most $(2D + 1)(4D + 1)^n$.*

Proof.   Let $A \subset \mathbb{R}^n$ be a set containing exactly one point in each nonempty set defined by a tuple of sign conditions. Thus, the cardinality of A is the number we want to bound. We denote by A_i' the subset of A containing those points on which the sign of f_i is either $>$ or $<$. We then choose $\delta > 0$ such that for all $i \leq k$ and all $x \in A_i'$ one has $|f_i(x)| > \delta$ and consider the polynomial $\tilde{f} = \prod_{i=1}^k (f_i - \delta)(f_i + \delta)$ whose zero set is the dotted lines in Figure C.

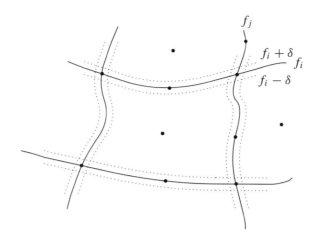

Figure C   The zero set of $\widetilde{f}$.

We claim that each connected component of W, the complement of the zero set of $\widetilde{f}$, contains at most one point of A. To prove this claim, consider two different points x, y in A. Then there exists $i \leq k$ such that, for instance, $f_i(x) > 0$ and $f_i(y) \leq 0$. Because of the choice of δ we then have that $f_i(x) - \delta > 0$ and since $\delta > 0$ that $f_i(y) - \delta < 0$. If x and y were in the same connected component of W, then it would be a path γ joining them inside W. But as in Example 2, any such path crosses the set of zeros of $f_i - \delta$ and therefore cannot be contained in W.

Finally, the number of connected components of $\widetilde{f} \neq 0$ is the same as the number of connected components in $\mathbb{R}^{n+1}$ of the zero set of the polynomial $Y\widetilde{f} - 1$. Applying Theorem 1 to this polynomial we deduce the statement of the proposition.

$\square$

Proposition 7   *Let $S \subseteq \mathbb{R}^n$ be defined by*

$$\begin{cases} f_i(x) = 0, & i = 1, \ldots, p \\ f_i(x) \geq 0, & i = p+1, \ldots, p+\ell \\ f_i(x) > 0, & i = p+\ell+1, \ldots, k \end{cases}$$

and let $d = \max\{\text{degree } f_1, \ldots, \text{degree } f_k\}$. Then the number of connected components of S is bounded by $(kd + 1)(2kd + 1)^{n+1}(4kd + 1)^n$.

Proof.   The set S is the union of sets S_j given by inequalities of the form $f_i \sigma_i 0$ for $i = 1, \ldots, k$ where $\sigma_i \in \{=, >\}$. By Proposition 6 there are at most $(2kd+1)(4kd+1)^n$ nonempty sets S_j. And by Proposition 5, each has at most $(kd + 1)(2kd + 1)^n$ connected components. Multiplying both bounds finishes the proof. $\square$

16.4 Additional Comments and Bibliographical Remarks

The bounds on the number of connected components of algebraic sets established in Section 16.1 are due to Oleinik and Petrovski [1949] with an important early special case done by Petrovski (see also [Oleinik 1951]) and, independently, to Milnor [1964] and Thom [1965].

The use of the number of connected components to provide complexity lower bounds originates with Dobkin and Lipton [1979] who considered linear trees, that is, trees performing only linear operations. To the best of our knowledge, the first use of the bounds obtained in Section 16.1 is due to Steele and Yao [1982]. Part (i) of Theorem 2 is due to Ben-Or [1983]; its modification in Part (ii) is taken from [Cucker, Montaña, and Pardo 1992].

There are several results providing bounds for the number of satisfiable sign conditions of a polynomial family. A classic reference is [Warren 1968]. Proposition 6 is taken from [Meyer auf der Heide 1985b].

17
Probabilistic Machines

Probabilistic methods are widely used in the design and analysis of algorithms. We have already seen several examples of this. In Chapter 11 we computed bounds for the expected loss of precision in solving large linear systems. This expected value suggests why, in general, we do not lose much precision solving linear systems.

A different use of randomness was seen in Chapter 9 in our first algorithm for computing approximate zeros of polynomials. In Theorem 4 of that chapter, it was shown that if we randomly choose the starting point of our homotopy, the expected *running* (i.e., halting) *time* is decreased by a factor of d. In this case, randomness is introduced in the algorithm itself and is used to get an *expected* running time better than the *worst case* running time. In spite of this randomness, the algorithm is always correct. The worst case running time does not improve.

In this chapter, we deal with a third use of randomness. We show how algorithms, by making random choices, can cut the running time drastically. We pay for this gain by producing an output that is in error with some (small) probability. We give an example of this type of algorithm for computing the approximate volume of a bounded semi-algebraic set.

On the other hand, we show that probabilistic machines may be simulated by deterministic ones, thus providing theoretical limitations for the power of randomization.

17.1   Machines That Flip Coins

Let us call a *straight-line program* any finite-dimensional machine $\mathcal{P}$ without branching nodes whose computation nodes only perform addition, subtraction, or multiplication between two elements. If the input space of this machine has dimension n and machine constants $\{c_i\}_{i \in I}$, we say that the straight-line program has n variables and program constants $\{c_i\}_{i \in I}$. It is clear that the only function such a machine can compute is a polynomial p in n variables. Note that the τ defined in Chapter 7 for an integer polynomial p is just the size of a minimal straight-line program computing p and having only one as program constant.

Since a straight-line program $\mathcal{P}$ has no branching there is only one computation path, and this path traverses all the nodes once and only once. If N is the number of nodes, we say that N is the *size* of $\mathcal{P}$. Notice that, up to a constant factor, N is the dimension of a real vector coding $\mathcal{P}$ since any node of $\mathcal{P}$ can be described with a constant number of reals.

We have seen in Chapter 5 that 4-FEAS, the problem of deciding whether a fourth-degree polynomial has a real root, is $NP_{\mathbb{R}}$-complete. Now it is simple to design a machine that, with input a fourth-degree polynomial f, produces in polynomial time in the size of f a straight-line program $\mathcal{P}_f$ that evaluates f. Therefore the problem:

> with input a straight-line program $\mathcal{P}$ with n variables, decide whether there is a point $x \in \mathbb{R}^n$ such that $\mathcal{P}$ applied to x is 0

is also $NP_{\mathbb{R}}$-complete. If we replace the preceding existential condition with a universal one we obtain the following problem, which we denote by SLP0:

> with input a straight-line program $\mathcal{P}$ with n variables, decide whether for all points $x \in \mathbb{R}^n$ one has that $\mathcal{P}$ applied to x is 0.

In other words, SLP0 is the problem of deciding for a given $\mathcal{P}$ whether the polynomial computed by $\mathcal{P}$ is identically zero. If this is the case, we simply write $\mathcal{P} \in$ SLP0.

A first remark is that if we consider the problem SLP0 over the field of real algebraic numbers, the Witness Theorem (Theorem 4 of Chapter 7) allows us to solve the problem in polynomial time. But over $\mathbb{R}$, the machine constants of the straight-line program can be arbitrary real numbers and the Witness Theorem does not apply in this case. However, one can easily prove that SLP0 belongs to $coNP_{\mathbb{R}}$. To show that $\mathcal{P} \notin$ SLP0 one just guesses a point $x \in \mathbb{R}^n$ and checks that $\mathcal{P}$ applied to x is not zero.

But one can do better. Let p be the polynomial computed by a straight-line program $\mathcal{P}$. If $p \neq 0$, its degree is bounded by 2^s, where s is the size of $\mathcal{P}$. Assume for the moment that such a p has only one variable. Then if $p \neq 0$, the number of real roots of p is at most 2^s. Consider the set $\{0, 1, \ldots, 4 \cdot 2^s - 1\}$ containing the first $4 \cdot 2^s$ nonnegative integers. Since at most one-fourth these numbers are roots of p each has probability at most $\frac{1}{4}$ of being a root of p. If the number of variables of p is $n > 1$, we can generalize the preceding argument to show that for a point ζ in $\{0, 1, \ldots, N-1\}^n$ one has

$$\Pr(p(\zeta) = 0) \leq \frac{2^s n}{N}.$$

So letting $N = 4n2^s$, the probability again is at most $\frac{1}{4}$.

Let us now suppose that a uniform machine M over $\mathbb{R}$ is provided with the ability to flip coins. That is, M can produce a random element in $\{0, 1\}$ (each with probability $\frac{1}{2}$) at unit cost. Then M can generate the random element ζ with $2 + s + \log n$ coin flips and test if it is a root of p in the following way.

> input $\mathcal{P}$
> $s := \text{size}(\mathcal{P})$
> $n := $ number of variables of p
> randomly choose $\zeta \in \{0, \ldots, 4n2^s - 1\}^n$
> evaluate $\mathcal{P}$ at ζ
> **if** the result is 0 ACCEPT
> **else** REJECT
> **end if** .

If $\mathcal{P} \in \text{SLP0}$, then the machine correctly accepts $\mathcal{P}$. Otherwise, the machine can either accept or reject it (depending on the choice of ζ) but the probability that it rejects $\mathcal{P}$ is at least $\frac{3}{4}$. In fact, the actual probability depends on the number of elements in the set $\{0, \ldots, 4n2^s - 1\}^n$ that are roots of p.

One could think that the risk of a bad answer is still high. In this case, it is worth noting that iterating the choice of ζ and its evaluation k times (and accepting if all the k evaluations produce 0) reduces the probability of a wrong answer to at most $1/4^k$.

We now formally define probabilistic machines and some probabilistic complexity classes. We do so both for ordered and unordered rings R with the understanding that machines over R branch on the order relation if R is ordered, unless otherwise specified. Some results, such as Lemma 1 and Proposition 1, hold in both contexts.

Definition 1 Let R be an arbitrary ring. A *probabilistic machine* over R is a machine over R as defined in Chapter 3 with an additional type of node. These new nodes, called *probabilistic*, have two next nodes and no associated maps. When a computation reaches a probabilistic node it randomly chooses the next node from the two possible ones, each with probability $\frac{1}{2}$.

In what follows, we consider time bounded machines, that is, machines M such that for some function $t : \mathbb{N} \to \mathbb{N}$ and for every input of size n, the computation halts after at most $t(n)$ steps either rejecting or accepting. For a given $x \in R^n$, different computations of M with input x can lead to either result depending on the outcome of the probabilistic nodes. Nevertheless, for any $x \in R^\infty$, the probability of the event "M accepts x" is well defined. Let $\{\gamma_{x,i}\}_{i \in I}$ be the set of possible paths traversed by input x. For each path $\gamma_{x,i}$, the probability of $\gamma_{x,i}$ is $2^{-\#(i)}$ where $\#(i)$ is the number of probabilistic nodes in $\gamma_{x,i}$. The probability that M accepts (rejects) x is the sum of the probabilities of the paths leading to acception (rejection).

We now define what it means for a set to be decided by a probabilistic machine.

Definition 2 Let $S \subseteq R^\infty$. The *error probability* of a probabilistic machine M with respect to the set S is the function $e_M : R^\infty \to \mathbb{R}$ given by

$$e_M(x) = \Pr(M \text{ accepts } x \ \& \ x \notin S) + \Pr(M \text{ rejects } x \ \& \ x \in S).$$

If neither of the two events in the preceding formula holds, we say that M *treats* x *correctly*.

Definition 3 Let $S \subseteq R^\infty$ and let M be a probabilistic machine. We say that M *decides* the set S if for every $x \in R^\infty$ one has $e_M(x) \le \frac{1}{4}$. We say that S belongs to BPP_R if there is a probabilistic machine M working in polynomial time that decides S.

Remark 1

(i) Note that we have used the expression "M decides S" both for probabilistic and deterministic machines (i.e., machines as defined in Part I). This is done to avoid having to say that a probabilistic machine probabilistically decides a set.

(ii) The acronym BPP comes from Bounded error Probabilistic Polynomial time. It is widely used in classical complexity theory.

(iii) We have given the definition of a probabilistic machine for the case of uniform machines. Similar definitions with the obvious changes can be given for finite-dimensional machines and for the algebraic decision trees introduced in Chapter 16.

We can now summarize our earlier discussion by saying that the problem SLP0 belongs to $\mathrm{BPP}_\mathbb{R}$. Moreover, it is clear that $\mathrm{P}_\mathbb{R} \subseteq \mathrm{BPP}_\mathbb{R}$. A surprising result in this chapter is that this inclusion is indeed an equality. Our first step in proving this result is a simulation of probabilistic machines by deterministic ones in the restricted context of finite-dimensional machines.

17.2 Simulating Probabilistic Trees

The goal of this section is to show that a probabilistic algebraic decision tree can be simulated by a deterministic tree with only a polynomial increase in depth. By this we mean that if T is a probabilistic tree deciding S, there is a deterministic tree T^* also deciding S whose depth is bounded by a polynomial in the depth of T. The computation of T^* is essentially the computation of T repeated a small number of times replacing the random choices in T by a *deterministic sample* consisting of a fixed set of coin-tossing results.

Let T be a probabilistic algebraic decision tree over a ring R with input space R^n and let $s \in \mathbb{N}$. We define the probabilistic algebraic decision tree T_s in the following way. On input $x \in R^n$, T_s performs s *runs* (computations) of the machine T with input x. The machine T_s accepts x if at least $s/2$ of these runs lead to acceptance; otherwise T_s rejects x. The next lemma bounds the error probability of T_s.

Lemma 1 *Let $s \geq 10$. If T decides a set $S \subset \mathbb{R}^n$, then T_s decides the same set. Moreover, if the error probability of T is at most ε, then that of T_s is at most $[4(\varepsilon - \varepsilon^2)]^{s/2}$.*

Proof.   Given $x \in \mathbb{R}^n$ let X_i be the event "the ith run of T does not treat x correctly" for $i = 1, \ldots, s$. Then X_i is a Bernoulli random variable with parameter $p < \varepsilon$; that is, $X_i = 1$ with probability p and $X_i = 0$ with probability $1 - p$. If we denote by S the random variable $X_1 + \ldots + X_s$, then we want to estimate

$$\Pr(S \geq s/2).$$

In Theorem 1 in the next section we give an upper bound for this probability, namely, that for $p < a < 1$,

$$\Pr(S \geq ak) \leq \left[\left(\frac{p}{a} \right)^a \left(\frac{1-p}{1-a} \right)^{1-a} \right]^k.$$

We delay this result to that section to avoid breaking our arguments here. Applying Theorem 1 for $a = \frac{1}{2}$, we deduce that $\Pr(S \geq s/2) \leq [4(p - p^2)]^{s/2}$. Since $p \leq \varepsilon < \frac{1}{4}$ and the function $f(x) = x - x^2$ is increasing in $\left[0, \frac{1}{4}\right]$, the bound on the error probability of T_s follows. And since $\varepsilon \leq \frac{1}{4}$ one has that $(4(\varepsilon - \varepsilon^2))^{s/2} \leq \left(4\frac{3}{16}\right)^{s/2} = \left(\frac{3}{4}\right)^{s/2}$ which is less than $\frac{1}{4}$ for $s \geq 10$.   □

Our next step is to show how the simulation works for finite input spaces.

Let R be any ring, A a finite subset of R^n, and A_{yes} a subset of A. We say that a probabilistic algebraic decision tree T *decides* the structured decision problem (A, A_{yes}) if $e_T(a) \leq \frac{1}{4}$ for all $a \in A$.

Proposition 1 *Let R be any ring, A a finite subset of R^n, and A_{yes} a subset of A. Let T be a probabilistic algebraic decision tree with depth d that decides (A, A_{yes}). Then there is a deterministic algebraic decision tree with depth $O(d \log |A|)$ that decides (A, A_{yes}). Here $|A|$ is the cardinality of A.*

Proof.   Since the depth of T is d, the number of probabilistic nodes that a path in T_s can have is bounded by sd. Thus, any sequence $b = (b_1, \ldots, b_{sd})$ of 0s and 1s determines a deterministic tree T^b as follows. For any probabilistic node η in T_s denote by η_0 and η_1 the next nodes of η. Also, let ℓ be the number of probabilistic nodes traversed in the path from the input node to η. Then we remove from T_s the edge between η and η_0 if $b_{\ell+1} = 1$ and the edge between η and η_1

otherwise. We define T^b to be the subtree of the resulting graph that contains the input node. Briefly, T^b is the tree obtained by replacing the "coin flips" in T_s by the "deterministic sample" b.

For any $a \in A$ and any sequence b as in the preceding, we say that b treats a correctly when T^b accepts a if and only if $a \in A_{\text{yes}}$. Given $a \in A$, denote by B_a the set of sequences b such that b does not treat a correctly. Then

$$e_{T_s}(a) = \frac{|B_a|}{2^{sd}}.$$

Now suppose that $e_{T_s}(a) \leq \delta$ for all $a \in A$ and $\delta|A| < 1$. The first inequality implies that for each $a \in A$, the cardinality of B_a is at most $\delta 2^{sd}$. Thus, the set

$$W = \{(a, b) \mid b \in B_a\}$$

has at most $|A|\delta 2^{sd}$ elements. But since $\delta|A| < 1$, the number of elements in W is less than 2^{sd}. We conclude that there exists a sequence b^* that treats all the elements in A correctly and thus, the deterministic tree T^* determined by b^* satisfies the statement of the proposition.

To finish the proof we show that for $s = O(\log|A|)$ we have $e_{T_s}(a)|A| < 1$ for all $a \in A$. To do so, let $\varepsilon < \frac{1}{4}$ be a bound of the error probability of T. Then by Lemma 1, $e_{T_s}(a) \leq [4(\varepsilon - \varepsilon^2)]^{s/2}$ and we have the following chain of equivalences

$$[4(\varepsilon - \varepsilon^2)]^{s/2}|A| < 1 \iff \frac{s}{2}\log[4(\varepsilon - \varepsilon^2)] + \log|A| < 0$$

$$\iff \frac{s}{2} > -\frac{\log|A|}{\log[4(\varepsilon - \varepsilon^2)]}$$

$$\iff s > \frac{2\log|A|}{\log[\frac{1}{4(\varepsilon - \varepsilon^2)}]},$$

where the logarithm in the denominator is positive and well defined since $\varepsilon \in (0, \frac{1}{4})$. This finishes the proof.   □

A direct application of Proposition 1 yields a general simulation result for finite rings.

Corollary 1 *If R is a finite ring, then any probabilistic algebraic decision tree T over R can be simulated by a deterministic tree with depth $O(dn \log|R|)$, where d is the depth of T, n the dimension of its input space, and $|R|$ the cardinality of R.*   □

We can now proceed with the real case which also follows from Proposition 1 but in a less direct manner.

Proposition 2 *Let T be a probabilistic algebraic decision tree over $\mathbb{R}$ of depth d and input space $\mathbb{R}^n$ and let S be the set decided by T. Then there is a deterministic tree T^* that decides S with depth $O(nd^2)$.*

Proof.   At each branching node i of the tree T, a test is performed of the form $g_i(x) \geq 0$, where g_i is a rational function that is evaluated on the input $x \in \mathbb{R}^n$. For any random sequence y of 0s and 1s, the signs of these evaluations select the path that the input x traverses. In particular, for any such y, two inputs x and x' that give the same signs to the g_i are either both accepted or both rejected for the deterministic run of T that follows y in the probabilistic nodes. Thus, x and x' are either both in S or both in the complement of S.

The degree of g_i, as well as the number m of branching nodes of T, are bounded by 2^d and thus, because of Proposition 6 of Chapter 16, we have that the number h of satisfiable m-tuples of sign conditions for $g_1, \ldots, g_m$ is bounded by

$$(2D + 1)(4D + 1)^n,$$

where $D = 2^d 2^d = 2^{2d}$. Therefore $h \leq 2^{O(dn)}$.

Let H be the set of m-tuples of sign conditions that are satisfied. For each $\sigma = (\sigma_1, \ldots, \sigma_m) \in H$ we consider the set $I_\sigma = \{x \in \mathbb{R}^n \mid g_i \sigma_i 0 \text{ for } i = 1, \ldots, m\}$. Then the sets I_σ for $\sigma \in H$ partition $\mathbb{R}^n$. Let us select a representative point $x_\sigma \in I_\sigma$ for each one of these sets.

By Proposition 1 there is a deterministic tree T^* of depth $O(d \log h) = O(nd^2)$ that restricted to the set of the x_σs simulates T. Moreover, the set of rational functions at the branching nodes of T^* (except for the last one) are in the set $\{g_i\}$ by the construction of T^*. We claim that T^* simulates T over all $\mathbb{R}^n$. To prove the claim we first note that by the construction of T^* and the preceding argument two inputs x and x' that give the same signs to the g_i are either both accepted or both rejected by T^*. Let us consider now any point $x \in \mathbb{R}^n$. Then there is a $\sigma \in H$ such that $x \in I_\sigma$. By definition of I_σ, the result of the test at any branching node of T will be the same for x and x_σ. Thus, x belongs to S if and only if x_σ does. Thus, we have

$$x \in S \iff x_\sigma \in S$$
$$\iff T^* \text{ accepts } x_\sigma$$
$$\iff T^* \text{ accepts } x$$

and this concludes our proof.   □

17.3   Two Bounds for the Tail of the Binomial distribution

Let $X_1, \ldots, X_k$ be independent random variables associated with a Bernoulli distribution with parameter $p, 0 < p < 1$. That is, for each $i \leq k, \Pr(X_i = 0) = 1 - p$ and $\Pr(X_i = 1) = p$.

Under these conditions the random variable $S = X_1 + \ldots + X_k$ has the binomial distribution defined by

$$\Pr(S = j) = \binom{k}{j} p^j (1-p)^{k-j} \qquad \text{for } 0 \leq j \leq k$$

and its expectation is pk. The goal of this section is to bound the tail $\Pr(S \geq h)$ for h larger than pk. We provide two different bounds. The first one, for a given $a \in \mathbb{R}$ with $p < a < 1$, bounds the probability $\Pr(S \geq ak)$ as a function of a, k, and p. The second one is independent of p.

Theorem 1 *For any $p \in [0, 1)$ and any $a \in \mathbb{R}$ with $p < a < 1$ one has*

$$\Pr(S \geq ak) \leq \left[\left(\frac{p}{a} \right)^a \left(\frac{1-p}{1-a} \right)^{1-a} \right]^k.$$

Proof.   For any $\gamma > 0$ let us estimate $\Pr(S \geq \gamma)$. For any $t > 0$ one has

$$\Pr(S \geq \gamma) = e^{-t\gamma} e^{t\gamma} \Pr(e^{tS} \geq e^{t\gamma})$$
$$\leq e^{-t\gamma} \mathrm{E}(e^{tS}). \qquad (17.1)$$

But

$$\mathrm{E}(e^{tS}) = \mathrm{E}(e^{t(\sum_{i=1}^{k} X_i)}) = \mathrm{E}\left(\prod_{i=1}^{k} e^{tX_i} \right)$$

$$= \prod_{i=1}^{k} \mathrm{E}(e^{tX_i}) \quad \text{by the independence of the } X_i$$

$$= \prod_{i=1}^{k} (pe^t + (1-p))$$

$$= (pe^t + (1-p))^k$$

and thus, inequality (17.1) becomes

$$\Pr(S \geq \gamma) \leq e^{-t\gamma} (pe^t + (1-p))^k.$$

Replacing here γ by ak and t by $\ln(a(1-p)/p(1-a))$ we obtain

$$\Pr(S \geq ak) \leq \left(\frac{p(1-a)}{a(1-p)} \right)^{ak} \left(\frac{a(1-p)}{1-a} + (1-p) \right)^k$$

$$= \left(\frac{p(1-a)}{a(1-p)} \right)^{ak} \left(\frac{1-p}{1-a} \right)^k$$

$$= \left[\left(\frac{p}{a} \right)^a \left(\frac{1-p}{1-a} \right)^{1-a} \right]^k.$$

$\square$

We now give another bound for the tail of the binomial distribution. It will be helpful in the next section.

As previously, $X_1, \ldots, X_k$ are Bernoulli random variables with parameter p and $S = X_1 + \ldots + X_k$.

Theorem 2 *For any $p \in [0, 1]$ and any $\eta > 0$ one has*

$$\Pr(S - kp \geq \eta) \leq e^{-2(\eta^2/k)}.$$

Proof.   As in Theorem 1 we have that for any $\gamma > 0$ and any $t > 0$,

$$\Pr(S \geq \gamma) \leq e^{-t\gamma} \prod_{i=1}^{k}(pe^t + (1 - p)).$$

Thus, for $\gamma = kp + \eta$ we deduce that for any $t > 0$,

$$\Pr(S - kp \geq \eta) \leq e^{-tkp}e^{-t\eta} \prod_{i=1}^{k}(pe^t + (1 - p)). \qquad (17.2)$$

Let us denote by $L(t)$ the logarithm (base e) of the function inside the product symbol; that is, $L(t) = \ln(pe^t + (1 - p))$. Then, denoting $1 - p$ by q and taking derivatives one obtains

$$L'(t) = \frac{pe^t}{pe^t + q}$$

$$L''(t) = \frac{pqe^t}{(pe^t + q)^2} = \left(\frac{pe^t}{pe^t + q}\right)\left(1 - \frac{pe^t}{pe^t + q}\right).$$

Now, for any $t > 0$ one has that $x(t) = (pe^t/(pe^t + q))$ satisfies $0 \leq x(t) \leq 1$ and thus, since the function $x(1 - x)$ is bounded by $\frac{1}{4}$ for $x \in [0, 1]$, we deduce that for all $t > 0$ one has $L''(t) \leq \frac{1}{4}$.

Expanding $L(t)$ in Taylor series around 0 we have that for a certain t^*

$$L(t) = L(0) + tL'(0) + \frac{1}{2}t^2L''(t^*)$$

and since $L''(t^*) \leq \frac{1}{4}$ we deduce that

$$L(t) \leq tp + \frac{1}{8}t^2$$

and, since $e^{L(t)} = pe^t + q$, that

$$pe^t + q \leq e^{tp+(1/8)t^2}.$$

Replacing this bound in inequality (17.2) we obtain

$$\Pr(S - kp \geq \eta) \leq e^{-tkp}e^{-t\eta}\left(e^{tp+(1/8)t^2}\right)^k$$

$$= e^{-t\eta+(k/8)t^2}$$

always for all $t > 0$. Set now $t = 4\eta/k$ and the statement follows.   □

Corollary 2 *For any $p \in [0, 1]$ and any $\eta > 0$ one has*

$$\Pr\left(|S - kp| \geq \eta\right) \leq 2e^{-2(\eta^2/k)}.$$

Proof. For $i = 1, \ldots, k$ let us denote by $\overline{X}_i$ the Bernoulli random variable with parameter $q = 1 - p$. That is, $\overline{X}_i = 1$ with probability q and $\overline{X}_i = 0$ with probability p. Then $\overline{S} = \overline{X}_1 + \ldots + \overline{X}_k$ is a binomial distribution satisfying that for $j \leq k$

$$\Pr\left(S = j\right) = \Pr\left(\overline{S} = k - j\right).$$

By Theorem 2 one has that for any $\eta > 0$,

$$\Pr\left(\overline{S} - kq \geq \eta\right) \leq e^{-2(\eta^2/k)}.$$

But the preceding probability equals

$$\Pr\left(kp - S \geq \eta\right)$$

and the corollary follows.   □

17.4   Approximating the Volume of Bounded Semi-Algebraic Sets

In this section we exhibit a probabilistic algorithm for approximating the volume of a semi-algebraic set E contained in a cube in $\mathbb{R}^n$. Indeed, computing such an approximation is the best we can do since the volume may be transcendental over the input data. Thus, we find ourselves in a setting close to that of Part II. We therefore compute an ε-approximation to the volume of E, where $\varepsilon > 0$ is given. Our complexity estimates depend on ε as well. Moreover, our main algorithm is probabilistic; that is, it will output a value that is a correct approximation (i.e., differing from the volume of E by at most ε) with high probability but not with total certainty.

We use this algorithm in the next section to show that there is no significant gain in allowing our random choices to come from the interval $[0, 1] \subset \mathbb{R}$ with the uniform distribution rather than be the discrete 0 or 1 random choices used so far.

The idea for computing the volume of E is to count the number of points that belong to E in a certain grid. We begin by considering a geometric invariant that is useful in estimating how fine a grid we must use.

Definition 4 Let $E \subset \mathbb{R}^n$. We define $\kappa(E)$ to be the maximum number of connected components (intervals) of $E \cap L$, where L is a line parallel to some axis.

Our first result gives a bound for $\kappa(E)$ for the case of a semi-algebraic set E.

Proposition 3 *Let $E \subset \mathbb{R}^n$ be a semi-algebraic set defined by a Boolean combination of s polynomial inequalities of degree, with respect to any variable, at most d. Then $\kappa(E) \leq sd + 1$.*

Proof. Let L be any axis-parallel line. The intersection $E \cap L$ is given by a Boolean combination of s polynomial inequalities in one variable of degree at most d. The roots of these s polynomials are bounded in number by sd and determine at most $sd + 1$ intervals in L. Each of these roots or intervals is contained in E or in its complement. The conclusion now follows in a straightforward way. □

Now we consider a set $E \subset [0, 1]^n$. We want to approximate its volume $v(E)$. To do so, we may consider a natural number $N \geq 1$, its inverse $h = 1/N$, and the grid $\mathcal{G}$ in $[0, 1]^n$ given by the points $(y_1, \ldots, y_n)$ with $y_i \in \{0, h, 2h, \ldots, (N-1)h\}$. We say that this grid has *fineness h*. It contains N^n points and we want to estimate which proportion $\mu_h(E)$ of them belong to E. More precisely, we want to bound the difference $|\mu_h(E) - v(E)|$. Let us deal first with the one-dimensional case.

Lemma 2 *Let $E \subseteq [0, 1]$ be a measurable set such that $\kappa(E)$ is finite. Then $|\mu_h(E) - v(E)| \leq h\kappa(E)$.*

Proof. Since any connected set in $\mathbb{R}$ is an interval, the set E can be written as the union of $\kappa(E)$ disjoint intervals $I_1, \ldots, I_{\kappa(E)}$. Thus, it is sufficient for us to estimate the difference $|\mu_h(I_j) - v(I_j)|$ for $j = 1, \ldots, \kappa(E)$ and use the inequality

$$|\mu_h(E) - v(E)| \leq \sum_{j=1}^{\kappa(E)} |\mu_h(I_j) - v(I_j)|.$$

Let I be one of the connected components of E and let k be the number of points in $\mathcal{G}$ contained in I. Then the volume $v(I)$ of I satisfies $(k-1)h \leq v(I) < (k+1)h$. On the other hand, the number $\mu_h(I)$ is exactly kh. We deduce that $|\mu_h(I) - v(I)| \leq h$ and consequently that $|\mu_h(E) - v(E)| \leq h\kappa(E)$. □

Theorem 3 *Let $E \subseteq [0, 1]^n$ be a measurable set such that $\kappa(E)$ is finite. Then $|\mu_h(E) - v(E)| \leq nh\kappa(E)$.*

Proof. By the preceding lemma, we know that the statement holds for $n = 1$. So let us suppose that $n \geq 2$. If we denote by f_E the characteristic function of E we have that $|\mu_h(E) - v(E)|$ equals

$$\left| h^n \sum_{i_1, \ldots, i_n = 0}^{N-1} f_E(hi_1, \ldots, hi_n) - \int_{[0,1]^n} f_E(x_1, \ldots, x_n) dx_1, \ldots, dx_n \right|$$

which is at most

$$\left| \int_{[0,1]^n} f_E(x_1, \ldots, x_n) dx_1, \ldots, dx_n \right.$$

$$- h \sum_{i=0}^{N-1} \int_{[0,1]^{n-1}} f_E(hi, x_2 \ldots, x_n) dx_2, \ldots, dx_n \left. \right|$$

$$+ \left| h \sum_{i=0}^{N-1} \int_{[0,1]^{n-1}} f_E(hi, x_2 \ldots, x_n) dx_2, \ldots, dx_n \right.$$

$$- h^n \sum_{i=0}^{N-1} \sum_{i_2,\ldots,i_n=0}^{N-1} f_E(hi, hi_2, \ldots, hi_n) \left. \right|.$$

We are going to bound the two terms of the preceding sum. The second one is bounded by

$$h \sum_{i=0}^{N-1} \left| \int_{[0,1]^{n-1}} f_E(hi, x_2, \ldots, x_n) dx_2, \ldots, dx_n \right.$$

$$- h^{n-1} \sum_{i_2,\ldots,i_n=0}^{N-1} f_E(hi, hi_2, \ldots, hi_n) \left. \right|. \qquad (17.3)$$

For any $i \leq N - 1$ consider the hyperplane H_i of those points in $\mathbb{R}^n$ whose first coordinate is hi and the intersection $F = E \cap H_i$. Then $F \subseteq [0, 1]^{n-1}$ and satisfies that $\kappa(F) \leq \kappa(E)$. By the induction hypothesis, one has that $|\mu_h(F) - v(F)| \leq (n-1)h\kappa(F)$ and this implies that (17.3) is bounded by

$$hN[(n-1)h\kappa(F)] \leq (n-1)h\kappa(E).$$

For the first term, inverting the order of the summation and the integration symbols, we have the bound

$$\sup_{(x_2,\ldots,x_n)\in[0,1]^{n-1}} \left| \int_0^1 f_E(x_1, \ldots, x_n) dx_1 - h \sum_{i=0}^{N-1} f_E(hi, x_2, \ldots, x_n) \right|.$$

But for any $(a_2, \ldots, a_n) \in [0, 1]^{n-1}$ the intersection G of E with the line given by $\{x_2 = a_2, \ldots, x_n = a_n\}$ satisfies that $\kappa(G) \leq \kappa(E)$. Thus, by Lemma 2, the value of

$$\left| \int_0^1 f_E(x_1, a_2, \ldots, a_n) dx_1 - h \sum_{i=0}^{N-1} f_E(hi, a_2, \ldots, a_n) \right|$$

is at most $h\kappa(E)$ from where we deduce that the value of the preceding supremum is bounded by $h\kappa(E)$.

Adding up both bounds we get $|\mu_h(E) - v(E)| \leq (n-1)h\kappa(E) + h\kappa(E) = nh\kappa(E)$. $\qquad \square$

Theorem 3 provides bounds for the proportion $\mu_h(E)$ of points in the grid $\mathcal{G}$ that belong to E, namely,

$$v(E) - nh\kappa(E) \leq \mu_h(E) \leq v(E) + nh\kappa(E)$$

and this for any measurable E with $\kappa(E)$ finite. If E is a semi-algebraic set, Proposition 3 allows us to replace $\kappa(E)$ by $ds + 1$ where s is the number of polynomials used to define E and d is a bound for their degree. Altogether, this suggests a deterministic algorithm for approximating $v(E)$. Consider the problem:

with input a Boolean combination ψ of s polynomial inequalities of degree at most d, defining a semi-algebraic set $E \subseteq [0, 1]^n$ and an $\varepsilon > 0$, compute v such that $|v(E) - v| \leq \varepsilon$.

Consider also the algorithm:

input (ψ, ε)
$N := \lceil \frac{n(ds+1)}{\varepsilon} \rceil$
$h := \frac{1}{N}$
$v := 0$
for $i_1, \ldots, i_n = 0$ **to** N **do**
    if $(hi_1, \ldots, hi_n) \in E$ **then** $v := v + 1$ **end if**
end do
output $\frac{v}{N^n}$.

Theorem 3 and Proposition 3 ensure that this algorithm solves the preceding problem. However, the number of iterations is N^n and thus we have an exponential dependence on the dimension n. We can reduce the number of iterations by choosing some points in $\mathcal{G}$ at random instead of trying them all. In this way it is possible to eliminate the exponential dependence on n and at the same time achieve a good dependence on n in the error probability δ of our algorithm.

More precisely let us consider inputs of the form $(\psi, \varepsilon, \delta)$, where ψ and ε are as previously and $\delta \in \mathbb{R}$ is positive. Again, we want to approximate, up to ε, the volume of E, but now we want to do so with probability at least $1 - \delta$. Consider the following algorithm.

input $(\psi, \varepsilon, \delta)$
$m := \lceil \log(\frac{2n(ds+1)}{\varepsilon}) \rceil$
$N := 2^m$
$h := \frac{1}{N}$
$k := \lceil \frac{2}{\varepsilon^2} \ln \left(\frac{2}{\delta} \right) \rceil$
$v := 0$
for $j = 1$ **to** k **do**
    randomly choose $(i_1, \ldots, i_n) \in N^n$
    if $(hi_1, \ldots, hi_n) \in E$ **then** $v := v + 1$ **end if**
end do
output $\frac{v}{k}$.

The number of iterations k is polynomial in $\log(1/\delta)$ and the exponential dependence on n is no longer present. However, the dependence on ε is still large in the sense that k is proportional to $1/\varepsilon^2$ and thus, exponential in $\log(1/\varepsilon)$. The correctness of this algorithm is shown in the next theorem.

Theorem 4 *The preceding algorithm outputs a value v that satisfies $|v-v(E)| \leq \varepsilon$ with probability at least $1 - \delta$. The number of times it checks for membership of a point in E is bounded by $\lceil (2/\varepsilon^2) \ln(2/\delta) \rceil$ and each of these points is obtained with at most $n \lceil \log(2n(ds + 1))/\varepsilon \rceil$ random bit choices.*

Proof.   By construction, h satisfies the inequality

$$h \leq \frac{\varepsilon}{2n(ds + 1)}$$

from which, according to Theorem 3 for the grid $\mathcal{G}$ in $[0, 1]^n$ of fineness h, the proportion $\mu_h(E)$ of points belonging to E satisfies

$$|\mu_h(E) - v(E)| \leq \frac{\varepsilon}{2}.$$

On the other hand, if we choose k random points in $\mathcal{G}$ and we let S_k denote the number of them belonging to E, according to Corollary 2,

$$\Pr\left(\left|\frac{S_k}{k} - \mu_h(E)\right| \geq \frac{\varepsilon}{2}\right) = \Pr\left(|S_k - k\mu_h(E)| \geq \frac{k\varepsilon}{2}\right)$$

$$\leq 2e^{-k\varepsilon^2/2}.$$

Therefore, to ensure that this probability is at most δ, it is enough to take k satisfying

$$k \geq \frac{2}{\varepsilon^2} \ln\left(\frac{2}{\delta}\right).$$

Putting together both bounds we finish the proof. $\square$

17.5   Machines That Pick Real Numbers in [0, 1]

All the probabilistic algorithms we have seen thus far take their random elements from $\{0, 1\}$ (or uniformly from $\{0, 1, \ldots, 2^m - 1\}$ by making m choices from $\{0, 1\}$). Thus our model of probabilistic machines (as introduced in Definition 1) appears both natural and adequate. We might ask, however, if our computational power is limited by restricting our random choices to be "coin-tossing" outcomes. It could be that by making random choices according to some other measure μ, we could significantly improve the efficiency of some algorithms. In this section we show that this is not the case for μ equal to the uniform measure on [0, 1].

We first define probabilistic machines for arbitrary distributions. Since we can no longer represent random choices by a choice between a finite number of next nodes, we use a definition of acceptance similar to the one we used to define nondeterminism. Moreover, we restrict ourselves to algebraic trees.

Definition 5 Let T be an algebraic decision tree with input space $\mathbb{R}^{n+m}$, μ a probability measure on $\mathbb{R}^m$, and S a subset of $\mathbb{R}^n$. The *error probability* of T with respect to the set S is the function $e_M : R^n \to \mathbb{R}$ given by

$$e_M(x) = \Pr(M \text{ accepts } (x, y) \& x \notin S) + \Pr(M \text{ rejects } (x, y) \& x \in S),$$

where the coordinates of the element $y \in \mathbb{R}^m$ are randomly chosen according to the measure μ. We say that T *decides* S when $e_M(x) \leq \frac{1}{4}$ for all $x \in \mathbb{R}^n$ and that T is a *probabilistic algebraic decision tree with measure μ.*

Remark 2

(i) Note that the probabilistic algebraic decision trees defined in Section 17.1 coincide with the probabilistic algebraic decision trees with μ the Bernoulli measure with parameter $\frac{1}{2}$.

(ii) With the appropriate changes, Lemma 1 holds for probabilistic algebraic decision trees with measure μ.

Theorem 5 *Let T be a probabilistic algebraic decision tree with the uniform measure in $[0, 1]$ having depth t and deciding a set $S \subseteq \mathbb{R}^n$ with random choices from $\mathbb{R}^m$ and error probability at most ε. Denote by s the smallest integer satisfying*

$$s \geq 4 \log \left(\frac{1}{\varepsilon} \right) \log \left(\frac{1}{4(\varepsilon - \varepsilon^2)} \right).$$

Then there is a probabilistic algebraic decision tree T^* with random choices from $\{0, 1\}$ and depth $c(m(t + \log m))$ that decides S with error probability $2\varepsilon^2$ where the constant c depends only on ε.

Proof.   Let T_s be the tree that performs s runs of T and accepts if at least $s/2$ of these runs leads to acceptance in T. The tree T_s also decides S and by Lemma 1 it does so with error probability

$$[4(\varepsilon - \varepsilon^2)]^{s/2} \leq \varepsilon^2,$$

the inequality being a consequence of our choice of s.
     For any $x \in \mathbb{R}^n$ we consider the function

$$f_x : [0, 1]^{sm} \longrightarrow \{0, 1\}$$

that associates with each $y \in [0, 1]^{sm}$ the outcome of T_s with input (x, y). We also consider the set $E_x \subseteq [0, 1]^{sm}$ defined by $E_x = \{y \in \mathbb{R}^{sm} \mid f_x(y) = 1\}$. Then we have that $x \in S$ if and only if $v(E_x) \geq 1 - \varepsilon^2$.
     Now E_x is defined by a Boolean combination of at most 2st polynomial inequalities of degree bounded by 2^t. Therefore $\kappa(E_x) \leq 2^t 2^{st} + 1$ and by Theorem 3, for any h satisfying

$$h \leq \frac{1}{\varepsilon^2 (sm)(2^{t+st} + 1)}$$

we have for the grid of fineness h that $|\mu_h(E_x) - v(E_x)| \leq 1/\varepsilon^2$. We deduce that for such h we have that

$$\mu_h(E_x) \begin{cases} \geq 1 - 2\varepsilon^2 & \text{if } x \in S \\ \leq \varepsilon^2 & \text{if } x \notin S. \end{cases}$$

We conclude that for $r = \lceil \log\left(\varepsilon^2 (sm)(2^{t+st} + 1)\right)\rceil$, $N = 2^r$, and $h = 1/N$ the probabilistic tree T^* given by

    input $(x_1, \ldots, x_n)$
    randomly choose $(i_1, \ldots, i_{sm}) \in \{0, \ldots, N-1\}$
    run T_s with input $(x_1, \ldots, x_n, hi_1, \ldots, hi_{sm})$
    and ACCEPT iff T_s does,

decides S with error probability at most $2\varepsilon^2$. Note that to generate the random point $(i_1, \ldots, i_{sm})$ one needs smr random choices in $\{0, 1\}$. Since $r \leq c(t + \log m)$ for a constant c depending only on ε we deduce the bound on the depth of T^*. Finally, notice that since $\varepsilon \leq \frac{1}{4}$ one has that $2\varepsilon^2 < \varepsilon \leq \frac{1}{4}$. $\square$

Theorem 5 deals with nonuniform machines, namely, with algebraic trees. It is straightforward to extend Definition 5 to uniform machines and to define the class $\text{BPP}_\mathbb{R}^\mu$ for a fixed probability measure μ on $\mathbb{R}$. The following result states that there is no gain of power by allowing random choices in $[0, 1]$ instead of in $\{0, 1\}$. It is an easy consequence of Theorem 5.

Theorem 6 *Denote by U the uniform measure in $[0, 1]$. Then the equality* $\text{BPP}_\mathbb{R} = \text{BPP}_\mathbb{R}^U$ *holds.* $\square$

17.6   Simulating Probabilistic Machines

In this section we extend Proposition 2, that is, the simulation of probabilistic algebraic trees by deterministic algebraic trees, to uniform machines. This result has a nonconstructible aspect.

Theorem 7 *The equality* $\text{BPP}_\mathbb{R} = \text{P}_\mathbb{R}$ *holds.*

Proof.   We only need to show the inclusion $\text{BPP}_\mathbb{R} \subseteq \text{P}_\mathbb{R}$. To do so, let us consider a set $S \in \text{BPP}_\mathbb{R}$ together with a probabilistic machine M that decides S in time bounded by a polynomial function p. For any $n \geq 1$ the computation of M restricted to inputs of size n is described by a probabilistic algebraic decision tree T_n of depth bounded by $p(n)$. According to the proof of Proposition 2 we know that there exist vectors $b_1^n, \ldots, b_{s(n)}^n \in \mathbb{Z}_2^{p(n)}$, where $s(n) = O(np(n))$, such that the deterministic tree resulting from s runs of T_n using b_i^n instead of random elements in the ith run and then taking a majority vote simulates T_n.

We now consider a real number $\alpha \in [0, 1)$ whose binary expansion codes the concatenation of $b_1^n, \ldots, b_{s(n)}^n$ for all $n \geq 1$. That is,

$$\alpha = 0.b_1^1 \ldots b_{s(1)}^1 b_1^2 \ldots b_{s(2)}^2 \ldots b_1^n \ldots b_{s(n)}^n \ldots.$$

Note that the kth bit in the preceding expansion can be obtained from α by the procedure

> **for** $i = 1$ **to** $k - 1$ **do**
> $\quad \alpha := 2\alpha$
> $\quad$ **if** $\alpha \geq 1$ **then** $\alpha := \alpha - 1$ **end if**
> **end do**
> $\alpha := 2\alpha$
> **if** $\alpha \geq 1$ **then output** 1
> **else output** 0
> **end if** .

Let M^* be a machine that, given an input of size n, first obtains $b_1^n, \ldots, b_{s(n)}^n$ from α and then does $s(n)$ runs of M using the b_i^n as a deterministic sample and takes a majority vote. The machine M^* decides S and its running time is still polynomially bounded.   □

Corollary 3 $P_\mathbb{R} = BPP_\mathbb{R}^U$.   □

Remark 3 Theorem 7 (and with it Corollary 3) is a bit surprising and warrants some discussion. At first glance, the existence of a polynomial time algorithm for SLP0 may not seem natural. In fact, if we try to exhibit such an algorithm we will very likely fail and, indeed, no such algorithm has been exhibited so far. This may seem to contradict Theorem 7. It does not, the reason being that Theorem 7 is not constructive. The machine M^* shown to exist contains a built-in constant α whose value we do not know.

These considerations suggest that it may be more natural to state the problem of whether randomization helps in a different way. Instead of asking whether the equality $P_\mathbb{R} = BPP_\mathbb{R}$ holds, we may ask whether any probabilistic machine can be simulated by a deterministic machine having the same machine constants and with only a polynomial slowdown. This is currently an open problem.

17.7 Additional Comments and Bibliographical Remarks

Probabilistic methods are used widely in numerical analysis (see, for instance, [Hammersley and Handscomb 1964]). In computer science interest in randomization has its roots in the paper by Rabin [1976] which provided a one-sided error probabilistic algorithm for primality testing. Independently, Solovay and Strassen [1977] (and the correction, [Solovay and Strassen 1978]) found a similar probabilistic algorithm for the same problem.

The first simulation of probabilistic machines by deterministic nonuniform ones is due to Adleman [1978]. This was done in the classical case and only for one-sided error machines. The technique used in Proposition 1 comes from Bennett and Gill [1981]. The extension to the real case shown in Proposition 2 comes from Meyer auf der Heide [1985b].

Theorem 1 exhibits one of the several inequalities, collectively known as Chernoff bounds, originating in Chernoff's paper [1952]. They provide tight bounds for the tail of the binomial distribution. Our presentation follows [Hagerup and Rüb 1990]. The bound in Theorem 2, although referred to sometimes as a Chernoff bound, is a particular case of a result of Hoeffding [1963].

The definition of $\kappa(E)$ as well as its bound in Proposition 3 and Theorem 3 are due to Koiran [1995]. Theorem 4 is also suggested there. The application of Theorem 3 to the simulations in Theorems 5 and 6 comes from a discussion with P. Koiran.

The equality of $BPP_\mathbb{R}$ and $P_\mathbb{R}$ was shown by Cucker, Montaña, and Pardo [1995]. In this paper, an extension to Proposition 2 is carried out for probabilistic machines that draw random elements from $\mathbb{R}$ using very general distributions.

Straight-line programs are among the oldest algebraic models of computation. They were first defined by Ostrowski [1954] and have been used ever since. A simpler version of the straight-line program utilizing only additions was defined by Scholz in [1937]. The probabilistic algorithm for SLP0 given in Section 17.1 is due to Schwartz [1980].

18

Parallel Computations

If time is of the essence in solving a problem, we may be willing to devote the resources of many computers to its resolution. It is a challenge to formulate an appropriate mathematical notion of parallel machine over the reals.

For our first formal model of parallel computation, defined in Section 18.2, we imagine a large number of identical machines, called *processors*. Each processor performs its own computations but each can also read from the registers (or memory) of any other. Thus if there are N processors there will be $N(N - 1)$ connections between them. We begin this chapter with some examples to motivate the formal model.

An alternative model of parallel machines, widely used in algebraic complexity, considers families of algebraic circuits. We describe this model in Sections 18.4 and 18.5 where we also describe the relationship between the complexity classes resulting from these two models. The nature of this relationship should help to better understand the notion of parallelism, and its robustness.

18.1 Some Old Problems Revisited

In this section we describe parallel algorithms for some computational problems in an informal way. Our goal is to motivate the formal definition of parallel machines much as the seven examples of Section 1.2 were meant to motivate the formal definitions of machines in Chapters 2 and 3.

Example 1   Adding n Numbers

Here we exhibit a parallel procedure to add n numbers which uses at most $2n$ processors and $O(\log n)$ steps.

For $n \in \mathbb{Z}^+$, let $k = \lceil \log n \rceil$, so that 2^k is the first power of 2 greater than or equal to n. We use 2^k processors M_i for $i = 0, \ldots, 2^k - 1$. The procedure $\mathcal{P}$ for adding n numbers can then be described in the following way.

(1) Let $x_0, \ldots, x_{n-1}$ be the inputs to the problem and let $x_n, \ldots, x_{2^k-1}$ all be zero if n is not a power of 2.

(2) For $i = 0, \ldots, 2^k - 1$, input (i, x_i) into processor M_i as follows.

$$(y_{-1}, y_0) \leftarrow (i, x_i).$$

(3) Let $\ell = 0$.

(4) If $\ell = k$, output the content of the zeroth register of M_0 and HALT.

(5) For $i = 0, \ldots, 2^k - 1$, if $y_{-1} = (2j + 1)2^\ell$ with $j = 0, \ldots, 2^{k-\ell-1} - 1$, then processor M_i halts. Else, if $y_{-1} = (2j)2^\ell$ with $j = 0, \ldots, 2^{k-\ell-1} - 1$, then processor M_i

(a) reads the content of the zeroth register of processor M_{i+2^ℓ} and puts it in its first register;

(b)

$$y_0 \leftarrow y_0 + y_1.$$

(6) Increase ℓ by 1 and go to (4).

Here when we say that a processor *halts* we mean that its state space will not change anymore. However, the contents of this state space will still be accessible to other processors. On the other hand, when we write HALT we mean that the whole parallel procedure terminates.

We consider Steps (1) and (2) as one input instruction accomplished in unit time. This procedure then uses k loops of (4), (5), and (6) to output $\sum_{i=0}^{n-1} x_i$.

For $n = 16$ the sequence of additions performed by $\mathcal{P}$ is shown in Figure A.

If we write (i, j) for "processor i reads the contents of the first register of processor j," then we can similarly depict the sequence of communication instructions where each level represents another loop of the procedure $\mathcal{P}$.

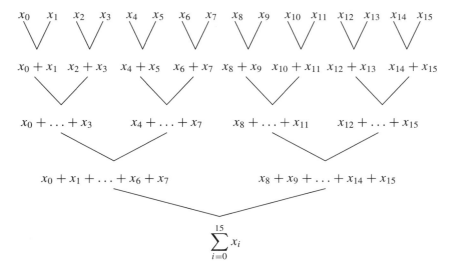

Figure A Sequence of operations in the addition of 16 numbers.

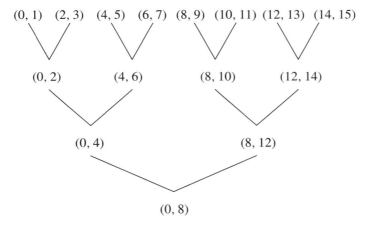

Figure B Sequence of communications in the addition of 16 numbers.

The computations done by all processors in (3) through (6) are clearly identical. Moreover, the read and output instructions are also identical. In the ℓth loop of the procedure processor M_i, for $i = 0, \ldots, 2^k - 1$, checks if y_{-1} is of the form $(2j)2^\ell$ or $(2j + 1)2^\ell$ for some integer j, $0 \le j \le 2^{k-\ell-1} - 1$. In the latter case processor M_i halts. In the former it reads the zeroth register in M_{i+2^ℓ} and adds the result to y_0 in M_i. Finally, at the end of the kth loop, processor M_0 outputs the contents of its zeroth register.

In the following, we refine the procedure $\mathcal{P}$ to carry out the checking in (5). Care need be taken to avoid introducing another loop of $\lceil \log n \rceil$ steps, thereby making the total number of steps of the order of $\log^2 n$.

First we need a lemma. For nonnegative integers n and k we say that $b_0, \ldots, b_{k-1}$ are the first k bits in the binary expansion of n, if $b_0, \ldots, b_q$, where $q = \max(k - 1, \lfloor \log n \rfloor)$, are all 0 or 1 and $n = \sum_{i=0}^{q} b_i 2^i$. Notice that n is of the form $(2j)2^\ell$ for some j if and only if $b_0, \ldots, b_\ell = 0$.

Lemma 1 *There is a machine M with input space $\mathbb{R}^2$ which for input $(k, y) \in \mathbb{N} \times \mathbb{N}$, $k \neq 0$, outputs $(b_{k-1}, \ldots, b_0)$ where $b_0, \ldots, b_{k-1}$ are the first k bits in the binary expansion of y in time $O(\ell)$, where $\ell = \max(k - 1, \lceil \log y \rceil)$.*

Proof.   We can assume that $y > 0$. In this case, we use the following algorithm, which computes the binary expansion of y and modify its output depending on how k compares with $\lceil \log y \rceil$.

$$
\begin{aligned}
&s := \lfloor \log y \rfloor + 1 \\
&r := 2^s \\
&\textbf{while } s > 0 \textbf{ do} \\
&\quad s := s - 1 \\
&\quad r := r/2 \\
&\quad \textbf{if } y \geq r \textbf{ then } b_s := 1 \\
&\quad\quad y := y - r \\
&\quad \textbf{else } b_s := 0 \\
&\quad \textbf{end if} \\
&\textbf{end do} \; .
\end{aligned}
$$

$\square$

Now we refine procedure $\mathcal{P}$. We drop the M_i in the instructions after Step (3) to emphasize that all the M_i are the same, only their inputs differ. Also, the registers of M_i are grouped in two classes to emphasize the fact that some computations are carried out over the integers and some over the reals. We denote registers in these two groups by z_i and y_i, respectively.

(1) Let $x_0, \ldots, x_{n-1}$ be the inputs to the problem and let $x_n, \ldots, x_{2^k-1}$ all be zero if n is not a power of 2.

(2) Input (i, k, x_i) into processor M_i as follows

$$(y_{-2}, y_{-1}, y_0) \leftarrow (i, k, x_i)$$
$$z_0 \leftarrow i.$$

(3) Use machine N in Lemma 1 to compute

$$(\ldots, y_{-1}, y_0 . y_1, \ldots) \leftarrow (\ldots, 0, b_{k-1}, \ldots, b_0, i, k, x_i . 0, \ldots),$$

where the b_i, $i = 0, \ldots, k - 1$ are the first k bits of i.

(4) $z_1 \leftarrow z_0 + 1$.

(5) If $y_{-1} = 0$, output the content of the zeroth register of M_0 and HALT.

(6) If $y_{-3} = 1$, halt. Else

    (a) read y_0 of processor M_{z_1} and let y_1 take this value;

    (b) $y_0 \leftarrow y_0 + y_1$;

    (c)

$$(\ldots, y_{-3}, y_{-2}, y_{-1}, y_0. y_1, \ldots) \leftarrow (\ldots, y_{-4}, y_{-2}, y_{-1}, y_0. y_1, \ldots).$$

(7) $z_1 \leftarrow 2z_1 - z_0$

    $y_{-1} \leftarrow y_{-1} - 1$.

(8) Go to (5).

Now if the first nonzero bit of i is the ℓth bit, processor M_i halts in the $(\ell + 1)$st loop of (5) through (8), and i is of the form $(2j + 1)2^{\ell+1}$. If processor M_i is still operating in the $(\ell + 1)$st loop of (5) through (8) and the ℓth bit of i is 0, then $z_2 = i + 2^{\ell+1}$ in Step (6) so the read and halt instructions in Step (6) here are the same as in (6) in $\mathcal{P}$. Instead of increasing ℓ by 1 and checking if $\ell = k$, we input k, decrease it by 1 in Step (7) and check for 0. Step (6c) is accomplished by a shift to the right followed by the computation

$$(y_{-4}, y_{-3}, y_{-2}, y_{-1}, y_0. y_1, y_2) \leftarrow (y_{-4}, y_{-2}, y_{-1}, y_0, y_1. y_2, 0).$$

Notice finally that, if we consider the communication step in (6a) as a single step, the number of steps performed at each iteration of the loop is constant.

Example 2   Computation of Dot Products.

We can now compute in parallel the dot product of vectors $(v_0, \ldots, v_{n-1})$ and $(w_0, \ldots, w_{n-1})$ in $\mathbb{R}^n$. As in the preceding example we may assume that $n = 2^k$. Then we use $2n$ processors M_i, $i = 0, \ldots, 2n - 1$ as follows.

(1) For $i = 0, \ldots, n - 1$ input v_i to processor M_i as follows,

$$(y_{-1}, y_0) \leftarrow (i, v_i)$$
$$(z_0, z_1) \leftarrow (i, n),$$

and input w_i to processor M_{n+i} as follows,

$$(y_{-1}, y_0) \leftarrow (n + i, w_i)$$
$$(z_0, z_1) \leftarrow (n + i, n).$$

(2) If $z_0 \geq z_1$ halt, else

    (a) $z_2 \leftarrow z_0 + z_1$;

    (b) read y_1 of processor M_{z_2} and let y_1 take this value.

(3) $y_0 \leftarrow y_0 y_1$.

(4) Proceed as in $\mathcal{P}$ to add registers y_0 of processors M_i, $i = 0, \ldots, n - 1$.

Again, we assume (1) as a single input instruction accomplished in unit time. Then, since (2) and (3) are carried out in a constant number of steps, the whole procedure is dominated by (4) and takes $O(\log n)$ steps.

Example 3   Product of Matrices.

We can also compute in parallel the product C of an $n \times m$ real matrix A with an $m \times q$ real matrix B. We may assume again that n, m, and q are powers of 2. Then we use $2nmq$ processors $M_{\ell jk}$ for $\ell = 0, \ldots, n - 1$, $j = 0, \ldots, q - 1$ and $k = 0, \ldots, 2m - 1$ in a procedure roughly described as follows.

For $\ell = 0, \ldots, n - 1$, and $k = 0, \ldots, m - 1$, we let $a_{\ell,k}$ be the entry of A in the $(\ell + 1)$st row and $(k + 1)$st column. For $j = 0, \ldots, q - 1$ and $k = 0, \ldots, m - 1$, let $b_{k,j}$ be the entry of B in the $(k + 1)$st row and $(j + 1)$st column.

(1) Input $a_{\ell k}$ to processor $M_{\ell 0k}$ for $\ell = 0, \ldots, (n - 1)$ and $k = 0, \ldots, (m - 1)$.

Input b_{kj} to processor M_{0jk+m} for $j = 0, \ldots, (q - 1)$ and $k = 0, \ldots, (m - 1)$.

(2)

(a) For $\ell = 0, \ldots, n - 1$, $j = 1, \ldots, q - 1$, and $k = 0, \ldots, m - 1$, processor $M_{\ell jk}$ copies $a_{\ell,k}$ from processor $M_{\ell 0k}$.

(b) For $\ell = 1, \ldots, n - 1$, $j = 0, \ldots, q - 1$, and $k = m, \ldots, 2m - 1$, processor $M_{\ell jk}$ copies $b_{j,k-m}$ from processor M_{0jk}.

(3) The set of processors $M_{\ell jk}$ for fixed (ℓ, j) with $0 \leq \ell < n$, $0 \leq j < q$, and $k = 0, \ldots, 2m - 1$ compute the dot product of the $(\ell + 1)$st row of A and the $(j + 1)$st column of B as in Example 2. The output of this computation, appearing in $M_{\ell j0}$, is the $(\ell + 1, j + 1)$st entry of C.

One may wish to have processors in the foregoing procedure indexed by a single index i running from 0 to $2nqm - 1$ and to have $M_{\ell 0k}$ for $\ell = 0, \ldots, n - 1$ and $k = 0, \ldots, m - 1$ and M_{0jk} for $j = 0, \ldots, q - 1$ and $k = m, \ldots, 2m - 1$ be the first $(n + q)m$.

To do this we introduce a pairing function $\psi_{r,s}$ which for natural numbers r, s bijectively maps the integer lattice points in the rectangle $[0, r - 1] \times [0, s - 1]$ to the nonnegative integers $0, \ldots, rs - 1$,

$$\psi_{r,s}(n_1, n_2) = n_1 s + n_2,$$

where $n_1 \in \{0, 1, \ldots, r - 1\}$ and $n_2 \in \{0, 1, \ldots, s - 1\}$. Moreover, the natural ordering in $\{0, 1, \ldots, rs - 1\}$ corresponds under $\psi_{r,s}$ with the lexicographic ordering in the lattice. An immediate consequence of the following lemma (whose proof is straightforward) is that the inverse of $\psi_{r,s}$ is computable in time $O(\log(rs))$.

Lemma 2 *Integer division may be implemented as a machine with no divisions and with integer constants that, given $(n, d) \in \mathbb{N}^2 \subset \mathbb{R}^2$, $d > 0$, computes q and r, the integer quotient, and remainder of the division of n by d, in time $O(\log(1+n))$.*

$\square$

Now we define for $\ell \in \{0, \ldots, n-1\}$, $j \in \{0, \ldots, q-1\}$, and $k \in \{0, \ldots, 2m-1\}$,

$$\phi(\ell, j, k) = \psi_{n,2qm}(\ell, \psi_{q,2m}(j, k))$$
$$= 2\ell qm + 2jm + k,$$

which bijectively maps the integer lattice points in the cube $[0, n-1] \times [0, q-1] \times [0, 2m-1]$ to the nonnegative integers $\{0, \ldots, 2nqm-1\}$. The inverse of ϕ is computable in time $O(\log(nqm))$ since the inverses of $\psi_{n,2qm}$ and $\psi_{q,2m}$ are.

Finally, given the set of indices

$$S = \{(\ell, 0, k) \mid \ell = 0, \ldots, n-1, k = 0, \ldots, m-1\}$$
$$\cup \{(0, j, k) \mid j = 0, \ldots, q-1, k = m, \ldots, 2m-1\},$$

we map S bijectively to $\{0, \ldots, (n+m)q - 1\}$ by

$$\phi'(a, b, c) = \begin{cases} \psi_{n,m}(a, c) & \text{if } (a, b, c) \in S \text{ and } c \leq m-1 \\ \psi_{q,m}(b, c) & \text{if } (a, b, c) \in S \text{ and } c \geq m. \end{cases}$$

Once again the inverse of ϕ' is computable in time $O(\log(nqm))$.

We now modify and expand (1) to use a single index for the processors and input the entries of the matrices A and B in the $(n+q)m$ first processors. The parallel procedure for matrix multiplication is then the following.

(1) For $\ell = 0, \ldots, n-1$ and $k = 0, \ldots, m-1$, input $a_{\ell k}$ to processor $M_{\phi'(\ell,0,k)}$ as follows.

$$(y_{-2}, y_{-1}, y_0) \leftarrow (\phi'(\ell, 0, k), 2nmq, a_{\ell k})$$
$$(z_0, z_1) \leftarrow (\phi'(\ell, 0, k), 2nmq).$$

For $k = 0, \ldots, m-1$ and $j = 0, \ldots, q-1$, input b_{kj} to processor $M_{\phi'(0,j,k)}$ as follows.

$$(y_{-2}, y_{-1}, y_0) \leftarrow (\phi'(0, j, k), 2nmq, b_{kj})$$
$$(z_0, z_1) \leftarrow (\phi'(0, j, k), 2nmq).$$

Otherwise, input 0 to processor M_i for $i \geq m(n+q)$ as follows.

$$(y_{-1}, y_0, y_1) \leftarrow (i, 2nmq, 0)$$
$$(z_0, z_1) \leftarrow (i, 2nmq).$$

(2) Compute $(\ell, j, k) = \phi^{-1}(i)$.

    (a) If $k \leq m - 1$, then processor M_i copies $a_{\ell,k}$ from processor $M_{\phi'(\ell,0,k)}$.

    (b) If $k \geq m$, then processor M_i copies $b_{k,j}$ from processor $M_{\phi'(0,j,k)}$.

(3) The set of processors $M_{\phi(\ell,j,k)}$ for fixed (ℓ, j) and $k = 0, \ldots, 2m - 1$ compute the dot product of the $(\ell + 1)$st row of A and the $(j + 1)$st column of B as in Example 2. The output of this computation is the $(i + 1, j + 1)$st entry of C.

As in the two previous examples, the total number of computational steps here is $O(\log n)$.

Example 4   Characteristic Polynomials and Matrix Inversion.

Here we just incorporate the parallel procedures given in the previous examples into the algorithm of Section 15.5. Note that Parts (1), (3), and and (5) in this algorithm require the computation of matrix powers of the form A^k with A an $n \times n$ matrix and $k \leq n$. These computations use $O(\log k)$ steps if we consider matrix multiplication as a single step. Computing these multiplications as in Example 3 yields a bound of $O(\log^2 n)$ for the total time.

18.2   A Parallel Model of Computation

We now define our parallel model of computation. We do so only for $R = \mathbb{R}$.

Definition 1   A *processor* M is a finite connected, directed graph containing nodes of the following eight types: input, output, integral branch, real branch, communication, computation, shift, and halt. A processor also has a state space $\mathcal{S}_M = \mathbb{Z}^k \times \mathbb{R}_\infty$ for some $k \in \mathbb{N}$, and a linear projection $\pi_M : \mathbb{Z}^k \to \mathbb{Z}$. The two components of $\mathcal{S}_M$ are denoted by $\mathcal{S}_M^{\mathbb{Z}}$ and $\mathcal{S}_M^{\mathbb{R}}$, respectively.

The nodes of a processor have associated maps and next nodes as described in the following. The input and output nodes are unique. The input node has no incoming edges and only one outgoing edge. The output and halt nodes have no outgoing edge. All other nodes have (possibly several) incoming edges and one outgoing edge except the branch nodes, which have two outgoing edges.

(1) Associated with a computation node η is a map $(z, y) \to (g_{\mathbb{Z},\eta}(z), g_{\mathbb{R},\eta}(y))$, where $g_{\mathbb{Z},\eta}$ is an integer polynomial and $g_{\mathbb{R},\eta}$ is a rational map as in the case of uniform machines.

(2) An integral branch node η (respectively, a real branch node) has an associated branching function which is a polynomial function $h_\eta : \mathbb{Z}^k \to \mathbb{Z}$ (respectively, $h_\eta : \mathbb{R}_\infty \to \mathbb{R}$ is a rational function). As in the case of uniform machines, the next nodes β_η^- and β_η^+ are associated with $h_\eta(z) < 0$ and $h_\eta(y) \geq 0$.

(3) A shift node η has an associated map $g_\eta : S_M^{\mathbb{R}} \to S_M^{\mathbb{R}}$, $g_\eta \in \{\sigma_l, \sigma_r\}$.

(4) A halt node η does not alter the state (z, y) and has no next node.

In order to describe the computation at an input, output, or communication node, we need to consider collections of processors.

A *uniform parallel machine* P is a sequence of identical processors $\{M_i\}, i \in \mathbb{N}$, together with an associated *activation function* $p : \mathbb{N} \to \mathbb{N}$, input space $\mathcal{I}_P = \mathbb{R}^\infty$, output space $\mathcal{O}_P = \mathbb{R}^\infty$, and state space

$$S_P = \bigsqcup_{h=1}^{\infty} (S_M)^h,$$

where each $M_i = M$ and $\sqcup$ is the disjoint union.

(5) Associated with the input node is a map

$$I_P : \mathcal{I}_P \to S_P$$

defined as follows. If $x \in \mathbb{R}^n \subset \mathcal{I}_P, x = (x_0, \ldots, x_{n-1})$, then $I_P(x) \in S_M^{p(n)}$. For $i = 0, \ldots, p(n) - 1$, the projection of $I_P(x)$ in S_{M_i} is (z, y), where $z = (i, n, 0, \ldots, 0) \in S_{M_i}^{\mathbb{Z}}$ and $y \in S_{M_i}^{\mathbb{R}}$ is given by

$$y = \begin{cases} (\ldots, 0, i, n, x_i.0, \ldots) & \text{if } i < n \\ (\ldots, 0, i, n, 0.0, \ldots) & \text{if } i \geq n. \end{cases}$$

(6) The output node has an associated function

$$O_P : S_P \to \mathcal{O}_P$$

defined as follows. If $x \in (S_M)^p \subset S_P$, then $O_P(x) = (w_0, \ldots, w_{s-1}) \in \mathbb{R}^s$. Here s is the content of register z_0 of M_0 and w_i is the content of register y_0 of processor M_i for $0 \leq i < s$. If $s \leq 0$, the function O_P is not defined.

(7) Given a state (z_i, y_i) in S_{M_i} for $0 \leq i < p(n)$, a communication node η replaces the zeroth coordinate of y_i with the zeroth coordinate of y_ℓ where $\pi_M(z_i) = \ell$ if $\ell < p(n)$. If $\ell \geq p(n)$, the computation is undefined. We say that M_i *reads* the contents of the zeroth register of processor M_ℓ.

Remark 1

(1) Notice that the activation function p in the preceding definition indeed "activates" the first $p(n)$ processors on an input of size n. In particular, computation at a communication node is only defined if the accessed processor M_ℓ is active. This situation is similar to the use of division in uniform machines. Here we can also suppose that a test is performed before each communication node to ensure that the computation at these nodes is always defined. To perform this test we need to have the value $p(n)$ stored in

a register in $\mathcal{S}_{M_i}^{\mathbb{Z}}$. This can be done by computing $p(n)$ at the beginning of the computation. Thus, although the cost of the comparison $\ell < p(n)$ can be assumed to be 1, we are adding the cost of computing $p(n)$ to the total cost of the computation.

(2) We generally omit the adjective "uniform" when we refer to uniform parallel machines.

(3) The integer part of the state space has a dual function. It simultaneously avoids noninteger addresses in the communication nodes and enables the computation of these addresses in an efficient way.

(4) A parallel machine P determines an input–output map $\varphi_P : \mathcal{I}_P \to \mathcal{O}_P$ in a fairly straightforward manner. Intuitively, an input $x \in \mathbb{R}^n \subseteq \mathcal{I}_P$ is plugged into the state space $(\mathcal{S}_M)^{p(n)} \subseteq \mathcal{S}_P$ by I_P and the processors M_i are activated for $i = 1, \ldots, p(n)$. These processors then follow the "program" given by their common flowchart. That is, at each computational step, processor M_i changes the contents of its state space according to the node instruction it is executing and then updates the current node. Thus, although all processors execute the same program, they do not necessarily execute the same instruction at a given moment. In particular, at a certain moment in the computation, processor M_i can modify y_0 while, simultaneously, processor M_j reads this value. In this case, we assume that the value read by M_j is the value of y_0 before being modified by M_i.

When the output node is reached by some processor, the whole procedure stops and the output is given by the image of O_P.

(5) All parallel machines considered here are *synchronous*. This means there is a global clock determining a time unit and at each such unit all processors do exactly one computational step. Equivalently, if p processors are active and

$$H_i : \mathcal{N} \times \mathcal{S}_P \to \mathcal{N} \times \mathcal{S}_{M_i}$$

is the "computing endomorphism" of processor M_i (one should say here transition function H_i being no longer an endomorphism), then the computing endomorphism of P

$$H : \prod_{i=0}^{p(n)-1} (\mathcal{N} \times \mathcal{S}_{M_i}) \to \prod_{i=0}^{p(n)-1} (\mathcal{N} \times \mathcal{S}_{M_i})$$

satisfies

$$H(\ldots, (\eta_i, x_i), \ldots) = (\ldots, H_i(\eta_i, (x_0, \ldots, x_{p(n)-1})), \ldots).$$

(6) Occasionally we use the adjective *sequential* to distinguish machines as defined in Chapters 2 and 3 from parallel machines.

(7) Parallel machines can be defined over other rings or fields too. We do not develop this direction here. In Section 18.6, references are given for the classical case.

One can now check that the parallel procedures in the preceding section give rise to parallel machines as previously defined. Some details need to be filled in to completely adjust to the definition. Mostly these are simple problems of reindexing. In Example 1, to add n numbers, we also have to adjust the input.

If we consider $x = (x_1, \ldots, x_n)$ as an element in $\mathcal{I}_P$, then the value $k = \lceil \log n \rceil$ is not part of the input according to Definition 1 and thus, we cannot suppose this value is stored in register y_{-1} as we did in Example 1. However, k can be computed from n in time $O(\log n)$ and stored in this register. Also, the assumption that n is a power of 2 needs to be justified more carefully. To adjust procedure $\mathcal{P}$ to comply with the preceding definition, we take $p(n) = 2n$ as the activation function and replace (1) and (2) by the following instructions which compute $\lceil n \rceil$ in y_1 and $2^{\lceil n \rceil}$ in y_2.

(1) Input $(x_1, \ldots, x_n)$ as in Definition 1.

(2a) $y_1 \leftarrow 0$

$\quad y_2 \leftarrow 1$

$\quad$ **while** $y_2 < y_{-1}$ **do**

$\quad\quad y_2 \leftarrow 2y_2$

$\quad\quad y_1 \leftarrow y_1 + 1$

$\quad$ **end do**

(2b) If $y_{-2} \geq y_2$ then halt.

Note that the computation in (2a) takes time $O(\log n)$.

Consider now the procedure for multiplying matrices given in Example 3. If we compare it with Definition 1, we note that the values ℓ, j, and k are obtained as $\varphi^{-1}(i)$ and stored in $\mathcal{S}_{M_i}^{\mathbb{R}}$, where i is the processor index. On the other hand, the comunication instructions in (3) of this procedure make use of the content of some register in $\mathcal{S}_{M_i}^{\mathbb{Z}}$ that will depend on ℓ, j, and k. To ensure a constant cost for these communication instructions, one possibility is to copy these three integers to three registers in $\mathcal{S}_{M_i}^{\mathbb{Z}}$ before the loop in (3). Note that we could compute ℓ, j, and k directly in $\mathcal{S}_{M_i}^{\mathbb{Z}}$ with the help of Lemma 2. But, in order to do so, we must have the values of n, m, and q in $\mathcal{S}_{M_i}^{\mathbb{Z}}$ and these three values, being part of the input, are fed into the real part of the state space of some processors. We are thus led to copy something from $\mathcal{S}_{M_i}^{\mathbb{R}}$ to $\mathcal{S}_{M_i}^{\mathbb{Z}}$. The next lemma shows that this copying can be done in logarithmic time.

Lemma 3 *The cost of copying a natural number y_0 in $\mathcal{S}_M^{\mathbb{R}}$ to a register in $\mathcal{S}_M^{\mathbb{Z}}$ is bounded by $O(\log y_0)$.*

Proof. We can assume that $y_0 > 0$. Then the following algorithm, where s also belongs to $\mathcal{S}_M^{\mathbb{R}}$, copies y_0 into z.

$$r := \lfloor \log y_0 \rfloor$$
$$s := 2^r$$
$$z := 0$$
while $s \geq 1$ **do**
    if $y_0 \geq s$ **then** $z := 2 * z + 1$;
           $y_0 := y_0 - s$;
    else $z := 2 * z$
    end if
    $s := \frac{s}{2}$
end do .

The number of steps used by this algorithm is $O(\log y_0)$.   □

As in Part I, we define the *halting time* of a parallel machine on input $x \in \mathbb{R}^\infty$ to be the number of nodes traversed in the computational path of x. We can then define the two main measures of parallel complexity.

Definition 2 Let $t : \mathbb{N} \to \mathbb{N}$. We say that a parallel machine P *works in time t* if for all $x \in \mathbb{R}^\infty$, the halting time of P for input x is at most $t(n)$, where n is the size of x. If $p : \mathbb{N} \to \mathbb{N}$ is the activation function of P, we say that P *uses p processors*.

Definition 3 For $k \geq 1$ we define the complexity class $\mathrm{PL}_{\mathbb{R}}^k$ (*parallel polylogarithmic time*) to be the class of sets $S \subseteq \mathbb{R}^\infty$ that are decided by a parallel machine that works in time $O(\log^k n)$ using a polynomial number of processors.

We summarize our discussion regarding the four problems in Section 18.1 in the following proposition.

Proposition 1

(1) *Addition of n numbers, dot product of vectors in $\mathbb{R}^n$, and matrix multiplication of $n \times n$ matrices can be computed in parallel time $O(\log n)$ with a linear number of processors.*

(2) *Determinants of square matrices and inverses of $n \times n$ matrices, when they exist, can be computed in parallel time $O(\log^2 n)$ with $O(n^3)$ processors. In particular, the problem of deciding whether a square matrix is singular is in $\mathrm{PL}_{\mathbb{R}}^2$.*   □

Remark 2 As we have remarked in Chapter 4, we pay more attention to the class $\mathrm{P}_{\mathbb{R}}$ than to classes defined by an upper bound of the kind $O(n^d)$ for a fixed d. Partly, this is because we think of two encodings of the same computational problem as equivalent when the vectors corresponding to the same input have polynomially related size. Consequently, although replacing the encoding of a problem by an equivalent one does not affect the membership of the problem in $\mathrm{P}_{\mathbb{R}}$, it can certainly modify the exponent of the polynomial time bound.

However, if a decision problem S belongs to $PL_{\mathbb{R}}^k$ for some $k \geq 1$, then any equivalent encoding of the same problem will also be in $PL_{\mathbb{R}}^k$. This justifies the closer look we give to each class $PL_{\mathbb{R}}^k$ rather than just the union of all these classes.

Example 5   A Parallel Solution for the Knapsack Problem.

In the examples of the preceding section we dealt with parallel procedures that work in parallel polylogarithmic time. We now consider an example that uses more resources. Recall that the Knapsack Problem is the subset KP of $\mathbb{R}^\infty$ defined by

$$\{x = (x_0, \ldots, x_{n-1}) \in \mathbb{R}^\infty \mid \exists b_0, \ldots, b_{n-1} \in \{0, 1\} \text{ s.t. } \sum_{i=0}^{n-1} b_i x_i = 1\}.$$

The following parallel procedure, which uses 2^n processors, decides KP.

(1) Input $(x_0, \ldots, x_{n-1}) \in \mathbb{R}^n$.

(2) Use machine N of Lemma 1 to compute

$$(y_{-j-2}, \ldots, y_{-3}, y_{-2}, y_{-1}) \leftarrow (b_{j-1}, \ldots, b_0, i, n),$$

where the $b_i, i = 0, \ldots, n-1$ are the first n bits of the processor index i.

(3) For $j = 0$ to $n - 1$ do

(a) read y_0 of processor M_j and put it in the second register;

(b) $y_3 \leftarrow y_{-(j+3)}$;

(c) $y_2 \leftarrow y_3 y_2$;

(d) $y_1 \leftarrow y_1 + y_2$.

(4) If $y_1 \neq 1$, then $y_1 \leftarrow 0$.

(5) Proceed as in procedure $\mathcal{P}$ of Example 1 to add the first registers of processors M_i for $i = 0, \ldots 2^n - 1$ and ACCEPT if the result is greater than zero.

This procedure takes time $O(n)$. Notice that Step (3b) is done in constant time since it can be performed with a combination of communication, shift, and computation nodes as in (6c) of procedure $\mathcal{P}$. Therefore, Step (3) takes time $O(n)$ as does Step (5).

Definition 4 We define the complexity class $PAR_{\mathbb{R}}$ (*parallel polynomial time*) to be the class of sets $S \subseteq \mathbb{R}^\infty$ that are decided by parallel machines that work in polynomial time using an exponentially bounded number of processors. That is, there are constants $c, d \in \mathbb{N}$ such that the number of active processors is bounded by 2^{cn^d}.

One first result concerning the classes just defined locates them relative to known complexity classes.

Proposition 2 *The following inclusions hold.*

(i) PL$_\mathbb{R}^k \subseteq$ P$_\mathbb{R}$ *for all* $k \in \mathbb{N}$.

(ii) PAR$_\mathbb{R} \subseteq$ EXP$_\mathbb{R}$.

Proof.   (i) Let $S \in$ PL$_\mathbb{R}^k$ for some $k \geq 1$. Then there exists a parallel machine P that decides S in time $c \log^k n$ using $p(n) = dn^q$ processors for some $c, d, q \geq 1$. The following roughly describes a sequential machine M^* that decides S in polynomial time.

> input $x \in \mathbb{R}^\infty$
> $n := \text{size}(x)$
> **for** $t = 1$ **to** $c \log^k n$ **do**
>   **for** $i = 0$ **to** $p(n) - 1$ **do**
>   simulate next step of processor M_i
>   **end do**
> **end do**
> ACCEPT if the output function of P returns 1.

Note that, in order to correctly perform the instruction in the loop, M^* needs to keep track of the current node for each processor and to organize a part of its own state space to allocate two copies of the state spaces of each of the $p(n)$ processors. The second copy is needed since otherwise, when a processor changes a coordinate of its state space, all other processors with larger index would be unable to read that coordinate.

(ii) The same idea applies with appropriate modifications for the bounds on the time and number of processors. $\square$

18.3   Deterministic Upper Bounds for NP$_\mathbb{R}$

As discussed in Remarks 3 and 5 of Chapter 5, a natural question arising from the consideration of NP$_\mathbb{R}$ is whether problems in NP$_\mathbb{R}$ are decidable. At first glance, it is not at all clear since, unlike the situation in the classical setting, the space of guesses now is infinitely large. However, by the decidability of the NP$_\mathbb{R}$-complete problem 4-FEAS, all problems in NP$_\mathbb{R}$ are decidable. Moreover, just as in the classical case, the best known deterministic algorithms for them are parallel polynomial time algorithms. The main difference is that the existing algorithms for deciding 4-FEAS, for instance, are unlikely to be described as "brute force." We close this section by stating a theorem concerning these algorithms. We do not attempt to prove this theorem, since all proofs we know are too long to be given in this book. References can be found in Section 18.6. On the other hand, this theorem is used only for two results in Chapter 20 with no further consequences.

Theorem 1 *There is an algorithm that, given a system ψ of k polynomial equalities and inequalities involving polynomials $p_1, \ldots, p_k \in \mathbb{R}[X_1, \ldots, X_n]$ whose* *degrees are bounded by $d \in \mathbb{N}$, decides whether there is a point $x \in \mathbb{R}^n$ satisfying ψ in parallel time $(n(\log k + \log d))^{O(1)}$ using $k^{n+1} d^{O(n)}$ processors.* $\square$

Corollary 1 $\mathrm{NP}_{\mathbb{R}} \subseteq \mathrm{PAR}_{\mathbb{R}}$.

Proof. The algorithm of Theorem 1 solves 4-FEAS in parallel polynomial time. Since any problem S in $\mathrm{NP}_{\mathbb{R}}$ reduces to 4-FEAS in polynomial time, the composition of this reduction with the stipulated algorithm yields a parallel algorithm for S that works in polynomial time. $\square$

18.4 Algebraic Circuits

Here and in the next section we describe an alternate model of parallel computation and compare the classes defined by polylogarithmic and polynomial bounds on the running time for this model with the classes $\mathrm{PL}_{\mathbb{R}}^k$ and $\mathrm{PAR}_{\mathbb{R}}$ just considered.

Let R be a ring. Recall that the sign function

$$\mathrm{sign} : R \to \{0, 1\}$$

is defined by $\mathrm{sign}(x) = 1$ if $x \geq 0$ and 0, otherwise, for ordered rings R and by $\mathrm{sign}(x) = 1$ if $x \neq 0$ and 0, otherwise, for unordered rings.

Definition 5 An *algebraic circuit* $\mathcal{C}$ over R is an acyclic directed graph where each node has indegree 0, 1, or 2. Nodes with indegree 0 are either labeled as *input nodes* or with elements of R (we call these *constant nodes*). Nodes with indegree 2 are labeled with the binary operators of R, that is, one of $\{+, \times, -\}$ (and division $/$ if R is a field). They are called *arithmetic nodes*. Nodes with indegree 1 are either *sign nodes* or *output nodes*. All the output nodes have outdegree 0. Otherwise, there is no upper bound on the outdegree of the other nodes. The nodes of an algebraic circuit are also called *gates*.

For an algebraic circuit $\mathcal{C}$, the *size* of $\mathcal{C}$ is the number of gates in $\mathcal{C}$. The *depth* of $\mathcal{C}$ is the length of the longest path from some input gate to some output gate.

Let $\mathcal{C}$ be an algebraic circuit with n input gates $g_1, \ldots, g_n$ and m output gates $h_1, \ldots, h_m$. Then with each gate g we inductively associate a function $f_g : R^n \to R$ as follows. If g is the ith input gate g_i, then $f_g(x_1, \ldots, x_n) = x_i$. If g is a constant gate labeled with $\alpha \in \mathbb{R}$, then $f_g(x_1, \ldots, x_n) = \alpha$. If g performs an operation (arithmetic or sign), then $f_g(x)$ is obtained by applying this operation to the functions associated with its parents (predecessors). If h is an output gate with predecessor g, f_h is simply f_g.

We refer to the function $\varphi_{\mathcal{C}} : R^n \to R^m$ associated with the output gates as the function *computed by the circuit*.

Remark 3

(1) In order for f_g and φ_C to be well-defined it is necessary to allow division by zero. This is due to the fact that there is no branching in the circuit. A procedure such as "if $x \neq 0$ output y/x, else output y" is implemented in a circuit by computing the formula $(1 - \text{sign}(x))y + (\text{sign}(x))x/y$ where $\text{sign}(x)$ is 1 if $x \neq 0$ and 0 otherwise. Notice that, unlike a f.d. machine, this simulation of the branching requires the computation of x/y beforehand.

By convention we understand that if a gate is labeled with / and its second argument is zero, the value computed by the gate is zero.

(2) Notice that the straight-line programs introduced in Section 17.1 are essentially algebraic circuits without divisions or sign gates.

A first relation between algebraic circuits and finite-dimensional machines is given in the following proposition.

Proposition 3 *Let R be unordered. For every algebraic circuit over R of size s there exists a finite-dimensional machine over R, having at most $3s$ nodes, which computes the same function in time at most $2s$. If R is ordered, the same holds now with $4s$ nodes and time at most $3s$.*

Proof. Let C be an algebraic circuit with gates $g_1, \ldots, g_n, g_{n+1}, \ldots, g_{n+r}$, $g_{n+r+1}, \ldots, g_{n+r+m}$, where $s = n + r + m$, the first n gates are input gates, and the last m are the outputs. The gates are numbered such that each g_j takes its inputs from gates numbered at most $j - 1$.

We construct a machine M with input space R^n, output space R^m and state space R^s computing the same function as C. To this end, with each gate g_{n+j} for $j = 1, \ldots, r + m$ we associate a collection G_j of at most three nodes of M and with one incoming and one outgoing edge. We do this in the following way. Let us denote by z_i the ith coordinate of the state space of M.

1. If g_{n+j} is a constant gate with associated constant $c \in R$, then G_j is a computation node. Its associated function has the form $z_{n+j} \leftarrow c$.

2. If g_{n+j} is a sign gate with entry g_k, then G_j consists of a group of three nodes connected in the following way

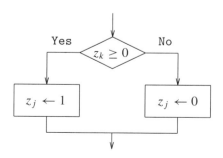

3. If g_{n+j} is an arithmetic gate with operation $\circ \in \{+, -, \times\}$ and left and right predecessors g_l and g_r, respectively, then G_j is a computation node. Its associated function has the form $z_{n+j} \leftarrow z_l \circ z_r$. If g_{n+j} is an arithmetic gate with operation $/$, then G_j consists of a group of three nodes as before (four, if R is ordered, with depth three) which checks z_r for zero and computes $z_{n+j} \leftarrow z_l/z_r$ if $z_r \neq 0$ and $z_{n+j} \leftarrow 0$ otherwise.

4. If g_{n+j} is an output gate with predecessor g_k, then G_j consists of a computation node with associated function $z_{n+j} \leftarrow z_k$.

Moreover, we define G_0 to be the input node of M with function I : $R^n \rightarrow R^s$ defined by $I(x_1, \ldots, x_n) = (x_1, \ldots, x_n, 0, \ldots, 0)$ and G_{r+m+1} to be the output node of M with function O : $R^s \rightarrow R^m$ defined by $O(x_1, \ldots, x_{n+r}, x_{n+r+1}, \ldots, x_{n+r+m}) = (x_{n+r+1}, \ldots, x_{n+r+m})$.

Finally, we connect each G_j to G_{j+1} for $j = 0, \ldots, r+m$. □

We now prove a converse to Proposition 3. The central idea of its proof is used again several times. We consider finite-dimensional machines M with the following special form. Let $\mathcal{S}_M = R^m$ be the state space of M. For any computation node η of M, its associated function is of the form

$$x_j \leftarrow x_k \circ x_g,$$

where $\circ$ is an arithmetical operation and $1 \leq j, k, g \leq m$ or of the form

$$x_j \leftarrow x_k \circ c,$$

where c is a machine constant. For any branch node the branch condition is of the form $x_i \geq 0$ with $1 \leq i \leq m$.

Proposition 4 *Let M be a finite-dimensional machine specialized as before with N nodes and state space R^m. If M decides a set $S \subseteq R^n$ in time T, then S can be decided by an algebraic circuit of size $O(mTN)$.*

Proof. We describe an algebraic circuit $\mathcal{C}$ with n input gates and one output gate which computes the characteristic function of S.

The circuit $\mathcal{C}$ is constructed from a family of smaller circuits $\mathcal{J}_t$ and $\mathcal{X}_{i,t}$, where $1 \leq t \leq T$ and $1 \leq i \leq m$. Consider a computation of M and let $x_{i,t}$ denote the value of the ith coordinate of the state space of M at step t. Let η_t denote the next node after this step. Then $\mathcal{J}_t$ is intended to compute η_t and $\mathcal{X}_{i,t}$ is intended to compute $x_{i,t}$. In both cases the input for the computation is η_{t-1} and $x_{j,t-1}$ for some values of j.

For $1 \leq i \leq n$, we can set $\mathcal{X}_{i,1}$ to be just one input node, and for $n+1 \leq i \leq m$, we set $\mathcal{X}_{i,1}$ to be a constant node with associated constant 0. Also, we set $\mathcal{J}_1$ to be a constant node with associated constant $\beta(1)$, the successor of the input node.

In this way, the circuits $\mathcal{X}_{i,1}$ and $\mathcal{J}_1$ perform the computation corresponding to the input node of M.

For $t \geq 2$ we have that η_t is given by

$$\sum_{r=2}^{N} [\eta_{t-1} = r]\beta(r),$$

where

$$[a = b] = \begin{cases} 1 & \text{if } a = b \\ 0 & \text{if } a \neq b. \end{cases}$$

Recall from Chapter 2 that if the rth node is not a branching node, then $\beta(r)$ is a well-defined value. Otherwise it is $\beta^+(r)$ or $\beta^-(r)$ according to whether sign $(x_{i_r}) = 1$ or sign $(x_{i_r}) = 0$ for a certain fixed $1 \leq i_r \leq m$. Therefore, in the preceding summation, each term can be computed by an algebraic circuit of constant size. We deduce that η_t can be computed by an algebraic circuit $\mathcal{J}_t$ of size $O(N)$.

Consider now an index $1 \leq i \leq m$ and let C_i be the set of computation nodes that modify the value of the ith coordinate of the state space of M. For $t \geq 2$, the value $x_{i,t}$ is then given by

$$\sum_{r \notin C_i} [\eta_{t-1} = r]x_{i,t-1} + \sum_{r \in C_i} [\eta_{t-1} = r](x_{k,t-1} \circ_r x_{g,t-1}),$$

where $\circ_r$ is the operation performed at node r on the kth and gth coordinates of the state space of M. (If node r operates on the kth such coordinate with a constant c, then $x_{g,t-1}$ is replaced by c.) Again, in this summation, each term can be computed by an algebraic circuit of constant size and therefore $x_{i,t}$ can be computed by an algebraic circuit $\mathcal{X}_{i,t}$ of size $O(N)$.

Wiring appropriately the circuits $\mathcal{J}_t$ and $\mathcal{X}_{i,t}$ for $1 \leq t \leq T$ and $1 \leq i \leq m$ with the necessary output nodes we obtain the claimed circuit $\mathcal{C}$ of size $O(mTN)$.

$\square$

18.5   The Class $NC_{\mathbb{R}}$

As said in the last section, we can consider algebraic circuits as a model of finite-dimensional parallel machines. The depth of a circuit $\mathcal{C}$ is a measure of parallel time and its size, that is, the number of its gates, of the amount of necessary hardware. The size thus corresponds to the number of processors used by a parallel machine. However, for algebraic circuits, a gate is no longer a machine executing a possibly complicated task but rather a very simple device. It can only perform an arithmetical operation or compute a sign.

Our next goal is to extend this finite-dimensional parallel model to a uniform one. We do so only for the case $R = \mathbb{R}$, although our next definition still applies to arbitrary rings.

First, we allow our model to take arbitrarily long inputs.

Definition 6 Let $f : R^\infty \to R^\infty$. We say that the family of algebraic circuits $\{C_n\}_{n\geq 1}$ computes f, if for all $n \geq 1$, the function computed by C_n is the restriction of f to $R^n \subset R^\infty$.

Secondly, we require a condition on the whole family $\{C_n\}_{n\geq 1}$ to ensure that its elements are not unrelated and to ensure a machine model with finite description. In order to define this condition, we note that gates of algebraic circuits can be described by four real numbers in the following way. If the gates of the circuit are $g_1, \ldots, g_k$, then gate g_j is described by the tuple $(j, t, i_\ell, i_r) \in \mathbb{R}^4$ where t represents the type of g_j according to the dictionary,

g_j	t
input	1
constant	2
$+$	3
$-$	4
$\times$	5
sign	6
output	7
$/$	8

For gates of indegree two, the numbers i_ℓ and i_r, respectively, denote the gates that provide left and right input to g_j. By convention, if g_j is an input gate, then i_ℓ and i_r are 0; if g_j is a constant gate, then i_ℓ equals its constant and i_r is zero; and finally, if g_j is a sign gate or an output gate, then i_ℓ denotes the gate that provides the input to g_j and i_r is zero. Thus, the whole circuit can then be described by a point in $\mathbb{R}^{4k}$. We suppose, without loss of generality, that when describing a circuit, the first gates are the input gates.

Definition 7 A family of circuits $\{C_n\}_{n\in\mathbb{N}}$ is said to be *uniform* if there exists a sequential machine M that outputs the description of the ith gate of C_n with input (n, i). If M works in time bounded by $O(\log n)$ we say that the family is *L-uniform*; if M works in time $O(n^k)$ for some positive integer k we say that the family is *P-uniform*.

We now define some new parallel complexity classes by bounding the depth and size of uniform families of circuits.

Definition 8 Let $\mathrm{NC}^k_\mathbb{R}$, $k \geq 1$ be the class of sets $S \subseteq \mathbb{R}^\infty$ such that there is an L-uniform family of algebraic circuits $\{C_n\}$ that computes the characteristic function of S and has size polynomial in n and depth $O(\log^k n)$. The union of the $\mathrm{NC}^k_\mathbb{R}$ is denoted by $\mathrm{NC}_\mathbb{R}$.

Relationships between the classes $\mathrm{PL}^k_\mathbb{R}$ and $\mathrm{NC}^k_\mathbb{R}$ are given in the next two theorems.

Theorem 2 *For every $k \geq 1$ we have* $\mathrm{PL}^k_\mathbb{R} \subseteq \mathrm{NC}^{k+1}_\mathbb{R}$.

Proof.   The proof relies on the idea used in the proof of Proposition 4 so we only sketch it here.

Let $S \in \mathrm{PL}_\mathbb{R}^k$ and let P be a parallel machine that decides S in time $t(n) = O(\log^k n)$ using $p(n)$ processors, where p is a polynomial. Recall that we denote by $\mathcal{S}_{M_j}^\mathbb{Z}$ and $\mathcal{S}_{M_j}^\mathbb{R}$ the integer and real parts of the state space of the jth processor of P. Let q be the dimension of $\mathcal{S}_{M_j}^\mathbb{Z}$.

For each $n \geq 1$ we produce a circuit $\mathcal{C}_n$ that decides $S \cap \mathbb{R}^n$. As in Proposition 4, we construct $\mathcal{C}_n$ from smaller circuits $\mathcal{J}_{j,t}$, $\mathcal{Z}_{i,j,t}$, and $\mathcal{Y}_{\ell,j,t}$, where $1 \leq j \leq p(n)$, $0 \leq i < k$, $1 \leq t \leq t(n)$, and $-p(n) \leq \ell \leq p(n)$. Those circuits are intended to do the following.

- $\mathcal{J}_{j,t}$ computes the current node $\eta_{j,t}$ corresponding to the jth processor after t steps of the computation.

- $\mathcal{Z}_{i,j,t}$ computes the value $z_{i,j,t}$ corresponding to the ith coordinate of $\mathcal{S}_{M_j}^\mathbb{Z}$ after t steps of the computation.

- $\mathcal{Y}_{\ell,j,t}$ computes the value $y_{\ell,j,t}$ corresponding to the ℓth coordinate of $\mathcal{S}_{M_j}^\mathbb{R}$ after t steps of the computation.

Circuits $\mathcal{J}_{j,t}$ and $\mathcal{Z}_{i,j,t}$ are obtained as in Proposition 4, also with size $O(N)$, where N is the number of nodes in the graph of the processors.

Circuits $\mathcal{Y}_{\ell,j,t}$ are designed in a similar way. However, for $\ell = 0$, they no longer have constant size with respect to n. Let $\mathcal{C}$ be the set of computation nodes that modify the value of the zeroth coordinate of $\mathcal{S}_{M_j}^\mathbb{R}$. Denote by $\mathcal{R}$ the set of communication nodes, by $\mathcal{S}_\ell$ the set of shift nodes that shift to the left, and by $\mathcal{S}_r$ the set of shift nodes that shift to the right. For $t \geq 2$, the value $y_{0,j,t}$ is then given by

$$\sum_{r \notin (\mathcal{C} \cup \mathcal{R} \cup \mathcal{S}_\ell \cup \mathcal{S}_r)} [\eta_{j,t-1} = r] y_{\ell,j,t-1} + \sum_{r \in \mathcal{C}} [\eta_{j,t-1} = r](y_{k,j,t-1} \circ_r y_{g,j,t-1}) \quad (18.1)$$

$$+ \sum_{r \in \mathcal{S}_\ell} [\eta_{j,t-1} = r] y_{-1,j,t-1} + \sum_{r \in \mathcal{S}_r} [\eta_{j,t-1} = r] y_{1,j,t-1} + \sum_{r \in \mathcal{R}} [\eta_{j,t-1} = r] \psi(j,t),$$

where the two first summations are as in Proposition 4. The last summation is more complicated since the communication function ψ is. The behavior of a communication node depends on the value of a coordinate z_a in $\mathcal{S}_{M_j}^\mathbb{Z}$ with $1 \leq a \leq q$. Recall that at such node, processor M_j reads the zeroth register of processor M_{z_a}. Therefore

$$\psi(j,t) = y_{0,\bar{z},t-1},$$

where $\bar{z} = z_{a,j,t-1}$. We can suppose that $0 \leq z_a < p(n)$. The bound follows from the definition of parallel machine.

Therefore to compute $\psi(j,t)$ we need at most $p(n)$ input values. Now a little programming shows that $\psi(j,t)$ can be computed by an algebraic circuit with $O(p(n))$ gates and depth $O(\log n)$ and, consequently, so can $\mathcal{Y}_{0,j,t}$.

For $\ell \neq 0$ the last summation in (18.1) does not appear and the circuit $\mathcal{Y}_{\ell,j,t}$ has constant depth.

The overall circuit obtained by appropriately wiring the circuits $\mathcal{J}_{j,t}$, $\mathcal{Z}_{i,j,t}$, and $\mathcal{Y}_{\ell,j,t}$ with the necessary output node has polynomial size and depth $O(\log^{k+1} n)$. A close look at its description shows that it can be produced by a sequential machine in an L-uniform way. $\qquad\square$

Theorem 3 *For every $k \geq 1$, we have that $\mathrm{NC}_{\mathbb{R}}^k \subseteq \mathrm{PL}_{\mathbb{R}}^k$.*

Proof. Suppose we are given a set S in $\mathrm{NC}_{\mathbb{R}}^k$. Consider an L-uniform circuit family $\{\mathcal{C}_n\}$ of depth $c \log^k n$ for a constant c and size $O(n^q)$ that decides the set S. We construct a parallel machine P with $p(n) = O(n^q)$ processors working within time $t(n) = O(\log^k n)$ whose accepted set is exactly S.

Let U be the sequential machine that computes the gates of $\mathcal{C}_n$ in time $O(\log n)$. We describe P in the following way where i denotes the processor index.

       input $(x_1, \ldots, x_n)$
(1)   compute the codification (i, op, i_ℓ, i_r) of the gate number i
(2)   if $op \neq$ constant then copy i_ℓ and i_r in integer registers as follows
                $(z_2, z_3) \leftarrow (i_\ell, i_r)$
(3)   compute $d := c \cdot \log^k n$
(4)   **for** $j = 1$ **to** d **do**
        if $op \in \{+, -, \times, /\}$ **then**
            read y_0 from M_{z_2} and let y_1 take this value
            read y_0 from M_{z_3} and let y_2 take this value
            $y_0 \leftarrow y_1 \, op \, y_2$
        elsif $op =$ sign **then**
            read y_0 from M_{z_2} and let y_1 take this value
            $y_0 \leftarrow \mathrm{sign}\,(y_1)$
        elsif $op =$ constant **then**
            $y_0 \leftarrow i_\ell$
        elsif $op =$ output **then**
            read y_0 from M_{z_2} and let y_0 take this value
        end if
      end do
(5)   **if** $op =$ output **then**
      output y_0
   end if .

Step (1) is accomplished in time $O(\log n)$ using the machine U. According to Lemma 3, the copying in Step (2) is also accomplished within time $O(\log n)$. Step (4) iterates the loop $O(\log^k n)$ times and each of these iterations takes constant time. We conclude that P works within time $O(\log^k n)$. $\qquad\square$

Remark 4

(1) As a consequence of Theorems 2 and 3 we deduce the following chain of inclusions.

$$\text{NC}_\mathbb{R}^1 \subseteq \text{PL}_\mathbb{R}^1 \subseteq \text{NC}_\mathbb{R}^2 \subseteq \text{PL}_\mathbb{R}^2 \subseteq \dots.$$

We also deduce that the union of the $\text{PL}_\mathbb{R}^k$ for $k \geq 1$ yields the class $\text{NC}_\mathbb{R}$.

(2) Consider the class of sets decided by P-uniform families of circuits with polynomial depth and exponential size. The same arguments of Theorems 2 and 3 can be used to prove that this class coincides with $\text{PAR}_\mathbb{R}$.

18.6   Additional Comments and Bibliographical Remarks

Classical complexity theory has dealt with several models of parallel machines. The one most resembling the parallel machine introduced in Section 18.2 is the PRAM, which exists in several variants (see, for instance, [Fortune and Wyllie 1978; Goldshlager 1982; or Savitch and Stimson 1979]). Our presentation is a development of [Cucker, Montaña, and Pardo 1995].

Algebraic circuits are probably the most widely used computational model in algebraic complexity. In the classical case, that is, when the ground field is $\mathbb{Z}_2$, algebraic circuits are equivalent to *Boolean circuits*. These are circuits whose gates compute the Boolean connectives $\vee$, $\wedge$, and $\neg$.

The consideration of uniform families of Boolean circuits was first suggested by Borodin [1977]. The classes NC^k were formally defined by Nicholas Pippenger [1979]. The name NC was coined by Cook [1979] as a mnemonic for "Nick's class." The classes $\text{NC}_\mathbb{R}^k$ are just the real version of the classes NC^k. Theorems 2 and 3 are well known in the classical case and can be found in [Stockmeyer and Vishkin 1984].

The name $\text{PAR}_\mathbb{R}$ does not appear in the classical theory since, classically, parallel polynomial time coincides with polynomial space. Denote by PSPACE the class of subsets of $\mathbb{Z}_2^\infty$ decidable by machines M using polynomial space. That is, for each input x of size n, the number of coordinates of the state space of M used during the computation with input x is bounded by a polynomial in n. The following is a consequence of results in [Borodin 1977].

Theorem 4   *The class of subsets of $\mathbb{Z}_2^\infty$ decided in parallel polynomial time coincides with* PSPACE.   □

Theorem 1 is the state-of-the-art answer to an old question: what is the complexity of the existential theory of the reals? Its origins lie in the introduction of the concept of *real closed field* by Artin and Schreier [1926] to solve Hilbert's 17th problem. In the 1940s Tarski [1951] proved[1] that there is an algorithm for

---

[1]The war years intervened in its publication.

quantifier elimination in the theory of real closed fields, that is, an algorithm that given as input a formula of this theory outputs an equivalent formula in which all the quantified variables have been eliminated. This algorithm allows one to solve the problem in Theorem 1. Given the system ψ of polynomial equalities and inequalities one decides whether there is a point $x \in \mathbb{R}^n$ satisfying ψ by checking the validity of a quantifier-free sentence equivalent to

$$\exists x_1 \ldots \exists x_n \ \psi(x_1, \ldots, x_n).$$

The quantifier-free sentence is just a Boolean combination of polynomial relations (polynomial equalities and inequalities) of the coefficients occurring in the system ψ. In the light of today's concerns, Tarski's algorithm is not feasible. Its complexity is of the order of

$$2^{2^{2^{\cdot^{\cdot^{2}}}}} \left. \right\} n \text{ times.}$$

In the 1970s, Collins [1975] and Wüthrich [1976] independently devised algorithms for quantifier elimination whose complexity has a doubly exponential dependence on n. A breakthrough was made later by Grigoriev and Vorobjov [1988] and Grigoriev [1988] with the introduction of the *critical points method*. The algorithm given in [Grigoriev 1988] allows one to eliminate quantifiers at a cost that is singly exponential in the number of variables n and doubly exponential only in the number of quantifier alternations. In particular, if all quantifiers are existential, the algorithm works in single exponential time. This algorithm is sequential and only works for systems of polynomials with rational coefficients. The cost measure considered is the bit cost. Several improvements appear in [Canny 1988; Heintz, Roy, and Solerno 1990; Renegar 1992a, 1992b, 1992c; Basu, Pollack, and Roy 1994]. Some of these articles provide parallel algorithms that work with arbitrary real numbers at unit cost. In particular, [Renegar 1992a] is a good reference for Theorem 1 as stated here.

19
Some Separations of Complexity Classes

In this chapter we continue to study the nature of parallel computation. The focus now is the limits of the power of parallelism. Thus, in Section 19.1 we prove that there are problems in $P_\mathbb{R}$ that cannot be efficiently parallelized. These problems are not in $NC_\mathbb{R}$. Section 19.2 contains lower bound estimates for parallel time in terms of the number of connected components as in Chapter 16.

Subsequently, completeness theorems are proved in the context of our parallel machines, in the spirit of Chapter 5.

In the last section we separate complexity classes given by deterministic and nondeterministic time $O(n^d)$ for any $d \geq 1$. This is done for a model consisting of uniform families of circuits.

19.1   A Separation Result

A natural question arising from the inclusion $NC_\mathbb{R} \subseteq P_\mathbb{R}$ is whether equality holds. An affirmative answer would imply the existence of efficient parallelizations for all problems in $P_\mathbb{R}$. The goal of this section is to show that equality does not hold. We prove a more general theorem from which several separations of complexity classes follow.

We begin with a special case of the Nullstellensatz for real algebraic sets. It provides conditions a polynomial must fulfill in order to generate the ideal of polynomials vanishing at its zeros.

Denote by $\mathbb{R}[X_1, \ldots, X_n]$ the ring of polynomials in the indeterminates $X_1, \ldots, X_n$ with real coefficients and by $\mathbb{R}(X_1, \ldots, X_n)$ its quotient field. A polynomial $p \in \mathbb{R}[X_1, \ldots, X_n]$ is *irreducible* if whenever a polynomial $q \in \mathbb{R}[X_1, \ldots, X_n]$ divides p, then $q \in \mathbb{R}$ or degree $q =$ degree p.

Our first step is to prove a lemma which is of interest in its own right. It states that if p is irreducible, then the set of its zeros is a hypersurface if and only if it contains a regular point, that is, a point $x \in \mathbb{R}^n$ such that $p(x) = 0$ and grad $p(x) \neq 0$.

Let $V \subset \mathbb{R}^n$ be a semi-algebraic set and $x \in V$. We say that V *has dimension d at x* if there are open neighborhoods $W \subset V$ and $U \subset \mathbb{R}^d$ of x and the origin, respectively, and an analytic function f that maps U diffeomorphically onto W. We define the *dimension of V* to be

$$\max_{x \in V} \dim(V, x),$$

where $\dim(V, x)$ denotes the dimension of V at x and the maximum is taken over the points x such that $\dim(V, x)$ is defined.

Lemma 1 *Let $p \in \mathbb{R}[X_1, \ldots, X_n]$ be a nonzero irreducible polynomial and let $\mathcal{Z}(p) \subseteq \mathbb{R}^n$ be its zero set. Then $\mathcal{Z}(p)$ has dimension $n - 1$ if and only if $\mathcal{Z}(p)$ contains a regular point.*

Proof.   For the only if part, we can suppose that the projection $\pi : \mathbb{R}^n \to \mathbb{R}^{n-1}$ over the first $n - 1$ coordinates satisfies that the dimension of $\pi(\mathcal{Z}(p))$ is $n - 1$. Then we can write p in powers of X_n

$$p = a_m X_n^m + a_{m-1} X_n^{m-1} + \ldots + a_0,$$

where a_i is a polynomial in $X_1, \ldots, X_{n-1}$, $m \geq 1$ and $a_m \neq 0$. Let

$$p' = \frac{\partial p}{\partial X_n} = m a_m X_n^{m-1} + \ldots + a_1$$

and r be the resultant of p and p'. Adapting Lemma 3 of Chapter 10 to our context of polynomials in X_n with coefficients in $\mathbb{R}[X_1, \ldots, X_{n-1}]$ we have that r is identically zero if and only if p and p' have a common factor of degree at least 1 in X_n. But this is not possible because p is irreducible. Moreover, for all $(x_1, \ldots, x_{n-1})$ not annihilating the polynomial a_m, we have that $r(x_1, \ldots, x_{n-1})$ is the resultant of $p(x_1, \ldots, x_{n-1}, X_n)$ and $p'(x_1, \ldots, x_{n-1}, X_n)$.

Consider the product $r a_m$. This is a nonzero polynomial in the indeterminates $X_1, \ldots, X_{n-1}$. Therefore its zero set in $\mathbb{R}^{n-1}$ has dimension at most $n - 2$ and we deduce the existence of a point $x = (x_1, \ldots, x_{n-1})$ belonging to $\pi(\mathcal{Z}(p))$ such that $r(x) \neq 0$ and $a_m(x) \neq 0$. Let $x^* \in \mathcal{Z}(p)$ such that $\pi(x^*) = x$. Then, since the resultant $r(x)$ of $p(x, X_n)$ and $p'(x, X_n)$ is not zero, we deduce that these two univariate polynomials have no common roots and thus that x^* is not a root of p' and is therefore a regular point of $\mathcal{Z}(p)$.

Now, for the if part, let $a = (a_1, \ldots, a_n)$ be a regular point of $\mathcal{Z}(p)$ and assume that $(\partial p / \partial X_n)(a) \neq 0$. The implicit function theorem then provides an open neighborhood $U \subseteq \mathbb{R}^{n-1}$ of $(a_1, \ldots, a_{n-1})$ and an analytic function $\varphi : U \to \mathbb{R}$ such that for every $x \in U$ we have $p(x, \varphi(x)) = 0$. This proves that the dimension of $\mathcal{Z}(p)$ is at least $n - 1$. But the other inequality follows from the fact that p is not zero. $\square$

We can now proceed to our simple form of the real Nullstellensatz. Since its proof relies on divisibility arguments, we recall here some notions related to factorial rings.

Let A be an integral domain and $p \in A$, $p \neq 0$ and not invertible. We say that p is *irreducible* if $p = xy$ with $x, y \in A$ implies that x is invertible or y is invertible. We say that $x \in A$, $x \neq 0$ has a *unique factorization into irreducible elements* if there exist irreducible elements $p_1, \ldots, p_r \in A$ such that

$$x = \prod_{i=1}^{r} p_i,$$

and if given two such factorizations

$$x = \prod_{i=1}^{r} p_i = \prod_{i=1}^{s} q_i,$$

we have $r = s$ and after a permutation of the indices i, we have $p_i = u_i q_i$ for some invertible element $u_i \in A$, $i = 1, \ldots, r$. An integral domain A is a *factorial ring* (or a *unique factorization domain*) if every nonzero element has a unique factorization into irreducible elements. The canonical example of a factorial ring is $\mathbb{Z}$. The most important example in what follows is the ring $k[X_1, \ldots, X_n]$ of polynomials over a field k.

Let A be a factorial ring and $f \in A[X]$. We say that f is *primitive* if any common divisor of the coefficients of f is invertible.

Lemma 2 *Let A be a factorial ring and $f, g \in A[X]$ primitive. Then fg is also primitive.*

Proof.   It is enough to see that there is no irreducible element dividing all the coefficients of fg. Suppose there exists one such element, say p. Then

$$fg \equiv 0 \ (\text{mod } p).$$

Since f and g are primitive, one has that

$$f \not\equiv 0 \ (\text{mod } p) \qquad \text{and} \qquad g \not\equiv 0 \ (\text{mod } p).$$

Therefore $fg \not\equiv 0 \ (\text{mod } p)$ and we arrived at a contradiction.   □

Irreducibility of polynomials over a factorial ring is closely related to irreducibility over its field of fractions.

Proposition 1 (Gauss' Lemma) *Let A be a factorial ring and k its field of fractions. If $f \in A[X]$ is irreducible in $A[X]$, it is also in $k[X]$.*

Proof.   Suppose $f(X) = a(X)b(X)$ with $a, b \in k[X]$ both with degree at least 1. There are elements $r, s \in k$ such that $a'(X) = ra(X)$ and $b'(X) = sb(X)$ are both in $A[X]$ and primitive. By the preceding lemma, $rsf(X)$ is also in $A[X]$ and primitive.

Write $rs = p/q$ with $p, q \in A$ relatively prime. Then p divides all the coefficients of $srf(X)$. Since the latter polynomial is primitive, we deduce that p is invertible in A. Consequently, $f(X) = p^{-1}qa'(X)b'(X)$ which is in contradiction to the fact that f is irreducible in $A[X]$.   □

Proposition 2 *Let $p \in \mathbb{R}[X_1, \ldots, X_n]$ be an irreducible polynomial such that the dimension of $\mathcal{Z}(p) \subseteq \mathbb{R}^n$ is $n - 1$. Then, for any polynomial $q \in \mathbb{R}[X_1, \ldots, X_n]$, q vanishes on $\mathcal{Z}(p)$ if and only if q is a multiple of p.*

Proof.   It is immediate that if q is a multiple of p, then it vanishes on $\mathcal{Z}(p)$.

For the converse, denote by X the variables $X_1, \ldots, X_{n-1}$, by A the ring $\mathbb{R}[X_1, \ldots, X_{n-1}]$, and by k its field of fractions. Consider a regular point $a = (a_1, \ldots, a_n)$ of $\mathcal{Z}(p)$ and U and φ as in the proof of Lemma 1. Then the power series $p(X, \varphi(X))$ is identically zero. Since p is irreducible in $\mathbb{R}[X_1, \ldots, X_n]$, by Gauss' Lemma it is also irreducible considered as an element in $k[X_n]$. Therefore it is the minimal polynomial of φ in the sense that it annihilates φ and it has minimal degree amongst the polynomials in $k[X_n]$ doing so.

Since q vanishes on $\mathcal{Z}(p)$ it follows that $q(x, \varphi(x)) = 0$ for all $x \in U$. Thus, p divides q in the ring $k[X_n]$. Now since p is irreducible it is also primitive in the ring $A[X_n]$. Let $q = pr$ with $r \in k[X_n]$ and suppose that $r \notin A[X_n]$. Then we can write $(s/d)r = r'$ with $s, d \in A$ relatively prime and $r' \in A[X_n]$ primitive. The equality $q = pr$ implies $sq = dpr'$. Since p and r' are primitive so is their product. Now s divides dpr' and is relatively prime to d. Thus, s divides the coefficients of pr' and therefore it is invertible in A. We conclude that p divides q in $A[X_n]$, that is, in $\mathbb{R}[X_1, \ldots, X_n]$.   □

Proposition 3 *Let $f_n \in \mathbb{R}[X_1, \ldots, X_n]$, $n \in \mathbb{N}$, be a family of nonconstant irreducible polynomials such that for each n, the zero set $\mathcal{Z}(f_n)$ is a variety of dimension $n - 1$. Let $d(n) = \deg(f_n)$. Then any parallel machine deciding the set $S = \{x \in \mathbb{R}^\infty \mid f_{\mathrm{size}(x)}(x) = 0\}$ has running time greater than $\log d(n)$.*

Proof.   Let us consider a parallel machine P deciding S within time $r(n)$ and activation function $p(n)$.

For each $n \geq 1$ and each input of size n, the computation performed by each processor M of P takes place in the subspace

$$\mathcal{S}_M^{\mathbb{Z}} \times \mathbb{R}^{2r(n)} = \mathbb{Z}^k \times \mathbb{R}^{2r(n)}$$

of the state space of M. The whole computation may be considered to take place in $\mathbb{R}^N$ with $N = p(n)(k + 2r(n))$. We call any point in $\mathbb{R}^N$ a configuration. An initial configuration of the computation with input $x = (x_1, \ldots, x_n)$ denotes the point $I_P(x)$ where I_P is the input map of P considered as mapping into $\mathbb{R}^N$.

Our first goal is to unwind the computation of P for inputs of size n into a kind of algebraic computation tree $\mathcal{T}$ with state space $\mathbb{R}^N$, but with computation nodes that can modify several coordinates simultaneously and branch nodes that can have large arity. We now describe $\mathcal{T}$.

At each step of the computation of P two actions take place. Some of the coordinates of the current configuration are modified replacing their contents by the result of operating (via one of $(+, -, \times, /)$) on the contents of two other coordinates or by reading another coordinate. On the other hand, a certain number of tests, bounded by $p(n)$, are performed to determine the next step of P, that is, the next node for the $p(n)$ processors.

The root of $\mathcal{T}$ contains the initial configuration and one "outgoing edge." Otherwise, we associate with each step of P two layers in $\mathcal{T}$ by attaching to each outgoing edge new nodes such as those between the dashed lines in Figure A, where the first node performs the operations and the second corresponds to the branching. Since at most $p(n)$ tests can be done at a step of P, the largest possible arity of $\mathcal{T}$ is $2^{p(n)}$.

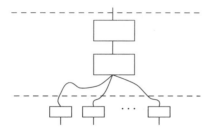

Figure A The layers associated with a step of P.

At the end of the computation, the first coordinate of the final configuration will be 0 or 1 and with one more question we can label the final leaf with Yes or No.

Now, at any node of $\mathcal{T}$ and for any coordinate y_j of the space $\mathbb{R}^N$, the contents of y_j can be expressed as a rational function $Q_j(X_1, \ldots, X_n)$ of the input $x_1, \ldots, x_n$. Let P_j/R_j be a representation of Q_j with P_j and R_j relatively prime. Then the degrees of P_j and R_j are at most $2^{r(n)}$.

Therefore, for each path γ in $\mathcal{T}$, the set S_γ of inputs $x \in \mathbb{R}^n$ that follow the path γ can be characterized by a system like

$$\bigwedge_{i=1}^{s_\gamma} Q_i(X_1, \ldots, X_n) \geq 0 \wedge \bigwedge_{i=s_\gamma+1}^{t_\gamma} Q_i(X_1, \ldots, X_n) < 0,$$

where the $Q_i(X_1, \ldots, X_n)$ are rational functions. Accordingly, the set S, which is the union of all the S_γ for all paths γ reaching a leaf labeled Yes, can be described by the disjunction

$$\bigvee_\gamma \left(\bigwedge_{i=1}^{s_\gamma} Q_{\gamma,i}(X_1, \ldots, X_n) \geq 0 \wedge \bigwedge_{i=1}^{r_\gamma} Q_{\gamma,i}(X_1, \ldots, X_n) < 0 \right).$$

By expressing the sign of Q_j in terms of the signs of P_j and R_j we can replace the rational functions by polynomials with degrees at most $2^{r(n)}$. Also, expressing an inequality like

$$F(X_1, \ldots, X_n) \geq 0$$

as the disjunction

$$F(X_1, \ldots, X_n) = 0 \vee F(X_1, \ldots, X_n) > 0$$

and then distributing, we can describe S as the disjunction

$$\bigvee_\eta \left(\bigwedge_{i=1}^{s_\eta} F_{\eta,i}(X_1, \ldots, X_n) = 0 \wedge \bigwedge_{i=1}^{r_\eta} G_{\eta,i}(X_1, \ldots, X_n) > 0 \right),$$

where now the $F_{\eta,i}(X_1, \ldots, X_n)$ and the $G_{\eta,i}(X_1, \ldots, X_n)$ are polynomials of degree at most $2^{r(n)}$ and the η are no longer paths in the tree.

For each η, let S_η be the set defined by the system

$$\bigwedge_{i=1}^{s_\eta} F_{\eta,i}(X_1, \ldots, X_n) = 0 \wedge \bigwedge_{i=1}^{r_\eta} G_{\eta,i}(X_1, \ldots, X_n) > 0.$$

Since the zero set $\mathcal{Z}(f_n)$ has dimension $n - 1$, there is an η such that S_η has dimension $n - 1$. Since the set described by the $G_{\eta,i}$s is open, it must be non-empty, and then it defines an open subset of $\mathbb{R}^n$. We must then have $s > 0$ since otherwise S_η would have dimension n.

Finally, all the polynomials F_i for $i = 1, \ldots, s$ vanish on S_η. But, since S_η has dimension $n - 1$, and the variety $\mathcal{Z}(f_n)$ is irreducible, this implies that every F_i must vanish on the whole variety. By Proposition 2 all the F_i are multiples of f_n. Thus their degrees are at least $d(n)$. We conclude that $2^{r(n)} \geq d(n)$ from which the statement follows. $\qquad\square$

Corollary 1 *The following inclusions of complexity classes are strict.*

(i) $\mathrm{PL}_{\mathbb{R}}^k \subset \mathrm{NC}_{\mathbb{R}}^{k+1}$ *for* $k \geq 0$.

(ii) $\mathrm{NC}_{\mathbb{R}} \subset \mathrm{P}_{\mathbb{R}}$.

(iii) $\mathrm{PAR}_{\mathbb{R}} \subset \mathrm{EXP}_{\mathbb{R}}$.

Proof. For (i) consider, for any $k \geq 0$, the family of polynomials $f_n = X_1 - X_2^{2^{\log n^{k+1}}}$ and the subset of $\mathbb{R}^\infty$ defined by this family as in the statement of Proposition 3. This set can be decided in $\mathrm{NC}_{\mathbb{R}}^{k+1}$ by an L-uniform family of circuits $\mathcal{C}_n$ computing, for any input $x \in \mathbb{R}^n$ the value $x_1 - x_2^{2^{\log n^{k+1}}}$ and then outputting 1 if this value is 0 and 0 otherwise. On the other hand, Proposition 3 shows that this set does not belong to $\mathrm{PL}_{\mathbb{R}}^k$.

Parts (ii) and (iii) are proved in a similar way, but now we take as f_n the polynomials $X_1 - X_2^{2^n}$ and $X_1 - X_2^{2^{2^n}}$, respectively. $\qquad\square$

19.2   Lower Bounds for Parallel Time

We prove a lower bound result in this section similar to Theorem 2 of Chapter 16 but for parallel time. Here we use the algebraic circuit as our model of a finite dimensional parallel machine.

Theorem 1 *There is a universal constant $\gamma > 0$ such that, for every $S \subseteq \mathbb{R}^n$ and every algebraic circuit C deciding S, the depth t of C satisfies the lower bound*

$$t \geq \gamma \left(\sqrt{\frac{\log(\# \text{ c.c. } (S))}{n}} \right),$$

where by # c.c. we mean number of connected components.

Proof.   Let C be an algebraic circuit that decides S. We can suppose that C has no division nodes. This can be shown, with the appropriate modifications, as in Proposition 3 of Chapter 7. We say that a gate g is at height h if the path from g to the only output gate of C has length h. Since the indegree of the gates of C is bounded by 2 there are at most 2^h gates with a given height h.

Now, for each $i \leq t$ let $g_{i,1}, \ldots, g_{i,s_i}$ be the sign gates of C whose height is $t - i$, and let $f_{i,1}, \ldots, f_{i,s_i}$ be the functions of $(x_1, \ldots, x_n)$ they respectively receive as input. Note that for each input $(x_1, \ldots, x_n) \in \mathbb{R}^n$, $f_{i,j}$ is a polynomial of $(x_1, \ldots, x_n)$. As just remarked, the number s_i is bounded by 2^{t-i}. On the other hand, the depth of any such gate is at most i and thus the degree of the input function $f_{i,j}$ is bounded by 2^i for $j = 1, \ldots, s_i$. Thus, applying Proposition 6 of Chapter 16, we see that at height $t - 1$ there are at most

$$(2^{t-1} \cdot 2)^{O(n)}$$

satisfiable s_1-tuples of sign conditions for $f_{1,1}, \ldots, f_{1,s_1}$. Let η_1 be this number.

Each s_1-tuple of satisfiable sign conditions determines the output of the s_1 sign gates at height $t - 1$ and thus determines specific polynomials $f_{2,1}, \ldots, f_{2,s_2}$ in the variables $(x_1, \ldots, x_n)$ as inputs for the sign gates at height $t - 2$. Since the number s_2 of these gates is bounded by 2^{t-2} and the degree of their input functions is bounded by 2^2, another application of Proposition 6 of Chapter 16 yields for each s_1-tuple σ at most

$$(2^{t-2} \cdot 2^2)^{O(n)}$$

satisfiable s_2-tuples of sign conditions for $f_{2,1}^\sigma, \ldots, f_{2,s_2}^\sigma$. (The superscript σ is written to recall that these input functions depend on the preceding tuple of sign conditions σ.)

We then conclude that the number of satisfiable $(s_1 + s_2)$-tuples of sign conditions is bounded by

$$\eta_1 \cdot (2^{t-2} \cdot 2^2)^{O(n)} = 2^{t \cdot O(n)} 2^{t \cdot O(n)}.$$

Iterating this argument t times we obtain a bound for the number of sign conditions satisfied for all the sign gates of C of

$$\overbrace{2^{t \cdot O(n)} \cdot \ldots \cdot 2^{t \cdot O(n)}}^{t \text{ times}} = 2^{t^2 \cdot O(n)}.$$

By Proposition 7 of Chapter 16 the sets defined by one of these $(s_1 + \ldots + s_n)$-tuples of sign conditions have at most

$$(2^t \cdot 2^t)^{O(n)} = 2^{t \cdot O(n)}$$

connected components.

Since S is a finite union of some of these sets, we deduce that

$$\# \text{ connected components } (S) \leq 2^{t^2 \cdot O(n)}.$$

Solving for t in this inequality we get the desired bound.   □

19.3   Completeness in $P_{\mathbb{R}}$

The goal of this section is to prove the existence of $P_{\mathbb{R}}$-complete problems. We have already answered the question "Does $NC_{\mathbb{R}} = P_{\mathbb{R}}$?" in the negative (Corollary 1). Our aim now is to show the existence of maximal elements in $P_{\mathbb{R}}$ with respect to the partial order given by polylogarithmic time reductions.

Definition 1 A decision problem $S \subseteq \mathbb{R}^\infty$ is $P_{\mathbb{R}}$-*complete* if it belongs to $P_{\mathbb{R}}$ and satisfies the following property. For every problem $A \in P_{\mathbb{R}}$ there is a function $\varphi_A : \mathbb{R}^\infty \to \mathbb{R}^\infty$ computable by a parallel machine in time $O(\log^k n)$ for some $k \geq 1$ such that for all $x \in \mathbb{R}^\infty$, $x \in A$ if and only if $\varphi_A(x) \in S$. The function φ_A is said to be a *polylogarithmic time reduction*.

Remark 1 Compare $P_{\mathbb{R}}$-completeness to NP-completeness of Chapter 5.

Definitions of completeness depend both on a class of decision problems and on a class of reductions. NP-completeness was defined for NP problems using polynomial time reductions and now $P_{\mathbb{R}}$-completeness is defined for $P_{\mathbb{R}}$-problems using parallel polylogarithmic time reductions.

The following is an immediate consequence of Corollary 1.

Proposition 4 *If $S \subseteq \mathbb{R}^\infty$ is $P_{\mathbb{R}}$-complete, then $S \notin NC_{\mathbb{R}}$.*   □

Let us consider the set of all pairs (C, a) consisting of an algebraic circuit C with n input gates and a point $a \in \mathbb{R}^n$. This is the set of admissible inputs for the following problem,

$$\text{CEP} = \{(C, a) \mid \text{the circuit } C \text{ with input } a \text{ outputs } 1\},$$

which we call the *Circuit Evaluation Problem*. The main theorem of this section asserts that the CEP is $P_\mathbb{R}$-complete. The first step of the proof is to show that the problem is indeed in $P_\mathbb{R}$. We show this in a lemma which is stated slightly more generally.

Lemma 3 *Let R be an arbitrary ring. Then there is a machine over R such that given an algebraic circuit C over R with n input gates and a point $a = (a_1, \ldots, a_n) \in R^n$, computes the value output by C on input a in time polynomial in the size of C.*

Proof.   The machine works by evaluating a gate when its predecessors have already been evaluated. Constant gates are considered as evaluated from the beginning and input gates are evaluated with the coordinates of a. A gate is said to be *ready* when its predecessors have already been evaluated. Note that circuit gates are coded by tuples (j, t, i_ℓ, i_r) and thus we can use i_ℓ to store the value resulting from the evaluation procedure and i_r to store the value minus one to indicate that the gate has been evaluated. The machine is then described in the following way.

> input (C, a)
> evaluate the input gates of C
> **while** C has unevaluated gates **do**
>   scan C for a ready gate g
>   evaluate g
> **end do**
> **output** the evaluation of the output gate.

$\square$

Given a set S in $P_\mathbb{R}$, a polylogarithmic time reduction from S to CEP is implicitly provided in the proof of Theorem 2 of Chapter 18 adapted to a sequential machine. From this proof the next two results follow.

Theorem 2 *The Circuit Evaluation Problem is $P_\mathbb{R}$-complete.* $\square$

Corollary 2 *If a set $S \subseteq \mathbb{R}^\infty$ can be decided by a uniform machine in time t, then it can be decided by a uniform family of algebraic circuits of size $O(t^2)$. Moreover, if $t(n) = n^{O(1)}$, then the family is indeed L-uniform.* $\square$

We close this section by exhibiting another $P_\mathbb{R}$-complete problem.

Let f_i, g_j, h_k be polynomials in n variables $x_1, \ldots, x_n$ for $i = 1, \ldots, r$, $j = 1, \ldots, s$, and $k = 1, \ldots, t$. We call a *clause* a system of the form

$$\bigvee_{i=1}^{r} f_i(x) = 0 \ \vee \ \bigvee_{j=1}^{s} g_j(x) \geq 0 \ \vee \ \bigvee_{k=1}^{t} h_k(x) > 0,$$

where $\vee$ means disjunction. The clause is satisfiable when any of its conditions is.

We now consider *semi-algebraic quadratic systems in conjunctive normal form*

$$\bigwedge_{\ell=1}^{q} \varphi_{\ell}(x), \qquad (19.1)$$

where $\varphi_{\ell}(x)$ is a clause all of whose polynomials have degree at most two and $\wedge$ means conjunction. That is, the system is satisfiable when all its clauses are.

We have seen in Chapter 5 that the problem SA-FEAS of deciding whether such a system is satisfiable is $NP_{\mathbb{R}}$-complete. There is, however, a straightforward approach for deciding satisfiability of such systems by means of substitutions whose success is not guaranteed but which is widely used in linear algebra (where it certainly does succeed).

Definition 2 If a clause C_1 consists of a single equation of the form $ax - b = 0$, and a second clause C_2 contains the variable x in some equation, then the *substitution* of C_1 in C_2 is the clause obtained by the following process.

(1) Substitute every ocurrence of x in C_2 by b/a.

(2) Perform the operations with real constants that are possible to do.

(3) If (2) yields an inequality without variables, decide if the inequality is TRUE or FALSE.

(4) Simplify the resulting clause according to the rules

- FALSE $\cup E_1 \cup \cdots \cup E_k = E_1 \cup \cdots \cup E_k$, and
- TRUE $\cup E_1 \cup \cdots \cup E_k =$ TRUE.

We say that the semi-algebraic system given by (19.1) is *solvable by substitution* if we can obtain a point $(y_1, \ldots, y_n) \in \mathbb{R}^n$ satisfying (19.1) after a finite number of substitutions.

Our computational problem can now be stated in the following way.

Given a semi-algebraic quadratic system, is it solvable by substitution?

Let us denote by 2-SUBS this problem.

Lemma 4 *The problem* 2-SUBS *is in* $P_{\mathbb{R}}$.

Proof. Let ϕ be a semi-algebraic system of quadratic polynomials, S the subset of clauses consisting of a single equation of the type $ax - b = 0$, and T the subset of the remaining clauses. We suppose that FALSE is not a clause in ϕ and that we have an order on S and T. The following procedure decides whether ϕ is solvable by substitution.

> **while** $S \neq \emptyset$ **do**
>   let E be the first equation in S
>   let x be the variable appearing in E
>   **for all** clause $C \neq E$ in $S \cup T$ **do**
>   **if** x appears in C **then**
>   let C' be the substitution of E and C
>   remove C
>   **if** C' =FALSE **then** REJECT and halt
>   **elsif** C' =TRUE **then** remove C'
>   **elsif** C' is a single equation on a single variable
>   **then** add C' at the end of S
>   **else** add C' at the end of T
>   **end if**
>   **end if**
>   **end do**
>   remove E
> **end do**
> **if** $T = \emptyset$ **then** ACCEPT **else** REJECT **end if** .

Notice that this algorithm works for any order on the sets S and T. □

Theorem 3 *The problem* 2-SUBS *is* $P_{\mathbb{R}}$-*complete.*

Proof.   We reduce CEP to 2-SUBS. Given an algebraic circuit $\mathcal{C} = (g_1, \ldots, g_n)$ with s input gates and one output gate and an input $a = (a_1, \ldots, a_s)$, we associate with the pair $(\mathcal{C}, a)$ the system

$$\phi := \bigwedge_{i=1}^{n} E_i \wedge E_{n+1},$$

where, for every i:

- if g_i is an input gate, $E_i := \{X_i - a_i = 0\}$;

- if g_i is an arithmetic gate that performs the operation $*$ on the outputs of g_j and g_k, then $E_i := \{X_i - (X_j * X_k) = 0\}$;

- if g_i is a constant gate with associated constant $c \in \mathbb{R}$, then $E_i := \{X_i - c = 0\}$;

- if g_i is a sign gate outputing 1 for an input ≥ 0 and 0 otherwise, whose input is the output of gate g_j, then $E_i := E_{i1} \vee E_{i2}$ where

$$E_{i1} := \{X_j \geq 0 \wedge X_i - 1 = 0\} \quad \text{and} \quad E_{i2} := \{X_j < 0 \wedge X_i = 0\}$$

(note that E_i can be readily transformed into the conjunction of four clauses);

- finally, $E_{n+1} := \{X_n - 1 = 0\}$.

This construction can be done with $n + 1$ processors (one for each gate and another for the equation E_{n+1}) each working in constant time.

Moreover, it is clear that the resulting system has a solution if and only if the circuit C outputs 1 with input a, and in this case, the solution can be found by substitution.

From these facts we get that 2-SUBS is $P_{\mathbb{R}}$-complete. □

19.4 Digital Nondeterminism

Nondeterminism for real machines, as introduced in Chapter 5, formalizes the idea of finding a witness for a yes-instance in a continuous space; problems such as 4-FEAS capture the difficulty of such a search. However, in many situations, witnesses already exist in a discrete space and thus we need only guess from among a finite set of possibilities. Let us define this more formally.

Definition 3 A set $S \subseteq \mathbb{R}^\infty$ belongs to $DNP_{\mathbb{R}}$ if it can be decided in polynomial time by a nondeterministic machine whose guesses are restricted to come from the set $\{0, 1\}^\infty$. Such a machine is called a *digital nondeterministic* machine.

We show that natural examples of problems in $DNP_{\mathbb{R}}$ exist.

Example 1 The Knapsack and the Traveling Salesman problems belong to $DNP_{\mathbb{R}}$.

In order to see that an input $(x_1, \ldots, x_n) \in \mathbb{R}^n$ for the Knapsack has a positive answer, one just guesses $(b_1, \ldots, b_n) \in \{0, 1\}^n$ and checks that the addition of the x_j such that $b_j = 1$ yields 1. Similarly, to see that the real matrix of distances (a_{ij}) has a tour with total distance at most $k \in \mathbb{R}$, it is enough to guess the order $i_2, \ldots, i_n$ in which the cities are visited (we agree that the tour begins in the first city), to check that this is a tour, and then that the addition of the distances $a_{i_j i_{j+1}}$ for $j = 2, \ldots, n - 1$ plus $a_{i_n 1} + a_{1 n_2}$ is at most k.

Notice that in the second algorithm, one guesses numbers $i_2, \ldots, i_n$ that are integers between 1 and n by guessing for a number i its binary expansion $b_0, \ldots, b_s$ and then performing the addition $\sum_{j=0}^{s} b_j 2^j$.

The class $DNP_{\mathbb{R}}$ also has complete problems. They are defined using polynomial time reductions as in Chapter 5. Let us consider the following problem called *Binary Circuit Satisfiability*.

> Given an algebraic circuit C with n input gates, decide whether there is a point $x \in \{0, 1\}^n$ such that the circuit C with input x outputs 1.

A trivial modification in the proof of Theorem 2 yields the following result.

Theorem 4 *The Binary Circuit Satisfiability is $DNP_{\mathbb{R}}$-complete.* □

19.5 The Complexity of the Knapsack Problem

In this section we give a lower bound for the Knapsack Problem. In fact, we define a family of problems that generalize Knapsack and obtain lower bounds for all of them.

In the following, let $k \in \mathbb{N}$, $k \geq 2$, and $K = \{0, 1, \ldots, k-1\}$.

We say that a set $T \subseteq K^n$ is a *threshold subset* of K^n if there exists a point $a = (a_1, \ldots a_n) \in \mathbb{R}^n$ such that the linear function

$$h_a = a_1 x_1 + \ldots + a_n x_n - 1$$

satisfies $h(x) > 0$ for any $x \in T$ and $h(x) < 0$ for any $x \in K^n - T$. In this case we also say that a *defines* the threshold subset T.

Note that for a given threshold set T there exist uncountably many choices of $a_1, \ldots, a_n$ satisfying the preceding condition. The following easy proposition characterizes the set of hyperplanes that separate a given threshold set.

Proposition 5 *Let $\mathcal{H}_n$ be the set $\{(x_1, \ldots, x_n) \in \mathbb{R}^n \mid \forall (b_1, \ldots, b_n) \in K^n, \sum_{i=1}^n x_i b_i \neq 1\}$. Two points $v, w \in \mathbb{R}^n$ define the same threshold set if and only if they belong to the same connected component of $\mathcal{H}_n$.*

Proof. The points v and w belong to the same connected component of $\mathcal{H}_n$ if and only if for all $(b_1, \ldots, b_n) \in K^n$ they lie on the same side of the hyperplane given by the equation $\sum_{i=1}^n x_i b_i = 1$. But this is equivalent to saying that the functions

$$\sum_{i=1}^n v_i Y_i - 1$$

and

$$\sum_{i=1}^n w_i Y_i - 1$$

define the same threshold subset of K^n. □

A trivial consequence of the preceding proposition is the following corollary.

Corollary 3 *For any threshold subset T of K^n, the set of points $v = (v_1, \ldots, v_n) \in \mathbb{R}^n$ defining T is an open set of $\mathcal{H}_n$.* □

Lemma 5 *Let H be an affine subspace of $\mathbb{R}^n$ of dimension $n-1$ containing K^{n-1} and T be a threshold subset of K^{n-1} in H. Also, let S be a finite set of points in $\mathbb{R}^n - H$. Then there exists an $(n-2)$-dimensional affine subspace L of H defining T such that every $(n-1)$-dimensional affine subspace of $\mathbb{R}^n$ passing through L contains at most one point of S.*

Proof. Let us consider a fixed injection

$$\mathbb{R}^n \hookrightarrow \mathbb{P}(\mathbb{R}^{n+1})$$

of the affine space of dimension n in its projective closure. For any affine subspace A of $\mathbb{R}^n$, we denote by $\tilde{A}$ its closure in $\mathbb{P}(\mathbb{R}^{n+1})$.

Now, for any two different points $x, y \in S$ we consider the intersection of the projective line defined by x and y with $\tilde{H}$. This defines a finite set of points F in $\tilde{H}$.

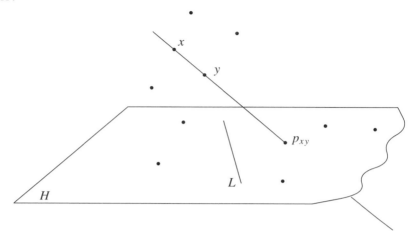

Figure B   The line given by x and y and the point p_{xy} it determines in H.

If we denote by $\widetilde{\mathcal{H}}_{n-1}$ the space of all the $(n-2)$-dimensional projective subspaces of $\tilde{H}$ we have that the set $\mathcal{H}_{n-1}$ defined in Proposition 5 is an open subset of $\widetilde{\mathcal{H}}_{n-1}$. Any element ℓ in $\widetilde{\mathcal{H}}_{n-1}$ is then given by a nonzero vector $(\ell_0, \ell_1, \ldots, \ell_{n-1}) \in \mathbb{R}^n$ together with all its nonzero multiples and thus, the space $\widetilde{\mathcal{H}}_{n-1}$ is isomorphic to $\mathbb{P}(\mathbb{R}^n)$. Now, for any point $p = (p_0, p_1, \ldots, p_{n-1}) \in F$, the condition $p \notin \ell$ can be written as

$$p_0 \ell_0 + p_1 \ell_1 + \ldots + p_{n-1} \ell_{n-1} \neq 0.$$

Thus, this condition defines a nonempty Zariski open set U_p in $\widetilde{\mathcal{H}}_{n-1}$. The requirement that no point of F belong to ℓ defines the finite intersection of the U_p for $p \in F$.

Since nonempty Zariski open sets are dense, the intersection of the U_p with the subset of $\mathcal{H}_{n-1}$ containing those $(n-2)$-dimensional affine subspaces of H that define T is nonempty. Any such subspace L in this intersection will fulfill our statement. $\qquad \square$

Theorem 5   *For any $k \geq 2$ and any $n \geq 1$ there are at least $k^{n^2/4}$ threshold subsets of K^n.*

Proof.   The statement is clearly true for $n = 1$ since in this case we have exactly k threshold subsets in K.

Let us then suppose $n \geq 2$ and let H be the affine subspace of $\mathbb{R}^n$ defined by the equation $X_n = 0$. There is a natural inclusion $K^{n-1} \subset H$ and, by induction hypothesis, the number of threshold subsets of K^{n-1} is greater than $k^{(n-1)^2/4}$. For each one of these threshold subsets, let us consider an $(n-2)$-dimensional affine subspace L satisfying the statement of Lemma 5. A hyperplane that rotates around the axis L will meet the points of K^n satisfying $X_n \neq 0$ one at a time. Therefore we can find $k^n - k^{n-1}$ hyperplanes through L defining different threshold subsets of K^n and, consequently, the total number of threshold subsets of K^n must be greater than

$$k^{(n-1)^2/4}(k^n - k^{n-1}) = k^{(n-1)^2/4}k^{n-1}(k-1)$$
$$\geq k^{(n^2+2n-2)/4}$$
$$\geq k^{n^2/4}.$$

$\square$

We call a *generalized Knapsack Problem of order d* the set

$$\mathrm{KP}_d = \{x \in \mathbb{R}^\infty \mid \exists b_1, \ldots, b_n \text{ s.t. } 0 \leq b_1, \ldots, b_n < 2^{n^d}$$
$$\text{with } n = \mathrm{size}(x) \text{ and } \sum_{j=1}^n b_j x_j = 1\}.$$

Note that the Knapsack Problem introduced in Section 1.2 is the generalized Knapsack Problem of order 0.

Lower bounds for generalized Knapsack problems follow.

Theorem 6 *For any $d \geq 0$, a lower bound of $\Omega(n^{d+2})$ holds for the number of arithmetical operations necessary to decide* KP_d.

Proof. According to Theorem 2 of Chapter 16, a lower bound of $\Omega(\log K_n)$ holds for KP_d where K_n is the number of connected components of the semi-algebraic subset of KP_d consisting of its points of size n. Now, by Proposition 5, this number coincides with the number of threshold subsets of K^n for $k = 2^{n^d}$. And by Theorem 5, this number is greater than $k^{n^2/4}$ and therefore greater than

$$(2^{n^d})^{n^2/4} = 2^{n^{d+2}/4}.$$

Taking logarithms the desired lower bound follows. $\square$

As a corollary, we deduce that the Knapsack Problem cannot be decided in parallel polylogarithmic time.

Corollary 4 *For any $d \geq 0$, the parallel time needed to decide KP_d is at least $\gamma\sqrt{n^{d+1}}$ for a universal constant $\gamma > 0$. Therefore, for all $d \geq 0$, one has that $\mathrm{KP}_d \notin \mathrm{NC}_\mathbb{R}$.* $\square$

Note that this Corollary cannot be deduced from the results in Section 19.1 since the algebraic set given by fixing the size n of the inputs fully decomposes into a union of hyperplanes.

We now use Theorem 6 to get a separation result for a specific model of computation.

Definition 4 Let $S \subseteq \mathbb{R}^{\infty}$. We say that S can be decided in *deterministic circuit time* $O(n^d)$, and we write $S \in \mathrm{DCTIME}_{\mathbb{R}}(O(n^d))$, if there is an L-uniform family of circuits $\{\mathcal{C}_n\}$ that decides S such that the size of $\mathcal{C}_n$ is $O(n^d)$.

We say that S can be decided in *nondeterministic circuit time* $O(n^d)$, and we write $S \in \mathrm{NCTIME}_{\mathbb{R}}(O(n^d))$, if there is an L-uniform family of circuits $\{\mathcal{C}_n\}$ and a polynomial $q(n)$ such that the size of $\mathcal{C}_n$ is $O(n^d)$ and the following holds. For each $n \geq 1$ the circuit $\mathcal{C}_n$ has $n + q(n)$ input gates and for every $x \in \mathbb{R}^n$, $x \in S$ if and only if there exists $y \in \mathbb{R}^{q(n)}$ such that $\mathcal{C}_n$ outputs 1 with input (x, y).

Proposition 6 *The union of the classes* $\mathrm{DCTIME}_{\mathbb{R}}(O(n^d))$ *for* $d \geq 1$ *is* $\mathrm{P}_{\mathbb{R}}$.

Proof.   This follows immediately from Lemma 3 and Corollary 2.   □

Remark 2 Deterministic circuit time as a measure of cost is close to the intuitive measure that counts the number of arithmetic operations performed by an algorithm. Thus, as we have just seen, polynomial circuit time yields the class $\mathrm{P}_{\mathbb{R}}$. Notice, however, that each class $\mathrm{DCTIME}_{\mathbb{R}}(O(n^d))$ contains decision problems that are unlikely to be solved by machines over $\mathbb{R}$ in time $O(n^d)$ due to management costs.

Theorem 7 *For all $d \geq 1$, the inclusion*

$$\mathrm{DCTIME}_{\mathbb{R}}(O(n^d)) \subset \mathrm{NCTIME}_{\mathbb{R}}(O(n^d))$$

is strict.

Proof.   According to Theorem 6, the number of arithmetic operations needed to decide the set KP_d has a lower bound of $\Omega(n^{d+2})$ for all $d \geq 0$. Therefore, $\mathrm{KP}_d \notin \mathrm{DCTIME}_{\mathbb{R}}(O(n^{d+1}))$.

On the other hand, the existence of an L-uniform family of circuits for KP_d of size $O(n^{d+1})$ follows from the algorithm,

> input $(x_1, \ldots, x_n)$
> **for** $j \in \{1, \ldots, n\}$ and $i \in \{0, \ldots, n^d - 1\}$ **do**
>   guess $b_{i,j} \in \{0, 1\}$
> **end do**
> compute $s := \sum_{i,j} b_{i,j} 2^i x_j$
> ACCEPT iff $s = 1$.

□

19.6   Additional Comments and Bibliographical Remarks

The main result of Section 19.1 is taken from [Cucker 1992b]. Proposition 2 is a special case of the real Nullstellensatz that characterizes the definition ideals for polynomial rings over the reals. The real Nullstellensatz was proved independently by Dubois [1969] and Risler [1970]; a weak form appeared in [Krivine 1964]. The books by Bochnak, Coste, and Roy [1987] and by Benedetti and Risler [1990] offer comprehensive expositions of algebraic geometry over the reals. Factorial rings and their properties can be found in [Childs 1995] and [Lang 1993a].

Theorem 1 is essentially due to Yao [1981]. Our presentation closely follows [Montaña and Pardo 1993]. Theorem 2 is a real version of a result of Ladner [1975]. Both Theorem 2 and Theorem 3 are taken from [Cucker and Torrecillas 1992].

Digital nondeterminism first appears in [Goode 1994] where Theorem 4 was also proved. Related results can be found in [Cucker and Matamala 1996]. The results of Section 19.5 are taken from [Cucker and Shub 1996]. They are also in [Meyer auf der Heide 1985a].

20
Weak Machines

In this chapter we study machines over $\mathbb{R}$ that penalize multiplication by modifying their cost in such a way that iterated multiplications become increasingly expensive. This situation is close to the classical where iterated multiplications can generate very large numbers and thus cause an increase in running time. We show that the complexity determined by this modified cost is closely related to the behavior of the algorithm on integer inputs in a very precise sense.

To do this, we introduce a cost measure called *weak* and study the properties of complexity classes defined using this cost. Our main results are: a P versus NP separation and the fact that the class NP_W obtained in this setting coincides with $NP_{\mathbb{R}}$. We also obtain a characterization of digital nondeterminism in terms of a bounded number of real guesses and prove that the linear programming problem over $\mathbb{R}$ can be solved in polynomial weak digital nondeterministic time.

20.1   A Restriction on the Number of Multiplications

For most of this chapter, only machines over $\mathbb{R}$ are considered.

Let M be a machine whose running time is bounded by a function t, and let $\alpha_1, \ldots, \alpha_k$ be its real constants. We have seen in Section 16.2 that for any input size n, the machine M determines an algebraic computation tree $T_{M,n}$ with depth $O(t(n))$. At a computation node v of this tree one coordinate of the state space of $T_{M,n}$, say x_ℓ, is modified. We associate with v a representation g_v/h_v, $g_v, h_v \in \mathbb{Z}[x, \alpha]$, of a rational function $f_v(x, \alpha)$ of the input variables $x = (x_1, \ldots, x_n)$ and the machine constants $\alpha = (\alpha_1, \ldots, \alpha_k)$. This function is just the kth coordinate of the composition of the arithmetic operations associated with the computation nodes along the path leading to v. The representation g_v / h_v is obtained by retaining numerators and denominators in this composition. For example, the representation of the product $g / h \cdot r / s$ is always gr / hs and the one of the addition $(g / h) + (r / s)$ always $(gs + hr) / hs$. We now use g_v and h_v to define the running time in the weak model.

Definition 1 The *weak cost* of any arithmetic node v is defined to be the maximum of $\deg(g_v)$, $\deg(h_v)$, and the maximum height of the coefficients of g_v and h_v, whereas the weak cost of any other node is 1. Here height denotes the height of an integer as defined in Chapter 6. For any $x \in \mathbb{R}^\infty$ of size n the *weak running time of M on x* is defined to be the sum of the costs of the nodes along its computational path in $T_{M,n}$. The *weak running time of M* is the function that associates with every n the maximum over all $x \in \mathbb{R}^n$ of the running time of M on x.

The classes P_W and NP_W of *weak deterministic* and *nondeterministic polynomial time*, respectively, are now defined using weak running time. Other classes, such as EXP_W of *weak exponential time*, can be similarly defined. In this chapter, and when necessary, we use the adjective *full* as oposed to weak when referring to the notions and complexity classes as introduced in Chapters 3 through 5.

Let us now consider a decision problem $S \subset \mathbb{R}^\infty$ over the reals. The set $S_\mathbb{Z} = S \cap \mathbb{Z}^\infty$ is the restriction of the problem S to the integers. For instance, let 4-FEAS be the set of points p in $\mathbb{R}^\infty$ such that p is the vector of coefficients of a polynomial of degree 4 having a real root. Accordingly, 4-FEAS$_\mathbb{Z}$ is the set of polynomials of degree 4 with integer coefficients that have a real root.

For most decision problems S over the reals, machines exist that do not perform any division and whose constants are all integers. Most algorithms presented in this book have this property,[1] in particular all in Part I. If M is such a machine, interpreting the arithmetical operations of M as operations over $\mathbb{Z}$ yields in a natural way a machine $M_\mathbb{Z}$ over $\mathbb{Z}$ for $S_\mathbb{Z}$. One can ask whether the cost bounds for M and $M_\mathbb{Z}$, considered with bit cost, are about the same. We now show that this is the case if the machine M is endowed with the weak cost.

Proposition 1 *Let S be a decision problem over $\mathbb{R}$ and let M be a machine with integer constants that solves S in weak polynomial time without performing divisions. Then $M_\mathbb{Z}$ solves $S_\mathbb{Z}$ in polynomial time over $\mathbb{Z}$.*

Proof. For an input in $\mathbb{Z}^\infty$, since M is weak, all the intermediately computed values have height polynomially bounded in terms of the heights of the inputs. Thus M, considered as a machine over $\mathbb{Z}$, works also in polynomial time. □

Remark 1 The converse of Proposition 1, that is, the assertion that if M decides $S_\mathbb{Z}$ in P over $\mathbb{Z}$, then it decides S in P_W, is false. Consider the problem of deciding whether an element in $\mathbb{R}^\infty$ belongs to $\mathbb{N}^\infty$. By Lemma 1 of Chapter 15, there exists

---

[1] An exception is the algorithm exhibited in Theorem 7 of Chapter 17.

a machine for solving this problem in $O(nm)$ arithmetic operations where n is the input length and m is the maximum of $\lceil \log(1 + |x_i|) \rceil$ over the input coordinates x_i. Considered over the integers, this machine works in polynomial time since the size of the input is exactly nm. However, as a machine over the reals, it has no time bound depending purely on the input size. The point here is that, although weak machines have a cost that is close to the bit cost, they measure the input size using unit height.

Remark 2 Proposition 1 and Remark 1 also hold over the rationals instead of integers with the appropriate modifications, that is, by allowing divisions and rational constants in M.

With the appropriate modifications, weakness can also be considered for parallel models of computation. Some care needs to be taken in working out the details. Then the class PAR$_W$ of sets decidable in weak parallel polynomial time can also be defined.

The main result of this section states some relations between weak and full complexity classes. Here the class DNP$_W$ of weak digital nondeterministic polynomial time is defined by requiring the guesses in NP$_W$ to be elements in $\{0, 1\}^\infty$ (cf. Section 19.4).

Theorem 1 *The relations in the following diagram hold*

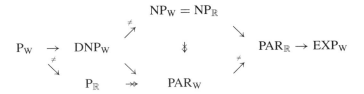

where an arrow $\rightarrow$ means inclusion, an arrow $\overset{\neq}{\rightarrow}$ means strict inclusion, and a crossed arrow $\nrightarrow$ means that the inclusion does not hold.

Before proving Theorem 1 we need some preliminary results. The following weak version of Proposition 3 of Chapter 19 is easily obtained by noting that the degrees of the polynomials in its proof are a bound for the weak running time.

Proposition 2 *Let $f_n \in \mathbb{R}[X_1, \ldots, X_n]$, $n \in \mathbb{Z}^+$, be a family of irreducible polynomials such that the zero set $\mathcal{Z}(f_n) \subset \mathbb{R}^n$ has real dimension $n - 1$. Let $d(n) = \deg(f_n)$. Then any parallel machine deciding the set $S = \{x \in \mathbb{R}^\infty \mid f_{\text{size}(x)}(x) = 0\}$ has weak running time greater than $d(n)$.* □

The following lemma is also useful.

Lemma 1 *If a set $S \subseteq \mathbb{R}^\infty$ can be decided in full parallel time t, then it can be decided in weak sequential time $2^{O(t)}$.*

Proof. The weak machine simply simulates the parallel one as in Proposition 2 of Chapter 18. The number of arithmetic operations done by the weak machine is $2^{O(t)}$ and, since the degrees and coefficient lengths of the rational functions associated with the parallel machine are bounded by 2^t, the result follows. $\quad\square$

Proof of Theorem 1. We first prove that $NP_\mathbb{R} = NP_W$. This is an immediate consequence of the NP_W-completeness of 4-FEAS. Thus, we need to prove two things: the hardness of 4-FEAS with respect to NP_W and the membership of 4-FEAS in this class.

For the first, we observe that the p-reductions given in Chapter 5 reducing any problem in $NP_\mathbb{R}$ to 4-FEAS, work in weak polynomial time. This can be seen by checking the weakness of these reductions.

To show membership of 4-FEAS in NP_W, note that the nondeterministic algorithm that checks whether a guess ζ is a root of the input polynomial only generates polynomials of degree 5 in the coefficients of the input and ζ and therefore runs in weak polynomial time.

We thus have an NP_W algorithm for solving all problems S in $NP_\mathbb{R}$ by composing the reduction of S to 4-FEAS with the algorithm for solving this latter problem.

Now we prove that $P_\mathbb{R} \not\subseteq PAR_W$. To do so, consider the set $S = \{x \in \mathbb{R}^\infty \mid x_1 = x_2^{2^n}$, where $n = \text{size}(x)\}$. By Proposition 2, the set S cannot be decided in weak polynomial parallel time. On the other hand, it clearly belongs to $P_\mathbb{R}$.

That $NP_W \not\subseteq PAR_W$ now follows since $P_\mathbb{R} \subset NP_\mathbb{R} = NP_W$. Note that it can also be shown observing that the following sentence

$$\exists y_1 \exists y_2 \ldots \exists y_{n-1} \left(x_2^2 = y_1 \ \& \ y_1^2 = y_2 \ \& \ \ldots \ \& \ y_{n-1}^2 = x_1 \right)$$

is equivalent to $x_1^{2^n} = x_2$ and that it can be checked in weak nondeterministic polynomial time.

We now show that the inclusions $DNP_W \subset NP_W$ and $PAR_W \subset PAR_\mathbb{R}$ are strict. The first assertion follows from the inclusion $DNP_W \subseteq PAR_W$ (the parallel machine just tests the exponential number of possible guesses independently) and the fact that $NP_W \not\subseteq PAR_W$ which we have just proved. The second one follows from the inclusion $P_\mathbb{R} \subseteq PAR_\mathbb{R}$ and the inequality $P_\mathbb{R} \not\subseteq PAR_W$ proved previously.

The inclusion $NP_\mathbb{R} \subset PAR_\mathbb{R}$ is Corollary 1 of Chapter 18 and the inclusion $PAR_\mathbb{R} \subset EXP_W$ follows from Lemma 1.

We finally prove that $NP_W \subseteq EXP_W$. According to Corollary 1 of Chapter 18, the problem 4-FEAS can be solved in parallel polynomial time in the full model. Thus, it can be solved in exponential time in the weak model by Lemma 1. Since 4-FEAS is NP_W-complete, the inclusion $NP_W \subseteq EXP_W$ follows. $\quad\square$

Corollary 1 *The problem* 4-FEAS *cannot be solved in weak polynomial time even allowing parallelism or digital nondeterminism.* $\quad\square$

Remark 3 Note that the weak setting contrasts with the classical and the full real settings since, in both the latter cases, the classes NP are included in their corresponding PAR.

Remark 4 Lemma 1 cannot be applied to show that $NC_\mathbb{R} \subseteq P_W$. This is because in time $\log^k n$ one can generate polynomials of degree $2^{\log^k n}$. A straightforward application of Proposition 2 shows that $NC_\mathbb{R}^k \not\subseteq P_W$ for every $k \geq 2$.

Nevertheless, several natural problems in $NC_\mathbb{R}$ do belong to P_W, for instance, the computation of characteristic polynomials as in Section 15.5. One can check this by inspecting the proof of Proposition 20 of Chapter 15.

Theorem 1 asserts that the classes NP_W and $NP_\mathbb{R}$ coincide. However, one does not know if $NPC_W = NPC_\mathbb{R}$, where NPC_W and $NPC_\mathbb{R}$ denote the classes of complete problems in NP_W and $NP_\mathbb{R}$, respectively. In the first case, the reductions considered are in P_W and, in the second, in $P_\mathbb{R}$. Actually, it is trivial that $NPC_W \subseteq NPC_\mathbb{R}$. The reverse inclusion, however, seems less trivial according to the next result.

Theorem 2 *If* $NPC_W = NPC_\mathbb{R}$, *then* $P_\mathbb{R} \neq NP_\mathbb{R}$.

Proof. Let $S \in DNP_W$ be a nonconstant decision problem; that is, $S \neq \mathbb{R}^\infty$ and $S \neq \emptyset$.

Let us suppose that $P_\mathbb{R} = NP_\mathbb{R}$. Then all nonconstant problems in $NP_\mathbb{R}$ are $NP_\mathbb{R}$-complete and, in particular, so is S. Therefore, by our hypothesis, S is NP_W-complete. Since $S \in DNP_W$ this implies that $DNP_W = NP_W$, contradicting Theorem 1. $\square$

Let us denote by $P_W^=$ and $DNP_W^=$ the classes of weak polynomial and nondeterministic polynomial time for machines over $\mathbb{R}$ that branch only on equality comparisons.

Theorem 3 *The Knapsack Problem belongs to* $DNP_W^=$ *but not to* $P_W^=$.

Proof. This is a straightforward consequence of Theorem 6 of Chapter 2. $\square$

20.2   A Characterization of Digital Nondeterminisnm

In this section we give a characterization of digital weak nondeterminism in terms of a bounded version of weak nondeterminism.

Definition 2 For any natural number $k \geq 1$ denote by $NP_W^{[k]}$ the subclass of NP_W containing those sets S such that there is a weak nondeterministic machine that works in polynomial time and accepts all $x \in S$ with guesses of size at most k.

Remark 5 It follows from Theorem 1 of Chapter 18 that the feasibility problem for a system of polynomial equations and inequalities in a constant number of variables is in $NC_\mathbb{R}$. One can also prove that it also belongs to P_W.[2] We rely on this fact to prove the next theorem.

---

[2]This can be checked by analyzing the complexity of the algorithms given in [Renegar 1992a; Heintz, Roy, and Solerno 1990; Basu, Pollack, and Roy 1994]. These algorithms

Theorem 4

(i) $DNP_W \subseteq NP_W^{[1]}$.

(ii) *For any* $k \geq 1$, $NP_W^{[k]} \subseteq DNP_W$.

Proof.　In order to prove (i) we note that for any $S \in DNP_W$ and any $x \in S$ we can encode the polynomial number of guesses in $\{0, 1\}$ necessary to decide the membership of x to S in the first digits of the binary expansion of a real number $\xi \in (0, 1)$. The machine in $NP_W^{[1]}$ then guesses ξ and pumps these digits as long as they are needed.

To prove (ii) let us consider a set S in $NP_W^{[k]}$. Then there is a polynomial p and a weak nondeterministic machine M that decides S using time at most $p(n)$ for every input of size n and with at most k real guesses. At each branching node of M, a test is done of the form $f(\alpha, x, z) \geq 0$, where $\alpha = (\alpha_1, \ldots, \alpha_h)$ are the constants of M, $x = (x_1, \ldots, x_n)$ is the input, $z = (z_1, \ldots, z_k)$ are the guessed values, and f is a rational function with integer coefficients. Moreover, if r and q are the numerator and denominator of f, then, since M is weak, we know that the degrees of r and q as well as the heights of their coefficients are also bounded by $p(n)$.

Consider a weak digital nondeterministic machine M_D that performs the following operations. With input $x_1, \ldots, x_n$ the machine M_D performs the computations of M in the ring $\mathbb{R}[Z_1, \ldots, Z_k]$ by working with pairs of numerators and denominators. At each step i of the computation in which a branching node is reached, M_D guesses $b_i \in \{0, 1\}$ and takes β^+ if $b = 1$ and β^- otherwise. Moreover, if the rational function whose positivity is tested at this branching node is f_i, M_D stores the f_i as well as the result b_i of the test. The weakness of M ensures the weakness of M_D.

At the end of this simulation, M_D rejects an input if M does. On the other hand, the sequence of guesses determines a path whose feasibiliy is given by a system of inequalities of the form $f_i(Z_1, \ldots, Z_k) \geq 0$ or $f_i(Z_1, \ldots, Z_k) < 0$, where the number of inequalities is at most $p(n)$. Because of the preceding remark, and since the degrees of the f_i are also bounded by $p(n)$, feasibility of this system is decidable in weak polynomial time. Thus, if the sequence of guesses leads to acceptance by M, the machine M_D tests the preceding system for feasibility and accepts if it is feasible.　□

Corollary 2　*For all* $k, q \geq 1$, $NP_W^{[k]} = NP_W^{[q]} = DNP_W$.　□

---

have only integer built-in constants and, when applied to a system of integer polynomials, they work within essentially the same time bounds for the classical cost. But unlike the example of Remark 1, this bound for the classical cost translates to a weak cost for the polynomial time algorithm over the reals.

20.3   Linear Programming and Digital Nondeterminism

The main result of this section is that $LPF \in DNP_W \cap coDNP_W$.

Let A be an $m \times n$ nonzero real matrix and $b \in \mathbb{R}^m$ and consider the polyhedron P defined as the set of $x \in \mathbb{R}^n$ such that $Ax \geq b$. Thus, P is defined by the inequalities $A_i x \geq b_i$ for $i = 1, \ldots, m$.

Suppose that P is nonempty. We say that a nonempty set $F \subseteq \mathbb{R}^n$ is a *minimal face* of P if there is a set $J \subseteq I = \{1, \ldots, m\}$ such that F is the set of $x \in \mathbb{R}^n$ satisfying $A_i x = b_i$ for all $i \in J$ and, for every $i \notin J$, the affine function $A_i - b_i$ is strictly positive on F. The set J *defines* F. It is easy to see that minimal faces are included in P. We now show that minimal faces exist.

Lemma 2 *Let A, b and P be as in the preceding and assume that P is not empty. Then there exist minimal faces of P.*

Proof.   For $m = 1$, the statement holds trivially. For $m > 1$, consider the matrix A' and the vector b' consisting of the first $m - 1$ rows of A and the first $m - 1$ components of b, respectively. Denote by P' the polyhedron in $\mathbb{R}^n$ defined by $A'x \geq b'$. If $A_m - b_m$ is strictly positive on P', then a minimal face of P', which exists by the induction hypothesis, will be a minimal face of P.

Otherwise, we must have $H \cap P' \neq \emptyset$, where H is the hyperplane given by the equation $A_m - b_m = 0$. Change coordinates such that H becomes the hyperplane given by $x_n = 0$ and consider A'' to be the matrix resulting from eliminating the last column in A'. If A'' is the zero matrix, then $J = \{i < m \mid b_i = 0\} \cup \{m\}$ defines a minimal face of P. Otherwise, consider the nonempty polyhedron P'' in $\mathbb{R}^{n-1}$ defined by the system $A''x \geq b'$. By the induction hypothesis, there is a minimal face F'' of P''. Let $J'' \subset \{1, \ldots, m-1\}$ define F''. The set $J = J'' \cup \{m\}$ defines a minimal face of P. $\square$

Theorem 5 *The Linear Programming Feasibility problem belongs to the class $DNP_W \cap coDNP_W$.*

Proof.   The fact that LPF belongs to DNP_W is shown by considering the following digital nondeterministic algorithm that decides LPF.

> input (A, b)
> **if** $A = 0$ and b contains some positive coordinate **then** REJECT
> **elsif** $A = 0$ and $b \leq 0$ **then** ACCEPT
> **else** guess $J^* \subseteq I$ defining a minimal face F
>   compute any solution x^* of the system $A_i x = b_i$ with $i \in J^*$
>   check that for all $i \notin J^*$ the inequality $A_i x^* \geq b_i$ holds
>   and ACCEPT in that case; otherwise REJECT
> **end if** .

Note that, by definition of minimal face, we can choose any x^* in F. Moreover, the main step in the algorithm solves a linear system and thus it is performed in weak polynomial time.

Now, in order to show that LPF also belongs to coDNP$_W$ we apply Farkas's Lemma. Our aim is to design a digital nondeterministic algorithm working in weak polynomial time such that on input (A, b) accepts if and only if (A, b) is not feasible. By Farkas's Lemma, we know that (A, b) is not feasible if and only if the system

$$
\begin{aligned}
y &\geq 0 \\
yA &= 0 \\
y \cdot b &> 0
\end{aligned}
$$

(20.1)

is feasible. Now we claim that this latter system is feasible if and only if

$$
\begin{aligned}
z &\geq 0 \\
zA &= 0 \\
z \cdot b &= 1
\end{aligned}
$$

(20.2)

is feasible. In fact, any solution of (20.2) proves feasibility of (20.1). On the other hand, if $y \in \mathbb{R}^m$ is any solution of (20.1), then $z = y/(y \cdot b)$ satisfies system (20.2).

Now we apply the preceding algorithm to (20.2) and we are done.   □

Theorem 5 has some immediate consequences. Since DNP$_W$ $\neq$ NP$_W$ we deduce the following corollary.

Corollary 3 *The set* LPF *is not complete in* NP$_W$ *nor is it complete in* coNP$_W$ *for reductions in* P$_W$.

It is not known if LPF is NP$_\mathbb{R}$-complete, nor is it likely to be by the following result.

Corollary 4 *If* LPF *is complete in* NP$_\mathbb{R}$ *or complete in* coNP$_\mathbb{R}$, *then we have the following collapse of complexity classes*

$$
\text{coDNP}_\mathbb{R} = \text{DNP}_\mathbb{R} = \text{NP}_\mathbb{R} = \text{coNP}_\mathbb{R}.
$$

20.4   Additional Comments and Bibliographical Remarks

Weak machines and complexity classes were introduced by Koiran [1993] where some evidence of the separation P$_W$ $\neq$ NP$_W$ was given in terms of classical complexity. More concretely, it was shown that if P$_W$ $=$ NP$_W$, then the classical polynomial hierarchy collapses at its second level. Subsequently, Cucker, Shub, and Smale [1994] showed that indeed P$_W$ $\neq$ NP$_W$. Section 20.1 closely follows this article. Theorem 4 is taken from [Meer 1994] where it is also shown that the quadratic programming problem belongs to DNP$_W$. The first part of Theorem 5 is then a special case of this result.

21
Additive Machines

A number of computational problems considered thus far have algorithms that make no use of multiplication or division. This is the case for the Knapsack or Traveling Salesman problems.

The goal of this chapter is to pursue the study of complexity classes of machines that do not perform multiplication or division. Additive machines rely on linear algebra and linear inequalities instead of the more complicated semi-algebraic geometry. Consequently, we are able to prove theorems about these restricted machines that are unproved in the unrestricted model.

More precisely, this chapter attempts a classification of complexity problems between P_{add} and EXP_{add} (the subscript add denotes the absence of multiplications). We introduce a hierarchy of classes of increasing complexity between NP_{add} and EXP_{add} by allowing "existential" guesses to alternate with "universal" assertions. The main results are a characterization of this hierarchy in terms of oracle machines and the fact that the guesses can be restricted to be 0 or 1 without loss in computational power. It is this feature that allows us to prove that all problems in this hierarchy belong to EXP_{add}.

21.1 Additive Machines and Complexity Classes

Only machines over $\mathbb{R}$ are considered in this chapter.

Definition 1 We define *additive machines* in the same way we defined machines in Section 3.2, but now the functions allowed at the computational nodes must have the form $x_i = y \circ z$, where $\circ$ is either addition or subtraction and y, z are either coordinates of the state space or machine constants belonging to the base ring R.

Again with appropriate modification, we define additive versions of P and NP. We denote by P_{add}, NP_{add}, and EXP_{add} the classes of problems decided in polynomial, nondeterministic polynomial, and exponential time, respectively. We also consider machines over $(\mathbb{R}, =)$, that is, machines which branch on equality comparisons only. We distinguish the complexity classes they define by the superscript $=$. Thus, $P_{add}^=$, $NP_{add}^=$, and $EXP_{add}^=$ denote the corresponding versions of the three complexity classes just mentioned.

Natural examples of sets in these classes exist. One can check that the Knapsack Problem belongs to $NP_{add}^=$ and the Traveling Salesman Problem belongs to NP_{add}. Moreover, as we have seen in Chapter 18, one can restrict the nondeterministic guesses to be sequences of 0s and 1s. This is not a particular feature of these examples. It turns out that all problems in NP_{add} or $NP_{add}^=$ can be decided with digital nondeterminism. Showing this is one of the main goals of this chapter.

Our first result deals with the amount of memory needed by additive machines. It shows that a computation of an additive machine can be done using a fixed finite amount of memory (besides that used for storing the input) without an exponential increase in the running time. Its proof strongly relies on the presence of the order relation, and therefore does not apply to machines over $(\mathbb{R}, =)$.

Definition 2 Let M be a machine and $t : \mathbb{N} \to \mathbb{N}$ a function. We say that M *works in space t* if, for any input of size n, the computation of M uses at most $t(n)$ coordinates of its state space in addition to those containing the input.

Theorem 1 *Let $S \subseteq \mathbb{R}^\infty$ and M be an additive machine over $(\mathbb{R}, <)$ that decides S in time bounded by a function t with $t : \mathbb{N} \to \mathbb{N}$, $t(n) \geq n$. Then there exists another additive machine M' over $(\mathbb{R}, <)$ and a constant $d \in \mathbb{N}$ such that M' decides S in time bounded by $O(t(n)^d)$ and works in constant space.*

Proof. Our argument relies on a property that is essential for many results in this chapter. Let $\alpha_1, \ldots, \alpha_k \in \mathbb{R}$ be the constants of M. For an input $u = (u_1, \ldots, u_n) \in \mathbb{R}^n$ and after t steps of the computation of M on this input, the content x of a coordinate of the state space of M has the form

$$x = \sum_{i=1}^k a_i \alpha_i + \sum_{i=1}^n b_i u_i, \qquad (21.1)$$

where a_i, b_i are integers of height at most t. For fixed t, u, α we denote by $\mathcal{H}_{u,\alpha}^t$ the subset of $\mathbb{R}$ consisting of the real numbers x for which a summation (21.1) exists.

This property enables us to "efficiently" encode the contents of the $t(n)$ coordinates that are used into a single real number (in fact, a positive integer). Denote by $\overline{a_i}$ and $\overline{b_i}$ the binary expansions of the integers a_i and b_i, respectively. We can encode an element $x \in \mathcal{H}_{u,\alpha}^t$ by the sequence over $\{0, 1, 2\}$,

$$s_x = \overline{a_1}\, 2\, \overline{a_2}\, 2 \ldots 2\, \overline{a_k}\, 2\, \overline{b_1}\, 2\, \overline{b_2}\, 2 \ldots 2\, \overline{b_n}.$$

Suppose that at a certain moment of the computation the first q coordinates of the state space have been used. Let $x_1, \ldots, x_q$ be their contents. This sequence can be encoded by a sequence s of elements from $\{0, 1, 2, 3\}$

$$s = s_1 3 s_2 3 \ldots 3 s_q,$$

where s_i is the sequence corresponding to x_i, and a 3 is used to separate s_i and s_{i+1} for $1 \leq i < q$. Let ζ be the integer whose expansion in base-4 is s. We say that ζ codes the sequence $x_1, \ldots, x_q$ with respect to the pair (u, α). The number x_i is the ith number encoded by ζ. For $i > q$ we agree that the ith number encoded by ζ is 0. Note that the sequence s has $O(nqt(n))$ elements and consequently, $\zeta < 4^{O(nqt(n))}$.

Given any finite sequence s over $\{0, 1, 2, 3\}$ denote by $h(s)$ the number of digits of s, that is, the number of its elements, and by ζ_s the integer whose expansion in base-4 is s.

The following problems can be solved by additive machines that use a constant number of registers in addition to the ones occupied by $u_1, \ldots, u_n$.

(i) Compute the integer ζ that codes the sequence $u_1, \ldots, u_n$ with respect to (u, α).

(ii) Given $\zeta \in \mathbb{N}$ coding a sequence $x_1, \ldots, x_q$ of elements in $\mathcal{H}_{u,\alpha}^{t(n)}$ and $i \in \mathbb{N}$, output $\zeta_i \in \mathbb{N}$ coding the value x_i of the ith real number encoded in ζ.

This is done as in Lemma 1 of Chapter 18 using only multiplications by four (which are done with two additions) and comparisons.

Let ℓ, s, r be sequences over $\{0, 1, 2, 3\}$ such that the expansion in base-4 of ζ has the form $\ell s r$, where s denotes the digits encoding x_i. Note that while computing ζ_i one can also compute $\zeta_\ell, \zeta_r, h(\ell),$ and $h(r)$.

(iii) Given $\zeta \in \mathbb{N}$ coding a sequence $x_1, \ldots, x_q$ of elements in $\mathcal{H}_{u,\alpha}^{t(n)}$ and $i \in \mathbb{N}$, output the value x_i of the ith real number encoded in ζ.

(iv) Given $\zeta \in \mathbb{N}$ which codes a sequence $x_1, \ldots, x_q$ of numbers in $\mathcal{H}_{u,\alpha}^{t(n)}$, ζ' coding an element x' in $\mathcal{H}_{u,\alpha}^{t(n)}$, and $i \in \mathbb{N}$, replace the value x_i of the ith real number encoded in ζ by x' and output the resulting ζ.

To do so, just note that the new value of ζ is now

$$r + 4^{h(r)} x' + 4^{h(r)+h(x')} \ell,$$

where ℓ and r are as in (ii) and $h(x')$ is the number of digits of the expansion in base-4 of ζ'.

(v) Given ζ_1 and ζ_2 coding real numbers $x_1, x_2 \in \mathcal{H}_{u,\alpha}^{t(n)}$, output ζ which codes $x_1 + x_2$. Do the same for $x_1 - x_2$, for $x_1 + \alpha_j$, or for $x_1 - \alpha_j$ with $1 \leq j \leq k$.

Just note that adding or subtracting x_1 and x_2 reduces to adding or subtracting their corresponding coefficients a_i and b_i.

The machine M' we are looking for behaves as follows. With input $u = (u_1, \ldots, u_n)$ it first computes $\zeta \in \mathbb{N}$ encoding $u_1, \ldots, u_n$ with respect to (u, α). This is (i) in the preceding. Then, it simulates M by executing the computations of M on ζ. At a branch node of M which tests the sign of a coordinate x_i of its state space, M' gets x_i from ζ as in (iii) and performs the required comparison. At a computation node, M' gets the terms of the operation from ζ and performs the operation as in (v). Then, it modifies ζ as in (iv) by appropriately storing the result of the operation.

Since (i) through (v) have cost polynomial in n and $t(n)$, the theorem follows.
□

Our next result separates P from NP in the most restricted setting we consider in this chapter.

Theorem 2 $P_{add}^= \neq NP_{add}^=$.

Proof.   As we have already remarked, the Knapsack Problem belongs to $NP_{add}^=$. On the other hand, we proved in Theorem 3 of Chapter 20 that it does not belong to $P_W^=$ and therefore does not belong to $P_{add}^=$. □

Our last result in this section is an extension of Theorem 5 in Chapter 6. It proves the existence of "small" points in (not necessarily closed) polyhedra when the defining matrix has "small" entries. A remarkable feature of this result is that it holds independently of the number of inequalities defining the polyhedron. This plays a central role in the next section.

Theorem 3 *Let P be a nonempty polyhedron of $\mathbb{R}^n$ defined by a system*

$$A_1 x \leq b_1; \; A_2 x < b_2 \tag{21.2}$$

of N_1 nonstrict inequalities and N_2 strict inequalities, where the entries of A_1 and A_2 are integers of height bounded by L. Then there is an $x \in P$ with the following description,

$$x_i = \sum_{j \in I_1} u_{ij} b_{1j} + \sum_{j \in I_2} v_{ij} b_{2j} + w_i, \; i = 1, \ldots, n,$$

where $I_1 \subset \{1, \ldots, N_1\}$, $I_2 \subset \{1, \ldots, N_2\}$, $\mathrm{card}(I_1) + \mathrm{card}(I_2) \leq n$, and the coefficients u_{ij}, v_{ij}, w_i are rationals of height at most $(Ln)^c$ for some constant c.

Proof.   Since P is not empty, for some $\epsilon_0 > 0$ and for any $\epsilon \in (0, \epsilon_0)$, the system

$$A_1 x \leq b_1; \; A_2 x \leq b_2 - \epsilon \tag{21.3}$$

is satisfiable. Conversely, for any $\epsilon > 0$ a solution to (21.3) is also a solution to (21.2). In order to obtain a solution of the desired form, we thus apply Theorem 5 of Chapter 6 to (21.3) for a well-chosen value of ϵ.

According to Theorem 5 in Chapter 6, for any $\epsilon \in (0, \epsilon_0)$ the new system (21.3) has a solution $x(\epsilon)$ of the form

$$x_i(\epsilon) = \sum_{j \in I_1} u'_{ij} b_{1j} + \sum_{j \in I_2} v'_{ij}(b_{2j} - \epsilon), \tag{21.4}$$

where u'_{ij} and v'_{ij} are rationals of height $(Ln)^{O(1)}$, and $|I_1| + |I_2| \le n$. Replacing the variables x_i in (21.3) by the right-hand side of (21.4) shows that every $\epsilon \in (0, \epsilon_0)$ is a solution of a system $\mathcal{S}$ of $N_1 + N_2$ inequalities of the form $z_i \epsilon + y_i \ge 0$, where

$$y_i = \sum_{j \in I_1} u''_{ij} b_{1j} + \sum_{j \in I_2} v''_{ij} b_{2j}$$

and the coefficients z_i, u''_{ij}, v''_{ij} are rationals of height $(Ln)^{O(1)}$.

Note that for different ϵ we may obtain different vectors of coefficients u''_{ij}, v''_{ij}. Nevertheless, the set $\mathcal{F}$ of possible vectors is finite since they are obtained as determinants of submatrices of the matrix A' in Theorem 5 in Chapter 6. Let us consider for any $f \in \mathcal{F}$ the set U_f containing those ϵ whose coefficient vector in the preceding is f. By reducing ϵ_0 if necessary, we can suppose that 0 belongs to the closure of U_f for all $f \in \mathcal{F}$.

Now, any $f \in \mathcal{F}$ gives rise to a system $\mathcal{S}_f$ as previously. For $\mathcal{S}_f$ to be satisfied by all $\epsilon \in U_f$, every y_i has to be nonnegative since 0 belongs to the closure of U_f. If z_i is also nonnegative, any $\epsilon > 0$ is a solution to the constraint $z_i \epsilon + y_i \ge 0$. If $z_i < 0$, any $\epsilon \le -y_i/z_i$ is a solution. Let

$$J = \{(i, f); z_i < 0 \text{ in the system corresponding to } f\}.$$

If $J = \emptyset$, we set $\epsilon = 1$. Otherwise, we set $\epsilon = \min_{i \in J} -y_i/z_i$. The desired solution to (21.2) is $x(\epsilon)$. $\qquad\square$

21.2   The Polynomial Hierarchy

Let us consider the following version of the Traveling Salesman problem, which we call the *Unique Solution Traveling Salesman Problem* and we denote UTSP.

> Given n cities, the distances (a_{ij}) between them, and a positive number k, does there exist a *unique* tour through all the cities with total distance at most k?

It is unlikely that UTSP $\in$ NP$_{\mathrm{add}}$ since, in order to check that a tour τ has the desired property, we need to check that the tour has total distance at most k and that all other possible tours (and there are $(n-1)! - 1$ of them) have total distance greater than k. Although the first condition can be tested in NP$_{\mathrm{add}}$, it seems unlikely that the second one can be. If we write the property defining UTSP in the following way,

$$\exists x \in \mathbb{R}^{n-1} \, \forall y \in \mathbb{R}^{n-1}(x \text{ is a tour and distance}(x) \leq k) \, \& \qquad (21.5)$$
$$(y \text{ is a tour and } x \neq y \Rightarrow \text{distance}(y) > k),$$

we notice that although the quantifier-free part of this formula is certainly a relation in P_{add}, the alternation of quantifiers acts as an obstruction for the membership in NP_{add}. On the other hand, we note that all the variables appearing under the scope of a quantifier can be restricted to belong to $\{0, 1\}$. This fact permits us to decide UTSP in exponential time by a brute force search on the $2^{2(n-1)}$ possible choices of x and y.

Thus, although both TSP and UTSP can be decided in exponential time, we might say that UTSP is more complex than TSP because of the quantifier alternation in the description (21.5).

Classifying objects according to the number of quantifier alternations in formulas describing them is a classical technique used in mathematical logic. This classification leads to a hierarchy of classes

$$C_0 \subseteq C_1 \subseteq \ldots C_i \subseteq C_{i+1} \subseteq \ldots ,$$

where the class C_i contains all those objects that can be described with not more than $i - 1$ alternations.[1] A primary concern is to show that the inclusions are all strict.

In this section we introduce the polynomial hierarchy for additive machines. The objects described here are subsets of $\mathbb{R}^\infty$ and the descriptions are like the one for the UTSP given in (21.5) in the sense that they allow a bounded number of quantifier alternations. We classify sets inside this hierarchy according to the number k of these alternations. Thus UTSP is a natural example of a problem in the class corresponding to $k = 2$ and problems in NP_{add}, such as TSP, examples for $k = 1$.

Recall that a decision problem is a subset of $\mathbb{R}^\infty$. We can also consider subsets of $(\mathbb{R}^\infty)^k$ for some $k \geq 2$ as decision problems (see Section 3.5).

Definition 3 Define $\Sigma_0^{add} = \Pi_0^{add} = P_{add}$. For $k \geq 1$, a set $L \subset \mathbb{R}^\infty$ is in Σ_k^{add} if there exists a set $A \subseteq (\mathbb{R}^\infty)^{k+1}$, $A \in P_{add}$, and polynomial functions $p_1, \ldots, p_k$ such that for any $u \in \mathbb{R}^n$, the following condition holds.

$$u \in L \Leftrightarrow Q_1 x_1 \in \mathbb{R}^{p_1(n)} \ldots Q_k x_k \in \mathbb{R}^{p_k(n)}(x_1, \ldots, x_k, u) \in A. \qquad (21.6)$$

The quantifiers $Q_i \in \{\exists, \forall\}$ alternate, starting with $Q_1 = \exists$, and the values $p_i(n)$ are natural numbers.

---

[1] Perhaps the most influential of these hierarchies has been the arithmetical hierarchy of Kleene that classifies undecidable problems over $\mathbb{Z}_2$. Here Σ_0 are the decidable problems, Σ_1 the recursively enumerable ones, and the remaining Σ_k for $k \in \mathbb{N}$ contain problems increasingly difficult under a certain well-defined measure.

The class Π_k^{add} is defined by the same condition except that the first quantifier is universal ($Q_1 = \forall$). The *polynomial hierarchy* PH$_{\mathrm{add}}$ is defined as the union of Σ_k^{add} for all k.

Remark 1 One can also define the classes $\Sigma_k^{\mathrm{add}=}$ and $\Pi_k^{\mathrm{add}=}$ by assuming that $A \in \mathrm{P}_{\mathrm{add}}^=$, and this leads to the hierarchy PH$_{\mathrm{add}}^=$. Unless otherwise stated, all the results in this chapter hold over both $(\mathbb{R}, <)$ and $(\mathbb{R}, =)$. We restrict the statements to the ordered case.

We already know the classes Σ_1^{add} and Π_1^{add}.

Proposition 1 $\Sigma_1^{\mathrm{add}} = \mathrm{NP}_{\mathrm{add}}$ *and* $\Pi_1^{\mathrm{add}} = \mathrm{coNP}_{\mathrm{add}}$ □

Now we introduce some notation that is useful in the next proposition and in Section 21.3. Let $\mathcal{C}$ be a class of subsets of $\mathbb{R}^\infty$ and denote by $\exists \mathcal{C}$ the class of sets $A \subseteq \mathbb{R}^\infty$ such that there is a $B \in \mathcal{C}$ and a polynomial p such that

$$A = \{x \in \mathbb{R}^\infty \mid \exists y \in \mathbb{R}^{p(\mathrm{size}(x))} \, (x, y) \in B\}.$$

Analogously, one defines the class $\forall \mathcal{C}$. It is clear that for every $k \geq 0$, $\exists \Pi_k^{\mathrm{add}} = \Sigma_{k+1}^{\mathrm{add}}$, and $\forall \Sigma_k^{\mathrm{add}} = \Pi_{k+1}^{\mathrm{add}}$.

Proposition 2 *If* $\Sigma_k^{\mathrm{add}} = \Pi_k^{\mathrm{add}}$ *for some* $k \geq 1$, *then* $\Sigma_{k+\ell}^{\mathrm{add}} = \Pi_{k+\ell}^{\mathrm{add}} = \Sigma_k^{\mathrm{add}}$ *for all* $\ell \geq 0$.

Proof. For $\ell = 0$ the statement trivially holds. Suppose it holds for $\ell \geq 0$. Then $\Sigma_{k+\ell}^{\mathrm{add}} = \Pi_{k+\ell}^{\mathrm{add}}$ and consequently $\exists \Sigma_{k+\ell}^{\mathrm{add}} = \exists \Pi_{k+\ell}^{\mathrm{add}}$. But $\exists \Sigma_{k+\ell}^{\mathrm{add}} = \Sigma_{k+\ell}^{\mathrm{add}}$ and $\exists \Pi_{k+\ell}^{\mathrm{add}} = \Sigma_{k+\ell+1}^{\mathrm{add}}$. Therefore $\Sigma_{k+\ell}^{\mathrm{add}} = \Sigma_{k+\ell+1}^{\mathrm{add}}$ and this finishes the proof. □

Corollary 1 *If* $\mathrm{P}_{\mathrm{add}} \neq \Sigma_k^{\mathrm{add}}$ *for some* $k \geq 1$, *then* $\mathrm{P}_{\mathrm{add}} \neq \mathrm{NP}_{\mathrm{add}}$.

We give now a definition of the polynomial hierarchy with digital guesses. Our main theorem in this section is the coincidence, level by level, of the polynomial hierarchies just defined with their digital counterparts.

Definition 4 The classes $\mathrm{D}\Sigma_k^{\mathrm{add}}$, $\mathrm{D}\Sigma_k^{\mathrm{add}=}$, $\mathrm{D}\Pi_k^{\mathrm{add}}$, and $\mathrm{D}\Pi_k^{\mathrm{add}=}$ are obtained from Σ_k^{add}, $\Sigma_k^{\mathrm{add}=}$, Π_k^{add}, and $\Pi_k^{\mathrm{add}=}$ by replacing condition (21.6) in Definition 3 by the condition

$$u \in L \Leftrightarrow Q_1 x_1 \in \{0, 1\}^{p_1(n)} \ldots Q_k x_k \in \{0, 1\}^{p_k(n)}(x_1, \ldots, x_k, u) \in A.$$

We can now proceed with our main theorem.

Theorem 4 *For all* $k \geq 0$ *the following identities hold.*

(i) $\mathrm{D}\Sigma_k^{\mathrm{add}} = \Sigma_k^{\mathrm{add}}$;

(ii) $\mathrm{D}\Pi_k^{\mathrm{add}} = \Pi_k^{\mathrm{add}}$.

Proof.   Both statements are trivially true when $k = 0$. Moreover, for each $k \geq 1$ the second identity follows from the first since the classes Π_k^{add} are the classes of sets whose complements are in Σ_k^{add} and the same holds for $\text{D}\Pi_k^{\text{add}}$ and $\text{D}\Sigma_k^{\text{add}}$.

We therefore prove both statements jointly by induction on k. To do so, it is enough to prove that $\Sigma_k^{\text{add}} = \text{D}\Sigma_k^{\text{add}}$.

Assume that this identity holds for every $i < k$, and let $L \in \Sigma_k^{\text{add}}$. Condition (21.6) holds for some set $A \in \text{P}_{\text{add}}$ and polynomials $p_1, \ldots, p_k$. The $k - 1$ internal quantifiers can be eliminated by induction hypothesis: there exists a set $B \in \text{P}_{\text{add}}$ and polynomial functions $q_2, \ldots, q_k$ such that for all $u \in \mathbb{R}^n$,

$$u \in L \Leftrightarrow$$
$$\exists x_1 \in \mathbb{R}^{p_1(n)} \; Q_2 x_2 \in \{0, 1\}^{q_2(n)} \ldots Q_k x_k \in \{0, 1\}^{q_k(n)}(x_1, \ldots, x_k, u) \in B,$$

where the quantifiers $Q_i \in \{\exists, \forall\}$ alternate, starting with $Q_2 = \forall$. Let M be an additive machine recognizing B in polynomial time. At any time during a computation of M with input $(x_1, x_2, \ldots, x_k, u)$, the content z of any coordinate of its state space is a linear function

$$z = \sum_{i=1}^{p_1(n)} a_i x_{1i} + \sum_{i=1}^{n} b_i u_i + \sum_{i}^{p} c_i \alpha_i + d \tag{21.7}$$

of its real inputs $x_1 = (x_{11}, \ldots, x_{1p_1(n)})$ and $u = (u_1, \ldots, u_n)$, and of its real constants $\alpha_1, \ldots, \alpha_p$. The coefficients a_i, b_i, c_i, and d are integers of height polynomial in n.

By enumerating all the accepting computation paths of M, the condition $(x_1, \ldots, x_k, u) \in B$ can be replaced by a disjunction of an exponential number of conditions of the form

$$\bigwedge_{j=1}^{r} z_j \geq 0 \; \& \; \bigwedge_{j=1}^{s} z_j < 0$$

with $r + s \leq t(n)$ which describe the answers at the branching nodes that define the path, and where z_j has the form (21.7).

Now, existential quantifiers in

$$Q_2 x_2 \in \{0, 1\}^{q_2(n)} \ldots Q_k x_k \in \{0, 1\}^{q_k(n)}(x_1, \ldots, x_k, u) \in B \tag{21.8}$$

can be replaced by disjunctions since $\exists x \in \{0, 1\} \; \psi(x)$ is equivalent to $\psi(0)$ or $\psi(1)$. In the same way, universal quantifiers can be replaced by conjunctions. This procedure eliminates the variables $x_2, \ldots, x_k$. Therefore the whole expression (21.8) can be replaced by a finite combination of conjunctions and disjunctions of inequalities with the form $z_j \geq 0$ or $z_j < 0$. By distributing, one can rewrite this combination in disjunctive normal form, that is, as a disjunction

$$\bigvee_{i=1}^{\ell} S_i,$$

where for $i = 1, \ldots, \ell$, S_i is a conjunction of inequalities

$$S_i = \bigwedge_{j=1}^{r'} z_j \geq 0 \ \& \ \bigwedge_{j=1}^{s'} z_j < 0.$$

The numbers ℓ, r', and s' may be exponential in the number of original inequalities.

We deduce from the preceding arguments that $u \in L$ if and only if one of the S_i is satisfiable in x_1, that is, if there exist $i \leq \ell$ and $a_1 \in \mathbb{R}^{p_1(n)}$ such that when replacing x_1 by a_1 in S_i this expression is true. The linear systems of inequalities in x_1 determined by the S_i may be very large with respect to the number of inequations, but their coefficients are of height polynomial in n. By Theorem 3, if one of the systems is satisfiable, it must have a solution of the form

$$x_{1i} = \sum_{j=1}^{n} b_{ij} u_j + \sum_{j=1}^{p} c_{ij} \alpha_j + d_i, \qquad (21.9)$$

where the coefficients b_{ij}, c_{ij}, and d_i are rationals of height polynomial in n. Hence the quantification $\exists x_1 \in \mathbb{R}^{p_1(n)}$ in (21.6) can be in fact restricted to those x_1 of the form (21.9). This shows that $\Sigma_k^{\text{add}} \subset D\Sigma_k^{\text{add}}$, since all the candidate solutions for x_1 can be described with a polynomial number of elements in $\{0, 1\}$. The inclusion $D\Sigma_k^{\text{add}} \subset \Sigma_k^{\text{add}}$ is obvious. $\qquad\square$

Remark 2 We observe that the inclusion $\text{DNP} \subseteq \text{NP}$ changes character depending on the use of multiplication. In the additive case, we have just seen that $\text{DNP}_{\text{add}} = \text{NP}_{\text{add}}$. On the other hand, in the weak case, we saw in the preceding chapter that $\text{DNP}_{\text{W}} \neq \text{NP}_{\text{W}}$. An open question is whether $\text{DNP}_{\mathbb{R}} = \text{NP}_{\mathbb{R}}$. We conjecture that the equality does not hold. Note, however, that to prove this would immediately imply that $\text{P}_{\mathbb{R}} \neq \text{NP}_{\mathbb{R}}$.

21.3 On the Definition of the Polynomial Hierarchy

In Section 21.2 we introduced the polynomial hierarchy by means of quantifier alternations. In classical complexity theory, the hierarchy is introduced in a recursive way using oracle machines; the characterization of its classes in terms of quantifier alternations is a fundamental theorem.

Oracle machines were introduced by Turing as a tool to define reductions. To get some idea of how this works, let us focus on the p-reductions that are given by the $\text{NP}_{\mathbb{R}}$-completeness of 4-FEAS. That is, for any set S in $\text{NP}_{\mathbb{R}}$, a polynomial time computable function φ_S exists such that for every $x \in \mathbb{R}^{\infty}$, $x \in S$ if and only if $\varphi_S(x) \in$ 4-FEAS. Thus, we might imagine a machine M_S that, given an input x from S, computes an element $\varphi_S(x)$ in $\mathbb{R}^{\infty}$ in polynomial time and then "queries" the set 4-FEAS as to whether this element belongs to it. If we charge unit cost for each query, this machine works in polynomial time. Thus, if we knew a polynomial time algorithm for 4-FEAS we would have one for S.

Oracle machines formalize this idea of machines that can make queries about membership in a fixed set. Such machines are more powerful a priori than the machine M_S in that rather than being allowed only one query, they can make as many queries as their time bounds permit.

In this section we define oracle additive machines; using them we characterize the levels of the polynomial hierarchy. We only consider the ordered case, but the same results hold in the unordered case as well.

Definition 5 An *oracle machine* is a machine as defined in Chapter 3 which, in addition to its input, output, and state space, possesses an oracle space $\mathcal{R} = \mathbb{N} \times \mathbb{R}^\infty$, an oracle set $A \subseteq \mathbb{R}^\infty$, and two additional types of nodes: *write* and *query*.

The input and output spaces are $\mathbb{R}^\infty$ and the state space, $\mathbb{R}_\infty$. Denote the content of the state space by $(\ldots, x_{-1}, x_0 . x_1, \ldots)$ and the content of the oracle space by $(j, y_1, y_2, \ldots)$. The input map works as in Chapter 3. Also, at the beginning of the computation, the content of the oracle space $\mathcal{R}$ is $(0, 0, \ldots)$; that is, all the coordinates are zero.

Both write and query nodes have a unique next node. Their behavior is described as follows.

- *Write nodes.* When one of these nodes is reached, the content of the coordinate x_0 of the state space is written in the jth coordinate of the real part of the oracle space, where j is the content of its natural number coordinate. Then the natural number coordinate is incremented by 1. That is,

$$\begin{array}{c} (\ldots, x_{-1}, x_0 . x_1, \ldots) \\ (j+1, y_1, \ldots, y_j, x_0, y_{j+2}, \ldots). \end{array} \leftarrow \begin{array}{c} (\ldots, x_{-1}, x_0 . x_1, \ldots) \\ (j, y_1, \ldots, y_j, y_{j+1}, y_{j+2}, \ldots). \end{array}$$

- *Query nodes.* When one of these nodes is reached the machine writes 1 in x_0 if the content in $\mathbb{R}^\infty$ in the oracle space belongs to A and 0 otherwise. Once the answer of the query has been written, all the coordinates of the oracle space are reset to 0. That is,

$$\left. \begin{array}{c} (\ldots, x_{-1}, 1 . x_1, \ldots) \\ (0, 0, \ldots) \end{array} \quad \text{if } (y_1, \ldots, y_j) \in A \\[2em] \begin{array}{c} (\ldots, x_{-1}, 0 . x_1, \ldots) \\ (0, 0, \ldots) \end{array} \quad \text{if } (y_1, \ldots, y_j) \notin A \end{array} \right\} \leftarrow \begin{array}{c} (\ldots, x_{-1}, x_0 . x_1, \ldots) \\ (j, y_1, \ldots, y_j, \ldots). \end{array}$$

Remark 3

(1) We refer to the oracle set A in the preceding definition simply as the oracle.

(2) Intuitively, oracle machines are machines as defined in Chapter 3 with an additional capacity to make queries about membership in a fixed set $A \subseteq \mathbb{R}^\infty$. This added feature corresponds to providing a "subroutine" for deciding membership in unit cost and enhances the power of such machines to solve problems as difficult as deciding membership in A.

(3) Halting time for an oracle machine is defined as the number of nodes traversed during the computation from input to output. It makes sense then to talk about sets $S \subseteq \mathbb{R}^\infty$ decided in polynomial time, or nondeterministic polynomial time, by machines that query an oracle A.

Oracle machines can be used to define new complexity classes from known ones.

Definition 6 Let $\mathcal{C}$ be a class of subsets of $\mathbb{R}^\infty$. We denote by $P_{add}(\mathcal{C})$ the class

$$\{X \subseteq \mathbb{R}^\infty \mid X \text{ is decided by an additive machine}$$
$$\text{in polynomial time with an oracle belonging to } \mathcal{C}\}.$$

The class $NP_{add}(\mathcal{C})$ is defined analogously.

Let us look at a couple of examples of problems in $P_{add}(NP_{add})$ and in $NP_{add}(NP_{add})$.

Example 1 Recall that KP is the set of points $(x_1, \ldots, x_n)$ such that there is a subset I of $\{1, \ldots, n\}$ satisfying $\sum_{i \in I} x_i = 1$. Consider the set S of points $x = (x_1, \ldots, x_n)$ such that if x belongs to KP then $x/2 = (x_1/2, \ldots, x_n/2)$ also belongs to KP. Unlike KP, the set S does not seem to belong to NP_{add}. We show that it belongs to $P_{add}(NP_{add})$.

To do so, note that $x \in \mathbb{R}^n$ belongs to S if and only if

$$\frac{x}{2} \in KP \quad \text{or} \quad x \notin KP.$$

Thus the algorithm

> compute $\frac{x}{2}$
> query whether $\frac{x}{2}$ belongs to KP
> query whether x belongs to KP
> ACCEPT if and only if the first query answers YES
>   or the second answers NO

decides in polynomial time, using KP as oracle, whether $x \in S$. We conclude that $S \in P_{add}(NP_{add})$.

Example 2 The set UTSP belongs to $NP_{add}(NP_{add})$.

Consider the following set in NP_{add}.

$$A = \{((a_{ij}), x, k) \in \mathbb{R}^\infty \mid (a_{ij}) \text{ is a matrix of distances, } x \text{ is a tour,}$$
$$k \in \mathbb{R}, \text{ and there exists a tour } y \text{ different from } x \text{ having total}$$
$$\text{distance at most } k\}.$$

To prove that $((a_{ij}), k) \in UTSP$ it suffices to guess the tour x, verify that the total distance of x is at most k —this is done by an NP_{add} machine— and then query whether $((a_{ij}), x, k)$ belongs to A. Thus $UTSP \in NP_{add}(NP_{add})$.

In the following, for each $k \in \mathbb{N}$, we denote by $\widetilde{\Sigma}_k^{\mathrm{add}}$ and $\mathrm{D}\widetilde{\Sigma}_k^{\mathrm{add}}$ the classes defined by

(i) $\widetilde{\Sigma}_0^{\mathrm{add}} = \mathrm{D}\widetilde{\Sigma}_0^{\mathrm{add}} = \mathrm{P}_{\mathrm{add}}$.

(ii) For $k > 0$, $\widetilde{\Sigma}_k^{\mathrm{add}} = \mathrm{NP}_{\mathrm{add}}(\widetilde{\Sigma}_{k-1}^{\mathrm{add}})$, and $\mathrm{D}\widetilde{\Sigma}_k^{\mathrm{add}} = \mathrm{DNP}_{\mathrm{add}}(\mathrm{D}\widetilde{\Sigma}_{k-1}^{\mathrm{add}})$.

Also, for each $k \in \mathbb{N}$, we define $\widetilde{\Pi}_k^{\mathrm{add}}$ to be $\mathrm{co}\widetilde{\Sigma}_k^{\mathrm{add}}$ and $\mathrm{D}\widetilde{\Pi}_k^{\mathrm{add}}$ to be $\mathrm{coD}\widetilde{\Sigma}_k^{\mathrm{add}}$.

The following results show that the preceding definition yields the same classes for digital and full nondeterminism and, moreover, that these classes coincide with the ones already defined in Section 21.2. We begin with the case of arbitrary guesses.

Theorem 5 *For every $k \geq 0$ we have*

(a) $\exists \widetilde{\Pi}_k^{\mathrm{add}} = \widetilde{\Sigma}_{k+1}^{\mathrm{add}}$.

(b) $\forall \widetilde{\Sigma}_k^{\mathrm{add}} = \widetilde{\Pi}_{k+1}^{\mathrm{add}}$.

Proof. We prove (a) and (b) together by induction on k.

For $k = 0$, (a) says that $\exists \mathrm{P}_{\mathrm{add}} = \mathrm{NP}_{\mathrm{add}}$ and this is the definition of $\mathrm{NP}_{\mathrm{add}}$. Part (b) follows by taking complements.

Now suppose that $k \geq 1$ and suppose that the statement holds for all $l < k$.

Let A be a set in $\exists \widetilde{\Pi}_k^{\mathrm{add}}$. Then, for some $B \in \widetilde{\Pi}_k^{\mathrm{add}}$ and any $x \in \mathbb{R}^\infty$, we have that $x \in A$ if and only if

$$\exists y \in \mathbb{R}^{p(|x|)} \ (x, y) \in B.$$

Then the nondeterministic additive machine that on input x guesses $y \in \mathbb{R}^{p(|x|)}$ and queries B for (x, y) decides A in polynomial time proving the membership of A in $\widetilde{\Sigma}_{k+1}^{\mathrm{add}}$.

Conversely, let us consider $A \in \widetilde{\Sigma}_{k+1}^{\mathrm{add}}$. Then there exists some $B \in \widetilde{\Sigma}_k^{\mathrm{add}}$ and a nondeterministic additive machine M that decides A in polynomial time using B as oracle. Therefore, for any $x \in \mathbb{R}^\infty$, x belongs to A if and only if $\exists y \in \mathbb{R}^{(p|x|)}$ such that M accepts (x, y) using the oracle B. Equivalently,

$$\exists y \in \mathbb{R}^{(p|x|)} \quad \exists z_1, \ldots, z_r \in \mathbb{R}^{(p|x|)} \ \exists w_1, \ldots, w_s \in \mathbb{R}^{(p|x|)}$$
$$[M \text{ accepts } (x, y) \text{ with queries } z_1, \ldots, z_r \text{ answered Yes}$$
$$\text{and with queries } w_1, \ldots, w_s \text{ answered No}]$$
$$\text{and } z_1, \ldots, z_r \in B$$
$$\text{and } w_1, \ldots, w_s \notin B.$$

Now the problem between square brackets can be decided in polynomial time by running M on input (x, y) and checking that the queries and their answers are the given ones. Consequently, this problem is in $\widetilde{\Pi}_k^{\mathrm{add}}$. Since $B \in \widetilde{\Sigma}_k^{\mathrm{add}}$, the problems

$z_1, \ldots, z_r \in B$ and $w_1, \ldots, w_s \notin B$ are in $\widetilde{\Sigma_k^{\text{add}}}$ and $\widetilde{\Pi_k^{\text{add}}}$, respectively. By the induction hypothesis, we can then decide $z_1, \ldots, z_r \in B$ in $\exists \Pi_{k-1}^{\text{add}}$. Therefore $z_1, \ldots, z_r \in B$ if and only if $\exists u \in \mathbb{R}^{q(n)}$ $(z_1, \ldots, z_r, u) \in D$ where q is a polynomial and D belongs to Π_{k-1}^{add} and thus, also to Π_k^{add}. This shows that $x \in A$ if and only if

$$\exists y \in \mathbb{R}^{(p|x|)} \quad \exists z_1, \ldots, z_r \in \mathbb{R}^{(p|x|)} \exists w_1, \ldots, w_s \in \mathbb{R}^{(p|x|)} \exists u \in \mathbb{R}^{q(n)}$$
$$[M \text{ accepts } (x, y) \text{ with queries } z_1, \ldots, z_r \text{ answered Yes}$$
$$\text{and with queries } w_1, \ldots, w_s \text{ answered No}]$$
$$\text{and } (z_1, \ldots, z_r, u) \in D$$
$$\text{and } w_1, \ldots, w_s \notin B.$$

Since now the set defined by the quantifier-free part of this expression belongs to $\widetilde{\Pi_k^{\text{add}}}$, we deduce that A belongs to $\exists \Pi_k^{\text{add}}$.

Again, Part (b) is obtained by taking complements. □

Corollary 2 *For every $k \geq 0$ we have*

(a) $\widetilde{\Sigma_k^{\text{add}}} = \Sigma_k^{\text{add}}$.

(b) $\widetilde{\Pi_k^{\text{add}}} = \Pi_k^{\text{add}}$. □

We now pass to the case of digital guesses.

Theorem 6 *For every $k \geq 0$ we have*

(a) $D\widetilde{\Sigma_k^{\text{add}}} = \Sigma_k^{\text{add}}$.

(b) $D\widetilde{\Pi_k^{\text{add}}} = \Pi_k^{\text{add}}$.

Proof. Both parts are shown together by induction on k. If $k = 0$, the result is certainly true. Let us then consider $k > 0$ and let L be a set in Σ_k^{add}. We then know that $L \in D\Sigma_k^{\text{add}}$ by Theorem 4 and thus, there exists a relation R decidable in P_{add} and polynomials $p_1, \ldots, p_k$ such that for every $u \in \mathbb{R}^\infty$,

$$u \in L \Leftrightarrow Q_1 x_1 \in \{0, 1\}^{p_1(n)} \ldots Q_k x_k \in \{0, 1\}^{p_k(n)} R(x_1, \ldots, x_k, u),$$

where $Q_1 = \exists$ and the Q_i alternate. Let us define the set B to be the subset of elements $(u, x_1) \in \mathbb{R}^\infty \times \mathbb{R}^\infty$ satisfying

$$Q_2 x_2 \in \{0, 1\}^{p_2(n)} \ldots Q_k x_k \in \{0, 1\}^{p_k(n)} R(x_1, \ldots, x_k, u) \wedge x_1 \in \{0, 1\}^\infty.$$

Clearly, $B \in \Pi_{k-1}^{\text{add}}$ and so, by induction hypothesis, B belongs to $D\widetilde{\Pi_{k-1}^{\text{add}}}$. Now the additive machine given by

input u
guess $x_1 \in \{0, 1\}^{p_1(n)}$
ACCEPT iff $(u, x_1) \in B$

decides L in digital nondeterministic polynomial time using B as oracle. Thus, $L \in D\Sigma_k^{add}$.

Let us prove the reverse inclusion. To do so, we consider a set $L \in \widetilde{D\Sigma_k^{add}}$. Such a set can be decided in digital nondeterministic polynomial time using an oracle B in $\widetilde{D\Sigma_{k-1}^{add}}$. Now, by induction hypothesis, $B \in \Sigma_{k-1}^{add}$ and since a digital guess is a guess we deduce that $L \in \Sigma_k^{add}$. Applying Corollary 2(a) we finally deduce that $A \in \Sigma_k^{add}$.

Part (b) follows by taking complements.   □

We close this section with a simple application of the characterization of the levels of the polynomial hierarchy by oracle machines.

We define parallel additive machines by restricting the functions allowed at the computational nodes of parallel machines to be additions or subtractions. The class PAR_{add} is then defined as in Section 18.2 with the appropriate changes. Notice that PAR_{add} can be characterized as the class of subsets of $\mathbb{R}^\infty$ that can be decided by a uniform family of additive circuits of polynomial depth.

Proposition 3

$$PH_{add} \subseteq PAR_{add} \subseteq EXP_{add}.$$

Proof.   The second inclusion is proved as in Proposition 2 of Chapter 18. We prove the first one by induction on k. Let us consider a set $S \in \Sigma_k^{add}$.

If $k = 0$, this means that $S \in P_{add}$ and we have that $S \in PAR_{add}$.

Now assume that $S \in \Sigma_{k+1}^{add}$. Then there is a set $A \in \Sigma_k^{add}$ and a nondeterministic machine M using A as an oracle and deciding S in polynomial time. By Theorem 6, we can suppose the machine M is restricted to use digital nondeterminism. Therefore it can be simulated by a parallel machine working in polynomial time, just by independently testing all the possible guesses (there are an exponential number of them) and then accepting if one of them leads to accept. On the other hand, by induction hypothesis, $S \in PAR_{add}$ and therefore we can replace the oracle queries of M by a parallel machine working within polynomial time.   □

21.4   The Polynomial Hierarchy for Unrestricted Machines

In Sections 21.2 and 21.3 we introduced the polynomial hierarchy for additive machines and characterized its levels by means of oracle machines. Moreover, we proved that the levels of the polynomial hierarchy remain unchanged if we require the quantified variables to range over the set $\{0, 1\}$. If we consider unrestricted machines, that is, machines which can also multiply (and divide, if the base ring is a field), much still holds.

For a ring R, define $\Sigma_0^R = \Pi_0^R$ to be P_R. For every $k \geq 1$, define Σ_k^R, Π_k^R, and PH_R as in Definition 3. In the case $R = \mathbb{Z}_2$, we omit the superscript R.

It is straightforward to check that the proofs given for Propositions 1 and 2, and for Theorem 5 (and Corollary 2) also apply in the unrestricted case. We can also prove the existence of complete problems. For every $k \geq 1$, consider the subset $\mathcal{S}_k$ of $\mathbb{R}^\infty$ defined by

$$\{(f, n_1, n_2, \ldots, n_k) \mid f \text{ is a degree-4 polynomial in } n_1 + \ldots + n_k$$
$$\text{variables such that } \exists x_1 \in \mathbb{R}^{n_1} \, \forall x_2 \in \mathbb{R}^{n_2} \ldots \exists x_k \in \mathbb{R}^{n_k} \, f(x) = 0\},$$

if k is odd, and by

$$\{(f, n_1, n_2, \ldots, n_k) \mid f \text{ is a degree-4 polynomial in } n_1 + \ldots + n_k$$
$$\text{variables such that } \exists x_1 \in \mathbb{R}^{n_1} \, \forall x_2 \in \mathbb{R}^{n_2} \ldots \forall x_k \in \mathbb{R}^{n_k} \, f(x) \neq 0\}$$

if k is even. In both cases, $f(x)$ denotes $f(x_1, \ldots, x_k)$.

Notice that $\mathcal{S}_1$ is just 4-FEAS and that the complement of $\mathcal{S}_2$ is a parameterized version of 4-FEAS. That is, we have two groups of variables, x and t, and the problem is to decide whether for all values of the "parameter" $t \in \mathbb{R}^{n_1}$ there is an $x \in \mathbb{R}^{n_2}$ such that $f(t, x) = 0$. Using the results of Section 5.4 we get, in a straightforward manner, the following.

Theorem 7 *For all $k \geq 1$, the set $\mathcal{S}_k$ is $\Sigma_k^{\mathbb{R}}$-complete.* $\square$

We close this section by stating a result that extends Proposition 3 to the unrestricted case. We do not prove this result here. References are in the next section.

Theorem 8 $\mathrm{PH}_{\mathbb{R}} \subseteq \mathrm{PAR}_{\mathbb{R}}$. $\square$

21.5   Additional Comments and Bibliographical Remarks

Additive machines have been considered in algebraic complexity since their early introduction by Scholz [1937]. The paper by Dobkin and Lipton [1979] gives more recent examples of problems studied within this setting.

Close to the additive model is the linear model in which multiplication by machine constants is allowed. Most of the results in this chapter hold for linear machines as well. The origin of Theorem 2 is the separation of P from NP in the linear case [Meer 1992]. The problem used there to establish the separation consists of several equations with binary variables. The version with only one equation (i.e., the Knapsack problem) used in Theorem 2 is originally proved in [Koiran 1994] where a systematic study of additive machines was initiated. Theorem 1 also appears in that paper.

Hierarchies are commonly studied in recursive function theory. The most influential has been the arithmetical hierarchy introduced by Kleene [1943] classifying nonrecursive problems. A hierarchy over the reals analogous to the arithmetical hierarchy was studied in [Cucker 1992a]. The polynomial hierarchy in the classical setting was introduced by Meyer and Stockmeyer [1973]. More detailed expositions with further results are in Stockmeyer [1977] and in Chapter 8 of the book by Balcázar, Díaz, and Gabarró [1988]. Investigation of the polynomial hierarchy in the additive setting appears in [Cucker and Koiran 1995] where the existence of complete problems in PAR_{add} is also proved. The results of Sections 21.2 and 21.3 are taken from this reference. We have not proved here the existence of complete problems in the classes Σ_k^{add}. Complete problems in NP_{add} are shown in [Cucker and Matamala 1996]. The existence of complete problems in Σ_k^{add} for $k \geq 2$ easily follows.

The argument used in Proposition 3 to show that a problem in the polynomial hierarchy can be decided in parallel polynomial time relies only on guesses from $\{0, 1\}^{\infty}$. Thus, it also works over finite rings such as $\mathbb{Z}_2$, or in the additive setting by Theorem 4. To prove the inclusion $PH_{\mathbb{R}} \subseteq PAR_{\mathbb{R}}$ stated in Theorem 8 requires a subtler argument. For references see Section 18.6.

Oracle machines were introduced by Turing [1936]. He used such machines to define reductions between problems. More specifically, and restricted to decision problems, we say that a set $S \subseteq R^{\infty}$ is *Turing reducible* to $A \subseteq R^{\infty}$ if there is a machine M over R deciding S and using A as oracle. If M belongs to a class $\mathcal{C}$, we write $S \in \mathcal{C}^A$ (or $S \in \mathcal{C}(A)$) and we say that S *belongs to $\mathcal{C}$ relative to A*. Since the class $\mathcal{C}$ is arbitrary, we may analyze complexity classes in the "relativized world of A." That is, we place ourselves in a setting in which A is given for free (much the same way the empty set has been in our development thus far). The techniques used to separate classes of undecidable problems since the 1930s, mainly variants of diagonalization, have the property that they relativize; that is, for every set A, the arguments given to separate $\mathcal{C}$ and $\mathcal{D}$ also yield a proof of the separation of $\mathcal{C}^A$ and $\mathcal{D}^A$.

In 1975, Baker, Gill, and Solovay [1975] proved the existence of sets A and B in $\mathbb{Z}_2^{\infty}$ such that

$$P^A = NP^A \qquad \text{and} \qquad P^B \neq NP^B. \tag{21.10}$$

This result showed that the efforts carried out at the time to separate P from NP, relying mainly on classical diagonalization techniques, were bound to fail. Some years later, Bennett and Gill [1981] proved that sets like A in (21.10) are rare in a precise sense. Namely, for a random oracle A, one has $P^A \neq NP^A$ with probability one. This same paper raised the *Random Oracle Hypothesis* according to which if $\mathcal{C}^A \neq \mathcal{D}^A$ with probability 1 for a random oracle A, then $\mathcal{C} \neq \mathcal{D}$. Results that disprove this conjecture (at least in some form) can be found in, among others, [Kurtz 1983]. All of the preceding are for $R = \mathbb{Z}_2$. The results in [Baker, Gill, and Solovay 1975] were extended to the reals (and other ordered rings) in [Emerson 1994].

Related to the theme of this chapter, there is an early paper of Sontag [1985] dealing with the classical polynomial time hierarchy and the theory of the reals with addition.

22
Nonuniform Complexity Classes

In Part I (Chapters 2 and 3) we introduced finite-dimensional machines with inputs from R^n for some $n \in \mathbb{N}$ and (uniform) machines whose inputs are from R^n for all $n \in \mathbb{N}$. On the other hand, one can imagine a machine where inputs are taken from R^∞ but where the program to be executed depends on the size of the given input. The simplest example of such a machine is a family of finite-dimensional machines containing one machine for each input dimension. These machines are called *nonuniform*.

The concept of nonuniformity naturally appears when we consider the restriction to $\mathbb{Z}_2$ of machines over $\mathbb{R}$. As we did in Chapter 6, one can consider elements of $\mathbb{Z}_2^\infty$ as inputs for real machines. Thus machines over $\mathbb{R}$ may be used to decide subsets of $\mathbb{Z}_2^\infty$. As we show, the computational power of this model depends on whether branching is done over $<$ or over $=$. With branching over $=$, resource bounds in real machines define uniform complexity classes over $\mathbb{Z}_2$. With branching over $<$, the power gained using real constants is described in terms of nonuniform complexity classes.

22.1 Complexity Classes Defined by Advice Functions

In Definition 6 of Chapter 18 we introduced the concept of a set decided by a family of algebraic circuits. We then investigated complexity classes that appear by imposing on the family a uniformity condition. Our first example of a nonuniform complexity class is obtained by considering the class of sets $S \subseteq R^\infty$ that can be decided by a family of algebraic circuits $\{C_n\}_{n \in \mathbb{N}}$ having size polynomial in n.

That is, each C_n has n input gates and there is a polynomial p such that for each n, size$(C_n) \leq p(n)$. The next theorem characterizes this class of sets in terms of uniform machines that are given some additional information together with the input.

Theorem 1 *Let R be a ring and $S \subseteq R^\infty$. The following conditions are equivalent.*

(i) *The set S can be decided by a family of algebraic circuits $\{C_n\}_{n \in \mathbb{N}}$ having size polynomial in n.*

(ii) *There exist a set $A \subseteq R^\infty \times R^\infty$ decidable in polynomial time, a function $f : \mathbb{N} \to R^\infty$, and a polynomial p such that*

> *1.* size$(f(n)) \leq p(n)$ *for each $n \in \mathbb{N}$, and*
>
> *2.* $S = \{x \mid (x, f(\text{size}(x))) \in A\}$.

Proof.   Let us consider a set $S \subseteq R^\infty$ satisfying (i). Then there is a polynomial q and circuits C_n for $n \geq 1$ with size$(C_n) \leq q(n)$ such that the subset S_n of elements of S having size n is decided by C_n.
   If $s_n \in R^\infty$ encodes the circuit C_n, we define

$$f : \mathbb{N} \to R^\infty$$

$$n \to s_n.$$

The size of $f(n)$ is bounded by a polynomial in n and therefore f satisfies condition 1 of (ii). We now consider a machine M taking inputs in $R^\infty \times R^\infty$ described as follows.

> input (x, s)
> let $n = $ size(x)
> **if** s encodes an algebraic circuit C with n input gates **then**
> > evaluate C with input x
> > **if** the evaluation yields 1 **then** ACCEPT
> > **else** REJECT
> > **end if**
> **else** REJECT
> **end if** .

According to Lemma 3 of Chapter 18, the machine M works in polynomial time. Let A be the subset of $R^\infty \times R^\infty$ decided by M. Then the pair (A, f) satisfies the conditions of (ii).
   Conversely, assume that (ii) holds for S. Then there exist a set $A \in P_R$ and a function f such that for all $x \in R^\infty$, $x \in S$ if and only if $(x, f(\text{size}(x))) \in A$.
   Let M be a machine deciding A in polynomial time. According to the proof of Theorem 2 of Chapter 18, for each $m \geq 1$ there is an algebraic circuit C'_m of size polynomial in m that decides the set of inputs of size m accepted by M. Now, for each $n \geq 1$, let $s = $ size$(f(n))$ and $m = n + s$. We can replace the last s input

gates of C'_m by constant gates with associated constants given by the coordinates of $f(n)$. If C_n is the circuit obtained in that way, we have that C_n has n input gates, has size polynomial in n, and decides exactly S_n.   □

Theorem 1 suggests a way of defining more general nonuniform complexity classes. Denote by *poly* the class of functions $f : \mathbb{N} \to R^\infty$ such that for some polynomial p we have $\text{size}(f(n)) \le p(n)$ for each $n \in \mathbb{N}$.

Definition 1 Let R be a ring and $\mathcal{C}$ a class of subsets of $R^\infty \times R^\infty$. The class $\mathcal{C}/poly$ is defined to be the class of all subsets $S \subseteq R^\infty$ for which there exist a set $A \in \mathcal{C}$ and a function $f \in poly$ such that $S = \{x \mid (x, f(\text{size}(x))) \in A\}$.

The function f in this definition is said to be an *advice function*. Note that, since the only restriction we put on f is on its rate of growth, f need not be computable and thus, sets in $\mathcal{C}/poly$ are not necessarily decidable.

We can now restate Theorem 1 as follows. A set can be decided by a family of circuits $\{C_n\}_{n\in\mathbb{N}}$ having size polynomial in n if and only if it is in $P_R/poly$.

Remark 1 Although Theorem 1 holds for both $\mathbb{Z}_2$ and $\mathbb{R}$ there is a remarkable difference between their corresponding classes P/*poly*.

Circuits over $\mathbb{Z}_2$ can make use of only two constant values, namely, 0 and 1. The power of nonuniformity is then given by the wiring of the circuits of the family, that is, by their architecture. To the contrary, circuits over $\mathbb{R}$ can use an infinity of values for their constant gates and it is this diversity that gives to $P_\mathbb{R}/poly$ its power. On the other hand, any family $\{C_n\}$ of circuits over $\mathbb{R}$ using the same values $\alpha_1, \ldots, a_k$ for their constant gates decides a set in $P_\mathbb{R}$. The machine M that simulates this family has a built-in constant $\beta \in \mathbb{R}$ that codes the architecture of C_n for all $n \in \mathbb{N}$ and produces this circuit in polynomial time whenever an input of size n is given to M.

A natural question that arises is whether we can trade intractability for nonuniformity. More concretely and restricting ourselves to computations over $\mathbb{R}$, does any $NP_\mathbb{R}$-complete problem belong to $P_\mathbb{R}/poly$? This is unsolved but should the answer to this question be "yes," then $NP_\mathbb{R}$ would be included in $P_\mathbb{R}/poly$. The following result suggests that the answer is likely to be "no."

Proposition 1 *If* $NP_\mathbb{R} \subset P_\mathbb{R}/poly$, *then* $\Sigma_3^\mathbb{R} = PH_\mathbb{R}$.

Proof.   To show that $\Sigma_3^\mathbb{R} = PH_\mathbb{R}$ it is enough to see that $\Sigma_3^\mathbb{R} = \Sigma_4^\mathbb{R}$ and the latter assertion follows from the membership in $\Sigma_3^\mathbb{R}$ of a $\Sigma_4^\mathbb{R}$-complete problem. We have seen in Theorem 7 of Chapter 21 that the set $\mathcal{S}_4$ is $\Sigma_4^\mathbb{R}$-complete. We now show that if $NP_\mathbb{R} \subset P_\mathbb{R}/poly$, then $\mathcal{S}_4$ is in $\Sigma_3^\mathbb{R}$.

Suppose then that $NP_\mathbb{R} \subset P_\mathbb{R}/poly$ and consider an input

$$z = (f, n_1, n_2, n_3, n_4)$$

of $\mathcal{S}_4$. By definition, $z \in \mathcal{S}_4$ if and only if

$$\exists x_1 \in \mathbb{R}^{n_1} \; \forall x_2 \in \mathbb{R}^{n_2} \; \exists x_3 \in \mathbb{R}^{n_3} \; \forall x_4 \in \mathbb{R}^{n_4} \; f(x_1, x_2, x_3, x_4) \ne 0, \qquad (22.1)$$

where $x = (x_1, x_2, x_3, x_4)$.

Since $\mathrm{NP}_\mathbb{R} \subset \mathrm{P}_\mathbb{R}/poly$, the set 4-FEAS is in $\mathrm{P}_\mathbb{R}/poly$. Therefore there is a machine M over $\mathbb{R}$ and an advice function $h \in poly$ such that for all $n \geq 1$ and for all polynomial f_n in n variables,

$$M \text{ accepts } (f_n, h(n)) \Leftrightarrow \exists z \in \mathbb{R}^n \ f_n(z) = 0.$$

Thus, if y_n is a point in $\mathbb{R}^{\text{size}(h(n))}$, we can express that y_n is a good advice with the formula

$$\forall f_n \left[M \text{ accepts } (f_n, y_n) \Leftrightarrow \exists z \ f_n(z) = 0 \right].$$

Given a point $(a_1, a_2, a_3) \in \mathbb{R}^{n_1 + n_2 + n_3}$, denote by $f(a_1, a_2, a_3)$ the degree-4 polynomial resulting from replacing in f the first $n_1 + n_2 + n_3$ variables by the constants (a_1, a_2, a_3). Then formula (22.1) is equivalent to

$$\exists y_n \ [\ y_n \text{ is a good advice } \& \exists x_1 \in \mathbb{R}^{n_1} \ \forall x_2 \in \mathbb{R}^{n_2} \ \exists x_3 \in \mathbb{R}^{n_3}$$
$$M \text{ rejects } (f(x_1, x_2, x_3), y_n) \ ].$$

Using standard formula equivalences one can rewrite this formula as a prefix of quantifiers followed by a quantifier-free part. The resulting formula has only three quantifier alternations, beginning with an existential one, and its quantifier-free part is a relation decidable in polynomial time. Therefore it shows the membership of $\mathcal{S}_4$ to $\Sigma_3^\mathbb{R}$.   □

As a consequence of Proposition 1 we notice that, although $\mathrm{P}_\mathbb{R}/poly$ contains undecidable problems, it is unlikely that $\mathrm{NP}_\mathbb{R}$-complete problems will be found in this class since this would imply the collapse of the polynomial hierarchy.

22.2   Boolean Parts

In this section we describe a setting where nonuniformity appears in a natural way, namely, the comparison of machines over $\mathbb{R}$ and over $\mathbb{Z}_2$. This comparison can only be done by feeding the machines over $\mathbb{R}$ with binary inputs since the other way round does not make sense. Thus the question we pose is: which kind of binary sets (i.e., subsets of $\mathbb{Z}_2^\infty$) can be decided by uniform real machines?

For simplicity, we restrict our exposition to the additive framework. References for similar results for the weak model and for the unrestricted one can be found in Section 22.3. Our first proposition shows that all binary sets can be decided by additive machines in exponential time. Interest then focuses on computational resources below $\mathrm{EXP}_{\text{add}}$. Our main results can then be summarized by saying that in this situation the increase in power of additive machines with respect to machines over $\mathbb{Z}_2$ is given by a polynomial advice. This is only true, however, for additive machines over $(\mathbb{R}, <)$. In the case of additive machines over $(\mathbb{R}, =)$, there is no gain of power for the use of real constants or unit cost arithmetic.

Proposition 2 *For each set $S \subseteq \mathbb{Z}_2^\infty$ there is an additive machine M that decides S in exponential time.*

Proof. It is very similar to the proof of Theorem 7 of Chapter 17. For any element $x = (x_1, \ldots, x_n) \in \mathbb{Z}_2^n$ denote by $\varphi(x)$ the positive natural number whose expansion in base-2 is $1x_1 \ldots x_n$. The function φ bijects $\mathbb{Z}_2^\infty$ with the set of natural numbers greater than or equal to 2.

For any set $S \subseteq \mathbb{Z}_2^\infty$, consider the real number

$$\alpha_S = 0.b_2 b_3 \ldots b_i \ldots ,$$

where $b_i = 1$ if $\varphi(i) \in S$ and 0 otherwise. The machine that with input i obtains b_i from α_S and accepts if b_i is 1 works in time exponential in the size of i.   □

Definition 2 Given a class $\mathcal{C}$ of subsets of $\mathbb{R}^\infty$ define

$$\text{BP}(\mathcal{C}) = \{S \cap \mathbb{Z}_2^\infty \mid S \in \mathcal{C}\}.$$

Here we are identifying $\mathbb{Z}_2$ with the subset $\{0, 1\}$ of $\mathbb{R}$. The class $\text{BP}(\mathcal{C})$ is called the *Boolean part* of $\mathcal{C}$.

Theorem 2 *The Boolean parts of* P_{add} *and* $\mathrm{P}_{\text{add}}^=$ *are P/poly and P, respectively.*

Proof. (1) Let us consider a set A in P/poly. There is a polynomial p and an advice function f such that $f(n)$ belongs to $\mathbb{Z}_2^{p(n)}$ for all n. Furthermore, there is a machine M over $\mathbb{Z}_2$ working in polynomial time and accepting the set $\{(x, f(\text{size}(x))) \mid x \in A\}$. Let us code in a single number $x_A \in \mathbb{R}$ the sequence of advices $f(1), f(2), \ldots$ as in Proposition 2. Then we can consider an additive machine $\tilde{M}$ that, for each input $x \in \mathbb{Z}_2^n$, first produces the digits of x_A and obtains $f(n)$, and then simulates M over $(x, f(\text{size}(x)))$. This shows that $P/poly \subseteq \text{BP}(\mathrm{P}_{\text{add}})$.

Conversely, let us consider a set A in the Boolean part of P_{add} and let M be an additive machine accepting A in time bounded by a polynomial p. The computation of M over inputs of size n is described by a decision tree T of depth $p(n)$ which performs only additions and subtractions. Therefore, if $\alpha_1, \ldots, \alpha_k$ are the real constants of M, then for each $x \in \mathbb{Z}_2^n$ the test performed by T at a node i has the form $f_i(x, \alpha) \geq 0$ with

$$f_i(x, \alpha) = \sum_{i=1}^n a_i x_i + \sum_{j=1}^k b_j \alpha_j \tag{22.2}$$

and where a_i and b_j are integers of polynomial height. For a given $x \in \mathbb{Z}_2^n$ and according to the outcome of the test (22.2), the point $\alpha \in \mathbb{R}^k$ then satisfies an inequality of the form $f_{i,x}(\alpha) \geq 0$ or $f_{i,x}(\alpha) < 0$, where $f_{i,x} \in \mathbb{Z}[Y_1, \ldots, Y_k]$ is defined by $f_{i,x}(y) = f_i(x, y)$. Let Φ be the system of all these polynomial inequalities obtained when i varies over all branching nodes of T and x varies over the 2^n possible points of $\mathbb{Z}_2^n$. The system Φ is satisfied by α. Then, according to Theorem 3 of Chapter 21, there is a point $\beta_n \in \mathbb{Q}^k$ all of whose coordinates have polynomial height that also satisfies Φ. Thus, if we replace $\alpha = (\alpha_1, \ldots, \alpha_k)$ by β_n in the tree T, the path followed by any $x \in \mathbb{Z}_2^n$ will not change and x will be accepted or rejected as in T.

Let us now consider the function $f : \mathbb{N} \to \mathbb{Q}$ defined by $f(n) = \beta_n$. Since the size of β_n is polynomial in n we have that $f \in poly$. The machine $\tilde{M}$ over $\mathbb{Z}_2$ which, with input $(x, f(\mathrm{size}(x)))$, simulates the behavior of M with the constants $\alpha_1, \ldots, \alpha_k$ replaced by $f(\mathrm{size}(x))$, shows that $A \in \mathrm{P}_\mathbb{R}/poly$.

2) The inclusion $\mathrm{P} \subseteq \mathrm{BP}(\mathrm{P}_{\mathrm{add}}^=)$ is trivial. To see the reverse inclusion let us consider a set $A \in \mathrm{BP}(\mathrm{P}_{\mathrm{add}}^=)$ and an additive machine M accepting A having constants $\alpha_1, \ldots, \alpha_k$. We recall that for any input $u \in \mathbb{Z}_2^n$ and at any time of the computation of M over u, the content of the coordinates of the state space of M is a real number of the form

$$x = \sum_{i=1}^k a_i \alpha_i,$$

where the a_i are integers of polynomial height. Let E_α be the vector space over $\mathbb{Q}$ spanned by $\alpha_1, \ldots, \alpha_k$ and let m be the dimension of E_α. We can assume that E_α is generated by $\alpha_1, \ldots, \alpha_m$. Then we consider the rational matrix (b_{ij}) for $1 \le i \le k$ and $1 \le j \le m$ such that

$$\alpha_i = \sum_{j=1}^m b_{ij} \alpha_j.$$

We therefore have that $x = \displaystyle\sum_{j=1}^m \left(\sum_{i=1}^k a_i b_{ij} \right) \alpha_j$ and thus, the test $x = 0$ can be done by a machine over $\mathbb{Z}_2$ just checking that $\displaystyle\sum_{i=1}^k a_i b_{ij} = 0$ for all $j = 1, \ldots, m$.

$\square$

The result of Theorem 2 easily extends to the polynomial hierarchy.

Corollary 1 *The following equalities hold.*

(a) $\mathrm{BP}(\Sigma_k^{\mathrm{add}}) = \Sigma_k/poly$.

(b) $\mathrm{BP}(\Pi_k^{\mathrm{add}}) = \Pi_k/poly$.

(c) $\mathrm{BP}(\Sigma_k^{\mathrm{add}\,=}) = \Sigma_k$.

(d) $\mathrm{BP}(\Pi_k^{\mathrm{add}\,=}) = \Pi_k$.

Proof. These identities follow from the equalities $\mathrm{BP}(\mathrm{P}_{\mathrm{add}}) = \mathrm{P}/poly$ and $\mathrm{BP}(\mathrm{P}_{\mathrm{add}}^=) = \mathrm{P}$ together with the fact, seen in Theorem 4 of Chapter 21, that we can restrict the guesses in the polynomial hierarchy for additive machines to belong to $\{0, 1\}^\infty$. When $x_1, \ldots, x_k, u$ are all binary variables, the conditions $(x_1, \ldots, x_k, u) \in A$ for $A \in \mathrm{P}_{\mathrm{add}}$ (respectively, $A \in \mathrm{P}_{\mathrm{add}}^=$) or for $A \in \mathrm{P}/poly$ (respectively, $A \in \mathrm{P}$) are equivalent. $\square$

Remark 2

(1) Letting $k = 1$ in Corollary 1 we get $BP(NP_{add}) = NP/poly$ and $BP(NP_{add}^=) = NP$.

(2) The second part of Theorem 2 not only applies to $P_{add}^=$. The same proof yields that $BP(EXP_{add}^=) = EXP$.

Our next result characterizes the Boolean part of PAR_{add}. We recall from Theorem 4 of Chapter 18 that over $\mathbb{Z}_2$ the class of sets decided in parallel polynomial time coincides with the class of sets decided in polynomial space. We do not use this fact in what follows. However, in order to be consistent with the standard notation in classical complexity theory, we denote the class of binary sets decided in parallel polynomial time by PSPACE.

Theorem 3 *The Boolean part of* PAR_{add} *is PSPACE/poly.*

Proof. Let S be in the Boolean part of PAR_{add}. We show that S belongs to PSPACE/*poly*.

Since S belongs to $BP(PAR_{add})$, there is a family of additive circuits $\{C_n\}$ having depth bounded by a polynomial $q(n)$ that decides S. Moreover, there is an additive machine M which given (n, i) produces the ith gate of C_n within a time that we can suppose to be also bounded by $q(n)$. Let $\alpha_1, \ldots, \alpha_k$ be the constants of M.

With the exception of the constant value for the constant gates, all the remaining values computed by M are positive integers with polynomial height. Without loss of generality we suppose that M first produces a base-2 representation of these numbers and then —without using the real constants $\alpha_1, \ldots, \alpha_k$— computes the corresponding integers. This property allows us to suppose that the integer value returned at the end of a computational path only depends on the path itself and not on the constants $\alpha_1, \ldots, \alpha_k$.

On the other hand the constant value $\gamma_{n,i}$ associated with a constant gate can be expressed as a linear combination

$$\gamma_{n,i} = \sum_{j=1}^{k} a_{i,j} \alpha_j$$

of $\alpha_1, \ldots, \alpha_k$ whose coefficients $a_{i,j}$ are integers of polynomial height. Also, let

$$\begin{cases} h_{n,i,j}(\alpha_1, \ldots, \alpha_k) \geq 0 \\ r_{n,i,j}(\alpha_1, \ldots, \alpha_k) < 0 \end{cases}$$

be the linear inequalities that determine the computation path followed by M on input (n, i) and let $\Upsilon_{n,i}$ be the linear system of inequalities resulting by replacing the α_l with indeterminates X_l for $l = 1, \ldots, k$.

For any n, and for any element $u \in \mathbb{Z}_2^n$, the computation done by C_n on input u can be described by a set of linear inequalities

$$\Phi_{n,u} = \begin{cases} f_{n,u,i}(\gamma_{n,1}, \ldots, \gamma_{n,i_n}) \geq 0 \\ g_{n,u,i}(\gamma_{n,1}, \ldots, \gamma_{n,i_n}) < 0, \end{cases}$$

where the $f_{n,u,i}$ and the $g_{n,u,i}$ are linear functions whose coefficients have polynomial height (because of the polynomial depth of C_n) and the third subindex runs over the sign gates of C_n.

Let us replace in $\Phi_{n,u}$ each occurrence of a $\gamma_{n,i}$ by its correspondent linear form

$$\ell_{n,i} = \sum_{j=1}^{k} a_{i,j} X_j$$

and let $\zeta_{n,u}$ be the resulting system of inequations. If we now define

$$\Phi_n = \bigcup_i \Upsilon_{n,i} \cup \left(\bigcup_{u \in \mathbb{Z}_2^n} \zeta_{n,u} \right),$$

we obtain a system of inequations that has the real solution $(\alpha_1, \ldots, \alpha_k)$.

From Theorem 3 of Chapter 21 we deduce the existence of a point $r = (r_1, \ldots, r_k) \in \mathbb{Q}^k$ satisfying the system Φ_n such that each component has height polynomial in n. Therefore if we consider the machine M^r resulting from replacing the constants $\alpha_1, \ldots, \alpha_k$ by $r_1, \ldots, r_k$, the computations done by M^r over binary inputs can be carried out by a machine over $\mathbb{Z}_2$ in polynomial time. The circuit C_n^r produced by M^r can be readily transformed into a binary circuit having polynomial depth.

Thus, we have seen that S belongs to PSPACE/*poly*.

On the other hand, one trivially shows that any set in PSPACE/*poly* can be accepted by a P-uniform family of additive circuits having polynomial depth.

$\square$

22.3   Additional Comments and Bibliographical Remarks

The use of finite-dimensional and nonuniform computational models is an ongoing theme in complexity theory. In the classical setting it goes back to the work of Shannon [1938] and the Russian mathematicians around Lupanov. One of the major results of Lupanov [1958] is that for almost all Boolean functions $f : \{0, 1\}^n \to \{0, 1\}$ one has

$$c(f) \le \left(1 + O\left(\frac{1}{\sqrt{n}} \right) \right) \cdot \left(\frac{2^n}{n} \right),$$

where $c(f)$ denotes the size of the smallest Boolean circuit computing f. This upper bound is tight [Riordan and Shannon 1942].

We have already remarked that these models are standard in algebraic complexity. They have been used mainly to obtain lower bound results. It is an open question, both in the classical and in our general setting, how to exploit uniformity to obtain sharper lower bound results.

The general definition of nonuniform complexity classes by means of advice functions appears first in [Karp and Lipton 1982]. In that paper classes of functions other than *poly* are also considered. Proposition 1 is proved over $\mathbb{Z}_2$ but in a slightly stronger form; arguments of combinatorial character allow one to prove that the polynomial hierarchy collapses to Σ_2 instead of Σ_3.

The computational power of real machines over binary inputs has been the object of several papers in the last years. Proposition 2 appears in [Blum, Shub, and Smale 1989]. In [Meer 1990] the complexity over $\mathbb{R}$ (i.e., using real machines) of some classical NP-complete problems is analyzed. Siegelman and Sontag [1992] characterized the power of neural networks with rational weights working on Boolean inputs and within polynomial time by showing that they compute exactly the sets in P. The same problem was then considered for neural networks with real weights; it was shown by Maass [1993] and by Siegelman and Sontag [1994] that the power of these nets working within polynomial time is exactly P/*poly*. Koiran [1993] extends this result to the uniform real machine with the weak cost. The main result of his article shows that $BP(P_W) = P/poly$.

Subsequently, several papers presented new results on Boolean parts. The results of Section 22.2 for additive machines appear in [Koiran 1994] and in [Cucker and Koiran 1995]. Also, it is shown in [Cucker, Shub, and Smale 1994] that $BP(PAR_W) = PSPACE/poly$; Cucker and Grigoriev [1997] further extend this result by showing that $BP(PAR_{\mathbb{R}}) = PSPACE/poly$.

Related results for machines that branch on equalities only can be found in [Cucker, Karpinski, Koiran, Lickteig, and Werther 1995].

Another result related to this chapter is an extension of Lupanov's Theorem to Boolean functions computed by algebraic circuits obtained by Gashkov [1980] and improved by Turán and Vatan [1994].

The book by Poizat [1995] offers a panorama of complexity theory from the point of view of model theory. It is centered on problems related to nonuniform complexity.

The final result we mention here concerning nonuniformity is related to the Knapsack Problem. Meyer auf der Heide [1984] proved that for any $n \geq 1$ the restriction of the Knapsack Problem to $\mathbb{R}^n$ can be solved by a linear decision tree with depth $O(n^4)$. This result does not prove that KP belongs to $P_{\mathbb{R}}/poly$ since polynomial depth decision trees, having exponential size, provide a more powerful computational model than polynomial size circuits. However, it does show that the Knapsack Problem can be solved in polynomial time for a well-defined nonuniform model of computation.

23
Descriptive Complexity

Complexity classes over a ring R are classes of problems categorized according to their solvability with respect to a given machine model and prescribed resource restrictions. We have seen various such classifications in the preceding chapters. In this chapter we show that a number of classes can be characterized in terms of the "complexity" of their descriptions. This provides machine-independent characterizations of important complexity classes and sheds some light on their comparison.

We have already seen how the register equations enable us to describe problems in $\mathrm{NP}_\mathbb{R}$ by means of degree-4 polynomial equations. That is, if S belongs to $\mathrm{NP}_\mathbb{R}$, then, for each n, we can associate a degree-4 polynomial $\Phi_n(y_1, \ldots, y_n, x_1, \ldots, x_m)$ with the property that $a \in S^n$ if and only if the equation $\Phi_n(a, x_1, \ldots, x_m) = 0$ is solvable over $\mathbb{R}$. Here $a = (a_1, \ldots, a_n)$ and $S^n = S \cap \mathbb{R}^n$. In other words, $a \in S^n$ if and only if

$$\exists x_1 \ldots \exists x_m \, (\Phi_n(a, x_1, \ldots, x_m) = 0) \text{ is a true statement over } \mathbb{R}.$$

Thus, for each n, we have a description of S^n —and there is a great deal of uniformity in these descriptions. For one, each statement is of a purely existential nature. For another, there is a polynomial p such that for each n, $m = p(n)$. And the polynomials Φ_n are related in structure since they all derive from the register equations of a single $\mathrm{NP}_\mathbb{R}$ decision machine M for S. A goal here is to understand the nature of this uniformity and construct a language where S is describable by a single sentence in a machine-independent way.

In this chapter we introduce a particular class of vocabularies and structures, the $\mathbb{R}$-structures, together with three logics for them: first-order, fixed point first-order

(FP$_\mathbb{R}$), and existential second-order ($\exists$SO$_\mathbb{R}$). We show that FP$_\mathbb{R}$ *captures* P$_\mathbb{R}$ and that $\exists$SO$_\mathbb{R}$ *captures* NP$_\mathbb{R}$.

Specifically, we show that $\mathbb{R}$-structures can be seen in a natural way as points in $\mathbb{R}^\infty$ and therefore we are able to associate sets of $\mathbb{R}$-structures with subsets of $\mathbb{R}^\infty$, that is, decision problems over $\mathbb{R}$. The main results of this chapter, Theorems 1 and 2, state that a set $S \subset \mathbb{R}^\infty$ is in P$_\mathbb{R}$ (or NP$_\mathbb{R}$) if and only if its associated sets of $\mathbb{R}$-structures are describable within fixed point first-order logic (respectively, existential second-order logic). In order to do so, we first provide the necessary definitions and concepts.

23.1   Vocabularies, Structures, and First-Order Logic

In this section we review some elementary notions from first-order mathematical logic. They are the building blocks for our machine-independent characterization of complexity classes.

Generally speaking, given a certain vocabulary L containing symbols for elements, functions, and relations, one builds up sentences from L according to specified "grammatical rules." These rules —called logics— have varying expressive capabilities depending on the constructions allowed.

Associated with a vocabulary is a class of sets —called structures— in which the symbols of L are interpreted and in which the sentences take a truth value. For instance, if L contains the symbol $<$, the sentence

$$\forall x \forall y \exists z \ (x < y \Rightarrow (x < z \ \& \ z < y))$$

is false in $\mathbb{Z}$ and true in $\mathbb{Q}$ when $<$ is interpreted in both rings as the standard ordering.

Definition 1 A *vocabulary* L is a set of symbols, each one being either relational or functional. To each symbol σ there is an associated natural number called its *arity* which is at least 1 for relation symbols. Function symbols of arity 0 are called *constant symbols* or simply *constants*. Given a vocabulary L and a set C we let $L(C)$ be the vocabulary L together with constant symbols for elements in C. (For each $c \in C$ we denote its associated symbol simply by c.)

Given a vocabulary L, a *structure* $\mathfrak{A}$ of L is a set A —called its *universe*— together with *interpretations* of the symbols of L. That is, with each relation symbol r of arity n there is an associated relation $r^{\mathfrak{A}} \subseteq A^n$, and with each function symbol f of arity n, an associated function $f^{\mathfrak{A}} : A^n \to A$. Note that interpretations of function symbols of arity 0 are elements in A.

In what follows, we assume that all vocabularies contain the binary relation $=$, which is interpreted as the diagonal in A^2 for all its structures.

Example 1 Let L be a vocabulary containing only one symbol e denoting a binary relation. Then the class of structures of L is exactly the class of graphs. For any structure $\mathfrak{A}$, the universe A is the set of vertices of the graph and the interpretation $e^{\mathfrak{A}}$ of e the relation characterizing the edges on A^2.

Example 2 Let us consider the vocabulary $L = \{<, 0, 1, +, -, \times\}$ where the first symbol is relational of arity two, the next two symbols are constants, and the last three are function symbols of arity 2. Possible structures for this vocabulary are the ordered field of real numbers and the ordered ring of integers. More precisely, in the case of the real numbers, the structure is $\langle \mathbb{R}, <^{\mathbb{R}}, 0^{\mathbb{R}}, 1^{\mathbb{R}}, +^{\mathbb{R}}, -^{\mathbb{R}}, \times^{\mathbb{R}} \rangle$, where $\mathbb{R}$ is the universe and $<^{\mathbb{R}}$, and so on are the natural interpretations of $<$, and so on in $\mathbb{R}$. In practice, we drop the superscript $\mathbb{R}$ and also denote this structure briefly by $\langle \mathbb{R}, < \rangle$.

We could also consider structures such as a set $A = \{a, b\}$ with $<^{\mathfrak{A}} = \{(a, a)\}$, $0^{\mathfrak{A}} = a$, $1^{\mathfrak{A}} = a$, and $f^{\mathfrak{A}}(x, y) = b$ for f being either $+, -$, or $\times$ and for $x, y \in A$.

In Example 2 one feels that the last structure was not intended. Indeed, if we want to consider only ordered fields or rings as structures, it is necessary to stipulate more. To do so, we express the axioms of the theory of ordered fields (or rings) using the given vocabulary and then consider only those structures satisfying these axioms. We do this formally as follows.

Definition 2 Let L be a vocabulary and V a countable set of variables. The set of *terms* is recursively defined in the following way: any variable or constant of L is a term and if $f \in L$ is a function symbol of arity k and $t_1, \ldots, t_a$ are terms, so is $f(t_1, \ldots, t_k)$.

An *atomic formula* is an expression of the form $r(t_1, \ldots, t_a)$, where $r \in L$ is a relation symbol of arity k and $t_1, \ldots, t_k$ are terms.

The set of *first-order formulas* over L is then recursively defined by the following conditions: any atomic formula is a formula, if ϕ and ψ are formulas, so are $\phi \vee \psi$, $\phi \,\&\, \psi$, and $\neg\phi$; and if ϕ is a formula containing a variable x, so are $(\exists x)\phi$ and $(\forall x)\phi$. The variables in a formula that are not under the scope of a quantifier are said to be *free*. Otherwise they are said to be *bound*. Formulas without free variables are called *first-order sentences*. Formulas without bound variables are said to be *quantifier-free*.

Example 3 Let L be the vocabulary in Example 2. Then 0, x, y^2 are terms, $0 < x$, $x = y^2$ are atomic formulas, $\exists y\, (x = y^2)$, $\neg(0 < x) \vee \exists y\, (x = y^2)$ are first-order formulas over L, and $\forall x\, (\neg(0 < x) \vee \exists y\, (x = y^2))$ is a sentence. The latter is sometimes written as $\forall x\, (0 < x \Rightarrow \exists y\, (x = y^2))$ using the shorthand $A \Rightarrow B$ for $\neg A \vee B$. An example of a sentence over $L(\mathbb{R})$ is $\exists y\, (\pi = y^2)$.

Definition 3 Let L be a vocabulary and $\mathfrak{A}$ a structure of L. For any term $t(x_1, \ldots, x_n)$ containing $n \geq 0$ variables and any point $a \in A^n$, the interpretation in $\mathfrak{A}$ of the function symbols in t extends to a natural interpretation $t^{\mathfrak{A}}(a)$ of t in $\mathfrak{A}$ where x_i is interpreted as a_i for $i = 1, \ldots, n$.

The interpretation of relation symbols of L in $\mathfrak{A}$ now enables us to associate *truth values* with atomic formulas *evaluated* at points in A^n. Thus, if $t_1, \ldots, t_k$ are terms containing the variables $x_1, \ldots, x_n$ and $a \in A^n$, we say $r(t_1(a), \ldots, t_k(a))$ is true in $\mathfrak{A}$ if $(t_1^{\mathfrak{A}}(a), \ldots, t_k^{\mathfrak{A}}(a)) \in r^{\mathfrak{A}}$ and false if not.

Finally, interpreting $\vee$, &, $\neg$, $\forall$, and $\exists$ in the natural way and proceeding recursively, we can associate with any formula φ with free variables $x_1, \ldots, x_n$ over L and any $a \in A^n$ a truth value. If this value is true we say that $\mathfrak{A}$ *satisfies* $\varphi(a)$ and we denote this by $\mathfrak{A} \models \varphi(a)$. If φ is a sentence (i.e., if $n = 0$), a truth value is associated with φ without any final appeal to elements of A. (Of course, such appeals come in recursively. Thus, for example, if $\exists x \, \varphi(x)$ is a sentence over L, we have $\mathfrak{A} \models \exists x \, \varphi(x)$ if and only if there is an $a \in A$ such that $\mathfrak{A} \models \varphi(a)$.)

If T is a set of sentences over L, we say that $\mathfrak{A}$ *satisfies* T when all the sentences of T are satisfied in $\mathfrak{A}$ and denote this fact by $\mathfrak{A} \models T$. The structures $\mathfrak{A}$ such that $\mathfrak{A} \models T$ are called *models* of T. The sentences in T are called *axioms* and T is called a *theory*.

Example 4 For $L = \{<, 0, 1, +, -, \times\}$ we can consider OF to be the set of axioms of the theory of ordered fields. For example, the axioms

$$\forall x \forall y \forall z \, [(x + y) + z = x + (y + z)],$$

$$\forall x \exists y \, (x \neq 0 \Rightarrow x \times y = 1),$$

and

$$\forall x \forall y \forall z \, (x < y \Rightarrow x + z < y + z)$$

require the models of OF to be associative under addition, to have multiplicative inverses for nonzero elements, and to preserve order under additions. Here $x \neq 0$ is a shorthand for $\neg(x = 0)$.

By adding the few standard remaining axioms, we note that OF can be given by a finite number of axioms. On the other hand, the theory RCF of real closed fields (see Remark 3.3 of Chapter 2) is an example of theory which needs an infinite set of first-order axioms. One adds to the axioms of OF the sentence φ

$$\forall x \, (0 < x \Rightarrow \exists y \, (x = y^2))$$

asserting that positive elements have square roots and, for each $k \in \mathbb{N}$, the sentence φ_k

$$\forall x_0 \forall x_1 \ldots \forall x_{2k} \exists y \, (y^{2k+1} + x_{2k} y^{2k} + \ldots + x_1 y + x_0 = 0)$$

asserting that monic polynomials of odd degree $(2k+1)$ have roots in the field. Thus we have that $\langle \mathbb{R}, < \rangle \models$ RCF and $\langle \overline{\mathbb{Q}} \cap \mathbb{R}, < \rangle \models$ RCF but not that $\langle \mathbb{Z}, < \rangle \models$ RCF.

It is customary to introduce new symbols to make the formulas easier to read. Thus, for instance, the term $1 + 1 + 1 + 1$ in the theory of real closed fields is generally written 4. We proceed formally and more generally as follows.

Definition 4 Let $\mathfrak{A}$ be a structure of vocabulary L with universe A and $R \subseteq A^n$ be an n-ary relation on A. We say that R is (first-order) *definable in $\mathfrak{A}$ over L* if there exists a formula φ over L with n free variables $x_1, \ldots, x_n$ such that for all $a \in A^n$

$$a \in R \text{ if and only if } \mathfrak{A} \models \varphi(a).$$

If T is a theory, we say that R is *definable in T over the vocabulary L* if there exists a formula φ over L with n free variables $x_1, \ldots, x_n$ such that for each model $\mathfrak{A}$ of T and for all $a \in A^n$,

$$a \in R \text{ if and only if } \mathfrak{A} \models \varphi(a).$$

A partial function $f : A^n \to A^k$ is *definable in $\mathfrak{A}$ over L* if its graph is definable over $\mathfrak{A}$ as a subset of A^{n+k}.

A function $f : A^n \to A$ is *term definable in $\mathfrak{A}$ over L* if there exists a term $t(x_1, \ldots, x_n)$ such that for all $a \in A^n$,

$$t^{\mathfrak{A}}(a) = f(a).$$

The two previous definitions also extend to definability in a theory T.

Example 5 Let $L = \{<, 0, 1, +, -, \times\}, \mathfrak{A} = \langle \mathbb{R}, < \rangle$, and consider the vocabulary $L(\mathbb{R})$. The subsets of $\mathbb{R}^n$ definable in $\mathfrak{A}$ over $L(\mathbb{R})$ are exactly the semi-algebraic sets. The term definable functions $f : \mathbb{R}^n \to \mathbb{R}$ are the polynomial functions. Notice that the function $\sqrt{\ } : [0, +\infty) \to \mathbb{R}$ is definable but not term definable.

We note that the apparently more geometric notion of definability over $\mathbb{R}$ given in Section 4.5 is equivalent to the one over $L(\mathbb{R})$ given here in that the definable sets in both cases are the same.

23.2 Logics on $\mathbb{R}$-Structures

Consider a decision problem over $\mathbb{R}$, say 4-FEAS. Our goal is to find a vocabulary L such that the admissible instances of 4-FEAS are structures of L. Moreover, we also want a finite set of axioms over L whose models are exactly the yes-instances of 4-FEAS.

The vocabularies we consider are split into two parts. One part is used for describing the finite, discrete "skeleton" of a structure, the other for describing the "arithmetic" or operations in the structure. If we are considering the problem 4-FEAS over $\mathbb{R}$, the first part is used to describe the structure of polynomials (e.g., the ordering of the monomials in a polynomial), and the second is used to describe the coefficients of a polynomial and the arithmetic in $\mathbb{R}$ (e.g., the evaluation of a polynomial at a real point). If we wish to describe machines or algebraic circuits, the first part of the vocabulary is used for describing the finite, discrete aspects of computation: the skeleton of flowcharts or circuits, tuples, registers, and time. The other part is used for describing the algebra of real computations.

Definition 5 Let L_s, L_f be finite vocabularies, where L_s may contain relation and function symbols, and L_f contains function symbols only. An $\mathbb{R}$-*structure of signature $\sigma = (L_s, L_f)$* is a pair $\mathfrak{D} = (\mathfrak{A}, \mathcal{F})$ consisting of

(i) a finite structure $\mathfrak{A}$ of vocabulary L_s, called the *skeleton* of $\mathfrak{D}$, whose (finite) universe A is also called the *universe* of $\mathfrak{D}$, and

(ii) a finite set $\mathcal{F}$ of functions $X : A^k \to \mathbb{R}$ interpreting the function symbols in L_f.

We denote the set of all $\mathbb{R}$-structures of signature σ by Struct(σ).

Definition 6 Let $\mathfrak{D}$ be an $\mathbb{R}$-structure of skeleton $\mathfrak{A}$. We denote by $|A|$ the cardinality of the universe A of $\mathfrak{A}$. An $\mathbb{R}$-structure $\mathfrak{D} = (\mathfrak{A}, \mathcal{F})$ is *ranked* if there is a unary function symbol r in L_f whose interpretation $r^{\mathfrak{D}} \in \mathcal{F}$ is a bijection between A and $\{0, 1, \dots, |A| - 1\}$. The function $r^{\mathfrak{D}}$ is called a *ranking*. A k-ranking on A is a bijection between A^k and $\{0, 1, \dots, |A|^k - 1\}$.

In the rest of this chapter, we only consider $\mathbb{R}$-structures of cardinality at least 2. Also, to simplify notation, we often denote the ranking $r^{\mathfrak{D}}$ by ρ.

Example 6 Let L_s be the empty set and L_f be $\{r, X\}$ with both function symbols of arity 1. Then a simple class of ranked $\mathbb{R}$-structures with signature $\sigma = (L_s, L_f)$ is obtained by letting A be a finite set, $r^{\mathfrak{D}}$ any ranking on A, and $X^{\mathfrak{D}}$ any unary function $X^{\mathfrak{D}} : A \to \mathbb{R}$. Since $r^{\mathfrak{D}}$ bijects A with $\{0, 1, \dots, n - 1\}$, where $n = |A|$, this $\mathbb{R}$-structure is a point $x_{\mathfrak{D}} = (x_1, \dots, x_n)$ in $\mathbb{R}^n \subset \mathbb{R}^{\infty}$. Here $x_i = X^{\mathfrak{D}}(a_i)$ with $r^{\mathfrak{D}}(a_i) = i - 1$. Conversely, for each point $x \in \mathbb{R}^{\infty}$ there is an $\mathbb{R}$-structure $\mathfrak{D}$ of signature σ such that $x = x_{\mathfrak{D}}$. Thus there is a bijection between $\mathbb{R}$-structures in this class and $\mathbb{R}^{\infty}$.

Again, we are interested in considering only some $\mathbb{R}$-structures for a given signature and again we do so by requiring certain axioms to be satisfied. We consequently extend first-order logic to first-order logic on $\mathbb{R}$-structures. Also, we enhance the expressive power of first-order logic by admitting additional grammatical rules.

So now, fix a countable set V of variables. These variables range only over the skeleton; we do not use variables ranging over $\mathbb{R}$.

Definition 7 The language of *first-order logic* for $\mathbb{R}$-structures, FO$_{\mathbb{R}}$, contains for each signature $\sigma = (L_s, L_f)$ a set of terms and formulas. We first define terms, of which there are two kinds. When interpreted in an $\mathbb{R}$-structure, each term t takes values either in the skeleton, in which case we call it an *index term*, or in $\mathbb{R}$, in which case we call it a *number term*. Terms are defined inductively as follows.

(i) The set of index terms is the closure of the set V of variables under the application of function symbols of L_s (in the natural way).

(ii) For each $c \in \mathbb{R}$, there is a number term, also denoted by c.

(iii) If $h_1, \dots, h_k$ are index terms and X is a k-ary function symbol of L_f, then $X(h_1, \dots, h_k)$ is a number term.

(iv) If t, t' are number terms, then so are $t + t', t - t', t \times t', t/t'$, and sign (t).

Atomic formulas are equalities $h_1 = h_2$ of index terms, equalities $t_1 = t_2$ of number terms, inequalities $t_1 < t_2$ of number terms, and expressions $P(h_1, \ldots, h_k)$, where P is a k-ary predicate symbol in L_s and $h_1, \ldots, h_k$ are index terms.

The set of *formulas* of $\mathrm{FO}_\mathbb{R}$ is the smallest set containing all atomic formulas and which is closed under Boolean connectives, $\vee$, &, and $\neg$, and quantification $(\exists v)\psi$ and $(\forall v)\psi$. Note that we do *not* consider formulas $(\exists x)\psi$ where x ranges over $\mathbb{R}$. Quantifier-free formulas and sentences are defined as in the preceding section.

Remark 1 The interpretation of formulas of $\mathrm{FO}_\mathbb{R}$ in an $\mathbb{R}$-structure $\mathfrak{D}$ is clear with the understanding that $x/0$ is interpreted as 0 for all $x \in \mathbb{R}$. Also, we define the concepts of truth value, satisfaction, model, axiom, and theory as in the preceding section.

Example 7 If $\mathfrak{D}$ is an $\mathbb{R}$-structure of signature (L_s, L_f) and $r \in L_f$ is a unary function symbol, we can express in first-order logic the requirement that r be interpreted as a ranking in $\mathfrak{D}$. This is done by the sentence

$$\forall x \forall y\, (r(x) = r(y) \Rightarrow x = y)$$
$$\&\exists o\, (r(o) = 0 \,\&\, \forall u\, [u \neq o \Rightarrow (r(o) < r(u) \,\&\, \exists v\, r(u) = r(v) + 1)]) \,.$$

As in the preceding section, given an $\mathbb{R}$-structure $\mathfrak{A}$ with universe A we may consider (first-order) *definable subsets* of A^n. Here we also consider *definable functions* taking values over the reals. We say that $F : A^n \to \mathbb{R}$ is *definable by a number term* if there exists a number term t with free variables $x_1, \ldots, x_n$ such that for all $a \in A^n$,

$$t^{\mathfrak{A}}(a) = F(a).$$

As in the preceding section, if T is a theory, we say that F is *definable in T* if the term t defines F in $\mathfrak{A}$ for every model $\mathfrak{A}$ of T. This notion plays a central role in this chapter.

Remark 2 If ρ is a ranking on A and $|A| = n$, then there are elements $o, \ell \in A$ such that $\rho(o) = 0$ and $\rho(\ell) = n - 1$. Note that these two elements are first-order definable since they are the only elements in A satisfying

$$\forall v\, (v \neq x \Rightarrow r(x) < r(v))$$

and

$$\forall v\, (v \neq x \Rightarrow r(v) < r(x)),$$

respectively. We use the symbols o and 1 as constant symbols that are to be interpreted as o and ℓ, respectively. Note that if $|A| = n$, then $n = \rho(\ell) + 1$. Thus, we may use n as a constant symbol to denote the cardinality of A.

Remark 3 Any ranking ρ induces, for all $k \geq 1$, a k-ranking ρ^k on A^k by lexicographical ordering. Note that for all $(a_1, \ldots, a_k) \in A^k$,

$$\rho^k(a_1, \ldots, a_k) = \rho(a_1)n^{k-1} + \rho(a_2)n^{k-2} + \ldots + \rho(a_k).$$

Again, ρ^k is definable and we freely use the symbol r^k which is to be interpreted as ρ^k.

Example 8 Let us see that we can describe algebraic circuits with first-order logic. To do so, we consider the vocabularies $L_s = \{f_l, f_r\}$, where f_l and f_r are function symbols of arity one and $L_f = \{r, C, F_t\}$, whose three symbols are also of arity one. If $\mathfrak{D}$ is an ℝ-structure with signature (L_s, L_f), we intend to interpret its universe A as the set of nodes of an algebraic circuit. Left and right predecessors are to be given by the functions f_l and f_r, respectively. The type of each node is given by F_t and the real numbers associated with the constant nodes by the function C. Finally, we require r to be interpreted as a ranking.

We now want to express all these requirements with first-order sentences. To interprete r as a ranking we use the sentence in Example 7. The sentence

$$\forall v \left[\bigvee_{k=1}^{8} F_t(v) = k \right] \tag{23.1}$$

ensures that the elements of A are of one of the eight kinds described in the table in Section 18.5.

The sentence

$$\forall v \forall w \; [(F_t(v) = 1 \; \& \; F_t(w) \neq 1) \Rightarrow r(v) < r(w)] \tag{23.2}$$

requires that input gates are the first gates of the circuit. Finally, the sentence

$$\forall v \; [(F_t(v) \neq 1 \; \& \; F_t(v) \neq 2) \Rightarrow (r(f_l(v)) < r(v) \; \& \; r(f_r(v)) < r(v))] \tag{23.3}$$

requires that when a gate v is not of input or constant type its predecessors are really so; that is, they are ranked before v. It also ensures that A, considered as a directed graph, is acyclic.

Denote by AC the theory with axioms (23.1) through (23.3), plus the sentence requiring r to be interpreted as a ranking. Then the models of AC are precisely the algebraic circuits.

We showed in Example 6 that there is a natural bijection between ℝ$^\infty$ and the ℝ-structures of a certain signature. In addition, for every signature σ and every ℝ-structure $\mathfrak{D} = (\mathfrak{A}, \mathcal{F})$ of signature σ, we can encode $\mathfrak{D}$ by a tuple $e(\mathfrak{D}) \in \mathbb{R}^\infty$ in the following way.

Choose a ranking ρ on A and replace all functions and relations in the skeleton by the appropriate characteristic functions $\chi : A^k \to \{0, 1\} \subseteq \mathbb{R}$. Thus we get a structure with skeleton a set A and functions $X_1, \ldots, X_t$ of the form $X_i : A^{k_i} \to \mathbb{R}$. Each of the functions X_i can be represented by a tuple $\xi_i = (x_0, \ldots, x_{m_i - 1}) \in \mathbb{R}^{m_i}$ with $m_i = |A|^{k_i}$ and $x_j = X_i(\bar{a}(j))$ where $\bar{a}(j)$ is the jth tuple in A^{k_i} with respect to the ranking on A^{k_i} induced by ρ. The concatenation

$$e(\mathfrak{D}) = \xi_1, \xi_2, \ldots, \xi_t$$

of these tuples gives the encoding $e(\mathfrak{D}) \in \mathbb{R}^{m_1 + \cdots + m_t}$. Note that $e(\mathfrak{D})$ depends on the ranking ρ chosen on A.

Obviously, for structures $\mathfrak{D}$ of a fixed finite signature, the length of $e(\mathfrak{D}) \in \mathbb{R}^\infty$ is bounded by some polynomial n^ℓ, where $n = |A|$ and ℓ depends only on the signature. Thus, appending zeros to $e(\mathfrak{D})$ if necessary, we can also view $e(\mathfrak{D}) = (x_0, \ldots, x_{n^\ell - 1})$ as a single function $X_\mathfrak{D} : A^\ell \to \mathbb{R}$, where $X(\bar{a}(i)) = x_i$ for all $i < n^\ell$. This means that one can encode an $\mathbb{R}$-structure in $\mathbb{R}^\infty$ representing the whole structure by a single function (of appropriate arity ℓ) from the ordered set $\{0, \ldots, n^\ell - 1\}$ into $\mathbb{R}$.

Furthermore this encoding is first-order definable in the following sense.

Lemma 1 *For every signature* $\sigma = (L_s, L_f)$*, there exists a sentence* $\kappa(r, x)$ *in first-order logic of signature* $(L_s, L_f \cup \{r, x\})$ *such that for all* $\mathbb{R}$*-structures* $\mathfrak{D}$ *of signature* σ*, all ranking* ρ*, and all functions* $X : A^\ell \to \mathbb{R}$*,*

$$(\mathfrak{D}, \rho, X) \models \kappa(r, x) \quad \text{iff} \quad X = e(\mathfrak{D}).$$

Proof.   Let $X_1, \ldots, X_t$ be as in the preceding discussion,

$$X_i : A^{k_i} \to \mathbb{R}.$$

The formula $\kappa(r, x)$ is

$$\forall v_1 \ldots \forall v_\ell \; \big[\; 0 \leq r^\ell(v) < \mathrm{n}^{k_1} \Rightarrow \exists w_1 \ldots \exists w_{k_1} \; (r^\ell(v) = r^{k_1}(w))$$
$$\& \; x(v) = X_1(w)) \; \&$$
$$\mathrm{n}^{k_1} \leq r^\ell(v) < \mathrm{n}^{k_2} \Rightarrow \exists w_1 \ldots \exists w_{k_2} \; (r^\ell(v) = \mathrm{n}^{k_1} + r^{k_1}(w)$$
$$\& \; x(v) = X_2(w)) \; \&$$
$$\vdots$$
$$\mathrm{n}^{k_{t-1}} \leq r^\ell(v) < \mathrm{n}^{k_t} \Rightarrow \exists w_1 \ldots \exists w_{k_t} \; (r^\ell(v) = \mathrm{n}^{k_1} + \ldots$$
$$+ \mathrm{n}^{k_{t-1}} + r^{k_1}(w) \; \& \; x(v) = X_t(w)) \; \&$$
$$\mathrm{n}^{k_t} \leq r^\ell(v) < \mathrm{n}^\ell \Rightarrow x(v) = 0 \; \big],$$

where, we recall, n stands for $r(1) + 1$. $\square$

We use this encoding to consider $\mathbb{R}$-structures as computational problems.

Definition 8 Given a signature $\sigma = (L_s, L_f)$, a *decision problem of* $\mathbb{R}$*-structures* of signature σ is a subset $S \subseteq \mathrm{Struct}(\sigma)$. The preceding lemma associates with a decision problem S of $\mathbb{R}$-structures the decision problem $(e(\mathrm{Struct}(\sigma)), e(S))$, where $e(S) = \{e(\mathfrak{D}) \mid \mathfrak{D} \in S\} \subset \mathbb{R}^\infty$.

If $\mathcal{C}$ is any complexity class over the reals, and S is a decision problem of $\mathbb{R}$-structures we say that S belongs to $\mathcal{C}$ whenever the problem $(e(\mathbb{R}_\sigma), e(S))$ belongs to $\mathcal{C}$. Also, if

$$S = \{\mathfrak{D} \in \mathrm{Struct}(\sigma) \mid \mathfrak{D} \models T\}$$

for a set of sentences T, we say that T *describes* the problem S. We are interested here in the case when T is given by a finite set of axioms; that is, $T = \{\varphi_1, \ldots, \varphi_n\}$. Then T can be replaced by the single sentence $\varphi_1 \wedge \ldots \wedge \varphi_n$.

With this convention, our first result shows that the expressive power of first-order logic is not too big. Recall from Section 18.5 the definition of the classes $NC_\mathbb{R}^k$.

Proposition 1 *Let σ be a signature and $S = \{\mathfrak{D} \in \text{Struct}(\sigma) \mid \mathfrak{D} \models \varphi\}$ with φ a first-order sentence of signature σ. Then $S \in NC_\mathbb{R}^1$.*

Proof. We can assume that φ is in prenex form, that is, in the form

$$Q_1 v_1 \ldots Q_s v_s \ \psi(v_1, \ldots, v_s),$$

where $Q_1, \ldots, Q_s$ are first-order quantifiers and $\psi(v_1, \ldots, v_s)$ is a quantifier-free formula.

Consider now an $\mathbb{R}$-structure $\mathfrak{D}$ and let A be its universe. Given a point $(a_1, \ldots, a_s) \in A^s$ it takes constant time to check whether $\psi^\mathfrak{D}(a_1, \ldots, a_s)$ holds over $\mathbb{R}$. On the other hand, if $|A| = n$, there are n^s possible tuples $(a_1, \ldots, a_s)$ in A^s. Thus, we can decide whether $\mathfrak{D}$ satisfies φ in parallel logarithmic time by independently checking the validity of ψ for all the n^s possibilities and accept if the resulting n^s truth values satisfy the prefix of quantifiers. □

23.3   Capturing $P_\mathbb{R}$ with Fixed Point First-Order Logic

Proposition 1 shows that sets of $\mathbb{R}$-structures describable within first-order logic are easy in the sense that they can be decided in $NC_\mathbb{R}^1$. It is not known whether the converse is true, that is, whether all sets of $\mathbb{R}$-structures decidable in $NC_\mathbb{R}^1$ can be described by first-order sentences. The goal of this section is to extend first-order logic with certain syntactic operators (i.e., grammatical rules) that enable us to build new number terms. The resulting logic, called fixed point first-order logic, and denoted by $FP_\mathbb{R}$, *captures* $P_\mathbb{R}$, that is, a set of $\mathbb{R}$-structures in $P_\mathbb{R}$ if and only if it can be described by a sentence in $FP_\mathbb{R}$. In the next section we describe another logic, existential second-order logic, which captures $NP_\mathbb{R}$.

A first-order number term $F(\bar{t})$ with free variables $\bar{t} = (t_1, \ldots, t_r)$ is interpreted in an $\mathbb{R}$-structure $\mathfrak{D}$ with universe A as a function $F^\mathfrak{D} : A^r \to \mathbb{R}$. As we have seen in Proposition 1, first-order number terms are easy to compute: they can be computed in constant time. Fixed point first-order logic extends first-order logic by introducing two new grammatical rules, the *maximization rule* and the *fixed point rule*. The first rule has the expressive power of first-order quantifiers but, unlike quantifiers, it yields a number term when applied to a number term. The second rule allows us to define $F^\mathfrak{D}$ in an inductive way.

For simplicity, in the rest of this chapter we restrict attention to *functional* $\mathbb{R}$-structures, that is, $\mathbb{R}$-structures whose signatures do not contain relation symbols. This represents no loss of expressive power since we can replace any relation $P \subseteq A^k$ by its characteristic function $\chi_P : A^k \to \mathbb{R}$.

We first define the maximization rule $MAX_\mathbb{R}$.

Definition 9 Let $F(\bar{u}, \bar{t})$ be a first-order number term with free variables $\bar{u} = (u_1, \ldots, u_r)$ and $\bar{t} = (t_1, \ldots, t_s)$. Then

$$\max_{\bar{u}} F(\bar{u}, \bar{t})$$

is also a number term with free variables $\bar{t}$. Its interpretation in any $\mathbb{R}$-structure $\mathfrak{D}$ and for any point $h \in A^s$ interpreting $\bar{t}$ is the maximum of $F^{\mathfrak{D}}(a, h)$ where a ranges over A^r.

Example 9 If the signature contains a symbol r which is interpreted as a ranking, then we can define the size n of the universe with the number term $\max_s r(s) + 1$.

In the rest of this section all $\mathbb{R}$-structures are ranked. As usual, r denotes the symbol interpreted as a ranking.

Definition 10 We denote by $\mathrm{MAX}_{\mathbb{R}}$ the logic obtained by adding to $\mathrm{FO}_{\mathbb{R}}$ the maximization rule.

The maximization rule enables us to express characteristic functions as number terms. If $\varphi(v_1, \ldots, v_r)$ is a first-order formula, we define its characteristic function $\chi[\varphi]$ on a structure $\mathfrak{D}$ by

$$\chi[\varphi](a_1, \ldots, a_r) = \begin{cases} 1 & \text{if } \mathfrak{D} \models \varphi(a_1, \ldots, a_r) \\ 0 & \text{otherwise,} \end{cases}$$

where $a_1, \ldots, a_r \in A$, the universe of $\mathfrak{D}$.

Proposition 2 *For every first-order formula $\varphi(v_1, \ldots, v_r)$ there is a number term in $\mathrm{MAX}_{\mathbb{R}}$ defining $\chi[\varphi]$.*

Proof. The proof is by induction on the construction of φ. If φ is atomic, it must have the form $F = G$ or $F < G$ with F and G number terms. This follows from the assumption that signatures do not contain relation symbols. Then

$$\chi[F = G] = \mathrm{sign}\,[-(F - G)^2]$$
$$\chi[F < G] = 1 - [\mathrm{sign}\,(F - G)].$$

If φ is $\psi_1 \,\&\, \psi_2$ or $\neg\psi$, then $\chi[\varphi]$ is $\chi[\psi_1]\chi[\psi_2]$ and $1 - \chi[\psi]$, respectively. Also, if φ is $\psi_1 \vee \psi_2$, then $\chi[\varphi]$ is $1 - (1 - \chi[\psi_1])(1 - \chi[\psi_2])$.

Finally, if φ is of the form $\exists x \psi(x)$ then

$$\chi[\varphi] = \max_x \chi[\psi(x)].$$

$\square$

The maximization rule can also be used to express some arithmetic in the universe. The following example is useful.

Example 10 Let $F(x, t)$ be a first-order number term with free variables $x = (x_1, \ldots, x_r)$ and $t = (t_1, \ldots, t_s)$. Consider the number term

$$\max_u \chi[r^s(u) = r^s(t) + 1]F(x, u) - \max_u \chi[r^s(u) = r^s(t) + 1](-F(x, u))$$

also with free variables x and t.

For every $\mathbb{R}$-structure $\mathfrak{D}$ with universe A and every element $h \in A^r$, $g \in A^s$ such that $\rho^s(g) \neq 0$, this new term is interpreted as $F^\mathfrak{D}(b, h)$ where b is the only element in A^s such that $\rho^s(b) = \rho^s(g) - 1$. If $\rho^s(g) = 0$, then it is interpreted as zero. We denote this term by $F(x, t - 1)$.

We now define the fixed point rule.

Definition 11 Fix a signature $\sigma = (L_s, L_f)$, an integer $r \geq 1$, and a pair (Z, D) of function symbols both of arity r and not contained in this signature. Let $F(Z, \bar{t})$ and $H(D, \bar{t})$ be number terms of signature $(L_s, L_f \cup \{Z, D\})$ with free variables $\bar{t} = (t_1, \ldots, t_r)$. We allow Z to appear several times in F but we do not require that its arguments are $t_1, \ldots, t_r$. The only restriction is that the number of free variables in F coincide with the arity of Z. A similar remark holds for H and D.

For any $\mathbb{R}$-structure $\mathfrak{D}$ of signature σ and any interpretation $\zeta : A^r \to \mathbb{R}$ of Z and $\Delta : A^r \to \mathbb{R}$ of D, respectively, the number terms $F(Z, \bar{t})$ and $H(D, \bar{t})$ define functions

$$F_\zeta^\mathfrak{D}, H_\Delta^\mathfrak{D} : A^r \to \mathbb{R}.$$

Let us consider the sequence of pairs $\{\Delta^i, \zeta^i\}_{i \geq 0}$ with $\Delta^i : A^r \to \mathbb{R}$ and $\zeta^i : A^r \to \mathbb{R}$ inductively defined by

$$\Delta^0(x) = 0 \quad \text{for all } x \in A^r$$
$$\zeta^0(x) = 0 \quad \text{for all } x \in A^r$$
$$\Delta^{i+1}(x) = \begin{cases} H_{\Delta^i}^\mathfrak{D}(x) & \text{if } \Delta^i(x) = 0 \\ \Delta^i(x) & \text{otherwise} \end{cases}$$
$$\zeta^{i+1}(x) = \begin{cases} F_{\zeta^i}^\mathfrak{D}(x) & \text{if } \Delta^i(x) = 0 \\ \zeta^i(x) & \text{otherwise.} \end{cases}$$

Since $\Delta^{i+1}(x)$ only differs from $\Delta^i(x)$ in case the latter is zero, one has that $\Delta^j = \Delta^{j+1}$ for some $j < |A|^r$. In this case, we also have that $\zeta^j = \zeta^{j+1}$. We denote these fixed points by Z^∞ and D^∞ and call them the *fixed points of* $F(Z, \bar{t})$ *and* $H(D, \bar{t})$ *on* $\mathfrak{D}$. We say that $F^\mathfrak{D}$ *updates* ζ.

Note that D plays the role of the characteristic function for the domain of Z; the different Δ^i determine the successive updatings of this domain. We say that E^∞ *is defined on* x if $D^\infty(x) \neq 0$.

The *fixed point rule* is now stated in the following way. If $F(Z, \bar{t})$ and $H(D, \bar{t})$ are number terms as previously, then

$$\mathbf{fp}[Z(\bar{t}) \leftarrow F(Z, \bar{t}), H(D, \bar{t})](\bar{u})$$

and

$$\mathbf{fp}[D(\bar{t}) \leftarrow H(D, \bar{t})](\bar{u})$$

are number terms of signature (L_s, L_f). Their interpretations on a given $\mathbb{R}$-structure $\mathfrak{D}$ are $Z^\infty(\bar{u})$ and $D^\infty(\bar{u})$, respectively.

An example of the fixed point rule is the evaluation of algebraic circuits.

Example 11 We consider the description of algebraic circuits given in Example 8 and extend this description by supposing that the constant $C(v)$ associated with an input gate v is the actual input to the circuit. We are then interested in defining the evaluation function $E : A \to \mathbb{R}$ that assigns to each gate of the circuit the value it computes for this input. To do so, let us denote by T the set $\{1, 2, \ldots, 8\}$ and by g_i the polynomial

$$g_i(x) = \prod_{\substack{j \in T \\ j \neq i}} \frac{x - j}{i - j}$$

that maps i to 1 and j to 0 for $i, j \in T, i \neq j$.

Now we consider the number term $F(E, v)$ given by

$$((g_1 + g_2)(F_t(v)))C(v) + g_3(F_t(v))(E(f_l(v)) + E(f_r(v)))$$
$$+ g_4(F_t(v))(E(f_l(v)) - E(f_r(v))) + g_5(F_t(v))(E(f_l(v)) \times E(f_r(v)))$$
$$+ g_6(F_t(v))(E(f_l(v))/E(f_r(v))) + g_7(F_t(v))\text{sign}(E(f_l(v)))$$
$$+ g_8(F_t(v))E(f_l(v))$$

as well as the number term $H(D, v)$ given by

$$((g_1 + g_2)(F_t(v))) + (g_7 + g_8)(F_t(v))(D(f_l(v)))$$
$$+ (g_3 + g_4 + g_5 + g_6)(F_t(v))(D(f_l(v)) \times D(f_r(v))).$$

The updating of D by H is clear. The value $\Delta^1(v)$ will be 1 if v is an input or constant node and 0 otherwise. Then Δ^{i+1} updates v from 0 to 1 if the $\Delta^i(x) = 1$ for the predecessors x of v. Eventually, all nodes are updated to 1 and D^∞ is the constant function 1.

The function $E^\infty : A \to \mathbb{R}$ defined by $F(E, v)$ is the evaluation function for which we are looking. Notice that since D^∞ is the constant function 1, the function E is total.

Definition 12 *Fixed point (first-order) logic* for $\mathbb{R}$-structures, denoted by $\text{FP}_\mathbb{R}$, is obtained by augmenting first-order logic $\text{FO}_\mathbb{R}$ with the maximization rule and the fixed point rule.

Now recall the Circuit Evaluation Problem (CEP). This is the set of pairs $(\mathcal{C}, b)$, where $\mathcal{C}$ is an algebraic circuit, $b \in \mathbb{R}^n$ (n is the number of input gates in $\mathcal{C}$), and $\varphi_{\mathcal{C}}(b) = 1$. We have seen in Section 19.3 that this is a $P_{\mathbb{R}}$-complete problem.

Example 12 Example 11 allows us to describe the Circuit Evaluation Problem within fixed point first-order logic for $\mathbb{R}$-structures. Since we only wish to consider circuits with a unique output gate we add to the axioms given in Example 8 the axiom

$$\forall u \ (F_t(u) = 8 \Rightarrow u = 1).$$

Also, we require the unique output gate to output 1, so we add the axiom

$$\mathbf{fp}[E(v) \leftarrow F(E, v), H(D, v)](1) = 1.$$

We can now proceed with our main result.

Theorem 1 *Let S be a decision problem of ranked $\mathbb{R}$-structures. Then the following two statements are equivalent.*

(i) $S \in P_{\mathbb{R}}$.

(ii) *There exists a sentence ψ in fixed point first-order logic such that $S = \{\mathfrak{D} \mid \mathfrak{D} \models \psi\}$.*

Proof.   Let M be a machine that decides S in polynomial time. We can assume without loss of generality that the functions associated with the computation nodes of M are single arithmetical operations between two coordinates of the state space and that the result of such an operation is stored in the zeroth coordinate of its state space. We also assume that all the branch nodes of M are of the form $x_0 \geq 0$, that is, that the value whose sign is tested is the content of the zeroth register of M.

Let m be such that for any $\mathbb{R}$-structure $\mathfrak{D}$ of size n the computation of M with input $\mathfrak{D}$ runs in time bounded by n^m. Thus, the time-n^m state space of M is $\mathbb{R}^{2n^m}$. Since $2n^m \leq n^{m+1}$ we consider the computations of M over $\mathbb{R}$-structures of size n to take place in the state space $\mathbb{R}^{n^{m+1}}$. A point in this space has coordinates $(x_{-n^m}, \ldots, x_{-1}, x_0, x_1, \ldots, x_{(n-1)n^m-1})$. Notice that the zeroth coordinate x_0 is in the $n^m + 1$th position in this vector.

Let $\mathfrak{D}$ be a ranked $\mathbb{R}$-structure of size n and A be its universe. Denote by x the point $e(\mathfrak{D})$.

Given an element $j \in A^k$ denote by $\bar{j}$ the integer $\rho^k(j)$, where ρ is the ranking on A. Also, given $t = (t_1, \ldots, t_k)$, where t_i is an index term for $i = 1, \ldots, k$, denote by $\bar{t}$ the integer $\rho^k(t_1^{\mathfrak{D}}, \ldots, t_k^{\mathfrak{D}})$.

Consider the function $\zeta : A \times A^{m+1} \times A^m \to \mathbb{R}$ such that

- $\zeta(0, v, t)$ is the node reached by M after $\bar{t} + 1$ steps,

- $\zeta(1, j, t)$ is the content of the $\bar{j}$th register of M after $\bar{t} + 1$ steps,

where we denote by 0 and 1 the two elements in A that are mapped to 0 and 1, respectively, by ρ. These two elements are definable by number terms so we freely use constant symbols 0 and 1 which are interpreted as 0 and 1, respectively. We denote by 0_m the sequence $(0, \ldots, 0)$ of m 0s.

If ζ can be defined in $\mathrm{FP}_\mathbb{R}$ by a number term Z, then the implication (i) $\Rightarrow$ (ii) follows. The $\mathbb{R}$-structure $\mathfrak{D}$ is accepted by M if and only if after n^m steps of the computation of M, the contents of its first register is a 1. That is, if $\zeta(1, \mathrm{M} + 1, \mathrm{T}) = 1$, where $\mathrm{M} + 1$ and T are index terms in $\mathrm{FP}_\mathbb{R}$ that are interpreted as the elements of A^{m+1} and A^m, respectively, which satisfy $\overline{\mathrm{M} + 1} = n^m + 2$ and $\overline{\mathrm{M} + 1} = n^m - 1$.

Our goal is to define ζ inductively. As a first attempt to do so we can write

$$Z(u, v, t) = \chi[\rho(u) = 0]Z(0, v, t) + \chi[\rho(u) = 1]Z(1, v, t). \qquad (23.4)$$

Note that for $u \notin \{0, 1\}$, $Z(u, v, t) = 0$. Moreover, we have that

$$Z(0, v, 0_m) = \beta(1),$$

where $\beta(1)$ is the successor of the input node, and that

$$Z(1, v, 0_m) = x_v,$$

where x_v is the $\overline{v}$th coordinate of the image of x under the input map of M. This value can be described with a first-order term according to Lemma 1.

Therefore we only need to find number terms defining $Z(0, v, t)$ and $Z(1, v, t)$ for $t \geq 1$. The way to do so resembles the idea used in the proof of Proposition 4 of Chapter 18.

The first function is simply dealt with since for $\overline{t} \geq 1$,

$$Z(0, v, t) = \sum_{r=2}^{N} \chi[Z(0, v, t - 1) = r]\beta(r),$$

where if the rth node is not a branching node, then $\beta(r)$ is a well-defined value and otherwise it is $\beta^+(r)$ or $\beta_-(r)$ according to whether $x_0 \geq 0$ or $x_0 < 0$ (notice the use of Example 10). The value $\beta(r)$ in this latter case can be described with the number term

$$\chi[Z(1, \mathrm{M}, t - 1) \geq 0]\beta^+(r) + \chi[Z(1, \mathrm{M}, t - 1) < 0]\beta^-(r).$$

Here N is the number of nodes of the machine M.

Thus, the term

$$\chi[\overline{t} = 0]\beta(1) + \chi[\overline{t} \neq 0] \sum_{r=2}^{N} \chi[Z(0, v, t - 1) = r]\beta(r) \qquad (23.5)$$

describes $Z(0, v, t)$.

We pass now to the value $Z(1, v, t)$. Denote by $\mathcal{C}$ the set of computation nodes of M. Then at a node $r \in \mathcal{C}$ the value $Z(1, v, t)$ is given by the term $Z_r(1, v, t)$,

$$\chi[\overline{v} = 0](Z(1, \mathrm{K}, t - 1) \circ_r Z(1, \mathrm{G}, t - 1)) + \chi[\overline{v} \neq 0]Z(1, v, t - 1),$$

where the $\circ_r$ is the operation performed at node r and it operates the $\overline{K}$th and $\overline{G}$th registers of M.

Denote now by $\mathcal{S}_\ell$ the set of shift nodes that shift to the left and by $\mathcal{S}_r$ the set of shift nodes that shift to the right. For $\overline{\iota} \geq 1$ the value $Z(1, v, t)$ is then given by

$$
\sum_{r \notin (\mathcal{C} \cup \mathcal{S}_\ell \cup \mathcal{S}_r)} \chi[Z(0, v, t-1) = r] Z(1, v, t-1)
$$
$$
+ \sum_{r \in \mathcal{C}} \chi[Z(0, v, t-1) = r] Z_r(1, v, t) \tag{23.6}
$$
$$
+ \sum_{r \in \mathcal{S}_\ell} \chi[Z(0, v, t-1) = r] Z(1, v-1, t-1)
$$
$$
+ \sum_{r \in \mathcal{S}_r} \chi[Z(0, v, t-1) = r] Z(1, v+1, t-1).
$$

Replacing terms (23.5) and (23.6) in (23.4) we get the number term that defines Z as a fixed point. The condition $Z(1, \text{M}, \text{T}) = 1$ is now expressed in fixed point logic thus proving (ii).

To prove that (ii) implies (i) is simpler. We only need to prove that number terms defined with the maximization and fixed point operator can be evaluated in polynomial time. For the first operator, one only evaluates the term to maximize over all the elements of A^s and selects the maximum. Since this number of elements is polynomial in $|A|$ the assertion follows. For the second operator, one applies the inductive definition. Again, since the number of updates is polynomially bounded, so is the total time required to do this.   □

23.4   Capturing NP$_\mathbb{R}$ with Existential Second-Order Logic

Second-order logic on $\mathbb{R}$-structures is the logic obtained by adding to FO$_\mathbb{R}$ the power to quantify over function symbols. Here we consider only a particular type of second-order formulas.

Definition 13 We say that ψ is an *existential second-order sentence* (of signature $\sigma = (L_s, L_f)$) if $\psi = \exists Y_1 \ldots \exists Y_r \phi$, where ϕ is a first-order sentence in FO$_\mathbb{R}$ of signature $(L_s, L_f \cup \{Y_1, \ldots, Y_r\})$. The function symbols $Y_1, \ldots, Y_r$ are called *function variables*. Existential second-order sentences are interpreted in $\mathbb{R}$-structures $\mathfrak{D}$ of signature σ in the natural way. A sentence ψ is true in $\mathfrak{D}$ if there exist functions $X_1, \ldots, X_r$, $X_i : A^{r_i} \to \mathbb{R}$ with r_i the arity of Y_i, such that the interpretation of ψ taking $Y_i^\mathfrak{D}$ to be X_i yields $\texttt{true}$. The set of second-order sentences together with this interpretation constitutes *existential second-order logic* and is denoted by $\exists$SO$_\mathbb{R}$.

Example 13 Let us see how to describe 4-FEAS with an existential second-order sentence.

Consider an $\mathbb{R}$-structure $\mathfrak{D} = (\mathfrak{A}, \mathcal{F})$, where $\mathcal{F}$ consists of a function $C : A^4 \to \mathbb{R}$ and a ranking $\rho : A \to \mathbb{R}$. Let $n = |A| - 1$ so that ρ is a bijection between A and $\{0, 1, \ldots, n\}$. Then $\mathfrak{D}$ defines an homogeneous polynomial $\widehat{g} \in \mathbb{R}[x_0, \ldots, x_n]$ of degree four, namely,

$$\widehat{g} = \sum_{(i,j,k,\ell) \in A^4} C(i, j, k, \ell) x_i x_j x_k x_\ell.$$

We obtain an arbitrary, not necessarily homogeneous, polynomial $g \in \mathbb{R}[x_1, \ldots, x_n]$ of degree four by setting $x_0 = 1$ in $\widehat{g}$. We also say that $\mathfrak{D}$ defines g. Notice that, conversely, for every polynomial g of degree four in n variables there is an $\mathbb{R}$-structure $\mathfrak{D}$ of size $n + 1$ such that $\mathfrak{D}$ defines g.

Denote by o, ℓ, $\bar{o}$, and $\bar{\ell}$ the first and last elements of A and A^4 with respect to ρ and ρ^4. These four elements are first-order definable and therefore we may use constant symbols o, $\mathsf{1}$, $\bar{\mathsf{o}}$, and $\bar{\mathsf{1}}$ which are interpreted as o, ℓ, $\bar{o}$, and $\bar{\ell}$, respectively. The following sentence quantifies two functions $X : A \to \mathbb{R}$ and $Y : A^4 \to \mathbb{R}$,

$$\psi \equiv (\exists X)(\exists Y) \Big(Y(\bar{\mathsf{o}}) = C(\bar{\mathsf{o}}) \,\&\, Y(\bar{\mathsf{1}}) = 0 \,\&\, X(\mathsf{o}) = 1 \,\&$$

$$\&\, \forall u_1 \ldots \forall u_4 \Big[u \neq \bar{\mathsf{o}} \Rightarrow \exists v_1 \ldots \exists v_4 \, (\rho^4(u) = \rho^4(v) + 1)$$

$$\&\, Y(u) = Y(v) + C(u)X(u_1)X(u_2)X(u_3)X(u_4) \Big] \Big),$$

where if $a_i = \rho^{-1}(i)$ for $i = 1, \ldots, n$, then $(X(a_1), \ldots, X(a_n)) \in \mathbb{R}^n$ describes the zero of g and $Y(u)$ is the partial sum of all its monomials up to $u = (u_1, \ldots, u_4) \in A^4$ evaluated at the point $(X(a_1), \ldots, X(a_n))$.

The sentence ψ describes 4-FEAS in the sense that for any $\mathbb{R}$-structure $\mathfrak{D}$ one has that $\mathfrak{D} \models \psi$ if and only if the polynomial g of degree four defined by $\mathfrak{D}$ has a real zero.

Existential second-order logic is at least as powerful as fixed point first-order logic.

Proposition 3 *For every sentence ψ in $\mathrm{FP}_{\mathbb{R}}$ there is a sentence $\widetilde{\psi}$ in $\exists\mathrm{SO}_{\mathbb{R}}$ such that for every $\mathbb{R}$-structure $\mathfrak{D}$*

$$\mathfrak{D} \models \psi \qquad \text{if and only if} \qquad \mathfrak{D} \models \widetilde{\psi}.$$

Proof. We must show that the maximization and fixed point rules can be defined within existential second-order logic. The maximization operator is defined in $\exists\mathrm{SO}_{\mathbb{R}}$ as follows. If $F(\bar{u}, \bar{t})$ is a number term, then we replace all occurrences of $\max_{\bar{u}} F(\bar{u}, \bar{t})$ by $G(\bar{t})$ in ψ. If ϕ denotes the resulting formula, then $\widetilde{\psi}$ is

$$\exists G \forall \bar{x} \, [\forall \bar{u} \, G(\bar{x}) \geq F(\bar{u}, \bar{x}) \,\&\, \exists \bar{u} \, G(\bar{x}) = F(\bar{u}, \bar{x}) \,\&\, \phi].$$

For the fixed point operator, if F, H, Z, D are as in Definition 11, we replace every occurrence of $\mathbf{fp}[Z(\bar{t}) \leftarrow F(Z, \bar{t}), H(D, \bar{t})](\bar{u})$ in ψ by $Z(\bar{u})$ and every occurrence of $\mathbf{fp}[D(\bar{t}) \leftarrow H(D, \bar{t})](\bar{u})$ by $D(\bar{u})$. Let ϕ be their resulting formula. Then $\widetilde{\psi}$ is

$$\exists Z \exists D \ [\forall \bar{v}(F(Z, \bar{v}) = Z(\bar{v}) \ \& \ H(D, \bar{v}) = D(\bar{v})) \ \& \ \phi].$$

$\square$

Our next goal is to prove that existential second-order logic captures NP$_\mathbb{R}$. To do so we begin with a lemma whose proof is immediate.

Lemma 2 *Let* $\sigma = (L_s, L_f)$ *be a signature and* S *a decision problem of* $\mathbb{R}$-*structures of signature* σ *which belong to* NP$_\mathbb{R}$. *Then there exists* $r \in \mathbb{N}$ *and a decision problem* H *of* $\mathbb{R}$-*structures of signature* $(L_s, L_f \cup \{Y\})$, *with* Y *of arity* r, *such that* H *belongs to* P$_\mathbb{R}$ *and*

$$S = \{\mathfrak{D} \in \mathbb{R}_\sigma \mid \exists Y \ (\mathfrak{D}, Y) \in H\}.$$

$\square$

Theorem 2 *Let* S *be a decision problem of* $\mathbb{R}$-*structures. Then the following statements are equivalent.*

(i) $S \in$ NP$_\mathbb{R}$.

(ii) *There exists an existential second-order sentence* ψ *such that* $S = \{\mathfrak{D} \mid \mathfrak{D} \models \psi\}$.

Proof. Suppose for the moment that we are considering ranked $\mathbb{R}$-structures. Since $S \in$ NP$_\mathbb{R}$, by Lemma 2 there is a decision problem H of ranked $\mathbb{R}$-structures such that $S = \{\mathfrak{D} \in \text{Struct}(\sigma) \mid \exists Y \ (\mathfrak{D}, Y) \in H\}$. By Theorem 1, there is a fixed point first-order formula φ that describes H and thus,

$$\mathfrak{D} \in S \quad \text{if and only if} \quad \mathfrak{D} \models \exists Y \ \varphi.$$

By Proposition 3, we can replace φ by an equivalent existential second-order formula $\exists Z \ \phi$ obtaining

$$\mathfrak{D} \in S \quad \text{if and only if} \quad \mathfrak{D} \models \exists Y \exists Z \ \phi$$

with ϕ a first-order formula.

We finally get rid of the assumption that our $\mathbb{R}$-structures are ranked. If no ranking is given, we can introduce one by existentially quantifying a function r and adding a first-order formula $\alpha(r)$ that requires r to be a ranking. It then follows that

$$\mathfrak{D} \in S \quad \text{if and only if} \quad \mathfrak{D} \models \exists r \exists Y \exists Z \ (\alpha(r) \ \& \ \psi).$$

To see that (ii) implies (i) consider a sentence $\psi = \exists Y_1 \ldots \exists Y_s \ \varphi$ with φ a first-order formula and Y_i a function variable for $i = 1, \ldots, s$. Given input $\mathbb{R}$-structure $\mathfrak{D}$, a nondeterministic machine M guesses assignments for $Y_1, \ldots, Y_s$ and then checks the validity of the resulting first-order formula. If $|A| = n$, the first part takes time $O(sn)$ and the second can be done in weak polynomial time according to Proposition 1. $\square$

23.5 Additional Comments and Bibliographical Remarks

Descriptive complexity can be viewed as a branch of finite model theory with the goal of characterizing the logical complexity of definable properties or queries. Its relationship to computational complexity became apparent with the work of Fagin [1974] who proved, in the classical setting, that the class NP can be characterized in a machine-independent way as the class of sets describable within existential second-order logic. Subsequently, a stream of results characterized in this way other complexity classes such as P [Vardi 1982; Immerman 1986] and introduced new concepts such as first-order reduction. An overview of the subject can be found in [Immerman 1989] and [Gurevich 1988].

This work, however, was restricted to the classical setting since only finite models were considered. In order to consider computations over more general structures, Grädel and Gurevich [1995] introduced the notion of metafinite models. A particular case, the $\mathbb{R}$-structures dealt with in this chapter, were developed by Grädel and Meer [1995]. Our presentation closely follows their work. Additional results along this line, characterizing the classes $PAR_{\mathbb{R}}$, $EXP_{\mathbb{R}}$, and $NC_{\mathbb{R}}^k$ among others, can be found in [Cucker and Meer TA].

The power of first-order logic in the discrete case is well known. The complexity class captured by this logic is AC_0. This is the class of sets decided by uniform families of Boolean circuits $\{C_n\}_{n \in \mathbb{N}}$ with unbounded fanin that have size polynomial in n and constant depth. It is known, moreover, that the strict inclusion $AC_0 \subset NC_1$ holds. This suggests that the converse of Proposition 1 mentioned in Section 23.3 is false.

For readers who are not familiar with the concepts and methods of mathematical logic, an elementary textbook is [Ebbinghaus, Flum, and Thomas 1994].

References

ABRAMOWITZ, M. and I. STEGUN (Eds.) (1964). *Handbook of Mathematical Functions with Formulas, Graphs, and Mathematical Tables*. National Bureau of Standards.

ABRAMSON, F. (1971). Effective computation over the real numbers. In *12th Annual IEEE Symp. on Switching and Automata Theory*, pp. 33–37.

ADLEMAN, L. (1978). Two theorems on random polynomial time. In *19th Annual IEEE Symp. on Foundations of Computer Science*, pp. 75–83.

AHO, A., J. HOPCROFT, and J. ULLMAN (1974). *The Design and Analysis of Computer Algorithms*. Addison-Wesley.

ALLGOWER, E. and K. GEORG (1990). *Numerical Continuation Methods*. Springer-Verlag.

ALLGOWER, E. and K. GEORG (1993). Continuation and path following. In A. Iserles (Ed.), *Acta Numerica*, pp. 1–64. Cambridge University Press.

ALLGOWER, E. and K. GEORG (1997). Numerical path following. In P. Ciarlet and J.-L. Lions (Eds.), *Handbook of Numerical Analysis*, Volume V, pp. 3–207. North-Holland.

ARTIN, E. and O. SCHREIER (1926). Algebraische Konstruktion reeller Körper. *Hamb. Abh. 5*, 85–99.

ATIYAH, M. and I. MACDONALD (1969). *Introduction to Commutative Algebra*. Addison-Wesley.

BACHMAN, P. (1894). *Die analytische Zahlentheorie*. Teubner.

BAKER, T., J. GILL, and R. SOLOVAY (1975). Relativizations of the P =?NP question. *SIAM J. on Computing 4*, 431–442.

BALCÁZAR, J., J. DÍAZ, and J. GABARRÓ (1988). *Structural Complexity I*. EATCS Monographs on Theoretical Computer Science, 11. Springer-Verlag.

BALCÁZAR, J., J. DÍAZ, and J. GABARRÓ (1990). *Structural Complexity II*. EATCS Monographs on Theoretical Computer Science, 22. Springer-Verlag.

BAREISS, E. (1968). Sylvester's identity and multistep integer-preserving Gaussian elimination. *Math. Comp. 22*, 565–578.

BARVINOK, A. and A. VERSHIK (1993). Polynomial-time computable approximation of families of semi-algebraic sets and combinatorial complexity. *American Mathematical Society Translation 155*, 1–17.

BASU, S., R. POLLACK, and M.-F. ROY (1994). On the combinatorial and algebraic complexity of quantifier elimination. In *35th Annual IEEE Symp. on Foundations of Computer Science*, pp. 632–641.

BEAUZAMY, B. and J. DÉGOT (1995). Differential identities. *Transactions of the Amer. Math. Soc. 347*, 2607–2619.

BEN-OR, M. (1983). Lower bounds for algebraic computation trees. In *15th Annual ACM Symp. on the Theory of Computing*, pp. 80–86.

BEN-OR, M., D. KOZEN, and J. REIF (1986). The complexity of elementary algebra and geometry. *J. of Computer and Systems Sciences 18*, 251–264.

BENEDETTI, R. and J.-J. RISLER (1990). *Real Algebraic and Semi-Algebraic Sets*. Hermann.

BENNETT, C. and J. GILL (1981). Relative to a random oracle A, $P^A \neq NP^A$ with probability 1. *SIAM J. on Computing 10*, 96–113.

BERKOWITZ, S. (1984). On computing the determinant in small parallel time using a small number of processors. *Information Processing Letters 18*, 147–150.

BINI, D. and V. PAN (1994). *Polynomial and Matrix Computations*. Birkhäuser.

BLUM, L. (1990). Lectures on a theory of computation and complexity over the reals (or an arbitrary ring). In E. Jen (Ed.), *Lectures in the Sciences of Complexity II*, pp. 1–47. Addison-Wesley.

BLUM, L. (1991). A theory of computation and complexity over the real numbers. In *Proceedings of the International Congress of Mathematicians*, pp. 1491–1507. Springer-Verlag.

BLUM, L., F. CUCKER, M. SHUB, and S. SMALE (1996a). Algebraic settings for the problem "P $\neq$ NP". In J. Renegar, M. Shub, and S. Smale (Eds.), *The Mathematics of Numerical Analysis*, Volume 32 of *Lectures in Applied Mathematics*, pp. 125–144. American Mathematical Society.

BLUM, L., F. CUCKER, M. SHUB, and S. SMALE (1996b). Complexity and real computation: A manifesto. *Int. J. of Bifurcation and Chaos 6*, 3–26.

BLUM, L. and M. SHUB (1986). Evaluating rational functions: Infinite precision is finite cost and tractable on average. *SIAM J. on Computing 15*, 384–398.

BLUM, L., M. SHUB, and S. SMALE (1989). On a theory of computation and complexity over the real numbers: NP-completeness, recursive functions and universal machines. *Bulletin of the Amer. Math. Soc. 21*, 1–46.

BLUM, L. and S. SMALE (1993). The Gödel incompleteness theorem and decidability over a ring. In M. Hirsch, J. Marsden, and M. Shub (Eds.), *From Topology to Computation: Proceedings of the Smalefest*, pp. 321–339. Springer-Verlag.

BLUM, M. (1967). A machine-independent theory of the complexity of recursive functions. *J. of the ACM 14*, 322–336.

BOCHNAK, J., M. COSTE, and M.-F. ROY (1987). *Géométrie algébrique réelle*. Springer-Verlag.

BORGWARDT, K. (1982). The average number of pivot steps required by the simplex method is polynomial. *Z. Oper. Res. 26*, 157–177.

BOROCH, L. and L. TREYBIG (1976). Bounds on positive integral solutions of linear diophantine equations. *Proceedings of the Amer. Math. Soc. 72*, 199–304.

BORODIN, A. (1977). On relating time and space to size and depth. *SIAM J. on Computing 6*, 733–744.

BORODIN, A. and I. MUNRO (1975). *The Computational Complexity of Algebraic and Numeric Problems*. American Elsevier.

BROCKETT, R. (1973). Lie algebras and Lie groups in control theory. In D. Mayne and R. Brockett (Eds.), *Geometric Methods in System Theory, Proceedings of the NATO Advanced Study Institute*, pp. 43–82. D. Reidel.

BROWNAWELL, W. (1987). Bounds for the degrees in the Nullstellensatz. *Annals of Math. 126*, 577–591.

BÜRGISSER, P., M. CLAUSEN, and A. SHOKROLLAHI (1996). *Algebraic Complexity Theory*. Springer-Verlag.

CAMPBELL, S. and C. MEYER (1979). *Generalized Inverses of Linear Transformations*. Pitman.

CANIGLIA, L., A. GALLIGO, and J. HEINTZ (1988). Borne simple exponentielle pour les degrés dans les théorèmes de zéros sur un corps de caractéristique quelconque. *C. R. Acad. Sci. Paris 307*, 255–258.

CANNY, J. (1988). Some algebraic and geometric computations in PSPACE. In *20th Annual ACM Symp. on the Theory of Computing*, pp. 460–467.

CHATELIN, F. and V. FRAYSSÉ (1993). Qualitative computing: Elements of a theory for finite precision computation. Lecture Notes for the Comett European Course, June 8–10, Thomson-CSF, LCR Corbeville, Orsay.

CHATELIN, F., V. FRAYSSÉ, and T. BRACONNIER (1995). Computations in the neighbourhood of algebraic singularities. *Num. Funct. Anal. Opt. 16*, 287–302.

CHEN, P. (1994). Approximate zeros of quadratically convergent algorithms. *Math. Comp. 63*, 247–270.

CHERNOFF, H. (1952). A measure of asymptotic efficiency for tests of a hypothesis based on the sum of observations. *Ann. Math. Statist. 23*, 493–507.

CHILDS, L. (1995). *A Concrete Introduction to Higher Algebra* (2nd ed.). Springer-Verlag.

CHURCH, A. (1936). An unsolvable problem of elementary number theory. *Amer. J. of Math. 58*, 354–363.

COBHAM, A. (1964). The intrinsic computational difficulty of problems. In International Congress for Logic, Methodology, and the Philosophy of Science, *Y. Bar-Hillel (Ed.), North-Holland*, pp. 24–30.

COLLINS, G. (1975). Quantifier elimination for real closed fields by cylindrical algebraic decomposition, Volume 33 of *Lect. Notes in Comp. Sci.*, pp. 134–183. Springer-Verlag.

COOK, S. (1971). The complexity of theorem proving procedures. In *3rd Annual ACM Symp. on the Theory of Computing*, pp. 151–158.

COOK, S. (1979). Deterministic CFL's are accepted simultaneously in polynomial time and log squared space. In *11th Annual ACM Symp. on the Theory of Computing*, pp. 338–345.

COPPERSMITH, D. and S. WINOGRAD (1990). Matrix multiplication via arithmetic progressions. *J. of Symbolic Computation 9*, 251–280.

CSANKY, L. (1976). Fast parallel matrix inversion algorithms. *SIAM J. on Computing 5*, 618–623.

CUCKER, F. (1992a). The arithmetical hierarchy over the reals. *J. of Logic and Computation 2*, 375–395.

CUCKER, F. (1992b). $P_\mathbb{R} \neq NC_\mathbb{R}$. *J. of Complexity 8*, 230–238.

CUCKER, F. (1993). On the complexity of quantifier elimination: The structural approach. *The Computer J. 36*, 400–408.

CUCKER, F., M. KARPINSKI, P. KOIRAN, T. LICKTEIG, and K. WERTHER (1995). On real Turing machines that toss coins. In *27th Annual ACM Symp. on the Theory of Computing*, pp. 335–342.

CUCKER, F. and P. KOIRAN (1995). Computing over the reals with addition and order: Higher complexity classes. *J. of Complexity 11*, 358–376.

CUCKER, F. and M. MATAMALA (1996). On digital nondeterminism. *Mathematical Systems Theory 29*, 635–647.

CUCKER, F. and K. MEER (TA). Logics which capture complexity classes over the reals. Preprint.

CUCKER, F., J. MONTAÑA, and L. PARDO (1992). Time bounded computations over the reals. *Int. J. of Algebra and Computation 2*, 395–408.

CUCKER, F., J. MONTAÑA, and L. PARDO (1995). Models for parallel computations with real numbers. In I. Shparlinski (Ed.), *Number Theoretic and Algebraic Methods in Computer Science*, pp. 53–63. World Scientific.

CUCKER, F. and F. ROSSELLÓ (1993). Recursiveness over the complex numbers is time bounded. In *13th Foundations of Software Technology and Theoretical Computer Science*, Volume 761 of *Lect. Notes in Comp. Sci.*, pp. 260–267. Springer-Verlag.

CUCKER, F. and M. SHUB (1996). Generalized Knapsack problems and fixed degree separations. *Theoretical Computer Science 161*, 301–306.

CUCKER, F. and M. SHUB (Eds.) (1997). *Foundations of Computational Mathematics. Selected Papers of a Conference Held at IMPA in Rio de Janeiro, January 1997.* Springer-Verlag.

CUCKER, F., M. SHUB, and S. SMALE (1994). Complexity separations in Koiran's weak model. *Theoretical Computer Science 133*, 3–14.

CUCKER, F. and A. TORRECILLAS (1992). Two P-complete problems in the theory of the reals. *J. of Complexity 8*, 454–466.

CURRY, J. (1989). On zero finding methods of higher order from data at one point. *J. of Complexity 5*, 219–237.

CUTLAND, N. (1980). *Computability*. Cambridge University Press.

DA COSTA, N. and F. DORIA (1990). Undecidability and incompleteness in classical mechanics. Preprint.

DAVIS, M. (1965). *The Undecidable*. Raven Press.

DE MELO, W. and B. SVAITER (1996). The cost of computing integers. *Proceedings of the Amer. Math. Soc. 124*, 1377–1378.

DEDIEU, J.-P. (1996). Approximate solutions of numerical problems, condition number analysis and condition number theorem. In J. Renegar, M. Shub, and S. Smale (Eds.), *The Mathematics of Numerical Analysis*, Volume 32 of *Lectures in Applied Mathematics*, pp. 263–283. American Mathematical Society.

DEDIEU, J.-P. (1997). Condition number analysis for sparse polynomial systems. In F. Cucker and M. Shub (Eds.), *Foundations of Computational Mathematics. Selected Papers of a Conference Held at IMPA in Rio de Janeiro, January 1997*, pp. 75–101. Springer-Verlag.

DEDIEU, J.-P. (TAa). Condition operators, condition numbers and condition number theorem for the generalized eigenvalue problem. To appear in *Linear Algebra and its Applic.*

DEDIEU, J.-P. (TAb). Estimations for the separation number of a polynomial system. To appear in *J. of Symbolic Computation*.

DEJON, B. and P. HENRICI (1969). *Constructive Aspects of the Fundamental Theorem of Algebra*. John Wiley & Sons.

DEMMEL, J. (1987a). On condition numbers and the distance to the nearest ill-posed problem. *Numer. Math. 51*, 251–289.

DEMMEL, J. (1987b). The probability that a numerical analysis problem is difficult. *Math. Comp. 51*, 251–289.

DEVANEY, R. (1989). *Chaotic Dynamical Systems*. Addison-Wesley.

DOBKIN, D. and R. LIPTON (1979). On the complexity of computations under varying sets of primitives. *J. of Computer and System Sciences 18*, 86–91.

DOUADY, A. and J. HUBBARD (1984). *Étude dynamique des polynômes complexes, I*, Volume 84-20 of *Publications Mathématiques d'Orsay*. Université de Paris-Sud, Dept. de Math., Orsay, France.

DOUADY, A. and J. HUBBARD (1985). *Étude dynamique des polynômes complexes, II*, Volume 85-40 of *Publications Mathématiques d'Orsay*. Université de Paris-Sud, Dept. de Math., Orsay, France.

DUBOIS, D. (1969). A Nullstellensatz for ordered fields. *Ark. Mat. 8*, 111–114.

EAVES, C. and H. SCARF (1976). The solution of systems of piecewise linear equations. *Math. of Oper. Research 1*, 1–27.

EBBINGHAUS, H.-D., J. FLUM, and W. THOMAS (1994). *Mathematical Logic* (2nd ed.). Springer-Verlag.

ECKART, C. and G. YOUNG (1936). The approximation of one matrix by another of lower rank. *Psychometrika 1*, 211–218.

EDELMAN, A. (1988). Eigenvalues and condition numbers of random matrices. *SIAM J. of Matrix Anal. and Applic. 9*, 543–556.

EDELMAN, A. (1989). *Eigenvalues and Condition Numbers of Random Matrices*. Ph. D. Thesis, M.I.T.

EDELMAN, A. (1992). On the distribution of a scaled condition number. *Math. Comput. 58*, 185–190.

EDELMAN, A. (1995). On the determinant of a uniformly distributed complex matrix. *J. of Complexity 11*, 352–357.

EDELMAN, A. and E. KOSTLAN (1995). How many zeros of a polynomial are real? *Bulletin of the Amer. Math. Soc. 32*, 1–37.

EDMONDS, J. (1965). Paths, trees, and flowers. *Canadian J. of Mathematics 17*, 449–467.

EDMONDS, J. (1967). Systems of distinct representatives and linear algebra. *J. of Research of the National Bureau of Standards, B. Math. and Mathematical Physics 71B*, 241–245.

EISENBUD, D. (1995). *Commutative Algebra with a View Toward Algebraic Geometry*. Springer-Verlag.

EMERSON, T. (1994). Relativization of the P $=$?NP question over the reals (and other ordered rings). *Theoretical Computer Science 133*, 15–22.

ENGELER, E. (1993). *Algorithmic Properties of Structures*. World Scientific.

ENGELER, E. (1995). *The Combinatory Programme*. Birkhäuser. In collaboration with K. Aberer, B. Amrhein, O. Gloor, M. von Mohrenschildt, D. Otth, G. Schwärzler, and T. Weibel.

FAGIN, R. (1974). Generalized first-order spectra and polynomial-time recognizable sets. *SIAM-AMS Proc. 7*, 43–73.

FEDERER, H. (1969). *Geometric Measure Theory*. Springer-Verlag.

FRIEDMAN, H. (1971). Algorithmic procedures, generalized Turing algorithms, and elementary recursion theory. In R. Gandy and C. Yates (Eds.), *Logic Colloquium 1969*, pp. 361–390. North-Holland.

FRIEDMAN, H. and R. MANSFIELD (1992). Algorithmic procedures. *Transactions of the Amer. Math. Soc. 332*, 297–312.

GARCIA, C. and F. GOULD (1980). Relations between several path following algorithms and local and global Newton methods. *SIAM Review 22*, 263–274.

GARCIA, C. and W. ZANGWILL (1981). *Pathways to Solutions, Fixed Points and Equilibria*. Prentice Hall.

GAREY, M. and D. JOHNSON (1979). *Computers and Intractability: A Guide to the Theory of NP-Completeness*. Freeman.

GASHKOV, S. (1980). The complexity of the realization of Boolean functions by networks of functional elements and by formulas in bases whose elements realize continuous functions. *Prob. Kibernetiki 37*, 52–118. (In Russian).

GÖDEL, K. (1931). Über formal unentscheidbare Sätze der Principia Mathematica und verwandter System I. *Monatshefte für Math. und Physik 38*, 173–198. An English translation by Elliot Mendelson, "On Formally Undecidable Propositions of Principia Mathematica and Related Systems I," appears in [Davis 1965].

GOLDSTINE, H. (1977). *A History of Numerical Analysis from the 16th Through the 19th Century*. Springer-Verlag.

GOLUB, G. and C. VAN LOAN (1989). *Matrix Computations*. John Hopkins Univ. Press.

GONZAGA, C. (1989). An algorithm for solving linear programming problems in $O(n^3 L)$ operations. In N. Megiddo (Ed.), *Progress in Mathematical Programming: Interior-Point and Related Methods*. Springer-Verlag.

GOODE, J. (1994). Accessible telephone directories. *J. of Symb. Logic 59*, 92–105.

GRÄDEL, E. and Y. GUREVICH (1995). Metafinite model theory. Preprint.

GRÄDEL, E. and K. MEER (1995). Descriptive complexity theory over the real numbers. In *27th Annual ACM Symp. on the Theory of Computing*, pp. 315–324.

GRIFFITHS, P. and J. HARRIS (1978). *Principles of Algebraic Geometry*. John Wiley & Sons.

GRIGORIEV, D. (1988). Complexity of deciding Tarski algebra. *J. of Symbolic Computation 5*, 65–108.

GRIGORIEV, D. and N. VOROBJOV (1988). Solving systems of polynomial inequalities in subexponential time. *J. of Symbolic Computation 5*, 37–64.

GUILLEMIN, U. and A. POLLACK (1974). *Differential Topology*. Prentice-Hall.

HAGERUP, T. and C. RÜB (1990). A guided tour to Chernoff bounds. *Information Processing Letters 33*, 305–308.

HAMMERSLEY, J. and D. HANDSCOMB (1964). *Monte Carlo Methods*. Methuen.

HARRINGTON, L., M. MORLEY, A. SEEDROV, and S. SIMPSON (Eds.) (1985). *Harvey Friedman's Research on the Foundations of Mathematics*. North-Holland.

HARTMANIS, J. and R. STEARNS (1965). On the computational complexity of algorithms. *Transactions of the Amer. Math. Soc. 117*, 285–306.

HARTSHORNE, R. (1977). *Algebraic Geometry*. Springer-Verlag.

HEINTZ, J., M.-F. ROY, and P. SOLERNO (1990). Sur la complexité du principe de Tarski-Seidenberg. *Bulletin de la Société Mathématique de France 118*, 101–126.

HENRICI, P. (1977). *Applied and Computational Complex Analysis*. John Wiley & Sons.

HERMAN, G. and S. ISARD (1970). Computability over arbitrary fields. *J. London Math. Soc. 2*, 73–79.

HERMANN, G. (1926). Die Frage der endlich vielen Schritte in der Theorie der Polynomideale. *Math. Ann. 95*, 736–788.

HIRSCH, M. (1976). *Differential Topology*. Springer-Verlag.

HIRSCH, M. and S. SMALE (1979). On algorithms for solving $f(x) = 0$. *Comm. Pure and Appl. Math 32*, 281–312.

HODGES, W. (1993). *Model Theory*, Volume 42 of *Encyclopedia of Mathematics and its Applications*. Cambridge University Press.

HOEFFDING, W. (1963). Probability inequalities for sums of bounded random variables. *J. of the American Statistical Association 58*, 13–30.

IMMERMAN, N. (1986). Relational queries computable in polynomial time. *Information and Control 68*, 86–104.

KANTOROVICH, L. and G. AKILOV (1964). *Functional Analysis in Normed Spaces.* MacMillan.

KARMARKAR, N. (1984). A new polynomial time algorithm for linear programming. *Combinatorica 4*, 373–395.

KARP, R. (1972). Reducibility among combinatorial problems. In R. Miller and J. Thatcher (Eds.), *Complexity of Computer Computations*, pp. 85–103. Plenum Press.

KARP, R. and R. LIPTON (1982). Turing machines that take advice. *L'Enseignement Mathématique 28*, 191–209.

KELLER, H. (1978). Global homotopic and Newton methods. In *Recent advances in Numerical Analysis*, pp. 73–94. Academic Press.

KELLOG, R., T. LI, and J. YORKE (1976). A constructive proof of Brouwer fixed-point theorem and computational results. *SIAM J. of Numer. Anal. 13*, 473–483.

KENDIG, K. (1977). *Elementary Algebraic Geometry.* Springer-Verlag.

KHACHIJAN, L. (1979). A polynomial algorithm in linear programming. *Dokl. Akad. Nauk SSSR 244*, 1093–1096. (In Russian, English translation in *Soviet Math. Dokl.*, 20:191–194, 1979).

KIM, M. (1988). On approximate zeros and rootfinding algorithms for a complex polynomial. *Math. Comp. 51*, 707–719.

KLEENE, S. (1936). General recursive functions of natural numbers. *Math. Annalen 112*, 727–742.

KLEENE, S. (1943). Recursive predicates and quantifiers. *Transactions of the Amer. Math. Soc. 53*, 41–73.

KNUTH, D. (1976). Big omicrom and big omega and big theta. *SIGACT News 8*, 18–24.

KO, K. (1991). *Complexity theory of real functions.* Birkhäuser.

KOIRAN, P. (1993). A weak version of the Blum, Shub & Smale model. In *34th Annual IEEE Symp. on Foundations of Computer Science*, pp. 486–495.

KOIRAN, P. (1994). Computing over the reals with addition and order. *Theoretical Computer Science 133*, 35–47.

KOIRAN, P. (1995). Approximating the volume of definable sets. In *36th Annual IEEE Symp. on Foundations of Computer Science*, pp. 134–141.

KOIRAN, P. (1996). Hilbert's Nullstellensatz is in the polynomial hierarchy. *J. of Complexity 12*, 273–286.

KOLLÁR, J. (1988). Sharp effective Nullstellensatz. *J. of Amer. Math. Soc. 1*, 963–975.

KOSTLAN, E. (1988). Complexity theory of numerical linear algebra. *J. of Computational and Applied Mathematics 22*, 219–230.

KOSTLAN, E. (1993). On the distribution of the roots of random polynomials. In M. Hirsch, J. Marsden, and M. Shub (Eds.), *From Topology to Computation: Proceedings of the Smalefest*, pp. 419–431. Springer-Verlag.

KRICK, T. and L. PARDO (1996). A computational method for diophantine approximation. In L. González Vega and T. Recio (Eds.), *Computational Algebraic Geometry*, Volume 143 of *Progress in Mathematics*, pp. 193–253. Birkhäuser.

KRIVINE, J.-L. (1964). Anneaux préordonnés. *J. Analyse Math. 12*, 307–326.

KUHN, T. (1957). *The Copernican Revolution: Planetary Astronomy in the Development of the Western Thought.* Harvard University Press.

KUNZ, E. (1985). *Introduction to Commutative Algebra and Algebraic Geometry.* Birkhäuser.

KURTZ, S. (1983). On the random oracle hypothesis. *Information and Control 57*, 40–47.

LADNER, R. (1975). The circuit value problem is log space complete for P. *SIGACT News 7*, 18–20.

LANG, S. (1991). *Diophantine Geometry.* Springer-Verlag.

LANG, S. (1993a). *Algebra* (3rd ed.). Addison-Wesley.

LANG, S. (1993b). *Complex Analysis* (3rd ed.). Springer-Verlag.

LEVIN, L. (1973). Universal sequential search problems. *Probl. Pered. Inform. IX 3*, 265–266. (In Russian, English translation in *Problems of Information Trans.* 9,3; corrected translation in Trakhtenbrot [1984]).

LI, T., T. SAUER, and J. YORKE (1987). Numerical solution of a class of deficient polynomial systems. *SIAM J. of Numer. Anal. 24*, 435–451.

LOOS, R. (1982). Generalized polynomial remainder sequences. In B. Buchberger, G. Collins, and R. Loos (Eds.), *Computer Algebra, Symbolic and Algebraic Computation*, pp. 115–138. Springer-Verlag.

LUPANOV, O. (1958). A method of circuit synthesis. *Izvestia V.U.Z. Radiofizika 1*, 120–140. (In Russian).

MAASS, W. (1993). Bounds for the computational power and learning complexity of analog neural nets. In *25th Annual ACM Symp. on the Theory of Computing*, pp. 335–344.

MACINTYRE, A. (1971). On ω_1-categorical theories of fields. *Fund. Math. 71*, 1–25.

MACINTYRE, A., K. MCKENNA, and L. VAN DEN DRIES (1983). Elimination of quantifiers in algebraic structures. *Adv. in Math. 47*, 74–87.

MALAJOVICH-MUÑOZ, G. (1993). *On the Complexity of Path-Following Newton Algorithms for Solving Systems of Polynomial Equations with Integer Coefficients.* Ph. D. thesis, University of California at Berkeley.

MATIYASEVICH, Y. (1993). *Hilbert's Tenth Problem.* The MIT Press.

MEER, K. (1990). Computations over $\mathbb{Z}$ and $\mathbb{R}$: A comparison. *J. of Complexity 6*, 256–263.

MEER, K. (1992). A note on a P $\neq$ NP result for a restricted class of real machines. *J. of Complexity 8*, 451–453.

MEER, K. (1993). Real number models under various sets of operations. *J. of Complexity 9*, 366–372.

MEER, K. (1994). On the complexity of quadratic programming in real number models of computation. *Theoretical Computer Science 133*, 85–94.

MEER, K. and C. MICHAUX (1997). A survey on real structural complexity theory. *Bulletin of the Belgian Math. Soc. 4*, 113–148.

MEYER, A. and L. STOCKMEYER (1973). The equivalence problem for regular expressions with squaring requires exponential time. In *13th IEEE Symp. on Switching and Automata Theory*, pp. 125–129.

MEYER AUF DER HEIDE, F. (1984). A polynomial linear search algorithm for the *n*-dimensional Knapsack problem. *J. of the ACM 31*, 668–676.

MEYER AUF DER HEIDE, F. (1985a). Lower bounds for solving diophantine equations on random access machines. *J. of the ACM 32*, 929–937.

MEYER AUF DER HEIDE, F. (1985b). Simulating probabilistic by deterministic algebraic computation trees. *Theoretical Computer Science 41*, 325–330.

MICHAUX, C. (1989). Une remarque à propos des machines sur $\mathbb{R}$ introduites par Blum, Shub et Smale. *C. R. Acad. Sci. Paris 309, Série I*, 435–437.

MICHAUX, C. (1991). Ordered rings over which output sets are recursively enumerable. *Proceedings of the Amer. Math. Soc. 112*, 569–575.

MICHAUX, C. (1994). $P \neq NP$ over the nonstandard reals implies $P \neq NP$ over $\mathbb{R}$. *Theoretical Computer Science 133*, 95–104.

MILNOR, J. (1964). On the Betti numbers of real varieties. *Proceedings of the Amer. Math. Soc. 15*, 275–280.

MILNOR, J. (1965). *Topology from the Differentiable Viewpoint*. University Press of Virginia.

MONTAÑA, J. and L. PARDO (1993). Lower bounds for arithmetic networks. *Applicable Algebra in Engineering, Communication and Computing 4*, 1–24.

MOREIRA, C. (TA). On asymptotical estimates for arithmetical cost functions. To appear in Proceedings of the Amer. Math. Soc..

MORGAN, A. (1987). *Solving Polynomial Systems Using Continuation for Scientific and Engineering Problems*. Prentice-Hall.

MORGAN, F. (1988). *Geometric Measure Theory, A Beginners Guide*. Academic Press.

MOSCHOVAKIS, Y. (1986). Foundations of the theory of algorithms. Draft.

MUMFORD, D. (1976). *Algebraic Geometry I, Complex Projective Varieties*. Springer-Verlag.

NEFF, C. (1994). Specified precision root isolation is in NC. *J. of Computer and System Sci. 48*, 429–463.

NEFF, C. and J. REIF (1996). An efficient algorithm for the complex roots problem. *J. of Complexity 12*, 81–115.

OLEINIK, O. (1951). Estimates of the Betti numbers of real algebraic hypersurfaces. *Mat. Sbornik (N.S.) 28*, 635–640. (In Russian).

OLEINIK, O. and I. PETROVSKI (1949). On the topology of real algebraic surfaces. *Izv. Akad. Nauk SSSR 13*, 389–402. (In Russian, English translation in *Transl. Amer. Math. Soc.*, 1:399–417, 1962).

OSTROWSKI, A. (1954). On two problems in abstract algebra connected with Horner's rule. In *Studies in Mathematics and Mechanics Presented to Richard von Mises*, pp. 40–48. Academic Press.

OSTROWSKI, A. (1973). *Solutions of Equations in Euclidean and Banach Spaces*. Academic Press.

PAN, V. (1966). Methods of computing values of polynomials. *Russian Math. Surveys 21*, 105–136.

PAN, V. (1987). Sequential and parallel complexity of approximate evaluation of polynomial zeros. *Comput. Math. Appl. 14*, 591–622.

PAN, V. (1995). Optimal (up to polylog factors) sequential and parallel algorithms for approximating complex polynomial zeros. In *27th Annual ACM Symp. on the Theory of Computing*, pp. 741–750.

PAN, V. (TA). Solving a polynomial equation: Some history and recent progress. To appear in *SIAM Review*.

PAPADIMITRIOU, C. (1981). On the complexity of integer programming. *J. of the ACM 28*, 765–768.

PAPADIMITRIOU, C. (1994). *Computational Complexity*. Addison-Wesley.

PEITGEN, H.-O. and D. SAUPE (Eds.) (1988). *The Science of Fractal Images*. Springer-Verlag.

PENROSE, R. (1991). *The Emperor's New Mind*. Penguin.

PIPPENGER, N. (1979). On simultaneous resource bounds. In *20th Annual IEEE Symp. on Foundations of Computer Science*, pp. 307–311.

POIZAT, B. (1995). *Les Petits Cailloux*. Aléa.

POST, E. (1943). Formal reductions of the general combinatorial decision problem. *Amer. J. of Math. 65*, 197–268.

PREPARATA, F. and M. SHAMOS (1985). *Computational Geometry: An Introduction*. Texts and Monographs in Computer Science, Springer-Verlag.

RABIN, M. (1960a). Computable algebra, general theory and theory of computable fields. *Transactions of the Amer. Math. Soc. 95*, 341–360.

RABIN, M. (1960b). Degree of difficulty of computing a function and a partial ordering of recursive sets. Technical Report 2, Hebrew University of Jerusalem.

RABIN, M. (1966). Mathematical theory of automata. In *19th ACM Symp. in Applied Mathematics*, pp. 153–175.

RABIN, M. (1976). Probabilistic algorithms. In J. Traub (Ed.), *Algorithms and Complexity: New Directions and Results*, pp. 21–39. Academic Press.

RENEGAR, J. (1987a). On the efficiency of Newton's method in approximating all zeros of systems of complex polynomials. *Math. of Oper. Research 12*, 121–148.

RENEGAR, J. (1987b). On the worst case arithmetic complexity of approximating zeros of polynomials. *J. of Complexity 3*, 90–113.

RENEGAR, J. (1992a). On the computational complexity and geometry of the first-order theory of the reals. Part I. *J. of Symbolic Computation 13*, 255–299.

RENEGAR, J. (1992b). On the computational complexity and geometry of the first-order theory of the reals. Part II. *J. of Symbolic Computation 13*, 301–327.

RENEGAR, J. (1992c). On the computational complexity and geometry of the first-order theory of the reals. Part III. *J. of Symbolic Computation 13*, 329–352.

RENEGAR, J. (1995a). Incorporating condition measures into the complexity theory of linear programming. *SIAM J. of Optimization 5*, 506–524.

RENEGAR, J. (1995b). Linear programming, complexity theory and elementary functional analysis. *Mathematical Programming 70*, 279–351.

RENEGAR, J. and M. SHUB (1992). Unified complexity analysis for Newton LP methods. *Math. Programming 53*, 1–16.

RENEGAR, J., M. SHUB, and S. SMALE (Eds.) (1996). *The Mathematics of Numerical Analysis*, Volume 32 of *Lectures in Applied Mathematics*. American Mathematical Society.

REZNICK, B. (1992). *Sums of Even Powers of Real Linear Forms*. Number 463 in Memoirs of the American Mathematical Society. AMS, Providence, RI.

RHEINBOLDT, W. (1988). On a theorem of S. Smale about Newton's method for analytic mappings. *Appl. Math. Lett. 1*, 69–72.

RIORDAN, J. and C. SHANNON (1942). The number of two-terminal series-parallel networks. *J. of Mathematics and Physics 21*, 83–93.

RISLER, J.-J. (1970). Une caractérisation des idéaux des variétés algébriques réelles. *C. R. Acad. Sci. Paris 271*, 1171–1173.

ROBINSON, J. (1949). Definability and decision problems in arithmetic. *J. of Symbolic Logic 14*, 98–114.

ROGERS, H. (1967). *Theory of Recursive Functions and Effective Computability*. McGraw-Hill.

ROYDEN, H. (1986). Newton's method. Preprint.

SCHOBER, G. (1980). Coefficient estimates for inverses of Schlicht functions. In D. Brannan and J. Clunie (Eds.), *Aspects of Contemporary Complex Analysis*, pp. 503–513. Academic Press.

SCHOLZ, A. (1937). Aufgabe 253. *Jahresber. Deutsch. Math.-Verein. 47*, 41–42.

SCHÖNHAGE, A. (1982). The fundamental theorem of algebra in terms of computational complexity. Technical report, Math. Institut der Univ. Tübingen.

SCHRIJVER, A. (1986). *Theory of Linear and Integer Programming*. John Wiley & Sons.

SCHWARTZ, J. (1980). Fast probabilistic algorithms for verification of polynomial identities. *J. of the ACM 27*, 701–717.

SHAFAREVICH, I. (1977). *Basic Algebraic Geometry*. Springer-Verlag.

SHAMIR, A. (1979). Factoring numbers in $O(\log n)$ arithmetic steps. *Information Processing Letters 8*, 28–31.

SHANNON, C. (1938). A symbolic analysis of relay and switching circuits. *Trans. AIEE 57*, 713–723.

SHEPHERDSON, J. and H. STURGIS (1963). Computability of recursive functions. *J. of the ACM 10*, 217–255.

SHISHIKURA, M. (1994). The boundary of the Mandelbrot set has Hausdorff dimension two. *Astérisque 222*, 389–405.

SHUB, M. (1993a). On the work of Steve Smale on the theory of computation. In M. Hirsch, J. Marsden, and M. Shub (Eds.), *From Topology to Computation: Proceedings of the Smalefest*, pp. 281–301. Springer-Verlag.

SHUB, M. (1993b). Some remarks on Bezout's theorem and complexity theory. In M. Hirsch, J. Marsden, and M. Shub (Eds.), *From Topology to Computation: Proceedings of the Smalefest*, pp. 443–455. Springer-Verlag.

SHUB, M. and S. SMALE (1985). Computational complexity: On the geometry of polynomials and a theory of cost I. *Ann. Sci. École Norm. Sup. 18*, 107–142.

SHUB, M. and S. SMALE (1986). Computational complexity: On the geometry of polynomials and a theory of cost II. *SIAM J. on Computing 15*, 145–161.

SHUB, M. and S. SMALE (1993a). Complexity of Bezout's Theorem I: Geometric aspects. *J. of the Amer. Math. Soc. 6*, 459–501.

SHUB, M. and S. SMALE (1993b). Complexity of Bezout's theorem II: Volumes and probabilities. In F. Eyssette and A. Galligo (Eds.), *Computational Algebraic Geometry*, Volume 109 of *Progress in Mathematics*, pp. 267–285. Birkhäuser.

SHUB, M. and S. SMALE (1993c). Complexity of Bezout's theorem III: Condition number and packing. *J. of Complexity 9*, 4–14.

SHUB, M. and S. SMALE (1994). Complexity of Bezout's Theorem V: Polynomial time. *Theoretical Computer Science 133*, 141–164.

SHUB, M. and S. SMALE (1995). On the intractability of Hilbert's Nullstellensatz and an algebraic version of "P = NP". *Duke Math. J. 81*, 47–54.

SHUB, M. and S. SMALE (1996). Complexity of Bezout's theorem IV: Probability of success; extensions. *SIAM J. of Numer. Anal. 33*, 128–148.

SIEGELMAN, H. and E. SONTAG (1992). On the computational power of neural nets. In *5th ACM Workshop on Computational Learning Theory*, pp. 440–449.

SIEGELMAN, H. and E. SONTAG (1994). Analog computation via neural networks. *Theoretical Computer Science 131*, 331–360.

SMALE, S. (1976). A convergent process of price adjustment and global Newton methods. *J. Math. Economy 3*, 107–120.

SMALE, S. (1981). The fundamental theorem of algebra and complexity theory. *Bulletin of the Amer. Math. Soc. 4*, 1–36.

SMALE, S. (1983). On the average number of steps of the simplex method of linear programming. *Mathematical Programming 27*, 241–262.

SMALE, S. (1985). On the efficiency of algorithms of analysis. *Bulletin of the Amer. Math. Soc. 13*, 87–121.

SMALE, S. (1986a). Algorithms for solving equations. In *Proceedings of the International Congress of Mathematicians*, pp. 172–195. American Mathematical Society.

SMALE, S. (1986b). Newton's method estimates from data at one point. In R. Ewing, K. Gross, and C. Martin (Eds.), *The Merging of Disciplines: New Directions in Pure, Applied, and Computational Mathematics*. Springer-Verlag.

SMALE, S. (1987). On the topology of algorithms I. *J. of Complexity 3*, 81–89.

SMALE, S. (1988). The Newtonian contribution to our understanding of the computer. In M. Stayer (Ed.), *Newton's Dream*. McGill-Queens University Press.

SMALE, S. (1990). Some remarks on the foundations of numerical analysis. *SIAM Review 32*, 211–220.

SOLOVAY, R. and V. STRASSEN (1977). A fast Monte-Carlo test for primality. *SIAM J. on Computing 6*, 84–85.

SOLOVAY, R. and V. STRASSEN (1978). Erratum on "A fast Monte-Carlo test for primality". *SIAM J. on Computing 7*, 118.

SONTAG, E. D. (1985). Real addition and the polynomial hierarchy. *Information Processing Letters 20*, 115–120.

STEELE, J. and A. YAO (1982). Lower bounds for algebraic decision trees. *J. of Algorithms 3*, 1–8.

STEIN, E. and G. WEISS (1971). *Introduction to Fourier Analysis on Euclidean Spaces*. Princeton University Press.

STOCKMEYER, L. (1977). The polynomial-time hierarchy. *Theoretical Computer Science 3*, 1–22.

STOCKMEYER, L. and U. VISHKIN (1984). Simulation of parallel random access machines by circuits. *SIAM J. on Computing 13*, 409–422.

STRASSEN, V. (1969). Gaussian elimination is not optimal. *Numer. Math. 13*, 354–356.

STRASSEN, V. (1976). Einige Resultate über Berechungskomplexität. *Jber. Deutsche Math. Verein. 78*, 1–8.

TARSKI, A. (1951). *A Decision Method for Elementary Algebra and Geometry*. University of California Press.

THOM, R. (1965). Sur l'homologie des variétés algébriques réelles. In S. Cairns (Ed.), *Differential and Combinatorial Topology*. Princeton University Press.

TIURYN, J. (1979). A survey of the logic of effective definitions. In E. Engeler (Ed.), *Logic of Programs*, Volume 125 of *Lect. Notes in Comp. Sci.*, pp. 198–245. Springer-Verlag.

TRAKHTENBROT, B. (1984). A survey of Russian approaches to perebor (brute-force search) algorithms. *Annals of the History of Computing 6*, 384–400.

TRAUB, J., G. WASILKOWSKI, and H. WOŹNIAKOWSKI (1988). *Information-Based Complexity*. Academic Press.

TRAUB, J. and H. WOŹNIAKOWSKI (1979). Convergence and complexity of Newton iteration for operator equations. *J. of the ACM 29*, 250–258.

TRAUB, J. and H. WOŹNIAKOWSKI (1982). Complexity of linear programming. *Oper. Research Letters 1*, 59–62.

TRIESCH, E. (1990). A note on a theorem of Blum, Shub, and Smale. *J. of Complexity 6*, 166–169.

TUCKER, J. (1980). Computing in algebraic systems. In F. Drake and S. Wainer (Eds.), *Recursion Theory, Its Generalizations and Applications*, London Math. Soc. Cambridge University Press.

TUCKER, J. and J. ZUCKER (1992). Examples of semicomputable sets of real and complex numbers. In M. O'Donnell and J. Myers Jr. (Eds.), *Constructivity in Computer Science*, Volume 613 of *Lect. Notes in Comp. Sci.*, pp. 179–198. Springer-Verlag.

TURÁN, G. and F. VATAN (1994). On the computation of Boolean functions by analog circuits of bounded fan-in. In *35th Annual IEEE Symp. on Foundations of Computer Science*, pp. 553–564.

TURING, A. (1936). On computable numbers, with an application to the Entscheidungsproblem. *Proc. London Math. Soc., Ser. 2 42*, 230–265.

VAN DER WAERDEN, B. (1949). *Modern Algebra*. F. Ungar Publishing Co.

VARDI, M. (1982). Complexity of relational query languages. In *14th Annual ACM Symp. on the Theory of Computing*, pp. 137–146.

VASSILIEV, V. (1992). *Complements of Discriminants of Smooth Maps: Topology and Applications*, Volume 98 of *Translations of Mathematical Monographs*. American Mathematical Society.

VAVASIS, S. (1991). *Nonlinear Optimization, Complexity Issues*. Oxford University Press.

VON NEUMANN, J. (1963). Collected Works, Volume A. *Taub (ed.)*. MacMillan.

WANG, X. (1993). Some results relevant to Smale's reports. In M. Hirsch, J. Marsden, and M. Shub (Eds.), *From Topology to Computation: Proceedings of the Smalefest*, pp. 456–465. Springer-Verlag.

WARREN, H. (1968). Lower bounds for approximation by non linear manifolds. *Transactions of the Amer. Math. Soc. 133*, 167–178.

WEIHRAUCH, K. (1987). *Computability*. EATCS Monographs on Theoretical Computer Science, 9. Springer-Verlag.

WEYL, H. (1932). *The Theory of Groups and Quantum Mechanics*. Dover.

WILKINSON, J. (1963). *Rounding Errors in Algebraic Processes*. Prentice Hall.

WILKINSON, J. (1965). *The Algebraic Eigenvalue Problem*. Clarendon Press.

WINOGRAD, S. (1967). On the number of multiplications required to compute certain functions. *Proc. National Acad. Sci. 58*, 1840–1842.

WINOGRAD, S. (1980). *Arithmetic complexity of computations*. SIAM Regional Conf. Ser. Appl. Math. 33.

WOŹNIAKOWSKI, H. (1977). Numerical stability for solving non-linear equations. *Numer. Math. 27*, 373–390.

WÜTHRICH, H. (1976). Ein Entscheidungsverfahren für die Theorie der reell-abgeschlossenen Körper. In E. Specker and V. Strassen (Eds.), *Komplexität von Entscheidungsproblemen*, Volume 43 of *Lect. Notes in Comp. Sci.*, pp. 138–162. Springer-Verlag.

YAO, A. (1981). On the parallel computation of the knapsack problem. In *13th Annual ACM Symp. on the Theory of Computing*, pp. 123–127.

YE, Y. (1994). Combining binary search and Newton's method to compute real roots for a class of real functions. *J. of Complexity 10*, 271–280.

ZARISKI, O. and P. SAMUEL (1979). *Commutative Algebra*, Volume 1 and 2. Springer-Verlag. Reprint of the 1958–1960 edition.

Index